PRENTICE-HALL INSTRUMENTATION AND CONTROLS SERIES

CHEN & HAAS *Elements of Control Systems Analysis: Classical and Modern Approaches*
KUO *Analysis and Synthesis of Sampled-Data Control Systems*
KUO *Discrete-Data Control Systems*
OGATA *Modern Control Engineering*
OGATA *State Space Analysis of Control Systems*

DISCRETE-DATA
CONTROL SYSTEMS

PRENTICE-HALL INTERNATIONAL, INC., *London*
PRENTICE-HALL OF AUSTRALIA, PTY. LTD., *Sydney*
PRENTICE-HALL OF CANADA, LTD., *Toronto*
PRENTICE-HALL OF INDIA PRIVATE LTD., *New Delhi*
PRENTICE-HALL OF JAPAN, INC., *Tokyo*

B. C. KUO

Professor of Electrical Engineering
University of Illinois

Kuo, Tsung-i

DISCRETE-DATA CONTROL SYSTEMS

PRENTICE-HALL, INC.

Englewood Cliffs, New Jersey

Current printing (last digit):

10 9 8 7 6 5 4 3 2 1

13-216002-1

Library of Congress Catalog Card No. 70-104174

Printed in the United States of America

To my three daughters

PREFACE

This project originally began as a revision of the author's previous text *Analysis and Synthesis of Sampled-Data Control Systems*, published in 1963. However, the revised work contained so much new material that a change of title seemed justified. Furthermore, the material covered in the original text, although conventional by present standards, is still quite useful for the practicing engineer.

The field of discrete-data control systems has been advancing so rapidly that to cover the entire subject with a book of any reasonable size is virtually out of the question. The topics included in the text are those which the author considers suitable for a one-semester course in discrete-data control systems. Much of the material has been class tested during a period of more than ten years. The book is also prepared with the needs of modern engineers in mind; it should therefore prove useful as a reference. The text has been designed so that it may be studied with a minimum amount of guidance.

It is assumed that the reader has a knowledge of the basic theory of feedback control systems. He need not have read the prior text, *Analysis and Synthesis of Sampled-Data Control Systems*, although it may be helpful to first get acquainted with some conventional techniques of treating discrete-data systems. In fact, some of the material, such as the topic on z-transformation, is repeated here to make the present text self-contained. The state variable method of analysis and design is emphasized in this book. Therefore, beyond the basic principles of the z-transform, the modified z-transform techniques and the submultiple sampling technique for the recovery of responses between sampling instants are not included.

The book can be divided into three parts: analysis, design, and simulation. The first five chapters are devoted essentially to analysis. Included in these chapters are discussions of the sampling and data-reconstruction processes, the z-transform method, the state variable technique, and stability. Chapters 6, 7, 8, and 9 are devoted to the design of discrete-data systems. Chapter 6 covers the time-optimal design which includes the concepts of controllability and observability, time-optimal control of single-variable and multivariable systems, as well as systems with input delays. Chapter 7 deals with the optimal design with respect to the minimization of a performance index. The summed-error-square criterion and the quadratic error criterion are considered. Chapters 8 and 9 discuss statistical analysis and design in the form of Wiener filter theory and Kalman filter theory. The last chapter, 10, is devoted to digital simulation of continuous-data systems. Illustrative examples are given for each topic discussed, and a number of problems are given for each chapter. The solutions to the problems are available, and qualified users may obtain the solutions manual from the publisher.

The author is grateful to many of his graduate students whose assistance in the form of thesis work, questions, and classroom dissussions have assisted in many ways in the preparation of this manuscript. He also wishes to acknowledge with gratitude the support of this project from the Electrical Engineering Department of the University of Illinois, especially the Publications Office which typed and produced part of the manuscript.

The encouragement and interest of Dr. E. C. Jordan, Head, and Professor W. E. Miller, Associate Head, of the Electrical Engineering Department of the University of Illinois, have been invaluable in the preparation of the manuscript. Finally, the author wishes to thank Professor A. P. Sage of the Southern Methodist University for reviewing the manuscript and his many valuable suggestions.

B. C. Kuo

CONTENTS

DISCRETE-DATA
CONTROL SYSTEMS

1

INTRODUCTION

1.1 DISCRETE-DATA CONTROL SYSTEMS

In recent years great advances have been made in the field of discrete-data control systems. These systems differ from the conventional continuous-data systems in that the signal in one or more parts of the system is in the form of either a pulse train or a numerical code. The terms, sampled-data systems and digital systems are often used in the control literature. However, usually, sampled-data systems refer to a general class of systems whose signals are in the form of pulsed data, whereas digital systems imply the use of a digital computer or digital transducers in a system so that the signals are numerically coded. In this text we shall use the term discrete-data systems in a broad sense to include all systems in which some form of digital or sampled signals take place either intentionally or inherently.

The use of sampled data in control systems can be traced back to at least seventy years ago.[1,2]† However, organized study and intensive research on the subject did not begin until the early nineteen fifties.[3] Today, it is generally recognized that discrete-data systems are definitely here to stay. The importance of discrete-data techniques in control systems is also exemplified by the rapid increase in the use of digital computers and digital controllers in control systems.

In general, discrete data may occur in a control system either inherently or intentionally. For instance, a radar tracking system is known to be inherently a discrete-data system; the signals transmitted and received by the system are

†Superscripts refer to numbered references at the end of the chapter.

in the form of pulse trains. On the other hand, intentional sampling operations are found in numerous modern control systems for the purposes of time sharing, better sensitivity, optimal performance and response, etc.

The signals found in a discrete-data system are not always functions of the continuous time variable t, but are in the form of pulse trains or digital-coded signals. No signal is supplied at all during the periods separating the information carrying pulses. Mathematically, the discrete-data signal may be described by a sampling operation which is carried out by a sampler. In general, the sampler converts a continuous-time signal into a discrete-time signal according to one of the many possible sampling schemes. Some of the well-known sampling schemes are:

1. Uniform-rate sampling
2. Multirate sampling
3. Skip sampling
4. Cyclic sampling
5. Finite pulsewidth sampling
6. Pulsewidth modulation
7. Random sampling

Figure 1-1 illustrates in a simple manner the basic elements of an open-

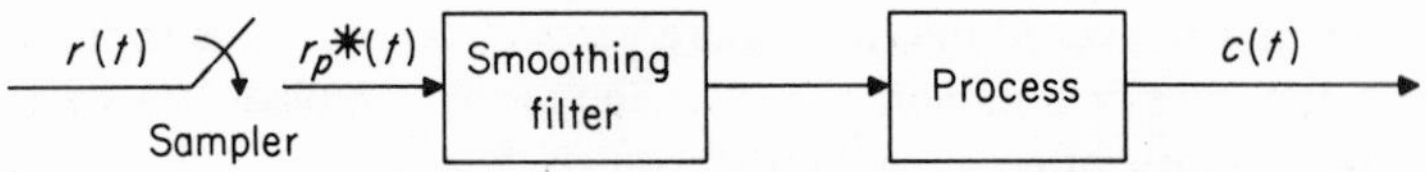

FIG. 1-1. An open-loop discrete-data system.

loop discrete-data system. A sampler samples the continuous-data signal $r(t)$ at discrete instants of time, and the output of the sampler is a sequence of pulses. Typical waveforms of the input and the output of the sampler are illustrated in Figure 1-2. In this case, the sampler is assumed to have a uniform sampling rate and is referred to as a uniform-rate or period sampler. If the sampler is a physical one, it must have a finite sampling duration so that the pulses at the output of the sampler have finite widths. However, as will be emphasized later, a great majority of the mathematical analysis of discrete-data systems assume that the sampling duration is infinitesimally small.

The smoothing filter that follows the sampler is used for the purpose of smoothing out the pulse signal before it enters the continuous-data process.

A unique characteristic of the discrete-data system of Figure 1-1 is that both sampled and continuous data exist in the system. Figure 1-3 shows the block diagram of a system in which all signals are in digital form. Digital computers have gained significant importance as control elements in control systems. The block diagram of a typical control system using a digital com-

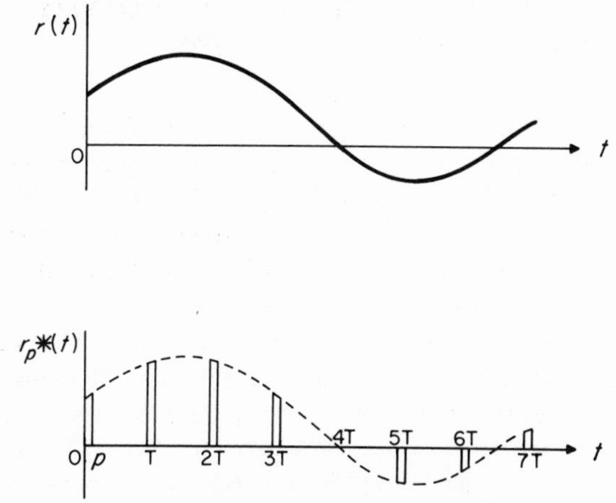

FIG. 1-2. Typical input and output signals of a sampler.

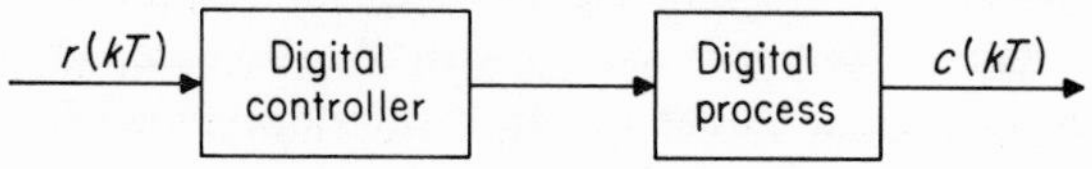

FIG. 1-3. An all-digital system.

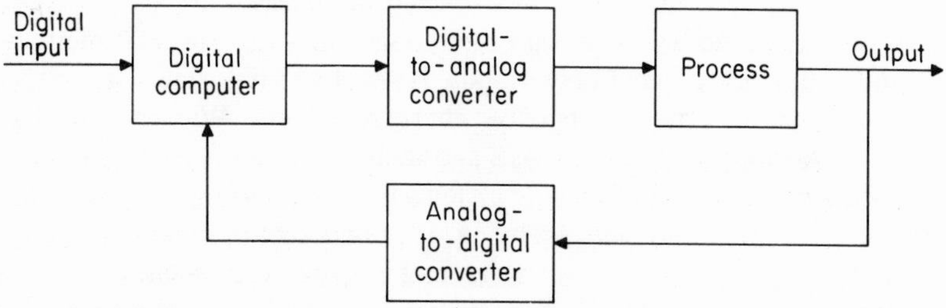

FIG. 1-4. A digital control system.

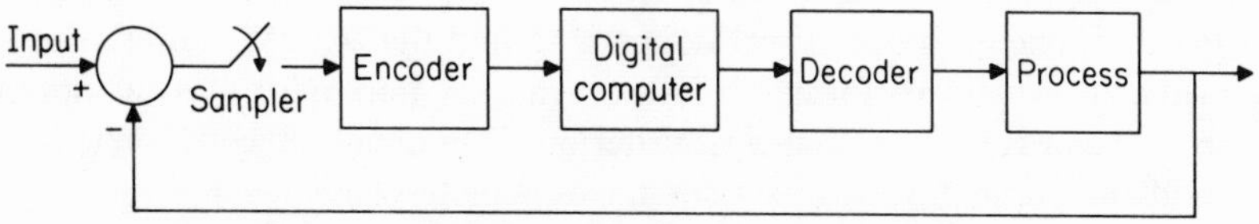

FIG. 1-5. A sampled-data system with a digital computer as controller.

puter is shown in Figure 1-4. The presence of digital coded signals in part of the system requires the use of digital-to-analog and analog-to-digital converters. It is also common to use a digital computer as a controller in a sampled-data control system. A typical arrangement is shown in Figure 1-5.

1.2 A BRIEF SURVEY OF THE METHODS OF ANALYSIS AND DESIGN OF DISCRETE-DATA SYSTEMS

Very little work was done on the analysis and design of discrete-data systems prior to 1950. Although early text books on servomechanisms written by MacColl[9] (1945), James, Nichols, and Phillips[10] (1947), and Oldenbourg and Sartorius[11] (1948), all have some coverage on topics of sampled-data systems, the treatments have mostly been limited to stability analysis and the derivation of transfer functions of sampled-data systems.

Since the early 1950's when digital computers were first used in control systems and when the need arose for complex automatic tracking systems and aerospace control systems, discrete-data control systems have been enjoying a most fruitful period of research and development. Not only have numerous papers been published in the United States and abroad, but several books[2–8] devoted solely to the subject of discrete-data systems have been written.

During the early stage of development, efforts were concentrated mostly on making extensions of the existing methods for continuous-data systems to discrete-data systems. For linear systems with discrete data, analysis and design techniques rely to a great extent on the z-transformation. During the period from 1950 to 1960, virtually all the standard techniques for the study of continuous-data control systems were successfully extended to the treatment of discrete-data systems. The control engineer was pleased to learn that with various degrees of modifications and limitations, such concepts and methods as the transfer functions, signal flow graphs, Bode plot, Routh-Hurwitz criterion, polar plot, Nyquist criterion, Nichols chart, root locus, Wiener filter design, etc., can all be applied to the analysis and design of discrete-data systems.

Since 1960, the field of control systems has gone through a rapid change in design techniques and philosophy. The principle of optimal control, made possible by the use of state variables and applied mathematics, have also revolutionalized the design of discrete-data systems. The objective of the modern control system design is to optimize the system in some prescribed sense. Usually, the quality of the system is measured by a performance index. For instance, one class of optimal design is to find the control to minimize the summed error-square performance index. Another important class of optimal design problem is the minimal-time criterion. The design objective in this case is to find the optimal control so that the system response reaches the desired value in the shortest possible time.

REFERENCES

1. Gouy, G., "On a Constant Temperature Oven," *J. Physique,* **6**, Series 3, 1897, pp. 479–483.

2. Kuo, B. C., *Analysis and Synthesis of Sampled-Data Control Systems.* Prentice-Hall, Englewood Cliffs, N.J., 1963.

3. Ragazzini, J. R., and Zadeh, L. H., "The Analysis of Sampled-data Systems," *Trans. AIEE,* **71**, Part II, 1952, pp. 225-234.

4. Tou, J. T., *Digital And Sampled-Data Control Systems.* McGraw-Hill, New York, 1959.

5. Ragazzini, J. R., and Franklin, G. F., *Sampled-Data Control Systems.* McGraw-Hill, New York, 1958.

6. Jury, E. I., *Sampled-Data Control Systems.* John Wiley & Sons, New York, 1958.

7. Lindorff, D. P., *Theory of Sampled-Data Control Systems.* John Wiley & Sons, New York, 1965.

8. Freeman, H., *Discrete-Time Systems.* John Wiley & Sons, New York, 1965.

9. MacColl, L. A., *Fundamental Theory of Servomechanisms.* D. Van Nostrand, Princeton, N. J., 1945.

10. James, H. M., Nichols, N. B., and Phillips, R. S., *Theory of Servomechanisms.* McGraw-Hill, New York, 1947.

11. Oldenbourg, R. C., and Sartorius, H., *The Dynamics of Automatic Control.* American Society of Mechanical Engineers, New York, 1948.

2

SAMPLING AND DATA RECONSTRUCTION PROCESSES

2.1 INTRODUCTION

It was pointed out in the last chapter that the signals found in a digital control system are discontinuous with respect to time. Digital signals, in general, may be in the form of a train of pulses introduced by a sampling operation, or a sequence of numbers representing the output of a digital controller or a digital computer.

When the digital signal is the output of a sampling device, we can visualize the operation as being that of a mechanical or electronic sampler which opens and closes once every so often, thus sending out a pulse train when a continuous-time signal is applied at the input. In this simple situation, the sampler closes for a very short interval of p second at discrete periodic instants $t = 0, T, 2T, \ldots, kT, \ldots$, and the sampler is classified as a periodic or uniform-rate sampler with finite pulsewidth. The diagram of a uniform-rate sampler with finite sampling duration is shown in Figure 2-1. For an input signal $f(t)$ which is a function of the continuous-time variable t, the output of the uniform-rate sam-

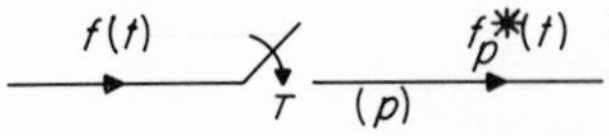

FIG. 2-1. A uniform-rate sampler with finite sampling duration.

pler is a train of finite-width pulses and is denoted by $f_p^*(t)$. Typical waveforms of the input and output signals of a uniform-rate sampler are shown in Figure 2-2.

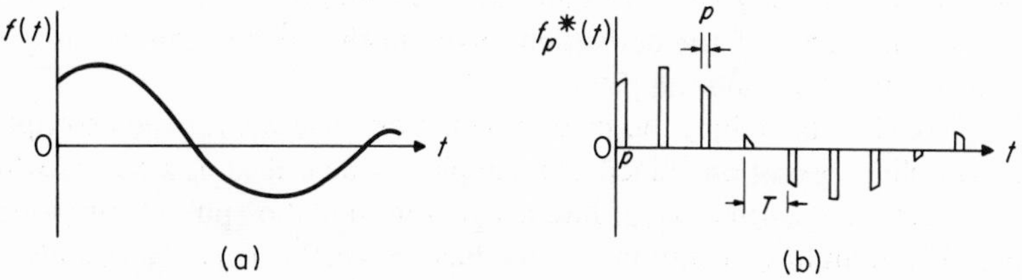

FIG. 2-2. Typical waveforms of the input and output signals of a uniform-rate sampler: (a) input; (b) output.

In general, the sampling scheme in digital control systems may take a great variety of forms. For instance, a sampler may have a nonuniform aperiodic or cyclic-variable sampling rate. It is also common to have digital systems which contain a multiple number of samplers with different sampling rates. The sampling rates of these samplers may or may not be synchronized, that is, operate at the same instants of time. This type of system is generally described as a multirate sampling system. In more complicated cases the sampling operation may be entirely random, i.e., the interval between successive samples may be thought of as following some random scheme.

A pulsewidth sampler, better known as the pulsewidth modulator, is a sampler whose output is a train of pulses with the pulsewidths varying as functions of the magnitude of the input signal at the sampling instants. The magnitudes of the output pulses are equal in this case. Typical input and output signals of a pulsewidth modulator are illustrated in Figure 2-3. A still

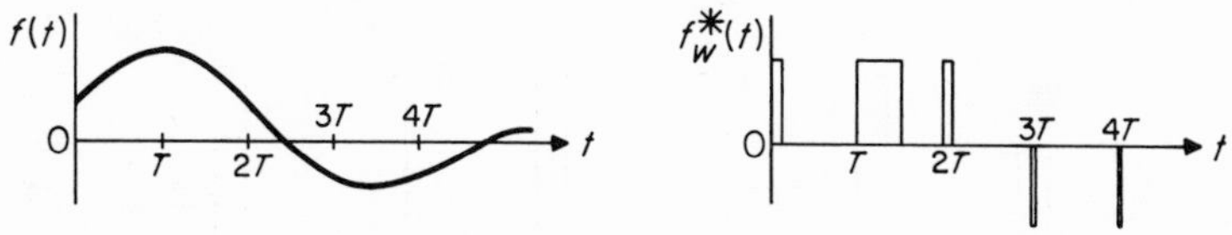

FIG. 2-3. Typical input and output signals of a pulsewidth modulator.

more complex situation is one in which the amplitudes of the output pulses also vary as some function of the input signal at the sampling instants. This type of sampler is known as a pulsewidth–pulse-amplitude modulator. A number of published works have shown that these and other nonuniform and unconventional types of sampling may be useful in improving the performance of a digital control system.

2.2 MATHEMATICAL DESCRIPTIONS OF THE UNIFORM-RATE SAMPLING PROCESS

For the sake of simplicity, only the uniform-rate sampling operation is discussed in this section. Some of the derivations given in this section can be easily extended to nonuniform-rate samplers.

First we shall introduce the pulse modulation concepts for the description of the sampling operation. Since the sampler is described as a device which converts a continuous-time signal into a signal in digital or pulse form, we can interpret the sampling operation as a modulator which generally operates on two inputs, the signal and the carrier. For the uniform-rate sampler shown in Figures 2-1 and 2-2, the signal is a continuous-time function $f(t)$, and the carrier is a unit pulse train with a period of T. If we designate the unit pulse train by $p(t)$, then

$$p(t) = \sum_{k=-\infty}^{\infty} [u(t - kT) - u(t - kT - p)] \quad (p < T) \tag{2-1}$$

where $u(t)$ is the unit step function, and it is assumed that the sampling operation begins at $t = -\infty$. The constant k takes on only integral values, and p is the sampling duration of the sampler. Then, the action of the uniform-rate sampler with finite sampler duration can be described by

$$f_p^*(t) = f(t) \times p(t) \tag{2-2}$$

or

$$f_p^*(t) = f(t) \sum_{k=-\infty}^{\infty} [u(t - kT) - u(t - kT - p)] \quad (p < T) \tag{2-3}$$

where the carrier signal $p(t)$ is shown in Figure 2-4.

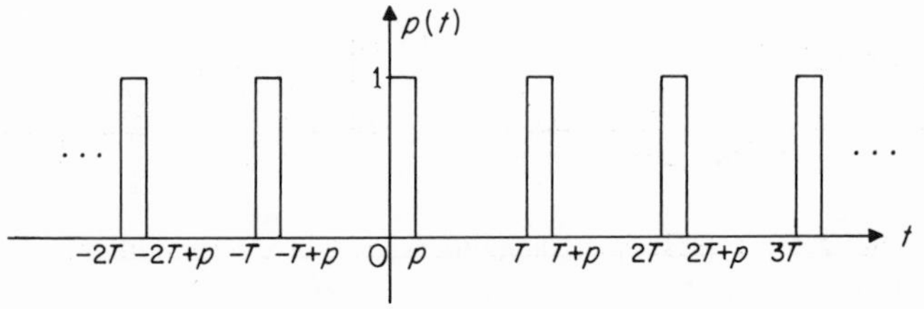

FIG. 2-4. The unit pulse train.

The block diagram for the pulse modulator is shown in Figure 2-5. It is easy to see that the pulse modulator representation of a sampler is applicable to many other types of sampling operations, and only the carrier signal $p(t)$ needs to be changed accordingly.

The operation of the sampler is described thus far by the time-domain characteristics of the input and output signals. It is of interest to investigate

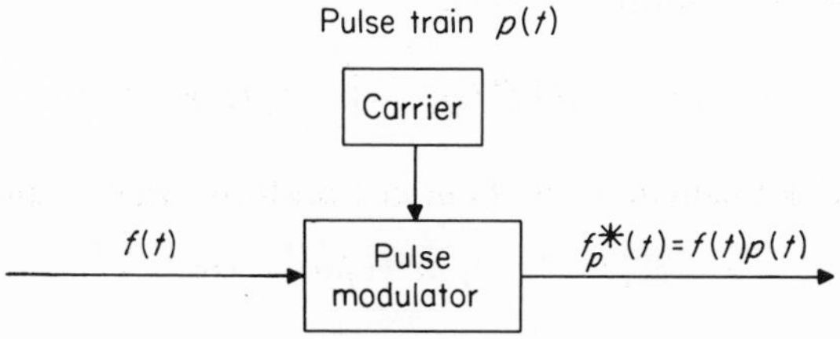

FIG. 2-5. Pulse modulator as a sampling device.

the frequency spectrum of the output signal of the sampler, $f_p^*(t)$. If we know the frequency content of the input signal $f(t)$, it would be interesting to determine what frequency components are contained in $f_p^*(t)$. It is not difficult to see that a sampler acts essentially as a harmonic generator, so that, in general, we expect to find much higher frequency components in $f_p^*(t)$ than in $f(t)$.

For a uniform-rate sampler, $p(t)$ is a periodic function with period T; we can express $p(t)$ as a Fourier series, i.e.,

$$p(t) = \sum_{n=-\infty}^{\infty} C_n e^{jn\omega_s t} \tag{2-4}$$

where ω_s is the *sampling frequency* in radians per second, and is related to the sampling period T by

$$\omega_s = \frac{2\pi}{T} \tag{2-5}$$

C_n denotes the complex Fourier series coefficient and is given by

$$C_n = \frac{1}{T}\int_0^T p(t)\, e^{-jn\omega_s t}\, dt \tag{2-6}$$

Since $p(t) = 1$ for $0 \leq t \leq p$, Eq. (2-6) becomes

$$C_n = \frac{1}{T}\int_0^p e^{-jn\omega_s t}\, dt = \frac{1 - e^{-jn\omega_s p}}{jn\omega_s T} \tag{2-7}$$

Using simple trigonometric identities, C_n is written as

$$C_n = \frac{p}{T}\frac{\sin(n\omega_s p/2)}{(n\omega_s p/2)} e^{-jn\omega_s p/2} \tag{2-8}$$

Substituting Eq. (2-8) into Eq. (2-4), we have

$$p(t) = \frac{p}{T}\sum_{n=-\infty}^{\infty} \frac{\sin(n\omega_s p/2)}{(n\omega_s p/2)} e^{-jn\omega_s p/2} e^{jn\omega_s t} \tag{2-9}$$

From Eq. (2-2) the output of the sampler is now written as

$$f_p^*(t) = \sum_{n=-\infty}^{\infty} C_n f(t)\, e^{jn\omega_s t} \tag{2-10}$$

where C_n is given by Eq. (2-8).

The Fourier transform of $f_p^*(t)$ is

$$F_p^*(j\omega) = \mathscr{F}[f_p^*(t)] = \int_{-\infty}^{\infty} f_p^*(t)\, e^{-j\omega t}\, dt \tag{2-11}$$

Using the shifting theorem of the Fourier transform which states that

$$\mathscr{F}[e^{jn\omega_s t} f(t)] = F(j\omega - jn\omega_s) \tag{2-12}$$

we have

$$F_p^*(j\omega) = \sum_{n=-\infty}^{\infty} C_n F(j\omega - jn\omega_s) \tag{2-13}$$

Equation (2-13) can also be written as

$$F_p^*(j\omega) = \sum_{n=-\infty}^{\infty} C_n F(j\omega + jn\omega_s) \tag{2-14}$$

since n extends from $-\infty$ to $+\infty$.

Taking the limit as $n \rightarrow 0$ in Eq. (2-8), we have

$$C_0 = \lim_{n \to 0} C_n = \frac{p}{T} \tag{2-15}$$

Hence from Eq. (2-13)

$$F_p^*(j\omega)|_{n=0} = C_0 F(j\omega) = \frac{p}{T} F(j\omega) \tag{2-16}$$

which shows an important fact that the frequency components contained in the original continuous-time input, $f(t)$, are still present in the sampler output $f_p^*(t)$, except that the amplitude is multiplied by the factor p/T. For $n \neq 0$, C_n is a complex quantity, but the magnitude of C_n may be written as

$$|C_n| = \frac{p}{T}\left|\frac{\sin(n\omega_s p/2)}{(n\omega_s p/2)}\right| \tag{2-17}$$

The magnitude of $F_p^*(j\omega)$ is

$$|F_p^*(j\omega)| = \left|\sum_{n=-\infty}^{\infty} C_n F(j\omega + jn\omega_s)\right| \tag{2-18}$$

The frequency spectrum of the unit pulse train $p(t)$ is simply the plot of the Fourier coefficient C_n as a function of ω, when n assumes various integral values between $-\infty$ and $+\infty$. The amplitude spectrum of C_n is shown in Figure 2-6(a). We see that the amplitude spectrum of C_n is not a continuous function but rather a line spectrum at discrete intervals of ω, with $\omega = n\omega_s$ for $n = 0, \pm 1, \pm 2, \ldots$. The amplitude of the spectrum is described by the right-hand side of Eq. (2-17).

Equation (2-18) can also be written as

$$|F_p^*(j\omega)| \leq \sum_{n=-\infty}^{\infty} |C_n|\, |F(j\omega - jn\omega_s)| \tag{2-19}$$

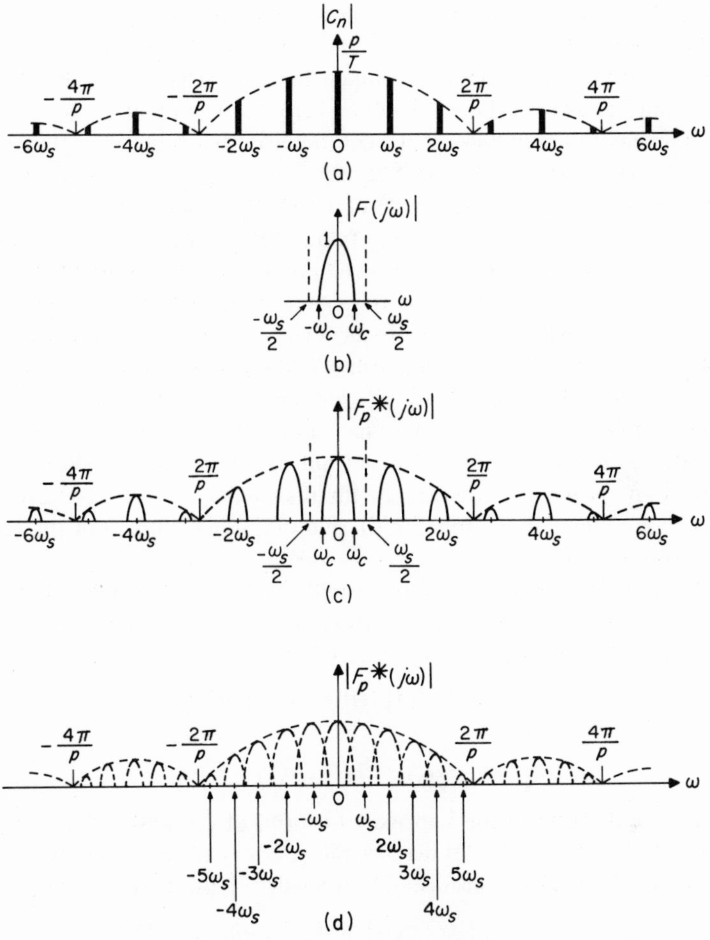

FIG. 2-6. Amplitude spectra of input and output signals of a finite-pulsewidth sampler: (a) amplitude spectrum of unit pulse train $p(t)$; (b) amplitude spectrum of continuous input signal $f(t)$; (c) amplitude spectrum of sampler output ($\omega_s > 2\omega_c$); (d) amplitude spectrum of sampler output ($\omega_s < 2\omega_c$).

which may be used to illustrate the amplitude spectrum of $F_p^*(j\omega)$. If the amplitude spectrum of the continuous input $f(t)$ is assumed to be as shown in Figure 2-6(b), by use of Eq. (2-19), the spectrum of $F_p^*(j\omega)$ will be of the form shown in Figure 2-6(c). It is interesting to note that $|F_p^*(j\omega)|$ contains not only the fundamental component of $F(j\omega)$ but also the harmonic or the complementary components, $F(j\omega + jn\omega_s)$, $n = \pm 1, \pm 2, \ldots$. The nth complementary component in the output spectrum is obtained by multiplying $|F(j\omega)|$ by its corresponding Fourier coefficient $|C_n|$ and shifting it by $n\omega_s$, n

$= \pm 1, \pm 2, \pm 3, \ldots$. Therefore, the sampler may be visualized as a harmonic generator whose output contains the weighted fundamental components, plus all the weighted complementary components at all frequencies separated by the sampling frequency. The band around zero frequency still carries essentially all the information contained in the continuous input signal, but the same information also repeats along the frequency axis; the amplitude of each component is weighted by the magnitude of its corresponding Fourier coefficient $|C_n|$. It should be pointed out that the frequency spectrum for $|F_p^*(j\omega)|$ shown in Figure 2-6(c) is sketched with the assumption that the sampling frequency ω_s is greater than twice the highest frequency contained in $f(t)$; that is, $\omega_s > 2\omega_c$. If $\omega_s < 2\omega_c$, distortion in the frequency spectrum of $|F_p^*(j\omega)|$ will appear because of the overlapping of the harmonic components. Figure 2-6(d) shows that if $\omega_s < 2\omega_c$, the spectrum of $|F_p^*(j\omega)|$ around zero frequency bears little resemblance to that of the original signal. Therefore, theoretically the original signal can be recovered from the spectrum shown in Figure 2-6(c) by means of an ideal low-pass filter, whereas the input signal cannot be recovered from the spectrum shown in Figure 2-6(d), due to the overlapping of the complementary components.

An alternate description of the sampled signal $f_p^*(t)$ can be obtained by means of the complex convolution method of Laplace transforms. With reference to Eq. (2-2), the Laplace transform of $f_p^*(t)$ is written

$$F_p^*(s) = \mathscr{L}[f_p^*(t)] = \mathscr{L}[f(t)p(t)]$$

or

$$F_p^*(s) = F(s) * P(s) \tag{2-20}$$

where $F(s)$ and $P(s)$ are the Laplace transforms of $f(t)$ and $p(t)$, respectively. The symbol, "$*$", in Eq. (2-20) denotes the complex convolution operation.

Taking the Laplace transform on both sides of Eq. (2-1), we have

$$P(s) = \sum_{k=0}^{\infty} \frac{1 - e^{-ps}}{s} e^{-kTs} \tag{2-21}$$

The summation on the right-hand side of Eq. (2-21) begins at $k = 0$ since the one-sided Laplace transform is defined for $0 \leq t < \infty$. The infinite series in Eq. (2-21) is written in closed form as

$$P(s) = \frac{1 - e^{-ps}}{s(1 - e^{-Ts})} \tag{2-22}$$

Then, Eq. (2-20) can be written

$$F_p^*(s) = F(s) * \frac{1 - e^{-ps}}{s(1 - e^{-Ts})} \tag{2-23}$$

It follows from the definition of the complex convolution that

$$F_p^*(s) = \frac{1}{2\pi j}\int_{c-j\infty}^{c+j\infty} F(\xi)P(s - \xi)\, d\xi \tag{2-24}$$

Hence

$$F_p^*(s) = \frac{1}{2\pi j}\int_{c-j\infty}^{c+j\infty} F(\xi)\frac{1 - e^{-p(s-\xi)}}{(s-\xi)[1 - e^{-T(s-\xi)}]}\,d\xi \tag{2-25}$$

where ξ is the variable of integration; $\sigma_1 < C < \sigma - \sigma_2, \sigma > \max(\sigma_1, \sigma_2, \sigma_1 + \sigma_2)$, where σ is the real part of s and σ_1 and σ_2 are the abscissas of convergence of $F(\xi)$ and $P(\xi)$, respectively. The path of integration of the integral of Eq. (2-25) is along the straight line from $\xi = C - j\infty$ to $C + j\infty$ in the complex ξ-plane as shown in Figure 2-7.

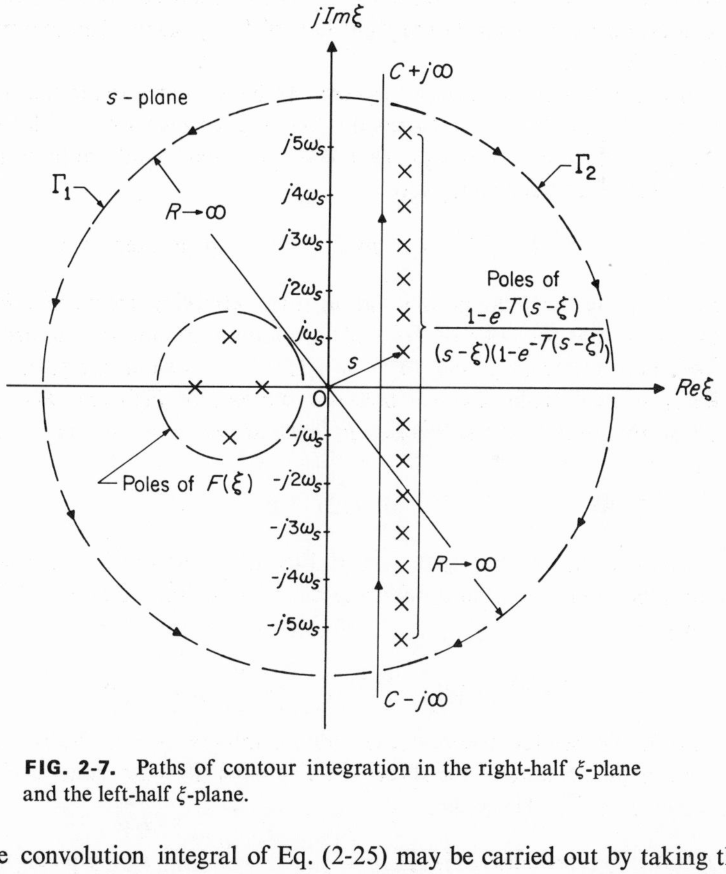

FIG. 2-7. Paths of contour integration in the right-half ξ-plane and the left-half ξ-plane.

The convolution integral of Eq. (2-25) may be carried out by taking the contour integration along a path formed by the $\xi = C - j\infty$ to $C + j\infty$ line, and a semicircle of infinitely large radius going around either the right-half or the left-half ξ-plane. The contour integral is then evaluated by means of the residue theorem of complex variables. In other words, Eq. (2-25) may be written as

$$1.\ F_p^*(s) = \frac{1}{2\pi j}\oint_{\Gamma_1} F(\xi)\frac{1 - e^{-p(s-\xi)}}{(s-\xi)[1 - e^{-T(s-\xi)}]}\,d\xi - \frac{1}{2\pi j}\oint F(\xi)\frac{1 - e^{-p(s-\xi)}}{(s-\xi)[1 - e^{-T(s-\xi)}]}\,d\xi \tag{2-26}$$

or

$$2.\ F_p^*(s) = \frac{1}{2\pi j}\oint_{\Gamma_2} F(\xi)\frac{1 - e^{-p(s-\xi)}}{(s-\xi)[1 - e^{-T(s-\xi)}]}\,d\xi - \frac{1}{2\pi j}\oint F(\xi)\frac{1 - e^{-p(s-\xi)}}{(s-\xi)[1 - e^{-T(s-\xi)}]}\,d\xi \tag{2-27}$$

where Γ_1 is a closed path enclosing the entire finite left half of the ξ-plane and Γ_2 is a closed path enclosing the right half of the ξ-plane. The two paths are shown in Figure 2-7.

In order to use the residue theorem, we have to investigate the poles and zeros of $F(\xi)$ and $P(s - \xi)$. Normally, the poles of $F(\xi)$ are considered to be in the left-half or on the imaginary axis of the ξ-plane and are finite in number, but $P(s - \xi)$ has simple poles at

$$\xi_n = s + \frac{2\pi nj}{T} = s + jn\omega_s \qquad -\infty < n \text{ (integer)} < \infty \tag{2-28}$$

where T is the sampling period and ω_s is the sampling frequency. Equation (2-28) shows that the poles of $P(s - \xi)$ are infinite in number and are located at frequency intervals of $n\omega_s$ ($n = 0, \pm 1, \pm 2, \ldots$) along the path of $\text{Re}(\xi) = \text{Re}(s)$ in the ξ-plane. Typical pole distributions of $F(\xi)$ and $P(s - \xi)$ are shown in Figure 2-7. If the function $F(\xi)$ has at least one more pole than zero, that is,

$$\lim_{\xi\to\infty} F(\xi) = 0 \tag{2-29}$$

the second terms on the right side of Eqs. (2-26) and (2-27) are zero, since the integrals along the infinite-radius semicircles vanish. Equation (2-26) then becomes

$$F_p^*(s) = \frac{2}{2\pi j}\oint_{\Gamma_1} F(\xi)\frac{1 - e^{-p(s-\xi)}}{(s-\xi)[1 - e^{-T(s-\xi)}]}\,d\xi \tag{2-30}$$

From the residue theorem the contour integral of Eq. (2-30) is equal to the sum of the residues of the integrand evaluated at those poles enclosed by the closed path Γ_1. Therefore,

$$F_p^*(s) = \sum \text{Residues of } \left\{F(\xi)\frac{1 - e^{-p(s-\xi)}}{(s-\xi)[1 - e^{-T(s-\xi)}]}\right\} \text{ at the poles of } F(\xi) \tag{2-31}$$

A more specific form can be arrived at for Eq. (2-31) if we assume that $F(\xi)$ is a rational function with k simple poles. Then Eq. (2-31) becomes

$$F_p^*(s) = \sum_{n=1}^{k} \frac{N(\xi_n)}{D'(\xi_n)} \frac{1 - e^{-p(s-\xi_n)}}{(s - \xi_n)[1 - e^{-T(s-\xi_n)}]} \tag{2-32}$$

where

$$F(\xi) = \frac{N(\xi)}{D(\xi)} \tag{2-33}$$

$$D'(\xi_n) = \left.\frac{dD(\xi)}{d\xi}\right|_{\xi=\xi_n} \tag{2-34}$$

and ξ_n denotes the nth pole of $F(\xi)$, $n = 1, 2, \ldots, k$.

Similarly, an expression can be written for $F_p^*(s)$ if $F(\xi)$ has some multiple-order poles.

Now consider Eq. (2-27). If the condition in Eq. (2-29) is satisfied, we have

$$F_p^*(s) = \frac{1}{2\pi j} \oint_{\Gamma_2} F(\xi) \frac{1 - e^{-p(s-\xi)}}{(s - \xi)[1 - e^{-T(s-\xi)}]} \tag{2-35}$$

Applying the residue theorem, Eq. (2-35) is written

$$F_p^*(s) = -\sum \text{Residues of } \left\{F(\xi) \frac{1 - e^{-p(s-\xi)}}{(s - \xi)[1 - e^{-T(s-\xi)}]}\right\}$$
$$\text{at the poles of } \left\{\frac{1 - e^{-p(s-\xi)}}{(s - \xi)[1 - e^{-T(s-\xi)}]}\right\} \tag{2-36}$$

where the minus sign comes from the fact that the contour integration along Γ_2 is taken in the clockwise direction. Since $P(s - \xi)$ has only simple poles that lie at periodic intervals along $\text{Re}(\xi) = \text{Re}(s)$ in the ξ-plane, Eq. (2-36) becomes

$$F_p^*(s) = -\sum_{n=-\infty}^{\infty} \frac{N(\xi_n)}{D'(\xi_n)} F(\xi_n) \tag{2-37}$$

where $\xi_n = s + jn\omega_s$ denotes the poles of $P(s - \xi)$, $n = 0, \pm 1, \pm 2, \ldots$. Also,

$$N(\xi_n) = 1 - e^{-p(s-\xi)}\big|_{\xi=\xi_n=s+jn\omega_s} = 1 - e^{-jn\omega_s p} \tag{2-38}$$

and

$$D'(\xi_n) = \frac{d}{d\xi}[(s - \xi)(1 - e^{-T(s-\xi)})]_{\xi=s+jn\omega_s} = -jn\omega_s T \tag{2-39}$$

Therefore, Eq. (2-37) becomes

$$F_p^*(s) = \sum_{n=-\infty}^{\infty} \frac{1 - e^{-jn\omega_s p}}{jn\omega_s T} F(s + jn\omega_s) \tag{2-40}$$

It is apparent that when s is replaced by $j\omega$, Eq. (2-40) is identical to Eq. (2-14) which is obtained using a different method.

2.3 FLAT-TOP APPROXIMATION AND THE IDEAL SAMPLER

In the analysis of the operation of a sampler, if the sampling duration p is very much smaller than the sampling period T and the largest time constant of the input signal, $f(t)$, the output of the finite-pulsewidth sampler, $f_p^*(t)$, can be approximated by a sequence of flat-topped pulses; that is,

$$\begin{aligned} f_p^*(t) &= f(kT) \qquad && \text{for } kT \leq t < kT + p \\ &= 0 && kT + p \leq t < (k+1)T \end{aligned} \tag{2-41}$$

where $k = 0, 1, 2, 3, \ldots$.

Then $f_p^*(t)$ is written

$$f_p^*(t) = \sum_{k=0}^{\infty} f(kT)[u(t - kT) - u(t - kT - p)] \tag{2-42}$$

where $u(t)$ is the unit step function.

Taking the Laplace transform on both sides of Eq. (2-25) gives

$$F_p^*(s) = \sum_{k=0}^{\infty} f(kT)\left(\frac{1 - e^{-ps}}{s}\right)e^{-kTs} \tag{2-43}$$

If the sampling duration p is very small,

$$1 - e^{-ps} = 1 - \left(1 - ps + \frac{(ps)^2}{2!} - \ldots\right) \cong ps \tag{2-44}$$

Then, Eq. (2-43) becomes

$$F_p^*(s) \cong p \sum_{k=0}^{\infty} f(kT)\, e^{-kTs} \tag{2-45}$$

or, equivalently,

$$f_p^*(t) \cong p \sum_{k=0}^{\infty} f(kT)\, \delta(t - kT) \tag{2-46}$$

where $\delta(t)$ is the unit impulse function.

The right-hand side of Eq. (2-46) is recognized as a train of impulses with the strength of the impulse at $t = kT$ equal to $pf(kT)$. This means that the finite-pulsewidth sampler can be approximated by an "impulse modulator" whose diagram is that shown in Figure 2-5 but with the carrier signal equal to the unit impulse train:

$$\delta_T(t) = \sum_{k=0}^{\infty} \delta(t - kT) \tag{2-47}$$

Or, the finite-pulsewidth sampler can be approximated by an *ideal sampler* connected in series with an attenuator with attenuation p. The equivalent situation is illustrated in Figure 2-8. Therefore, an ideal sampler is defined as a sampler which closes and opens instantaneously, for a zero time duration,

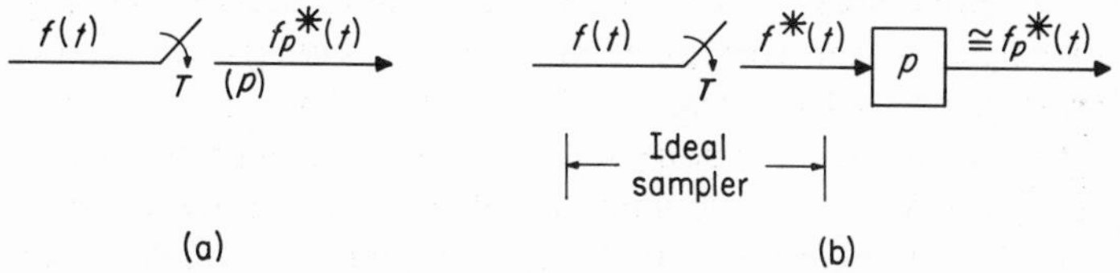

FIG. 2-8. (a) Finite pulsewidth sampler; (b) ideal sampler connected in series with an attenuator p. (a) and (b) are equivalent if p is much smaller than T and the largest time constant of $f(t)$.

every T seconds. It should be noted, however, that the attenuator is necessary only if the sampler is not followed by a data-hold device.

The output of the ideal sampler is written as

$$f^*(t) = \sum_{k=0}^{\infty} f(kT)\,\delta(t - kT) = f(t)\cdot\delta_T(t) \tag{2-48}$$

where $f(t)$ is the input of the sampler, and sampling is assumed to begin at $t = 0$. Taking the Laplace transform on both sides of Eq. (2-48), we get

$$F^*(s) = \sum_{k=0}^{\infty} f(kT)\,e^{-kTs} \tag{2-49}$$

which is the Laplace transform of the output of the ideal sampler.

Typical input and output signals of an ideal sampler are illustrated as shown in Figure 2-9. The output of the ideal sampler is a train of impulses

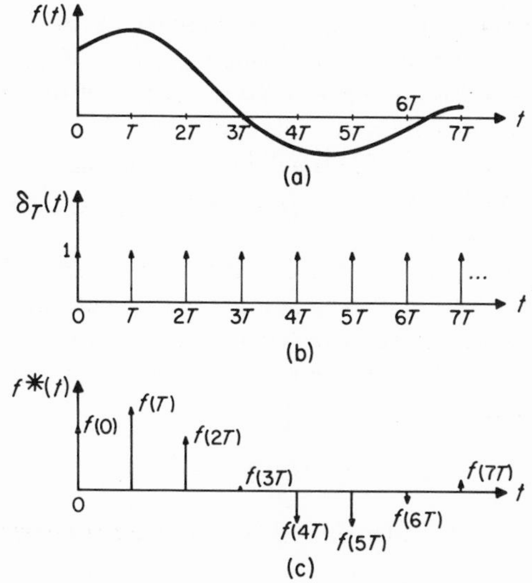

FIG. 2-9. (a) Input signal to ideal sampler; (b) unit impulse train; (c) output of ideal sampler. $f^*(t) = f(t)\delta_T(t)$.

with the respective areas (strengths) of the impulses equal to the magnitudes of the input signal at the corresponding sampling instants. Since an impulse function is defined to have zero pulsewidth and infinite pulse amplitude, in Figure 2-9 the impulses are represented by arrows with appropriate amplitudes representing the strengths of the impulses. We can also derive an alternate expression for $F^*(s)$ using Eq. (2-40). Since $\delta_T(t)$ and $p(t)$ are related by

$$\delta_T(t) = \lim_{p\to 0} \frac{1}{p} p(t) \tag{2-50}$$

$F^*(s)$ can be written as

$$\begin{aligned} F^*(s) &= \lim_{p\to 0} \frac{1}{p} F_p^*(s) = \lim_{p\to 0} \frac{1}{p} \sum_{n=-\infty}^{\infty} \frac{1 - e^{-jn\omega_s p/2}}{n\omega_s T} E(s + jn\omega_s) \\ &= \frac{1}{T} \sum_{n=-\infty}^{\infty} F(s + jn\omega_s) \end{aligned} \tag{2-51}$$

However, Eq. (2-40) is obtained from the contour integration around the path Γ_2 which is in the right-half ξ-plane. It is important to examine the convergence properties of the integral taken along the semicircle with infinite radius. Applying the limit as $p \to 0$ to $F_p^*(s)/p$ and using Eq. (2-35), we have

$$\begin{aligned} F^*(s) &= \lim_{p\to 0} \frac{1}{2\pi jp} \oint_{\Gamma_2} F(\xi) \frac{1 - e^{-p(s-\xi)}}{(s-\xi)(1 - e^{-T(s-\xi)})} d\xi \\ &= \frac{1}{2\pi j} \oint_{\Gamma_2} F(\xi) \frac{1}{1 - e^{-T(s-\xi)}} d\xi \end{aligned} \tag{2-52}$$

In this case since $1/1 - e^{-T(s-\xi)}$ has a simple pole at infinity in the ξ-plane, the part of the integral of Eq. (2-52) along the semicircle of infinite radius in the right half of the ξ-plane may not vanish. In fact, if the degree of the denominator of $F(\xi)$ in ξ is not higher than the degree of the numerator by at least two, the integral along the semicircle may have a finite value or may even not converge. Therefore Eq. (2-51) is valid only if $F(s)$ has a pole-zero excess greater or equal to two. In other words, the input signal $f(t)$ must not have a jump discontinuity at $t = 0$.

A more general expression than Eq. (2-51) can be derived for $F^*(s)$ by defining $\delta_T(t)$ as an even function so that the unit impulse at the time origin $t = 0$ is considered to be a pulse with amplitude $1/p$, width extending from $t = -p/2$ to $t = p/2$, and in the limit $p \to 0$. Let the Fourier series of $\delta_T(t)$ for $t > 0$ be designated by

$$\sum_{n=-\infty}^{\infty} C_n e^{jn\omega_s t}$$

where it can be easily shown that $C_n = 1/T$. Then the Fourier series for $\delta_T(t)$ for $t \geq 0$ is

$$\delta_T(t) = \frac{1}{2}\delta(t) + \sum_{n=-\infty}^{\infty} \frac{1}{T} e^{jn\omega_s t} \tag{2-53}$$

The $\delta(t)/2$ term on the right side of Eq. (2-53) is added because only half of the unit impulse was considered for $t > 0$. Now substituting Eq. (2-53) into Eq. (2-48) and taking the Laplace transform, we have

$$F^*(s) = \frac{f(0^+)}{2} + \frac{1}{T}\sum_{n=-\infty}^{\infty} F(s + jn\omega_s) \tag{2-54}$$

In a similar fashion we can show that for the ideal sampler case the expression for $F^*(s)$ which is analogous to Eq. (2-32) is (simple poles)

$$F^*(s) = \sum_{n=1}^{k} \frac{N(\xi_n)}{D'(\xi_n)} \frac{1}{1 - e^{-T(s-\xi_n)}} \tag{2-55}$$

where $N(\xi_n)$ and $D'(\xi_n)$ are defined in Eqs. (2-33) and (2-34), respectively.

The applications of Eqs. (2-49), (2-51), (2-54) and (2-55) will now be illustrated by an example.

EXAMPLE 2-1

Consider that a unit step function $f(t) = u(t)$ is sampled by an ideal sampler at intervals of T seconds. From Eq. (2-48), the output of the ideal sampler is

$$f^*(t) = \sum_{k=0}^{\infty} f(kT)\,\delta(t - kT) = \sum_{k=0}^{\infty} \delta(t - kT) \tag{2-56}$$

The Laplace transform of $f^*(t)$ is

$$\begin{aligned} F^*(s) &= \sum_{k=0}^{\infty} e^{-kTs} = 1 + e^{-Ts} + e^{-2Ts} + \dots. \\ &= \frac{1}{1 - e^{-Ts}} \text{ for } |e^{-Ts}| < 1 \end{aligned} \tag{2-57}$$

Now let us obtain the same answer using Eq. (2-55). We have

$$F(s) = \frac{N(s)}{D(s)} = \frac{1}{s} \tag{2-58}$$

Thus $N(s) = 1$ and $D'(s) = 1$, and Eq. (2-55) gives

$$F^*(s) = \frac{1}{1 - e^{-Ts}} \tag{2-59}$$

which is the same result as in Eq. (2-57).

Since the signal $f(t)$ has a jump discontinuity at $t = 0$, using Eq. (2-51) will lead to an erroneous result. In this case, Eq. (2-54) should be used. Therefore,

$$\begin{aligned} F^*(s) &= \frac{f(0^+)}{2} + \frac{1}{T}\sum_{n=-\infty}^{\infty} F(s + jn\omega_s) \\ &= \frac{1}{2} + \frac{1}{T}\sum_{n=-\infty}^{\infty} \frac{1}{s + jn\omega_s} \end{aligned} \tag{2-60}$$

which can be written

$$\begin{aligned} F^*(s) &= \frac{1}{2} + \frac{1}{Ts} + \sum_{n=1}^{\infty} \frac{2s}{s^2 + n^2\omega_s^2} \frac{1}{2} \coth\left(\frac{sT}{2}\right) \\ &= 1 + \frac{1}{e^{sT} - 1} = \frac{1}{1 - e^{-Ts}} \end{aligned} \tag{2-61}$$

In general, Eqs. (2-49) and (2-55) are more useful in the determination of the expression of the Laplace transform of the sampled signal (by an ideal sampler), whereas Eqs. (2-51) and (2-54) are useful in frequency response plots and analysis.

Referring to Eqs. (2-51) and (2-54) we see that the ideal sampler is again a harmonic generator. The ideal sampler reproduces in its output the spectrum of the continuous input $f(t)$ as well as the complementary components at integral multiples of the sampling frequency; and all components have the same amplitude of $1/T$. Assuming that the amplitude spectrum of the continuous input signal is as shown in Figure 2-10(a), the corresponding amplitude spectrum of the sampled signal $f^*(t)$ when $\omega_s > 2\omega_c$ is shown in Figure 2-10(b), where ω_s is the sampling frequency and ω_c is the highest frequency contained in $f(t)$. Again, if the sampling frequency is less than $2\omega_c$, distortion will occur in the output spectrum because of the overlapping of the sidebands in $|F^*(j\omega)|$.

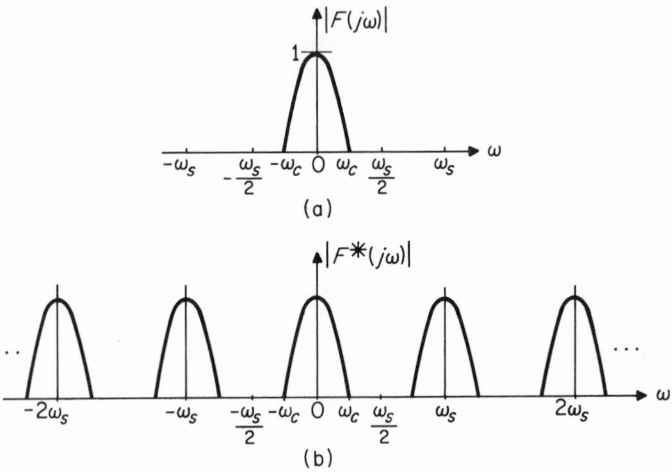

FIG. 2-10. Amplitude spectra of input and output signals of an ideal sampler: (a) amplitude spectrum of continuous input $f(t)$. (b) Amplitude spectrum of sampler output ($\omega_s > 2\omega_c$).

2.4 THE SAMPLING THEOREM

The simple physical reasoning made in the preceding section concerning the minimum sampling frequency required for full recovery of the continuous signal actually answers a very basic yet important question in regard to the proper rate of sampling if sampling is applied intentionally. If sampling is used in a given system, we may ask the question: What are the limitations on the rate of sampling? Theoretically, there is no upper limit on the sampling frequency, although any physical sampler must have a maximum sampling rate. When the sampling frequency reaches infinity the continuous case is approached. (We must treat the infinite sampling frequency carefully when ideal sampling is considered since impulses are involved.) As far as the lower limit of the sampling frequency is concerned, intuitively, we know that if the continuous signal changes rapidly with respect to time, sampling the signal at too low a rate may lose vital information on the signal between the sampling instants. Consequently, it may not be able to reconstruct the original signal from the information contained in the sampled data.

We can conclude from the amplitude spectra shown in Figure 2-10 that the lowest sampling frequency for possible signal reconstruction is $2\omega_c$, where ω_c is the highest frequency contained in $f(t)$. Formally, this is known as the *sampling theorem.*[1,2] The theorem states *that if a signal contains no frequency higher than ω_c radians per second, it is completely characterized by the values of the signal measured at instants of time separated by $T = \frac{1}{2}(2\pi/\omega_c)$ second.* In practice, however, stability of closed-loop systems and other factors also dictate the choice of the sampling frequency and may make it necessary to sample at a rate much higher than this theoretical minimum. Furthermore, strictly speaking, a band-limited signal does not exist physically in communication or control systems. All physical signals found in the real world do contain components covering a wide frequency range. It is because the amplitudes of the higher frequency components are greatly diminished that a band-limited signal is assumed. Therefore, in practice, these factors plus the unrealizability of an ideal low-pass filter make impossible the exact reproduction of a continuous signal from the sampled signal even if the sampling theorem is satisfied.

An interesting note on the sampling theorem may be inserted here. A signal can still be defined completely by sampling it at a rate less than $2\omega_c$ radians per second provided that the derivatives of the signal are known at the sampling instants as well as the amplitude information. L. J. Fogel and others[3,4] have proved that if a signal contains no frequency higher than ω_c radians per second, it is completely characterized by the values $f^{(n)}(kT), f^{(n-1)}(kT), \ldots, f^{(1)}(kT)$, and $f(kT)$, $(k = 0, 1, 2, \ldots)$ of the signal measured at instants of time separated by $T = \frac{1}{2}(n + 1)(2\pi/\omega_c)$ second, where

$$f^{(n)}(kT) = \left. \frac{d^n f(t)}{dt^n} \right|_{t=kT} \tag{2-62}$$

This means that when the values of the first derivative of $f(t)$ at $t = kT$, $f^{(1)}(kT)$, for $k = 0, 1, 2, \ldots$, are known in addition to the values of $f(kT)$, the maximum allowable sampling time is $T = 2\pi/\omega_c$, which is twice the time required when $f(kT)$ alone is measured. The addition of each succeeding derivative allows the time between samples to become larger according to $T = \frac{1}{2}(n + 1)(2\pi/\omega_c)$, where n is the order of the highest derivative when all lower ordered derivatives are observed for each sample.

2.5 SOME S-PLANE PROPERTIES OF $F^*(s)$

Two important properties of the output of the ideal sampler can be observed from Eq. (2-51).

1. $F^*(s)$ is a periodic function with period $j\omega_s$.

The property of $F^*(s)$ as a periodic function is apparent in view of Eq. (2-51) and Figure 2-10. Analytically, we can show this fact by substituting $s + jm\omega_s$ for s in Eq. (2-49), where m is an integer. We have

$$F^*(s + jm\omega_s) = \sum_{k=0}^{\infty} f(kT)\, e^{-kT(s+jm\omega_s)}$$
$$= \sum_{k=0}^{\infty} f(kT)\, e^{-kTs} \tag{2-63}$$

since $e^{-jkm\omega_s T} = 1$. Therefore,

$$F^*(s + jm\omega_s) = F^*(s) \tag{2-64}$$

whenever m is an integer. In other words, given any point $s = s_1$ in the s-plane, the function $F^*(s)$ has the same value at all periodic points

$$s = s_1 + jm\omega_s$$

for m = any integer. This equality is illustrated in Figure 2-11. Thus the s-plane is divided into an infinite number of periodic strips as shown in Figure 2-11. The strip between $\omega = -\omega_s/2$ and $\omega = \omega_s/2$ is called the *primary strip* and all others occurring at higher frequencies are designated as the complementary strips. The function $F^*(s)$ has the same value at all congruent points in the various periodic strips.

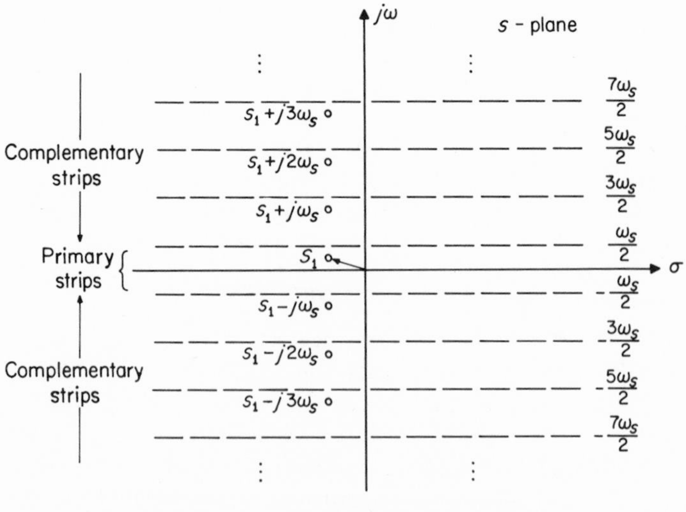

FIG. 2-11. Periodic strips in the s-plane.

2. If the function $F(s)$ has a pole at $s = s_1$, then $F^*(s)$ has poles at $s = s_1 + jm\omega_s$ for m = integers from $-\infty$ to $+\infty$.

The fact that this statement is true can be observed easily from Eq. (2-51). A typical set of pole configurations of $F(s)$ and the corresponding $F^*(s)$ is shown in Figure 2-12.

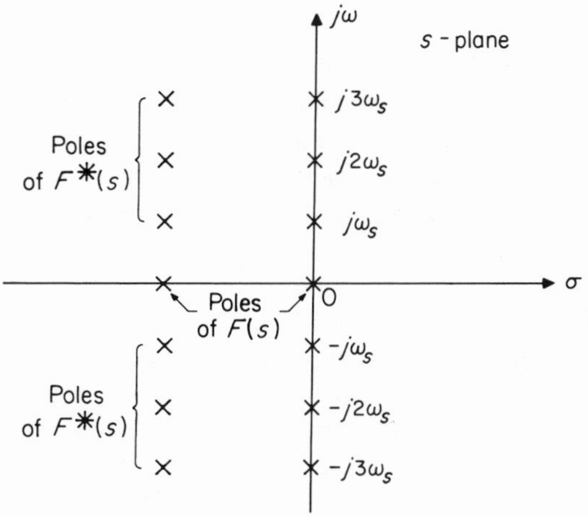

FIG. 2-12. Periodicity of the poles of $F^*(s)$.

2.6 RECONSTRUCTION OF SAMPLED SIGNALS

In most feedback control systems employing sampled data or digital data, the high frequency harmonic components in $f^*(t)$ that resulted from the sampling operation must be removed before the signal is applied to the continuous-data section of the system. In systems with components which are designed to be actuated by continuous-data signals the smoothing of the pulse signals is necessary since otherwise these components may be subject to excessive wear. Therefore, it is often necessary to consider a data-reconstruction device, or, simply, a filtering device for the removal of the high frequency components contained in $f^*(t)$, before the signal is applied to the controlled process.

The block diagram of a typical digital control system using a data-reconstruction device is shown in Figure 2-13. In this case the digital signal is first processed by a digital controller $D^*(s)$; the output of $D^*(s)$, which is still a digital signal, is then smoothed by the data-reconstruction device before the signal is applied to the continuous-data components of the controlled process. The output of the data-reconstruction device, $h(t)$, is considered to be a function of the continuous time variable t.

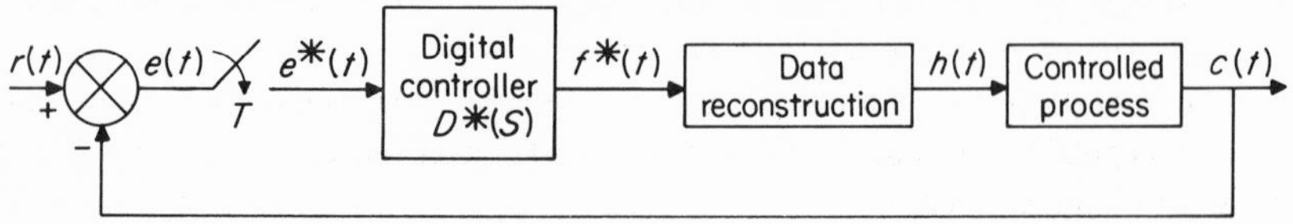

FIG. 2-13. Digital control system with data reconstruction device.

In order to study how data reconstruction may be accomplished, let us assume that an ideal sampler is sampling at a sampling frequency of ω_s which is at least twice as large as the highest frequency component contained in the continuous input being sampled. Figure 2-14 shows the amplitude spectrum of $F^*(s)$, from which it is apparent that in order to duplicate the continuous-data signal the sampled signal must be sent through an ideal low-pass filter with an amplitude characteristic as shown in Figure 2-15. Unfortunately, an ideal filter characteristic is physically unrealizable, since it is well known that its time response begins before an input is applied. However, even if we could realize the ideal filter response, we have mentioned previously that perfect reproduction of the continuous signal is still based on the assumption that $f(t)$ is band-limited. Therefore, for all practical considerations, it is impossible to recover a continuous signal perfectly once it is sampled. The best we can do in data

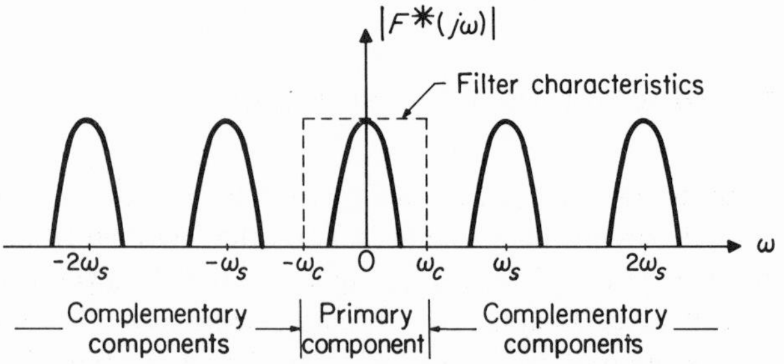

FIG. 2-14. Reconstruction of continuous signal from digital data using an ideal low-pass filter.

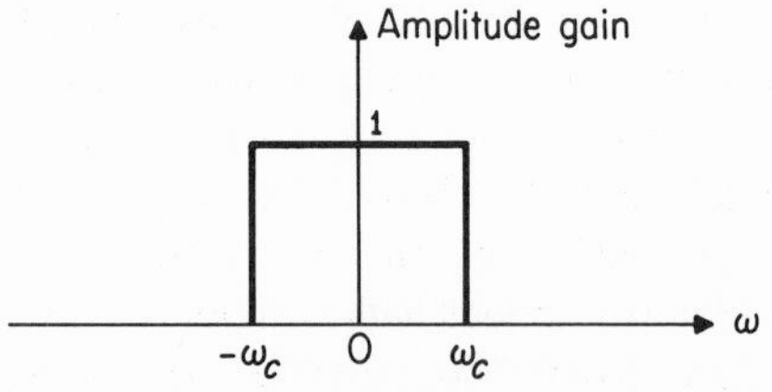

FIG. 2-15. Amplitude characteristic of an ideal filter.

reconstruction is to try to approximate the original time function as closely as possible. Furthermore, as will be shown later in this chapter, a better approximation of the original signal requires in general, a long time delay that is undesirable in view of its adverse effect on system stability. Consequently, the design of a data-reconstruction device usually involves a compromise between the requirements of stability and the desirability of a close approximation of the continuous signal.

The problem confronting us is that we have a sequence of numbers, $f(0), f(T), \ldots, f(kT), \ldots$, or a train of impulses with the strength of the impulse occurring at $t = kT$ equal to $f(kT)$, $k = 0, 1, 2, \ldots$, and the continuous-data signal $f(t)$, $t \geq 0$, must be reconstructed from the information contained in the pulsed data. This data-reconstruction process may be regarded as an extrapolation process, since the continuous signal is to be constructed based on information available only at past sampling instants. For instance, the original signal $f(t)$ between two consecutive sampling instants kT and $(k + 1)T$ is to be estimated based on the values of $f(t)$ at all preceding sampling instants of $kT, (k - 1)T, (k - 2)T, \ldots, 0$; that is, $f(kT), f(k - 1)T, f(k - 2)T, \ldots, f(0)$.

A well-known method[10] of generating this desired approximation is based on the power series expansion of $f(t)$ in the interval between sampling instants kT and $(k + 1)T$; that is

$$f_k(t) = f(kT) + f^{(1)}(kT)(t - kT) + \frac{f^{(2)}(kT)}{2!}(t - kT)^2 + \ldots \tag{2-65}$$

where

$$f_k(t) = f(t) \quad \text{for } kT \leq t < (k + 1)T \tag{2-66}$$

$$f^{(1)}(kT) = \left.\frac{df(t)}{dt}\right|_{t=kT} \tag{2-67}$$

and

$$f^{(2)}(kT) = \left.\frac{d^2f(t)}{dt^2}\right|_{t=kT} \tag{2-68}$$

In order to evaluate the coefficients of the series given by Eq. (2-65), the derivatives of the function $f(t)$ must be obtained at the sampling instants. Since the only available information concerning $f(t)$ is its magnitudes at the sampling instants, the derivatives of $f(t)$ must be estimated from the values of $f(kT)$. A simple expression involving only two data pulses gives an estimate of the first derivative of $f(t)$ at $t = kT$ as

$$f^{(1)}(kT) = \frac{1}{T}[f(kT) - f(k - 1)T] \tag{2-69}$$

An approximated value of the second derivative of $f(t)$ at $t = kT$ is

$$f^{(2)}(kT) = \frac{1}{T}\{f^{(1)}(kT) - f^{(1)}[(k - 1)T]\} \tag{2-70}$$

Substitution of Eq. (2-69) into Eq. (2-70) yields

$$f^{(2)}(kT) = \frac{1}{T^2}\{f(kT) - 2f[(k-1)T] + f[(k-2)T]\} \qquad (2\text{-}71)$$

From these approximated values of $f^{(1)}(kT)$ and $f^{(2)}(kT)$, we see that the higher the order of the derivative to be approximated, the larger will be the number of delay pulses required. In fact, we can easily show that the number of delay pulses required to approximate the value of $f^{(n)}(kT)$ is $n + 1$. Thus, the extrapolating device described above consists essentially of a series of time delays, and the number of delays depends on the degree of accuracy of the estimate of the time function $f(t)$. The adverse effect of time delay on the stability of feedback control systems is well known. Therefore, an attempt to utilize the higher order derivative of $f(t)$ for the purpose of more accurate extrapolation is often met with serious difficulties in maintaining system stability. Furthermore, a high-order extrapolation also requires complex circuitry and results in a high cost in construction. For these two reasons, quite frequently only the first term of Eq. (2-65) is used in practice. A device generating only $f(kT)$ of Eq. (2-65), for the time interval $kT \leq t < (k+1)T$, is normally known as the *zero-order extrapolator*, since the polynomial used is of the zeroth order. It is also commonly referred to as the *zero-order hold* device, because it holds the value of the previous sample during a given sampling period, until the next sample arrives. Similarly, a device that implements the first two terms of Eq. (2-65) is called a *first-order hold*, since the polynomial generated by this device is of the first order. Details of the zero-order hold and the first-order hold are discussed in the following sections.

2.7 THE ZERO-ORDER HOLD

When only the first term of the power series in Eq. (2-65) is used to approximate a signal between two conservative sampling instants, the polynomial extrapolator is called a *zero-order hold*. Then, Eq. (2-65) is simply

$$f_k(t) = f(kT) \qquad (2\text{-}72)$$

Equation (2-72) defines the impulse response of the zero-order hold to be as shown in Figure 2-16. In principle, the operation of a sampler and zero-order hold device is illustrated by the simple circuit shown in Figure 2-17. The capacitor is assumed to be charged up to the voltage $f(kT)$ at the sampling instant $t = kT$ instantaneously. (The actual rate of charging up the capacitor is, of course, determined by the capacitance and the source impedance.) As the sampling switch is opened during the sampling period T, the capacitor holds the charge until the next pulse comes along from the sampler. The input impedance of the amplifier is assumed to be infinite so that there is no path

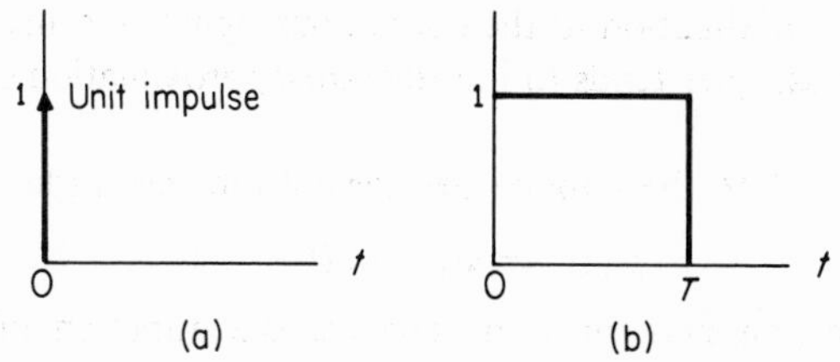

FIG. 2-16. (a) Unit impulse input to zero-order hold; (b) impulse response of zero-order hold.

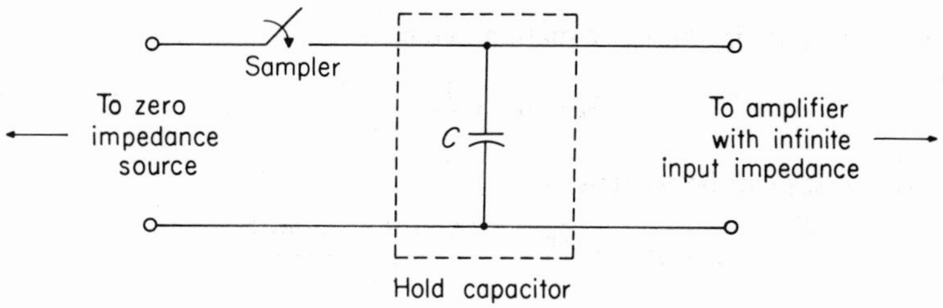

FIG. 2-17. Simplified diagram of a sampler and zero-order hold device.

for the capacitor to discharge. The zero-order hold circuit thus serves to stretch the input pulses into a series of rectangular pulses of width T. In practice, however, the amplifier does have a large finite input impedance. Therefore, the actual waveform of the output of the zero-order hold is not a train of exact rectangular pulses but rather, a series of exponentially decayed pulses having a large time constant. The input and output waveforms of an ideal zero-order hold device are shown in Figure 2-18. Observe that the output of the zero-order

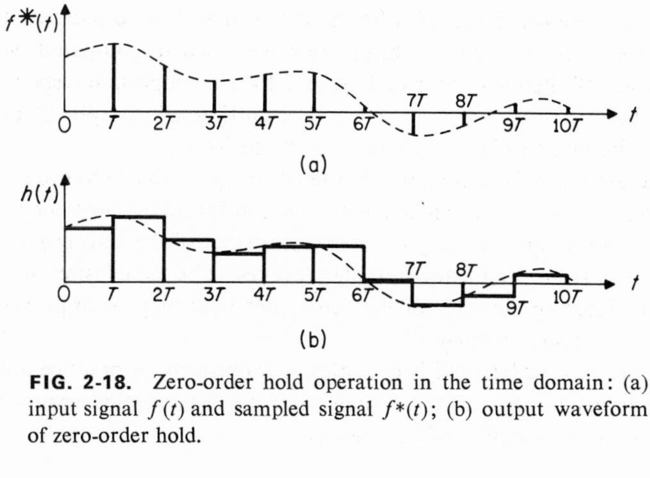

FIG. 2-18. Zero-order hold operation in the time domain: (a) input signal $f(t)$ and sampled signal $f^*(t)$; (b) output waveform of zero-order hold.

hold is a step approximation to the continuous signal, and increasing the sampling rate of the sampler tends to improve the approximation of the continuous signal.

From Figure 2-16 the impulse response of the zero-order hold is written

$$g_{h0}(t) = u(t) - u(t - T) \tag{2-73}$$

where $u(t)$ is the unit-step function. The transfer function of the zero-order hold is then

$$G_{h0}(s) = \frac{1 - e^{-Ts}}{s} \tag{2-74}$$

Replacing s by $j\omega$ in the last equation, we have

$$G_{h0}(j\omega) = \frac{1 - e^{-j\omega T}}{j\omega} \tag{2-75}$$

Equation (2-75) can be written as

$$G_{h0}(j\omega) = \frac{2e^{-j\omega T/2}(e^{j\omega T/2} - e^{-j\omega T/2})}{j2\omega}$$

$$= \frac{2 \sin (\omega T/2)\, e^{-j\omega T/2}}{\omega} \tag{2-76}$$

or

$$G_{h0}(j\omega) = T \frac{\sin (\omega T/2)}{(\omega T/2)} e^{-j\omega T/2} \tag{2-77}$$

Since T is the sampling period in seconds, and $T = 2\pi/\omega_s$, where ω_s is the sampling frequency in rad/sec, Eq. (2-77) becomes

$$G_{h0}(j\omega) = \frac{2\pi}{\omega_s} \frac{\sin \pi(\omega/\omega_s)}{(\omega\pi/\omega_s)} e^{-j\pi(\omega/\omega_s)} \tag{2-78}$$

The gain and phase characteristics of the zero-order hold as functions of ω are shown in Figure 2-19. This figure shows that the zero-order hold behaves essentially as a low-pass filter. However, when compared with the characteristics of the ideal filter of Figure 2-14, the amplitude response of the zero-order hold is zero at $\omega = \omega_s$, instead of cutting off sharply at $\omega_s/2$. At $\omega = \omega_s/2$, the magnitude of $G_{h0}(j\omega)$ is equal to 0.636.

Figure 2-18 clearly indicates that the accuracy of the zero-order hold as an extrapolating device depends greatly on the sampling frequency ω_s. The effect of the sampling frequency on the performance of the zero-order hold can also be shown from frequency response curves. The conclusion is that, in general, the filtering property of the zero-order hold may be improved by increasing the sampling frequency.

Since the zero-order hold is used almost exclusively in practice and in the ensuing chapters of this text, we shall refer to the sampler-zero-order hold combination as the *sample-and-hold* device.

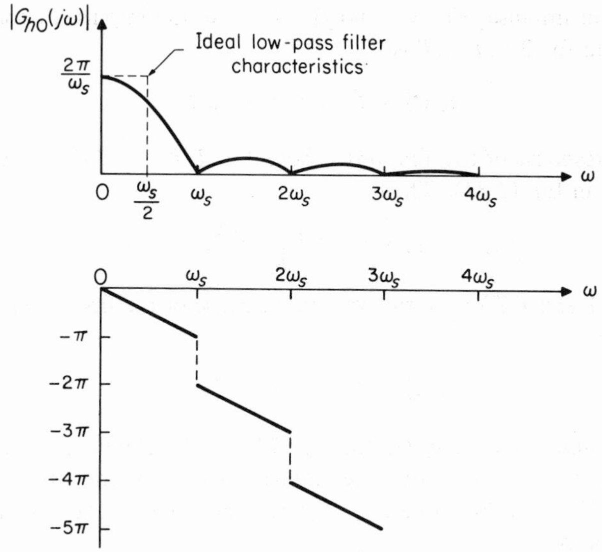

FIG. 2-19. Gain and phase characteristics of a zero-order hold device.

2.8 THE FIRST-ORDER HOLD

When the first two terms of the power series in Eq. (2-65) are used to approximate the time function $f(t)$ between two successive sampling instants kT and $(k+1)T$, the polynomial extrapolator is called a *first-order hold.* Equation (2-65) becomes

$$f_k(t) = f(kT) + f^{(1)}(kT)(t - kT) \quad kT \leq t < (k+1)T \tag{2-79}$$

The first derivative of $f(t)$ at $t - kT$ can be approximated by

$$f^{(1)}(kT) = \frac{f(kT) - f[(k-1)T]}{T} \tag{2-80}$$

since only the values of $f(t)$ at the sampling instants are known. Substituting Eq. (2-80) into Eq. (2-79) yields

$$f_k(t) = f(kT) + \frac{f(kT) - f[(k-1)T]}{T}(t - kT) \tag{2-81}$$

The impulse response of the first-order hold may be derived from Eq. (2-81) by applying an impulse of unit strength at $t = 0$ as the input $f(t)$. Therefore, when $k = 0$, Eq. (2-81) reads

$$f_0(t) = f(0) + \frac{f(0) - f(-T)}{T}t \tag{2-82}$$

Since for a unit impulse $f(0) = 1$ and $f(-T) = 0$, the impulse response of the first-order hold for $0 \leq t < T$ is

$$g_{h1}(t) = f_0(t) = 1 + \frac{1}{T}t \tag{2-83}$$

The impulse response of the first-order hold for $T \leq t < 2T$ is obtained by setting $k = 1$ in Eq. (2-81). Then

$$f_1(t) = f(T) + \frac{f(T) - f(0)}{T}(t - T) \tag{2-84}$$

Since $f(0) = 1$ and $f(T) = 0$, the impulse response of the first-order hold for $T \leq t < 2T$ is

$$g_{h1}(t) = f_1(t) = 1 - \frac{1}{T}t \tag{2-85}$$

The impulse response $g_{h1}(t)$ for $t > 2T$ can be shown to be zero since $f(kT) = 0$ for $k \geq 2$. Therefore, the impulse response of the first-order hold is as shown in Figure 2-20. For all $t \geq 0$, the impulse response of the first-order hold is written as

$$g_{h1}(t) = \left(1 + \frac{t}{T}\right)u(t) - 2\left(1 + \frac{t-T}{T}\right)u(t-T) + \left(1 + \frac{t-2T}{T}\right)u(t-2T) \tag{2-86}$$

and the transfer function is

$$G_{h1}(s) = \frac{Ts + 1}{Ts^2}(1 - e^{-Ts})^2 \tag{2-87}$$

FIG. 2-20. (a) Unit impulse input; (b) impulse response of first-order hold.

The time-domain characteristics of the first-order hold used to approximate a continuous-time function are illustrated in Figure 2-21. Figure 2-21 shows that for the sampling period chosen, the waveforms of the output of the hold device do not indicate any advantage of the first-order hold over the zero-order hold. As a matter of fact, the zero-order hold actually gives better approximations in certain sampling intervals. However, it can be shown that if the sampling frequency is increased, the first-order hold does give a

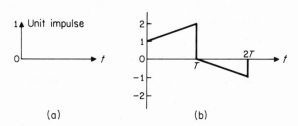

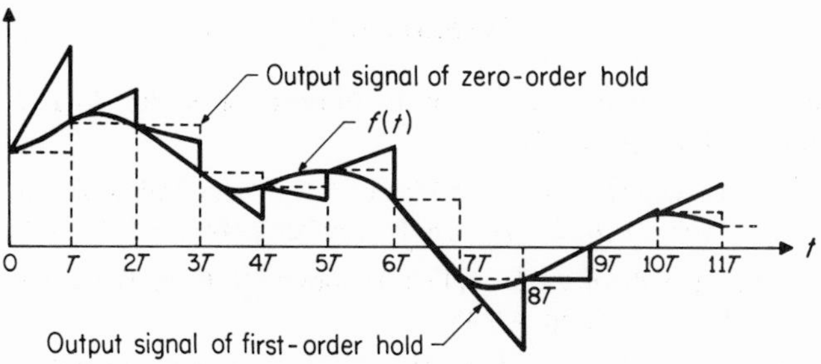

FIG. 2-21. Reconstruction of a continuous signal from sampled data by zero-order hold and first-order hold.

better approximation. One should not be concerned with the large errors in the output of the first-order hold during the first sampling period ($0 \leq t < T$); this is due to the assumption that $f(t)$ has a jump discontinuity at $t = 0$.

The frequency response curves for the first-order hold are shown in Figure 2-22. From the gain characteristic we see that the first-order hold cuts off at

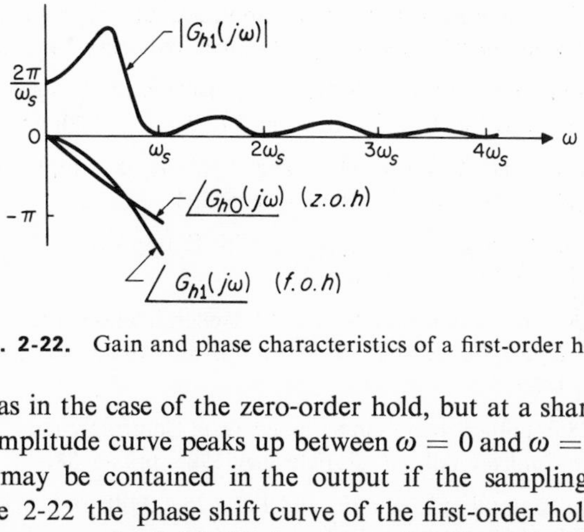

FIG. 2-22. Gain and phase characteristics of a first-order hold.

$\omega = \omega_s$, just as in the case of the zero-order hold, but at a sharper rate. The fact that the amplitude curve peaks up between $\omega = 0$ and $\omega = \omega_s$ shows that higher ripple may be contained in the output if the sampling frequency is low. In Figure 2-22 the phase shift curve of the first-order hold is compared with that of the zero-order hold. Notice that at low frequencies, the phase lag produced by the zero-order hold exceeds that of the first-order hold but, as the frequency becomes higher, the opposite is true. However, in feedback control systems, the high frequency characteristics of the one-loop transfer function normally control the system stability. Therefore, application of the first-order hold to control systems often complicates the problem of stability of the closed-loop system. Consequently, although the first-order hold in general does a better job in data reconstruction than the zero-order hold, the fact that it is more complex to implement and its adverse effect on system stability made it seldom practical.

REFERENCES

1. Shannon, C. E., "Communication in the Presence of Noise," *Proc. I. R. E,*. Vol. 37, January 1949, pp. 10-21.
2. Shannon, C. E., Oliver, B. M., and Pierce, J. R., "The Philosophy of Pulse Code Modulation," *Proc. I. R. E.*, Vol. 36, November 1948, pp. 1324-1331.
3. Fogel, L. J., "A Note of the Sampling Theorem," *I. R. E. Trans. on Information Theory*, **1**, March 1955, pp. 47-48.
4. Jagerman, and Fogel, L. J., "Some General Aspects of the Sampling Theorem," *I. R. E. Trans. on Information Theory*, **2**, December 1956, pp. 139-146.
5. Linden, D. A., "A Discussion of Sampling Theorems," *Proc. I. R. E.*, Vol. 47, July 1959, pp. 1219-1226.
6. Oliver, R. M., "On the Functions Which are Represented by the Expansions of the Interpolation-Theory," *Proc. Royal Society* (Edinburgh), **35**, 1914-1915, pp. 181-194.
7. Kuo, B. C., *Analysis and Synthesis of Sampled-Data Control Systems*. Prentice-Hall, Englewood Cliffs, N. J., 1963.
8. Barker, R. H., "The Reconstruction of Sampled-Data," *Proc. Conference on Data Processing and Automatic Computing Machines*, Salisbury, Australia, June 1957.
9. Porter, A., and Stoneman, F., "A New Approach to the Design of Pulse Monitored Servo Systems," *J. I. E. E.*, London, **97**, Part II, 1950, pp. 597-610.
10. Ragazzini, J. R., and Zadeh, L. H., "The Analysis of Sampled-Data Systems," *Trans. A. I. E. E.*, **71**, Part II, 1952, pp. 225-234.
11. Tsypkin, Y. Z., "Sampled-Data Systems With Extrapolating Devices," *Avtomat. Telemekh*, **19**, No. 5, 1958, pp. 389-400.
12. Linden, D. A., and Abramson, N. M., "A Generalization of the Sampling Theorem," *Tech. Report No. 1551-2*, Solid-State Elec. Laboratory, Stanford University, August, 1959.
13. Jury, E. I., "Sampling Schemes in Sampled-Data Control Systems," *I. R. E. Trans. on Automatic Control*, Vol. AC-6, February 1961, pp. 86-88.
14. Beutler, F. J., "Sampling Theorems and Bases in a Hilbert Space," *Information and Control*, **4**, 1961, pp. 97-117.

3

THE Z-TRANSFORM

3.1 DEFINITION OF THE Z-TRANSFORM

One of the mathematical tools devised for the analysis and design of digital systems is the z-transform. The role of the z-transform to digital systems is similar to that of the Laplace transform to continuous-data systems. In recent years, however, the popularity of the z-transform method has been gradually replaced by the state variable method (Chapter 4). The fact is that although analysis and design using z-transform methods are straightforward for linear single-variable digital systems with uniform sampling, in extensions of these methods to multivariable, nonlinear, time-varying, and nonuniform-rate sampling systems, we often encounter great difficulties. On the other hand, the state variable approach usually allows the formulation of an analysis or design problem in a more general sense.

The motivation of using the z-transform in the study of digital systems can be explained by referring to the expression of the sampled signal of Eq. (2-49).

$$\mathscr{L}[f^*(t)] = F^*(s) = \sum_{k=0}^{\infty} f(kT)e^{-kTs} \tag{3-1}$$

Since the expression of $F^*(s)$ normally contains the factor e^{-Ts}, it is not a rational function of s. We know that when factors involving e^{-Ts} appear in a transfer function, difficulties in taking the inverse Laplace transform may arise. Therefore, it is desirable to first transform the irrational function $F^*(s)$ into a rational function, say $F(z)$, where z and s are related through a transformation. An obvious choice of the relationship between z and s is

$$z = e^{Ts} \tag{3-2}$$

although $z = e^{-Ts}$ would be just as good for the purpose. Solving for s in Eq. (3-2), we get

$$s = \frac{1}{T} \ln z \tag{3-3}$$

In these last two equations T is the sampling period and z is a complex variable. When Eq. (3-2) is substituted into Eq. (3-1), we have

$$F^*\left(s = \frac{1}{T} \ln z\right) = F(z) = \sum_{k=0}^{\infty} f(kT)z^{-k} \tag{3-4}$$

which when written in closed form will be a rational function of z. Therefore, we can define $F(z)$ as the *z-transform of $f(t)$*; i.e.,

$$F(z) = z\text{-transform of } f(t) = \mathscr{Z}[f(t)] \tag{3-5}$$

where the symbol "$\mathscr{Z}$" denotes "the z-transform of." In view of the way the z-transform is defined in Eq. (3-4), we can also write

$$\begin{aligned} F(z) &= [\text{Laplace transform of } f^*(t)] \mid_{s=(1/T)\ln z} \\ &= [F^*(s)] \mid_{s=(1/T)\ln z} \end{aligned} \tag{3-6}$$

It appears that the z-transform of the time function $f(t)$ contains the same information as the Laplace transform of $f^*(t)$, $F^*(s)$, although the transformation is made from s to z. Furthermore, since the z-transform of $f(t)$ is obtained from the Laplace transform of $f^*(t)$ by means of the transformation $z = e^{Ts}$, we can say in general that any continuous function that possesses a Laplace transform also has a z-transform.

In summary, the operation of taking the z-transform of a continuous signal, $f(t)$, involves the following three steps:

1. $f(t)$ is sampled to give $f^*(t)$.
2. The Laplace transform of $f^*(t)$ is taken, and

$$F^*(s) = \mathscr{L}[f^*(t)] = \sum_{k=0}^{\infty} f(kT)e^{-kTs}$$

3. Replace e^{Ts} by z in $F^*(s)$ to get $F(z)$, and

$$F(z) = \sum_{k=0}^{\infty} f(kT)z^{-k}$$

We can also use the expression in Eq. (2-55) to find the z-transform of a function. Replacing e^{-Ts} by z^{-1} in Eq. (2-55), we have

$$F(z) = \sum_{n=1}^{k} \frac{N(\xi_n)}{D'(\xi_n)} \frac{1}{1 - e^{\xi_n T}z^{-1}} \tag{3-7}$$

indicating that if $F(s)$ has a finite number of simple poles, the z-transform of the function is a rational function in z. Furthermore, the degree of the denominator of $F(z)$ in z^{-1} is equal to that of the denominator of $F(s)$ in s.

3.2 EVALUATION OF Z-TRANSFORMS

In this section several examples of finding the z-transform of a continuous-data signal will be given.

EXAMPLE 3-1

Find the z-transform of a unit-step function $u(t)$.

1. When the unit-step function is sampled by an ideal sampler, the output of the sampler is a unit-impulse train described by

$$u^*(t) = \delta_T(t) = \sum_{k=0}^{\infty} \delta(t - kT) \tag{3-8}$$

2. Taking the Laplace transform on both sides of Eq. (3-8), we get

$$U^*(s) = \Delta_T(s) = \sum_{k=0}^{\infty} e^{-kTs} \tag{3-9}$$

which can be written in a closed form

$$U^*(s) = \Delta_T(s) = \frac{1}{1 - e^{-Ts}} \quad \text{for} \quad |e^{-Ts}| < 1 \tag{3-10}$$

3. Replacing e^{Ts} by z in Eq. (3-10), we have

$$U(z) = \mathscr{Z}[u(t)] = \frac{1}{1 - z^{-1}} = \frac{z}{z - 1} \tag{3-11}$$

for $|z^{-1}| < 1$.

EXAMPLE 3-2

Find the z-transform of the function $f(t) = e^{-at}$, where a is a real constant.

1. The sampled signal is described by

$$f^*(t) = \sum_{k=0}^{\infty} e^{-akT}\delta(t - kT) \tag{3-12}$$

2. The Laplace transform of $f^*(t)$ is

$$F^*(s) = \sum_{k=0}^{\infty} e^{-akT}e^{-kTs} \tag{3-13}$$

which can be written in closed form as

$$F^*(s) = \frac{1}{1 - e^{-(s+a)T}}, \quad |e^{-(s+a)T}| < 1 \tag{3-14}$$

3. Replacing e^{-Ts} by z^{-1} in Eq. (3-14), the z-transform of $f(t) = e^{-at}$ is obtained as

$$F(z) = \frac{1}{1 - e^{-aT}z^{-1}} = \frac{z}{z - e^{-aT}} \tag{3-15}$$

for $|z^{-1}| < e^{-aT}$.

The same answer as Eq. (3-15) can be obtained using Eq. (3-7). The Laplace transform of e^{-at} is

$$F(s) = \frac{1}{s + a} \tag{3-16}$$

Thus in Eq. (3-7), $N(\xi) = 1$ and $D(\xi) = \xi + a$. For the pole at $\xi_1 = -a$, $N(\xi_1) = 1$, and $D'(\xi_1) = 1$. Therefore, Eq. (3-7) gives

$$F(z) = \frac{N(\xi_1)}{D'(\xi_1)} \frac{1}{1 - e^{\xi_1 T} z^{-1}} = \frac{1}{1 - e^{-aT} z^{-1}} \tag{3-17}$$

EXAMPLE 3-3

Find the z-transform of $f(t) = \sin \omega t$.

1. The sampled version of $f(t)$ is

$$f^*(t) = \sum_{k=0}^{\infty} \sin \omega kT \, \delta(t - kT) \tag{3-18}$$

or

$$f^*(t) = \sum_{k=0}^{\infty} \frac{e^{j\omega kT} - e^{-j\omega kT}}{2j} \delta(t - kT) \tag{3-19}$$

2. The Laplace transform of $f^*(t)$ is

$$F^*(s) = \sum_{k=0}^{\infty} \frac{e^{j\omega kT} - e^{-j\omega kT}}{2j} e^{-kTs} \tag{3-20}$$

which can be written in closed form as

$$F^*(s) = \frac{1}{2j}\left[\frac{1}{1 - e^{-T(s-j\omega)}} - \frac{1}{1 - e^{-T(s+j\omega)}}\right] \tag{3-21}$$

After simplification, Eq. (3-21) becomes

$$F^*(s) = \frac{e^{-Ts} \sin \omega T}{e^{-2Ts} - 2e^{-Ts} \cos \omega T + 1} \tag{3-22}$$

3. From Eq. (3-22) the z-transform of $\sin \omega t$ is written

$$F(z) = \frac{z^{-1} \sin \omega T}{z^{-2} - 2z^{-1} \cos \omega T + 1} = \frac{z \sin \omega T}{z^2 - 2z \cos \omega T + 1} \tag{3-23}$$

Now using Eq. (3-7), with

$$F(s) = \mathscr{L}[\sin \omega t] = \frac{\omega}{s^2 + \omega^2} \tag{3-24}$$

we have

$$\begin{aligned} N(\xi) &= \omega \\ D(\xi) &= \xi^2 + \omega^2 \\ D'(\xi) &= 2\xi \\ \xi_1 = j\omega, \quad D'(\xi_1) &= 2j\omega \\ \xi_2 = -j\omega, \quad D'(\xi_2) &= -2j\omega \end{aligned}$$

and we have

$$F^*(s) = \sum_{n=1}^{2} \frac{N(\xi_n)}{D'(\xi_n)} \frac{1}{1 - e^{-T(s-\xi_n)}}$$
$$= \frac{\omega}{2j\omega}\left[\frac{1}{1 - e^{-T(s-j\omega)}} - \frac{1}{1 - e^{-T(s+j\omega)}}\right] \tag{3-25}$$

Since Eq. (3-25) is identical to Eq. (3-21), the same answer for $F(z)$ as in Eq. (3-23) can be obtained using Eq. (3-7).

In this section we have illustrated by examples how the z-transforms of some simple time functions can be obtained by use of Eq. (3-4) or Eq. (3-7). In general, just as working with the Laplace transformation, a table of transforms for the z-transforms associated with commonly used functions can be assembled. A short table of z-transforms is given in Table 3-1, and a more extensive one may be found in the literature.[12]

One important difference between the properties of the Laplace transform and the z-transform must be mentioned here. It is wellknown that the Laplace transform and its inverse transform are unique, so that if $F(s)$ is the Laplace transform of $f(t)$, then $f(t)$ is the inverse Laplace transform of $F(s)$. However, in the z-transformation, the inverse z-transform is not unique. If the z-trans-

Table 3-1

TABLE OF Z-TRANSFORMS

Time Function	*Laplace Transform*	*z-Transform*
Unit step $u(t)$	$\frac{1}{s}$	$\frac{z}{z-1}$
t	$\frac{1}{s^2}$	$\frac{Tz}{(z-1)^2}$
$\frac{t^2}{2}$	$\frac{1}{s^3}$	$\frac{T^2z(z+1)}{2(z-1)^3}$
$\frac{t^n}{n!}$	$\frac{1}{s^{n+1}}$	$\lim_{a\to 0} \frac{(-1)^n}{n!} \frac{\partial^n}{\partial a^n}\left(\frac{z}{z-e^{-aT}}\right)$
e^{-at}	$\frac{1}{s+a}$	$\frac{z}{z-e^{-aT}}$
te^{-at}	$\frac{1}{(s+a)^2}$	$\frac{Tze^{-aT}}{(z-e^{-aT})^2}$
$1-e^{-at}$	$\frac{a}{s(s+a)}$	$\frac{z(1-e^{-aT})}{(z-1)(z-e^{-aT})}$
$\sin \omega t$	$\frac{\omega}{s^2+\omega^2}$	$\frac{z \sin \omega T}{z^2 - 2z\cos \omega T + 1}$
$e^{-at}\sin \omega t$	$\frac{\omega}{(s+a)^2+\omega^2}$	$\frac{ze^{-aT}\sin \omega T}{z^2 - 2ze^{-aT}\cos \omega T + e^{-2aT}}$
$\cos \omega t$	$\frac{s}{s^2+\omega^2}$	$\frac{z(z-\cos \omega T)}{z^2 - 2z\cos \omega T + 1}$
$e^{-at}\cos \omega t$	$\frac{s+a}{(s+a)^2+\omega^2}$	$\frac{z^2 - ze^{-aT}\cos \omega T}{z^2 - 2ze^{-aT}\cos \omega T + e^{-2aT}}$

form of $f(t)$ is $F(z)$, the inverse z-transform is *not* necessarily equal to $f(t)$. A simple example would be the z-transform of unit-step function, which is $z/(z-1)$. However, the inverse z-transform of $z/(z-1)$ can be any time function which has a value of unity at $t = 0, T, 2T, \ldots$. The problem of not being unique in the inverse transform is one of the difficulties of the z-transform method, and this property should be remembered when using the z-transform table.

3.3 MAPPING BETWEEN THE S-PLANE AND THE Z-PLANE

The study of the mapping between the s-plane and the z-plane by the z-transformation $z = e^{Ts}$ is necessarily important. The design and analysis of continuous-data control systems often rely on the pole-zero configurations of system transfer functions in the s-plane. Similarly, the poles and zeros of the z-transform of a system transfer function may conceivably determine the sampled response of the system. Therefore, in this section some of the important contours frequently used in the s-plane, such as the constant-damping-ratio line, the constant-frequency line, or the imaginary axis, will be mapped into the z-plane by the transformation $z = e^{Ts}$.

The s-plane is first divided into an infinite number of periodic strips as shown in Figure 2-11. The primary strip extends from $\omega = -\omega_s/2$ to $+\omega_s/2$; and the other complementary strips extend from $-\omega_s/2$ to $-3\omega_s/2$, $-3\omega_s/2$ to $-5\omega_s/2, \ldots$, for negative frequencies, and from $\omega_s/2$ to $3\omega_s/2$, $3\omega_s/2$ to $5\omega_s/2, \ldots$, for positive frequencies. If we consider only the primary strip shown in Figure 3-1(a), the path described by (1)-(2)-(3)-(4)-(5)-(1) in the left half of the s-plane is mapped into a unit circle centered at the origin in the z-plane [Figure 3-1(b)] by the transformation $z = e^{Ts}$. Similarly, since

$$e^{(s + jn\omega_s)T} = e^{Ts} e^{2\pi jn} = e^{Ts} = z \tag{3-26}$$

all the other left-half s-plane complementary strips are also mapped into the same unit circle in the z-plane. In other words, all points in the left half of the s-plane are mapped into the region inside the unit circle in the z-plane. The points in the right half of the s-plane are thus mapped into the region outside the unit circle in the z-plane.

The loci of the constant damping factor, constant frequency, and constant damping ratio in the z-plane, for sampled-data systems, are given in the following discussion.

1. The Constant Damping Loci

For a constant damping factor σ_1 in the s-plane, the corresponding z-plane locus is a circle of radius $z = e^{\sigma_1 T}$ centered at the origin in the z-plane as shown in Figure 3-2.

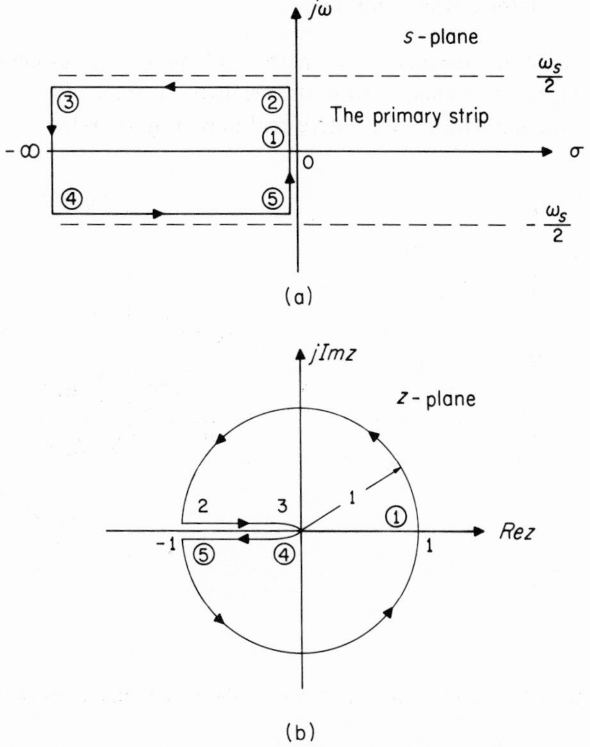

FIG. 3-1. Mapping of the primary strip in the left hand of the s-plane into the z-plane by the z-transformation.

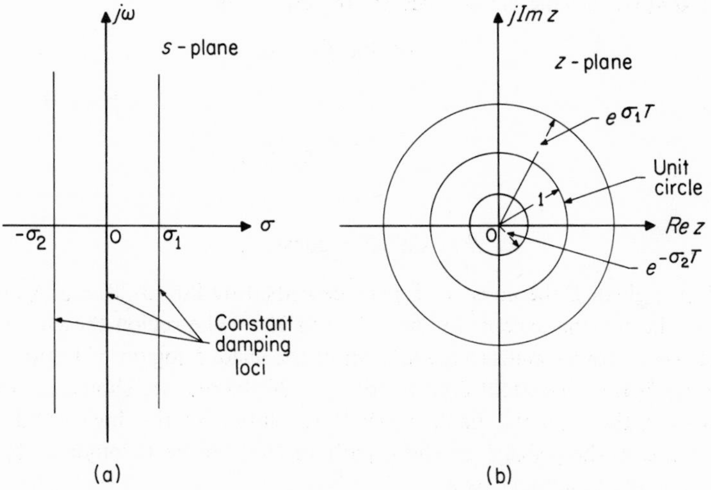

FIG. 3-2. Constant damping loci in the s-plane and the z-plane.

2. The Constant Frequency Loci

For any constant frequency ω_1 in the s-plane, the corresponding z-plane locus is a straight line emanating from the origin at an angle of $\theta = \omega_1 T$ radian; the angle is measured from the positive real axis (Figure 3-3).

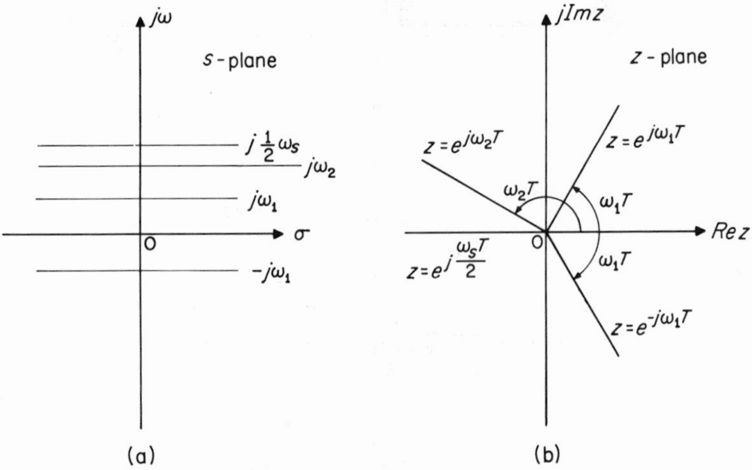

FIG. 3-3. Constant frequency loci in the s-plane and the z-plane.

3. The Constant Damping Ratio Loci

For a constant damping ratio, the constant line in the s-plane is shown in Figure 3-4(a); the line is described by the equation

$$s = -\omega \tan \beta + j\omega \tag{3-27}$$

Hence

$$z = e^{-2\pi\omega \tan \beta/\omega_s} \underline{/2\pi\omega/\omega_s} \tag{3-28}$$

where

$$\beta = \sin^{-1}\zeta = \text{constant} \tag{3-29}$$

For a given β the constant ζ path described by Eq. (3-28) is a logarithmic spiral in the z-plane except for $\beta = 0°$ and $90°$. The region shown shaded in Figure 3-4(a) corresponds to the interior of the shaded region of Figure 3-4(b). In Figure 3-5, the constant ζ paths for $\beta = 30$ degrees are shown in both the s-plane and the z-plane. Each one-half revolution of the logarithmic spiral corresponds to the passage of the ζ-path in the s-plane through a change of $\frac{1}{2}\omega_s$ along the imaginary axis.

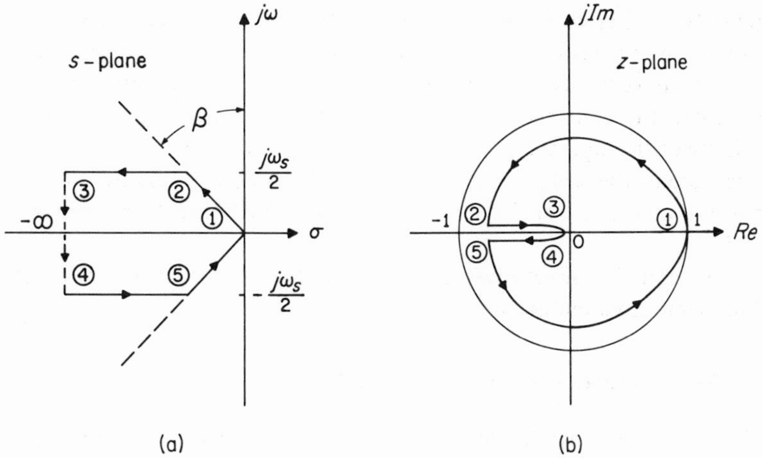

FIG. 3-4. Constant-damping-ratio loci in the s-plane and the z-plane.

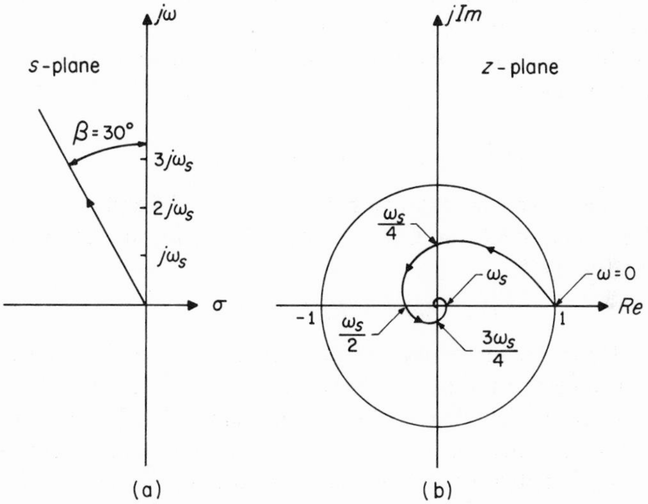

FIG. 3-5. Constant-damping-ratio loci for $\beta = 30°$ ($\zeta = 50\%$) in the s-plane and the z-plane.

3.4 THE INVERSE Z-TRANSFORM

Although the inverse z-transform is not unique, we can state that the transformation from $F(z)$ to $f^*(t)$ or $f(kT)$, $k = 0, 1, 2, \ldots$, is a unique one. Therefore, the inverse z-transform yields a time series which specifies the values of $f(t)$ only at discrete instants of time $t = 0, T, 2T, \ldots$, and nothing can be said about the values of $f(t)$ at all other times. However, for samplers which satisfy the sampling theorem, we can still use the discrete values of $f^*(t)$ to approximate the continuous function $f(t)$ with good accuracy by drawing a smooth curve through the points of $f(kT)$, $k = 0, 1, 2, \ldots$.

Because of the uniqueness problem, in the following discussion we shall refer to the inverse z-transform as an operation which goes from $F(z)$ to $f^*(t)$ or $f(kT)$.

In general, the inverse z-transformation can be carried out with one of the following three methods:

1. Partial Fraction Expansion Method

This method is parallel to the partial fraction expansion method used in Laplace transformation, with one slight modification.

To find the inverse z-transform, if $F(z)$ has one or more zeros at $z = 0$, $F(z)/z$ is expanded into a sum of simple first-order terms by partial fraction expansion, and the z-transform table is used to find the corresponding time functions of each of the expanded terms.

Recall that in the analysis of continuous-data functions, the inverse Laplace transform of a function $F(s)$ can be obtained by expanding $F(s)$ as

$$F(s) = \frac{A}{s+a} + \frac{B}{s+b} + \frac{C}{s+c} + \cdots \tag{3-30}$$

where a, b, and c, are the negative poles of $F(s)$, and simple poles are assumed in this illustration; A, B, and C are constants. The inverse Laplace transform of $F(s)$ is given by

$$f(t) = Ae^{-at} + Be^{-bt} + Ce^{-ct} + \cdots \tag{3-31}$$

for $t \geq 0$. However, in the case of the z-transform, we do not expand $F(z)$ into a form similar to Eq. (3-30), since the inverse z-transform of a term such as $A/(z + a)$ is not found in the z-transform table. Furthermore, it is known that the inverse z-transform of $Az/(z - e^{-aT})$ is Ae^{-akT}. Therefore, the expansion of $F(z)/z$ is called for. Of course, both sides of the expansion of $F(z)/z$ must be multiplied by the term z after the partial fraction expansion is made.

For functions which do not have any zeros at $z = 0$, the partial fraction expansion of $F(z)$ is performed in the usual manner, that is,

$$F(z) = \frac{A}{z+a} + \frac{B}{z+b} + \cdots \tag{3-32}$$

Then, let

$$F_1(z) = zF(z) = \frac{Az}{z+a} + \frac{Bz}{z+b} + \cdots \tag{3-33}$$

and the inverse z-transform of $F_1(z)$, $f_1(kT)$, is obtained with the help of the z-transform table. Therefore, the inverse z-transform of $F(z)$ is simply

$$\begin{aligned} f(kT) &= \mathscr{Z}^{-1}[F(z)] = \mathscr{Z}^{-1}[z^{-1}F_1(z)] \\ &= f_1[(k-1)T] \end{aligned} \tag{3-34}$$

where the notation $\mathscr{Z}^{-1}$ denotes inverse z-transformation.

EXAMPLE 3-4

Given the z-transform

$$F(z) = \frac{(1 - e^{-aT})z}{(z-1)(z-e^{-aT})} \tag{3-35}$$

where a is a constant and T is the sampling period, find the inverse z-transform $f^*(t)$ by means of the partial fraction expansion method.

The partial fraction expansion of $F(z)/z$ is written

$$\frac{F(z)}{z} = \frac{1}{z-1} - \frac{1}{z-e^{-aT}}$$

Therefore,

$$F(z) = \frac{z}{z-1} - \frac{z}{z-e^{-aT}}$$

From the z-transform table of Table 3-1, the inverse z-transform of $F(z)$ is found to be a time function whose values at the sampling instants are described by

$$f(kT) = 1 - e^{-akT}$$

Therefore,

$$f^*(t) = \sum_{k=0}^{\infty} (1 - e^{-akT})\delta(t - kT) \tag{3-36}$$

Notice that the time function $f(t)$ cannot be determined from the inverse z-transform, since no information on the behavior of the function is available between the sampling instants.

2. The Power Series Method

Expanding Eq. (3-4), we have

$$F(z) = f(0) + f(T)z^{-1} + f(2T)z^{-2} + \dots + f(kT)z^{-k} + \dots \tag{3-37}$$

which suggests that the coefficients of the infinite series expansion of $F(z)$ in powers of z^{-1} are the values of $f(t)$ at the sampling instants. Therefore, if $F(z)$ is given as a rational function of z

$$F(z) = \frac{a_m z^m + a_{m-1}z^{m-1} + \dots + a_1 z + a_0}{b_n z^n + b_{n-1}z^{n-1} + \dots + b_1 z + b_0} \tag{3-38}$$

then dividing the numerator by the denominator by long division yields

$$F(z) = A_0 z^{m-n} + A_1 z^{m-n-1} + \dots \tag{3-39}$$

Now comparing Eq. (3-39) with Eq. (3-37) we clearly see that the coefficients $A_0, A_1, \dots$ are related to the values of $f(t)$ at the sampling instants; i.e.,

$$A_0 = f[(m-n)T]$$
$$A_1 = f[(m-n-1)T]$$
$$\vdots$$

For a physically realizable system, $f(kT) = 0$ for $k < 0$; therefore, $n \geq m$.

The significance of the power series method is that the values of the function $f(t)$ at $t = 0, T, 2T, \dots, kT, \dots$ can be determined by expanding $F(z)$ into a power series of z^{-1}. The main difference between the partial fraction expansion method and the present one is that the former gives a closed form solution to $f(kT)$ whereas the latter usually gives a sequence of numbers.

EXAMPLE 3-5

Determine the inverse z-transform of $F(z)$ in Eq. (3-35) by means of the power series expansion method. Equation (3-35) is written

$$F(z) = \frac{(1 - e^{-aT})z}{z^2 - (1 + e^{-aT})z + e^{-aT}} \tag{3-40}$$

Dividing the denominator into the numerator yields

$$F(z) = (1 - e^{-aT})z^{-1} + (1 - e^{-2aT})z^{-2} + (1 - e^{-3aT})z^{-3} + \dots \tag{3-41}$$

From Eq. (3-41) it is easily observed that

$$f(kT) = 1 - e^{-akT} \tag{3-42}$$

and

$$f^*(t) = \sum_{k=0}^{\infty} f(kT)\delta(t - kT)$$
$$= \sum_{k=0}^{\infty} (1 - e^{-akT})\delta(t - kT) \tag{3-43}$$

which is identical to the result in Eq. (3-43) obtained by the partial fraction expansion method.

3. The Inversion Formula Method

It is of interest to compare the definitions of the Laplace transform and the z-transform. Given $f(t)$ as a function of t and Laplace transformable, the Laplace transform and the z-transform of $f(t)$ are respectively,

$$F(s) = \mathscr{L}[f(t)] = \int_0^{\infty} f(t)e^{-st}\,dt \tag{3-44}$$

and

$$F(z) = \mathscr{Z}[f(t)] = \sum_{k=0}^{\infty} f(kT)z^{-k} \tag{3-45}$$

The similarity between the two equations should not be surprising.

The inverse Laplace transform of $F(s)$ is

$$f(t) = \frac{1}{2\pi j}\int_{c-j\infty}^{c+j\infty} F(s)e^{ts}\, ds \tag{3-46}$$

where c denotes the abscissa of convergence and is chosen so that it is greater than the real parts of all the singularities of the integrand $F(s)e^{ts}$. We shall show that a similar expression can be derived for the inverse z-transform so that

$$f(kT) = \frac{1}{2\pi j}\oint_{\Gamma} F(z)z^{k-1}\, dz \tag{3-47}$$

where Γ is a closed path (usually a circle) in the z-plane which encloses all the singularities of $F(z)$.

We now proceed with the derivation of Eq. (3-47).

Substituting $t = kT$ in Eq. (3-46), we have

$$f(kT) = \frac{1}{2\pi j}\int_{c-j\infty}^{c+j\infty} F(s)e^{kTs}\, ds \tag{3-48}$$

The path of integration of the line integral of Eq. (3-48) is as shown in Figure 3-6(a). This path of integration is seen to pass through the periodic strips of the s-plane vertically, and thus the integral of Eq. (3-48) may be broken up

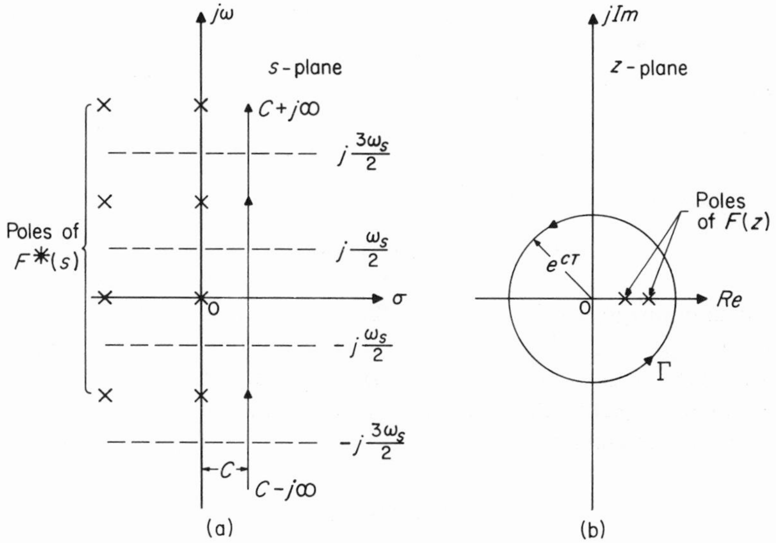

FIG. 3-6. Paths of integration of the inversion formula in the s-plane and the z-plane.

into a sum of integrals, each performed within one of the periodic strips. Equation (3-48) can now be written

$$f(kT) = \frac{1}{2\pi j}\sum_{k=-\infty}^{\infty}\int_{c+j\omega_s(k-1/2)}^{c+j\omega_s(k+1/2)} F(s)e^{kTs}\,ds \tag{3-49}$$

where $\omega_s = 2\pi/T$. Replacing s by $s + jk\omega_s$, where k is an integer, Eq. (3-49) becomes

$$f(kT) = \frac{1}{2\pi j}\sum_{k=-\infty}^{\infty}\int_{c-j\omega_s/2}^{c+j\omega_s/2} F(s + jk\omega_s)e^{kT(s+jk\omega_s)}\,d(s + jk\omega_s) \tag{3-50}$$

or

$$f(kT) = \frac{1}{2\pi j}\sum_{k=-\infty}^{\infty}\int_{c-j\omega_s/2}^{c+j\omega_s/2} F(s + jk\omega_s)e^{kTs}\,ds \tag{3-51}$$

Interchanging the summation and integration signs, Eq. (3-51) becomes

$$f(kT) = \frac{1}{2\pi j}\int_{c-j\omega_s/2}^{c+j\omega_s/2}\sum_{k=-\infty}^{\infty} F(s + jk\omega_s)e^{kTs}\,ds \tag{3-52}$$

Since from Eq. (2-51)

$$F^*(s) = \frac{1}{T}\sum_{k=-\infty}^{\infty} F(s + jk\omega_s) \tag{3-53}$$

Eq. (3-51) becomes

$$f(kT) = \frac{T}{2\pi j}\int_{c-j\omega_s/2}^{c+j\omega_s/2} F^*(s)e^{kTs}\,ds \tag{3-54}$$

Now substituting $z = e^{Ts}$ into Eq. (3-54), and since

$$F^*(s)\big|_{s=(1/T)\ln z} = F(z) \tag{3-55}$$

$$e^{kTs} = z^k \tag{3-56}$$

and

$$ds = d\left(\frac{1}{T}\ln z\right) = \frac{1}{T}z^{-1}\,dz \tag{3-57}$$

Eq. (3-54) is written

$$f(kT) = \frac{1}{2\pi j}\oint_{\Gamma} F(z)z^{k-1}\,dz \tag{3-58}$$

The path of integration from $s = c - j(\omega_s/2)$ to $s = c + j(\omega_s/2)$ is mapped onto the circle $|z| = e^{cT}$ in the z-plane, as shown in Figure 3-6(b). Since $F^*(s)$ does not have singularities on or to the right of the path of integration $s = c + j\omega$ in the s-plane, all singularities of $F(z)$ must lie inside the circle Γ, $|z| = e^{cT}$, in the z-plane.

EXAMPLE 3-6

Determine the inverse z-transform of $F(z)$ in Eq. (3-35) by means of the inversion formula of Eq. (3-47).

Substituting Eq. (3-35) into Eq. (3-47), we have

$$f(kT) = \frac{1}{2\pi j} \oint_{\Gamma} \frac{z(1 - e^{-aT})}{(z - 1)(z - e^{-aT})} z^{k-1}\, dz \tag{3-59}$$

where Γ is a circle which is large enough to enclose the poles of $F(z)$ at $z = 1$ and $z = e^{-aT}$. The contour integral of Eq. (3-59) can be evaluated using the residue theorem. Thus

$$f(kT) = \sum \text{residues of } F(z)z^{k-1} \text{ at the poles of } F(z) \tag{3-60}$$

Therefore,

$$\begin{aligned} f(kT) &= \sum \text{Residues of } \frac{z(1 - e^{-aT})}{(z - 1)(z - e^{-aT})} z^{k-1} \text{ at } z = 1 \text{ and } z = e^{-aT} \\ &= 1 - e^{-akT} \end{aligned} \tag{3-61}$$

It is apparent that this result agrees with that obtained earlier in Eqs. (3-36) and (3-42) by the two preceding methods.

3.5 THEOREMS OF THE Z-TRANSFORM

The derivations and applications of the z-transformation can often be simplified when theorems of the z-transform are referred to. The proofs and illustrations of some of the basic theorems in z-transforms are given below.

1. Addition and Subtraction

If $f_1(t)$ and $f_2(t)$ have z-transforms

$$F_1(z) = \mathscr{Z}[f_1(t)] = \sum_{k=0}^{\infty} f_1(kT)z^{-k} \tag{3-62}$$

and

$$F_2(t) = \mathscr{Z}[f_2(t)] = \sum_{k=0}^{\infty} f_2(kT)z^{-k} \tag{3-63}$$

respectively, then

$$\mathscr{Z}[f_1(t) \pm f_2(t)] = F_1(z) \pm F_2(z) \tag{3-64}$$

Proof. By definition

$$\begin{aligned} \mathscr{Z}[f_1(t) \pm f_2(t)] &= \sum_{k=0}^{\infty} [f_1(kT) \pm f_2(kT)]z^{-k} \\ &= \sum_{k=0}^{\infty} f_1(kT)z^{-k} \pm \sum_{k=0}^{\infty} f_2(kT)z^{-k} \\ &= F_1(z) \pm F_2(z) \end{aligned} \tag{3-65}$$

2. Multiplication by a Constant

If $F(z)$ is the z-transform of $f(t)$, then

$$\mathscr{Z}[af(t)] = a\mathscr{Z}[f(t)] = aF(z) \tag{3-66}$$

where a is a constant.

Proof. By definition

$$\mathscr{Z}[af(t)] = \sum_{k=0}^{\infty} af(kT)z^{-k} = a\sum_{k=0}^{\infty} f(kT)z^{-k} = aF(z) \tag{3-67}$$

3. Real Translation (Shifting Theorem)

If $f(t)$ has the z-transform $F(z)$, then

$$\mathscr{Z}[f(t - nT)] = z^{-n}F(z) \tag{3-68}$$

and

$$\mathscr{Z}[f(t + nT)] = z^{n}[F(z) - \sum_{k=0}^{n-1} f(kT)z^{-k}] \tag{3-69}$$

where n is a positive integer.

Proof. By definition

$$\mathscr{Z}[f(t - nT)] = \sum_{k=0}^{\infty} f(kT - nT)z^{-k} \tag{3-70}$$

which can be written as

$$\mathscr{Z}[f(t - nT)] = z^{-n}\sum_{k=0}^{\infty} f(kT - nT)z^{-(k-n)} \tag{3-71}$$

Since $f(t)$ is assumed to be zero for $t < 0$, Eq. (3-71) becomes

$$\mathscr{Z}[f(t - nT)] = z^{-n}\sum_{k=n}^{\infty} f(kT - nT)^{-(k-n)} = z^{-n}F(z) \tag{3-72}$$

To prove Eq. (3-69), we write

$$\begin{aligned}\mathscr{Z}[f(t + nT)] &= \sum_{k=0}^{\infty} f(kT + nT)z^{-k} \\ &= z^{n}\sum_{k=0}^{\infty} f(kT + nT)z^{-(k+n)} \\ &= z^{n}\left[F(z) - \sum_{k=0}^{n-1} f(kT)z^{-k}\right]\end{aligned} \tag{3-73}$$

EXAMPLE 3-7

Find the z-transform of a unit-step function which is delayed by one sampling period T. Using the shifting theorem stated by Eq. (3-68), we have

$$\mathscr{Z}[u(t - T)] = z^{-1}\mathscr{Z}[u(t)] = z^{-1}\frac{z}{z-1} = \frac{1}{z-1} \tag{3-74}$$

4. Complex Translation

If $f(t)$ has the z-transform $F(z)$, then

$$\mathscr{Z}[e^{\mp at}f(t)] = [F(s \pm a)]\,|_{z=e^{Ts}} = F(ze^{\pm aT}) \tag{3-75}$$

where a is a constant

Proof. By definition

$$\mathscr{Z}[e^{\mp at}f(t)] = \sum_{k=0}^{\infty} f(kT)e^{\pm akT}z^{-k} \tag{3-76}$$

Let $z_1 = ze^{\pm aT}$, then Eq. (3-76) is written

$$\mathscr{Z}[e^{\mp at}f(t)] = \sum_{k=0}^{\infty} f(kT)z_1^{-k} = F(z_1) \tag{3-77}$$

Therefore,

$$\mathscr{Z}[e^{\mp at}f(t)] = F(ze^{\pm aT}) \tag{3-78}$$

EXAMPLE 3-8

Find the z-transform of $f(t) = e^{-at}\sin \omega t$ with the help of the complex translation theorem.

From the z-transform table of Table 3-1, the z-transform of $e^{-at}\sin \omega t$ is found to be

$$\mathscr{Z}[e^{-at}\sin \omega t] = \frac{ze^{-aT}\sin \omega T}{z^2 - 2ze^{-aT}\cos \omega T + e^{-2aT}} \tag{3-79}$$

and the z-transform of $\sin \omega t$ is

$$\mathscr{Z}[\sin \omega t] = \frac{z \sin \omega T}{z^2 - 2z\cos \omega T + 1} \tag{3-80}$$

Apparently, the result in Eq. (3-79) may be obtained by substituting ze^{aT} for z in Eq. (3-80).

5. Initial Value Theorem

If the function $f(t)$ has the z-transform $F(z)$, and if the limit

$$\lim_{z\to\infty} F(z)$$

exists, then

$$\lim_{k\to 0} f(kT) = \lim_{z\to\infty} F(z) \tag{3-81}$$

The theorem simply states that the behavior of the discrete signal $f^*(t)$ in the neighborhood of $t = 0$ is determined by the behavior of $F(z)$ at $z = \infty$.

Proof. By definition, $F(z)$ is written

$$F(z) = \sum_{k=0}^{\infty} f(kT)z^{-k} = f(0) + f(T)z^{-1} + f(2T)z^{-2} + \dots \tag{3-82}$$

Taking the limit on both sides of the last equation as z approaches infinity, we get

$$\lim_{z\to\infty} F(z) = f(0) = \lim_{k\to 0} f(kT) \tag{3-83}$$

6. Final Value Theorem

If the function $f(t)$ has the z-transform $F(z)$ and if the function $(1 - z^{-1})F(z)$ does not have poles on or outside the unit circle $|z| = 1$ in the z-plane, then

$$\lim_{k\to\infty} f(kT) = \lim_{z\to 1} (1 - z^{-1})F(z) \tag{3-84}$$

Proof. Let us consider two finite sequences,

$$\sum_{k=0}^{n} f(kT)z^{-k} = f(0) + f(T)z^{-1} + \dots + f(nT)z^{-n} \tag{3-85}$$

and

$$\sum_{k=0}^{n} f[(k-1)T]z^{-k} = f(0)z^{-1} + f(T)z^{-2} + \dots + f[(n-1)T]z^{-n} \tag{3-86}$$

We assume that $f(t) = 0$ for all $t < 0$ so that the term $f(-T)$ that would have appeared in Eq. (3-86) is zero. Now comparing Eq. (3-85) with Eq. (3-86) we see that the latter can also be written as

$$\sum_{k=0}^{n} f[(k-1)T]z^{-k} = z^{-1}\sum_{k=0}^{n-1} f(kT)z^{-k} \tag{3-87}$$

Taking the limit as $z \to 1$ of the difference between Eqs. (3-86) and (3-87) yields

$$\lim_{z\to 1}\left[\sum_{k=0}^{n} f(kT)z^{-k} - z^{-1}\sum_{k=0}^{n-1} f(kT)z^{-k}\right]$$
$$= \sum_{k=0}^{n} f(kT) - \sum_{k=0}^{n-1} f(kT) = f(nT) \tag{3-88}$$

In the last equation if we take the limit as n approaches infinity, we have

$$\lim_{n-\infty} f(nT) = \lim_{n\to\infty}\lim_{z\to 1}\left[\sum_{k=0}^{n} f(kT)z^{-1} - z^{-1}\sum_{k=0}^{n-1} f(kT)z^{-k}\right] \tag{3-89}$$

Interchanging the limits on the right-hand side of the last equation, and since

$$\lim_{n\to\infty}\sum_{k=0}^{n} f(kT)z^{-k} = \lim_{n\to\infty}\sum_{k=0}^{n-1} f(kT)z^{-k} = F(z) \tag{3-90}$$

we have

$$\lim_{n\to\infty} f(kT) = \lim_{z\to 1} (1 - z^{-1})F(z) \tag{3-91}$$

which is the expression for the final-value theorem.

EXAMPLE 3-9

Given the z-transform

$$F(z) = \frac{0.792z^2}{(z - 1)(z^2 - 0.416z + 0.208)} \tag{3-92}$$

determine the final value of $f(kT)$ by use of the final-value theorem.

Since

$$(1 - z^{-1})F(z) = \frac{0.792z}{z^2 - 0.416z + 0.208} \tag{3-93}$$

does not have poles on or outside the unit circle $|z| = 1$, the final-value theorem may be applied. Thus from Eq. (3-84)

$$\lim_{k \to \infty} f(kT) = \lim_{z \to 1} \frac{0.792z}{z^2 - 0.416z + 0.208} = 1 \tag{3-94}$$

This result can be checked readily by expanding $F(z)$ into a power series in z^{-1}.

$$F(z) = 0.792z^{-1} + 1.12z^{-2} + 1.091z^{-3} + 1.01z^{-4} + 0.9832^{-5} + 0.989z^{-6} + 0.99z^{-7} + \dots \tag{3-95}$$

In this case when sufficient terms are carried out in the expansion we can see that the coefficient of the series converges rapidly to its final steady state value of unity.

7. Partial Differentiation Theorem

Let the z-transform of the function $f(t, a)$ be denoted by $F(z, a)$, where a is an independent variable or a constant. The z-transform of the partial derivative of $f(t, a)$ with respect to a is given by

$$\mathscr{Z}\left\{\frac{\partial}{\partial a}[f(t, a)]\right\} = \frac{\partial}{\partial a} F(z, a) \tag{3-96}$$

Proof. By definition

$$\begin{aligned} \mathscr{Z}\left\{\frac{\partial}{\partial a}[f(t, a)]\right\} &= \sum_{k=0}^{\infty} \frac{\partial}{\partial a} f(kT, a)z^{-k} \\ &= \frac{\partial}{\partial a} \sum_{k=0}^{\infty} f(kT, a)z^{-k} = \frac{\partial}{\partial a} F(z, a) \end{aligned} \tag{3-97}$$

The following example illustrates that the z-transforms of certain types of functions may be obtained easily if the partial differentiation theorem is applied.

EXAMPLE 3-10

Determine the z-transform of $f(t) = te^{-at}$ by use of the partial differentiation theorem.

The z-transform of $f(t)$ is

$$\mathscr{Z}[f(t)] = \mathscr{Z}[te^{-at}] = \mathscr{Z}\left[-\frac{\partial}{\partial a}e^{-at}\right] \tag{3-98}$$

From Eq. (3-96),

$$\mathscr{Z}\left[-\frac{\partial}{\partial a}e^{-at}\right] = \frac{-\partial}{\partial a}\mathscr{Z}[e^{-at}] = \frac{-\partial}{\partial a}\left[\frac{z}{z-e^{-aT}}\right]$$
$$= \frac{Tze^{-aT}}{(z-e^{-aT})^2} \tag{3-99}$$

8. Real Convolution

If the functions $f_1(t)$ and $f_2(t)$ have the z-transforms $F_1(z)$ and $F_2(z)$, respectively, and $f_1(t) = f_2(t) = 0$ for $t < 0$, then

$$F_1(z)F_2(z) = \mathscr{Z}\left[\sum_{n=0}^{k} f_1(nT)f_2(kT-nT)\right] \tag{3-100}$$

Proof. The right-hand side of Eq. (3-100) is written as

$$\mathscr{Z}\left[\sum_{n=0}^{k} f_1(nT)f_2(kT-nT)\right] = \sum_{k=0}^{\infty}\sum_{n=0}^{k} f_1(nT)f_2(kT-nT)z^{-k}$$
$$= \sum_{k=0}^{\infty}\sum_{n=0}^{\infty} f_1(nT)f_2(kT-nT)z^{-k} \tag{3-101}$$

Letting $m = k - n$ and interchanging the order of summation, we have

$$\mathscr{Z}\left[\sum_{n=0}^{k} f_1(nT)f_2(kT-nT)\right] = \sum_{n=0}^{\infty} f_1(nT)z^{-n}\sum_{m=-n}^{\infty} f_2(mT)z^{-m} \tag{3-102}$$

Since $f_2(t) = 0$ for $t < 0$, the last equation becomes

$$\mathscr{Z}\left[\sum_{n=0}^{k} f_1(nT)f_2(kT-nT)\right] = \sum_{n=0}^{\infty} f_1(nT)z^{-n}\sum_{m=0}^{\infty} f_2(mT)z^{-m}$$
$$= F_1(z)F_2(z) \tag{3-103}$$

The reader should recognize that the real convolution theorem of the z-transformation is analogous to that of the Laplace transformation. An important fact to remember is that the inverse transform (Laplace or z) of the product of two functions is *not* equal to the product of the two functions in the real domain which are the inverse transforms of the two functions in the complex domain respectively; that is

$$\mathscr{Z}^{-1}[F_1(z)F_2(z)] \neq f_1(kT)f_2(kT) \tag{3-104}$$

9. Complex Convolution

If the z-transforms of $f_1(t)$ and $f_2(t)$ are $F_1(z)$ and $F_2(z)$, respectively, then the z-transform of the product of the two functions is

$$\mathscr{Z}[f_1(t)f_2(t)] = \frac{1}{2\pi j}\oint_\Gamma \frac{F_1(\xi)F_2(z\xi^{-1})}{\xi}\,d\xi \tag{3-105}$$

where Γ is a circle which lies in the region (annulus) described by

$$\sigma_1 < |\xi| < \frac{|z|}{\sigma_2}$$

and

$$|z| > \max(\sigma_1, \sigma_2, \sigma_1\sigma_2)$$

where

$$\sigma_1 = \text{Radius of convergence of } F_1(\xi)$$
$$\sigma_2 = \text{Radius of convergence of } F_2(\xi)$$

Proof. By definition the z-transform of $f_1(t)f_2(t)$ is written

$$\mathscr{Z}[f_1(t)f_2(t)] = \sum_{k=0}^{\infty} f_1(kT)f_2(kT)z^{-k} \tag{3-106}$$

In order that this z-transform series converges absolutely, the magnitude of z must be greater than the largest value of σ_1, σ_2, and $\sigma_1\sigma_2$; that is, $|z| > \max(\sigma_1, \sigma_2, \sigma_1\sigma_2)$.

We may write $f_1(kT)$ as an inverse z-transform relationship,

$$f_1(kT) = \frac{1}{2\pi j}\oint_\Gamma F_1(\xi)\xi^{k-1}\,d\xi \tag{3-107}$$

where Γ denotes a circle which encloses all the singularities of $F_1(\xi)\xi^{k-1}$; therefore, $|\xi| > \sigma_1$. Substituting Eq. (3-107) in Eq. (3-106), we obtain

$$\mathscr{Z}[f_1(t)f_2(t)] = \frac{1}{2\pi j}\oint_\Gamma \frac{F_1(\xi)}{\xi}\,d\xi \sum_{k=0}^{\infty} f_2(kT)(\xi^{-1}z)^{-k} \tag{3-108}$$

Since

$$F_2(\xi^{-1}z) = \sum_{k=0}^{\infty} f_2(kT)(\xi^{-1}z)^{-k} \tag{3-109}$$

and is absolutely convergent for $|\xi^{-1}z| > \sigma_2$, or $|\xi| < |z|/\sigma_2$, Eq. (3-108) becomes

$$\mathscr{Z}[f_1(t)f_2(t)] = \frac{1}{2\pi j}\oint_\Gamma \frac{F_1(\xi)}{\xi}F_2(z\xi^{-1})\,d\xi \tag{3-110}$$

with

$$\sigma_1 < |\xi| < \frac{|z|}{\sigma_2}$$

EXAMPLE 3-11

Determine the z-transform of $f(t) = te^{-at}$ by use of the complex convolution theorem.

Let $f_1(t) = t$ and $f_2(t) = e^{-at}$. Then

$$F_1(z) = \frac{Tz}{(z-1)^2} \qquad |z| > 1 = \sigma_1 \tag{3-111}$$

$$F_2(z) = \frac{z}{z - e^{-aT}} \qquad |z| > e^{-aT} = \sigma_2 \tag{3-112}$$

Substituting Eqs. (3-106) and (3-107) into Eq. (3-105) yields

$$\mathscr{Z}[f_1(t)f_2(t)] = \frac{1}{2\pi j}\oint_\Gamma \frac{T\xi}{\xi(\xi-1)^2}\frac{z\xi^{-1}}{z\xi^{-1} - e^{-aT}}\,d\xi \tag{3-113}$$

where Γ is a circular path which lies in the annular ring,

$$1 < |\xi| < \frac{|z|}{e^{-aT}} = |z|e^{aT} \tag{3-114}$$

and $|z| > 1$.

Therefore, the integration path of Eq. (3-113) encloses only the poles of the integrand at $\xi = 1$. Applying the residues theorem to Eq. (3-113), we get

$$\mathscr{Z}[f_1(t)f_2(t)] = \text{residue of } \frac{Tz\xi^{-1}}{(\xi-1)^2(z\xi^{-1} - e^{-aT})} \quad \text{at} \quad \xi = 1$$

$$= \frac{\partial}{\partial\xi}\left[\frac{Tz\xi^{-1}}{(z\xi^{-1} - e^{-aT})}\right]_{\xi=1} = \frac{Tze^{-aT}}{(z - e^{-aT})^2} \tag{3-115}$$

which agrees with the result obtained earlier in Example 3-10.

3.6 LIMITATIONS OF THE Z-TRANSFORM METHOD

In the preceding sections we have seen that the z-transformation is a very convenient tool for the representation of linear digital systems. However, the z-tranform method has limitations; and in certain cases, care must be taken in the application and the interpretation of the results of the z-transforms.

The following considerations should be kept in mind when applying the z-transformation:

1. We must realize that the derivation of the z-transformation is based on the assumption that the sampled signal is approximated by a train of impulses whose areas are equal to the magnitude of the input signal of the sampler at the sampling instants. This assumption is considered valid only if the sampling duration of the sampler is small when compared with the significant time constant of the system.
2. The z-transform of the output of a linear system, $C(z)$, specifies only the values of the time function $c(t)$ at the sampling instants; $C(z)$ does not contain any information concerning the value of $c(t)$, between sampling instants. Therefore, given any $C(z)$, the inverse z-transform $c(kT)$ describes $c(t)$ only at the sampling instants $t = kT$.
3. In analyzing a linear system by the z-transform method, the transfer func-

tion of the system, $G(s)$, must have at least one more pole than zeros, or equivalently, the impulse response of $G(s)$ must not have any jump discontinuity at $t = 0$. Otherwise, the system response obtained by the z-transform method is misleading and sometimes even incorrect.[12]

A complete description of any system response almost always necessitates a knowledge of the waveform between sampling instants. Several methods using the z-transformation have been developed for the evaluation of response of digital systems between sampling instants. Among these methods, the submultiple method[12] and the modified z-transform method[12] are the most common. However, in recent years the state variable method has become perhaps the most widely used tool in the analysis and design of digital control systems.

3.7 THE PULSE TRANSFER FUNCTION

Thus far our examination of sampled-data systems has been confined to the discussions of the properties and mathematical representation of sampled signals only. The situation must be investigated when a sampled signal is applied to the input terminals of a linear system.

For the linear open-loop system with continuous-data input $r(t)$ as shown in Figure 3-7, the transfer function

$$G(s) = \frac{C(s)}{R(s)} \tag{3-116}$$

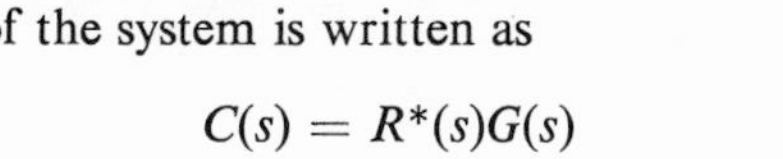

FIG. 3-7. A linear system with continuous-data input.

described the input-output relationship of the system. If the same system is now subjected to a sampled signal as shown in Figure 3-8, the Laplace transform of the output of the system is written as

$$C(s) = R^*(s)G(s) \tag{3-117}$$

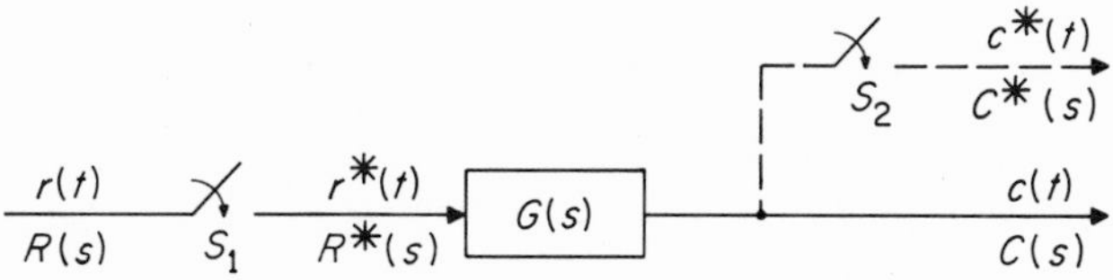

FIG. 3-8. A linear system with sampled data.

where $R^*(s)$ is the Laplace transform of the discrete-data input $r^*(t)$. Our objective here is to find a way of expressing the discrete-data system in terms of the z-transforms $C(z)$, $R(z)$, and $G(z)$. A simple way of accomplishing this is to form $C^*(s)$ using Eq. (2-51); we obtain

$$C^*(s) = \frac{1}{T}\sum_{n=-\infty}^{\infty} C(s + jn\omega_s) = \frac{1}{T}\sum_{n=-\infty}^{\infty} R^*(s + jn\omega_s)G(s + jn\omega_s) \quad (3\text{-}118)$$

Since

$$R^*(s + jn\omega_s) = R^*(s) \quad (3\text{-}119)$$

[Eq. (2-64)], Eq. (3-118) becomes

$$C^*(s) = R^*(s)\frac{1}{T}\sum_{n=-\infty}^{\infty} G(s + jn\omega_s) \quad (3\text{-}120)$$

Now defining

$$G^*(s) = \frac{1}{T}\sum_{n=-\infty}^{\infty} G(s + jn\omega_s) \quad (3\text{-}121)$$

we have from Eq. (3-120)

$$C^*(s) = R^*(s)G^*(s) \quad (3\text{-}122)$$

Upon transforming into z by $z = e^{Ts}$, Eq. (3-122) gives

$$C(z) = G(z)R(z) \quad (3\text{-}123)$$

which is the desired transfer relation for a linear system with discrete-data input. Notice that the output of the system may still be a continuous-data signal, but in Eq. (3-123) $C(z)$ defines the output only at the sampling instants.

An alternate way of arriving at Eq. (3-123) would be to use the impulse response method. Suppose that at $t = 0$, a unit impulse is applied as input to the system of Figure 3-8. The output of the system is simply the impulse response $g(t)$. If a fictitious sampler S_2 which is synchronized to S_1 and with the same sampling period is placed at the system output, the output of S_2 is described by

$$c^*(t) = g^*(t) = \sum_{k=0}^{\infty} g(kT)\delta(t - kT) \quad (3\text{-}124)$$

where $g(kT)$ for $k = 0, 1, 2, 3, \ldots$, is defined as the *weighting sequence* or the *impulse sequence* of the system.

When the discrete-data signal $r^*(t)$ is applied as input to the linear system the output of the system is written as

$$c(t) = r(0)g(t) + r(T)g(t - T) + r(2T)g(t - 2T) + \ldots \quad (3\text{-}125)$$

At $t = nT$, where n is a positive integer, Eq. (3-125) gives

$$c(kT) = r(0)g(kT) + r(T)g[(k - 1)T] + \ldots + r(kT)g(0)$$

$$= \sum_{n=0}^{k} r(nT)g(kT - nT) \quad (3\text{-}126)$$

Taking the z-transform on both sides of Eq. (3-126) and using the real convolution theorem of Eq. (3-100), we have

$$C(z) = R(z)G(z) \tag{3-127}$$

where

$$G(z) = \sum_{k=0}^{\infty} g(kT)z^{-k} \tag{3-128}$$

is defined as the *z-transfer function* of the linear system. Therefore, the z-transfer function $G(z)$ relates the z-transform of the input, $R(z)$, to the z-transform of the output $C(z)$ in the same way as the continuous-data transfer function $G(s)$ does for $R(s)$ and $C(s)$. However, if the linear system contains all continuous-data elements, the output signal is a function of the continuous t. The z-transform expression of Eq. (3-127) gives information on the continuous $c(t)$ only at the discrete time instants $t = kT$. This is one of the limitations of the z-transform method. In certain cases the loss of information between sampling instants may not be vital; e.g., if the value of $c(t)$ does not vary appreciably between sampling instants. On the other hand, if $c(t)$ contains large-amplitude oscillations between sampling instants, the z-transform method would often give misleading results. Still another comment is that in view of Eqs. (3-121) and (3-128) the z-transfer function $G(z)$ is derived from the impulse response $g(t)$ in exactly the same manner as $R(z)$ is derived from the signal $r(t)$.

When a discrete-data system contains cascaded elements, care must be taken in deriving the transfer function for the overall system. Figure 3-9 illus-

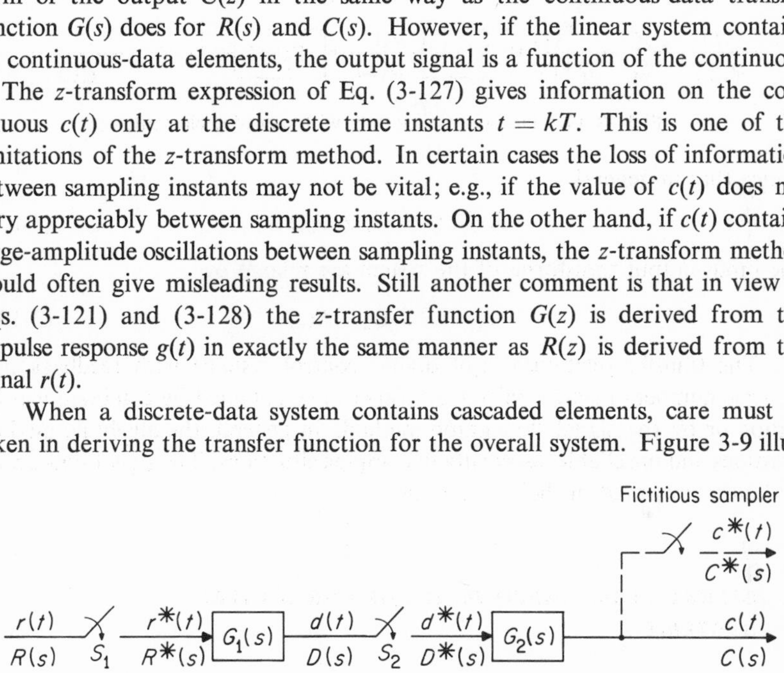

FIG. 3-9. Discrete-data system with sampler-separated elements.

trates a discrete-data system with cascaded elements G_1 and G_2. The two elements are separated by a second sampler S_2 which is identical and synchronized to the sampler S_1. The z-transfer function of the overall system is derived as follows: The transfer relationships of the two system G_1 and G_2 are

$$D(z) = G_1(z)R(z) \tag{3-129}$$

and

$$C(z) = G_2(z)D(z) \tag{3-130}$$

Substitution of Eq. (3-129) in Eq. (3-130) yields

$$C(z) = G_1(z)G_2(z)R(z) \tag{3-131}$$

Therefore, the z-transfer function of two linear elements separated by a sampler is equal to the product of the z-transfer functions of the two elements.

When a discrete-data system contains two cascade elements not separated by a sampler, as shown in Figure 3-10, the z-transfer function of the overall system should be written as

$$\mathscr{Z}[G_1(s)G_2(s)] = G_1G_2(z) = G_2G_1(z) \tag{3-132}$$

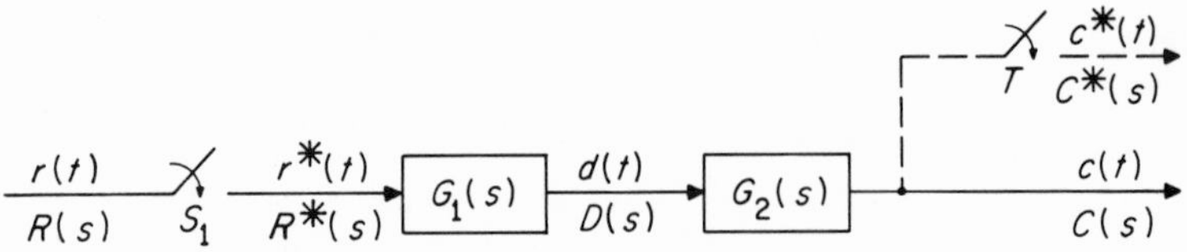

FIG. 3-10. Discrete-data system with cascaded elements.

Notice that in general

$$G_1G_2(z) \neq G_1(z)G_2(z) \tag{3-133}$$

The input-output transforms of the system are related by

$$C(z) = G_1G_2(z)R(z) \tag{3-134}$$

The transfer relationships of digital control systems with feedback and multiple number of synchronized samplers can be obtained by algebraic manipulation or by the signal flow graph method. In general, the algebraic method is tedious and unreliable, especially in complex situations. The signal flow graph method is presented in the next section.

3.8 SIGNAL FLOW GRAPH METHODS FOR DIGITAL SYSTEMS

It is well-known that transfer functions of linear systems with continuous-data can be determined from the signal flow graphs using Mason's gain formula.[20] However, since digital systems usually contain digital and continuous-data signals, Mason's gain formula cannot be applied directly in its original form. Two methods of extending the signal flow graph technique to the evaluation of input-output relationships of digital systems will be discussed. The first method[12,13,14,15,18] relies on the forming of a "sampled signal flow graph" in which all the node variables are discrete quantities. Once all the node variables represent either continuous-data or digital data uniformly, Mason's gain formula can again be used. The second method, which is due to Sedler and Bekey,[21] allows the derivation of the input-output transfer relations of a digital system directly from the system's signal flow graph. Since the signals in the system are usually mixed, a gain formula different from the original one by Mason has to be introduced.

The Sampled Signal Flow Graph Method

The method involves the following steps:

1. With the system's block diagram as the starting point, construct an equivalent signal flow graph for the system. This is also known as the signal flow graph of the system, since it is entirely equivalent to the block diagram representation. As an illustration, Figure 3-11 shows the block diagram of a system and its equivalent signal flow graph.

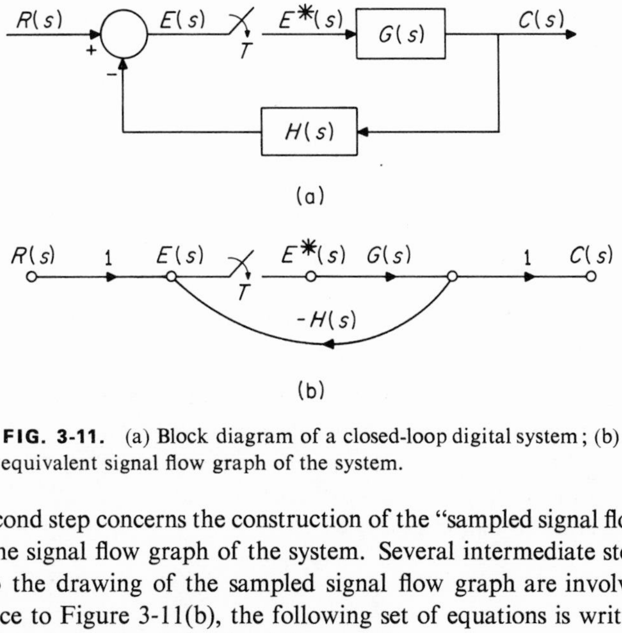

FIG. 3-11. (a) Block diagram of a closed-loop digital system; (b) equivalent signal flow graph of the system.

2. The second step concerns the construction of the "sampled signal flow graph" from the signal flow graph of the system. Several intermediate steps which lead to the drawing of the sampled signal flow graph are involved. With reference to Figure 3-11(b), the following set of equations is written for all the non-input nodes of the signal flow graph.

$$E(s) = R(s) - G(s)H(s)E^*(s) \tag{3-135}$$

$$C(s) = G(s)E^*(s) \tag{3-136}$$

Notice that in this step Mason's gain formula is applied to all the non-input nodes with the inputs of the system as well as the outputs of the samplers considered as input variables. The samplers may be considered as deleted from the flow graph once their output variables are defined.

Taking the pulse transform on both sides of Eqs. (3-135) and (3-136), we have

$$E^*(s) = R^*(s) - GH^*(s)E^*(s) \tag{3-137}$$

$$C^*(s) = G^*(s)E^*(s) \tag{3-138}$$

where it has been shown earlier that

$$[G(s)H(s)]^* = HG^*(s) = GH^*(s) = \frac{1}{T}\sum_{n=-\infty}^{\infty} G(s + jn\omega_s)H(s + jn\omega_s) \tag{3-139}$$

and

$$[E^*(s)]^* = \frac{1}{T}\sum_{n=-\infty}^{\infty} E^*(s + jn\omega_s) = E^*(s) \tag{3-140}$$

Now since Eqs. (3-137) and (3-138) contain only digital variables, the signal flow graph drawn using these equations is called the *sampled signal flow graph* of the system in Figure 3-11. The sampled signal flow graph is shown in Figure 3-12.

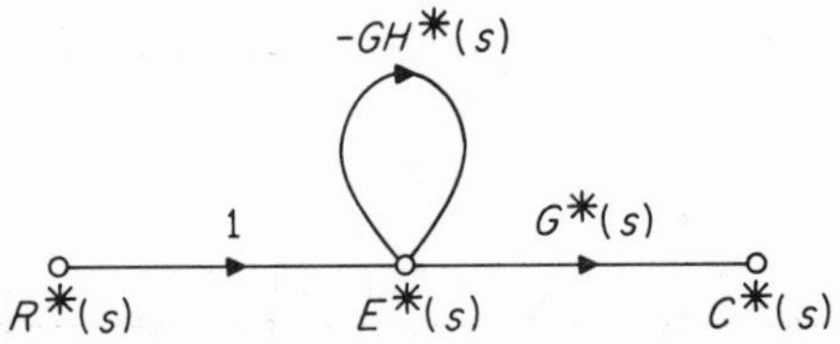

FIG. 3-12. Sampled signal flow graph of the system shown in Fig. 3-11.

3. Once the sampled signal flow graph is obtained, the transfer relation between any pair of input and output nodes on this flow graph can be determined by use of Mason's gain formula. For instance, applying Mason's gain formula to the sampled signal flow graph of Figure 3-12, with $C^*(s)$ and $E^*(s)$ regarded as output nodes, we get

$$C^*(s) = \frac{G^*(s)}{1 + GH^*(s)} R^*(s) \tag{3-141}$$

and

$$E^*(s) = \frac{1}{1 + GH^*(s)} R^*(s) \tag{3-142}$$

4. The transfer relationships between inputs and continuous-data outputs can be determined from the *composite signal flow graph.* The composite signal flow graph of a digital system is obtained by combining the equivalent and the sampled signal flow graphs. More specifically, the composite signal flow graph is formed by connecting the output nodes of samplers on the equivalent signal flow graph with unity-gain branches from the same nodes on the sampled signal flow graph. As an illustrative example, the composite signal flow graph of the system in Figure 3-11 is drawn as shown in Figure 3-13.

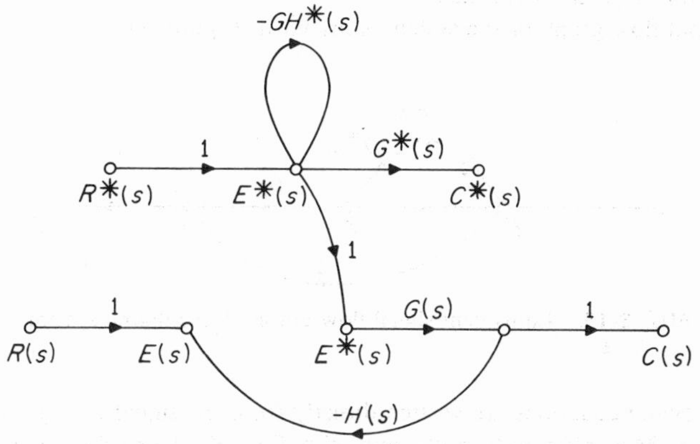

FIG. 3-13. Composite signal flow graph of the digital system in Fig. 3-11.

Application of Mason's gain formula to the composite signal flow graph yields the input-output transfer relations for all the digital and the continuous-data outputs. Therefore, from Figure 3-13, we get

$$C(s) = \frac{G(s)}{1 + GH^*(s)} R^*(s) \tag{3-143}$$

and

$$E(s) = R(s) - \frac{G(s)H(s)}{1 + GH^*(s)} R^*(s) \tag{3-144}$$

The procedure described above can be applied to linear multiloop, multisampler systems without difficulty, provided that the samplers are synchronized and of the same sampling frequency.

When one is more proficient with the method, equations such as (3-137) and (3-138) can be written directly from the equivalent signal flow graph.

EXAMPLE 3-12

In this example the discrete and continuous-data outputs of the digital system shown in Figure 3-14 will be determined by the sampled signal flow graph method.

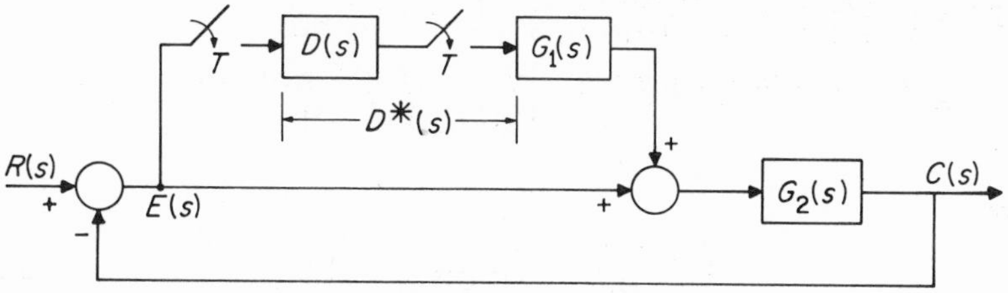

FIG. 3-14. Block diagram of digital control system.

The following steps are performed:

1. The signal flow graph of the system is drawn in Figure 3-15.

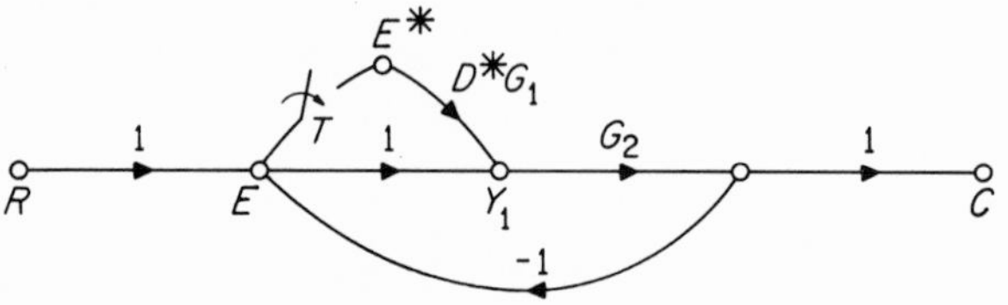

FIG. 3-15. Equivalent signal flow graph of the digital system in Fig. 3-14.

2. The following equations are written directly from the signal flow graph of Figure 3-15, using Mason's gain formula, and with $E(s)$, $Y_1(s)$, and $C(s)$ as output nodes, $R(s)$ and $E^*(s)$ as input nodes.

$$E = \frac{R}{1+G_2} - \frac{D^*G_1G_2}{1+G_2}E^* \tag{3-145}$$

$$Y_1 = \frac{R}{1+G_2} + \frac{D^*G_1}{1+G_2}E^* \tag{3-146}$$

$$C = \frac{RG_2}{1+G_2} + \frac{D^*G_1G_2}{1+G_2}E^* \tag{3-147}$$

For simplicity, all symbols of functions of s have been dropped, that is, $E^*(s)$ is denoted by E^*, $R(s)$ by R, etc.

Taking the pulse transform on both sides of Eqs. (3-145), (3-146), and (3-147), we get

$$E^* = \left(\frac{R}{1+G_2}\right)^* - D^*\left(\frac{G_1G_2}{1+G_2}\right)^* E^* \tag{3-148}$$

$$Y_1{}^* = \left(\frac{R}{1+G_2}\right)^* + D^*\left(\frac{G_1}{1+G_2}\right)^* E^* \tag{3-149}$$

$$C^* = \left(\frac{RG_2}{1+G_2}\right)^* + D^*\left(\frac{G_1G_2}{1+G_2}\right)^* E^* \tag{3-150}$$

3. Using these last six equations, the composite signal flow graph of the system is drawn as shown in Figure 3-16.
4. Now applying Mason's gain formula to the composite signal flow graph of Figure 3-16, the discrete and continuous-data outputs of the system are determined.

$$C^* = \left(\frac{RG_2}{1+G_2}\right)^* + \frac{D^*(G_1G_2/1+G_2)^*}{1+D^*(G_1G_2/1+G_2)^*}\left(\frac{R}{1+G_2}\right)^*$$
$$= \frac{(RG_2/1+G_2)^* + [(RG_2/1+G_2)^* + (R/1+G_2)^*]D^*(G_1G_2/1+G_2)^*}{1+D^*(G_1G_2/1+G_2)^*} \tag{3-151}$$

Since

$$\left(\frac{RG_2}{1+G_2}\right)^* + \left(\frac{R}{1+G_2}\right)^* = R^* \tag{3-152}$$

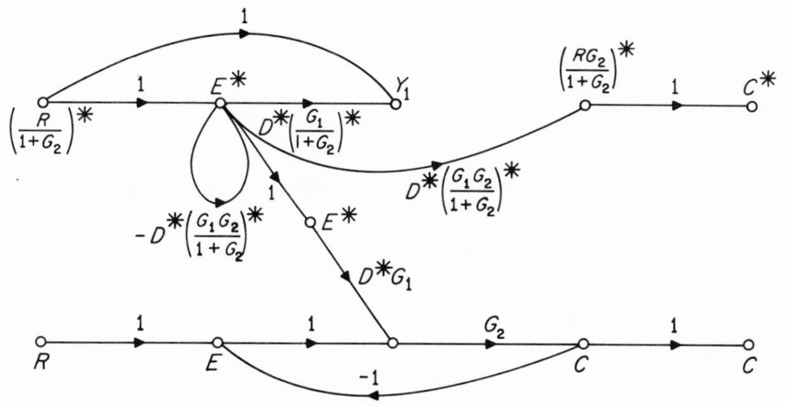

FIG. 3-16. Composite signal flow graph of the digital system shown in Fig. 3-14.

Eq. (3-151) is simplified to

$$C^* = \frac{(RG_2/1 + G_2)^* + R^* D^*(G_1 G_2/1 + G_2)^*}{1 + D^*(G_1 G_2/1 + G_2)^*} \tag{3-153}$$

In z-transform notation, Eq. (3-153) becomes

$$C(z) = \frac{(RG_2/1 + G_2)(z) + R(z)D(z)(G_1 G_2/1 + G_2)(z)}{1 + D(z)(G_1 G_2/1 + G_2)(z)} \tag{3-154}$$

where

$$\frac{G_1 G_2}{1 + G_2}(z) = \mathscr{Z}\left[\frac{G_1(s)G_2(s)}{1 + G_2(s)}\right] \tag{3-155}$$

Similarly,

$$\begin{aligned} C &= \frac{RG_2[1 + D^*(G_1 G_2/1 + G_2)^*] + (R/1 + G_2)^* G_1 G_2}{1 + G_2 + D^*(G_1 G_2/1 + G_2)^* + G_2 D^*(G_1 G_2/1 + G_2)^*} \\ &= \frac{RG_2}{1 + G_2} + \frac{(G_1 G_2/1 + G_2)D^*(R/1 + G_2)^*}{1 + D^*(G_1 G_2/1 + G_2)^*} \end{aligned} \tag{3-156}$$

The Direct Signal Flow Graph Method

The method proposed by Sedlar and Bekey[21] is a direct method which allows the evaluation of input-output relations from the signal flow graph of a digital control system without intermediate steps. This often makes the task of obtaining input-output relations for digital systems less tedious.

Since digital systems generally contain continuous and discrete signals, it is necessary to introduce a new node symbol which will be associated with the operation of sampling. Therefore, the following two types of nodes are defined:

White node. A white node ∘ is used to represent continuous variables.

Black node. A black node • is used to represent discrete-data or sampled-data variables.

With the introduction of the black node all samplers and digital operations in a digital system can be denoted by black nodes. Furtheremore, all the node variables of this new signal flow graph may be represented by unstarred quantities. It should be interpreted that the variable represented by a black node is equal to the sampled form of the sum of all signals entering the node.

As a simple illustration, the signal flow graph shown in Figure 3-11(b) is drawn as shown in Figure 3-17 with the sampling operation replaced by a black node. Therefore, the variable at the black node is given by

$$Y_1(s) = E^*(s) \tag{3-157}$$

As a matter of convenience, *the input variable is normalized to unity by introducing a branch with gain equal to the input* (see Figure 3-17).

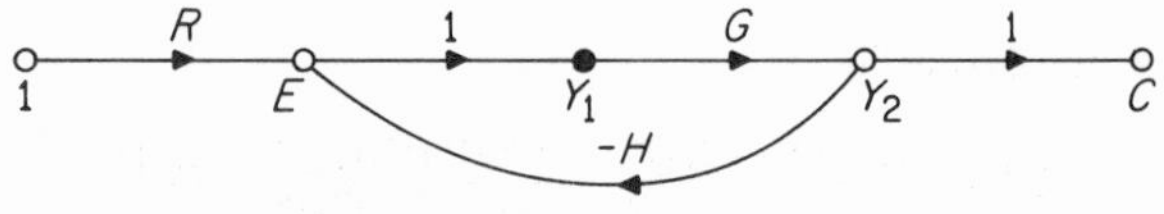

FIG. 3-17. Signal flow graph of Fig. 3-11(b) with sampling operation denoted by a black node.

In addition to the definitions such as nodes and branches which have been established for conventional signal flow graphs, the following terms are necessary for the signal flow graphs used here.

Segment. A segment, denoted by $\sigma(y_m, y_n)$, is a path connecting between two nodes y_m and y_n, where

y_m is either an input node or a black node

and

y_n is either an output node or a black node,

and all other nodes of σ are white. Not one node of a segment should be traversed more than once.

Type-1 paths and loops. Type-1 paths and loops contain only white nodes, and are denoted by $u^{(1)}$ and $v^{(1)}$, respectively.

Type-2 paths and loops. Type-2 refers to paths and loops which contain at least one black node and are denoted by $u^{(2)}$ and $v^{(2)}$, respectively.

Type-3 elementary path. A type-1 path is elementary if it does not traverse any node more than once.

Type-1 elementary loop. A type-1 loop is elementary if it does not traverse any node (other than the coincident input and output nodes of the loop) more than once.

Type-2 elementary path. A type-2 elementary path is one which is composed of distinct segments such that no black node is met more than once.

Type-2 elementary loop. A type-2 elementary loop is one which is composed of distinct segments such that no black node is met more than once.

Forward path. A path connected between an input node and an output node. A forward path can be type-1 or type-2.

Type-1 connected (in touch). A type-1 path or loop is connected (in touch) with another type-1 path or loop or a segment if and only if they contain a node in common.

Type-2 connected. A type-2 path or loop is connected (in touch) with another type-2 loop or path if and only if they contain a black node in common.

Path gain. The path gain of a path u is the product of the individual gains of all the branches of the path. For a path with black nodes, sampled form of the gains preceding the black nodes should be taken. We shall use $P_i^{(1)}$ to denote the gain of the ith type-1 elementary path, and $P_i^{(2)}$ the gain of the ith type-2 elementary path.

Segment gain. The segment gain of a segment σ_i, S_i, is the product of the individual gains of all the branches of the segment.

Loop gain. The loop gain of the loop v is the product of the individual gains of all the branches of the loop. We define $L_i^{(1)}$ = gain of the ith type-1 elementary loop, and $L_i^{(2)}$ = gain of the ith type-2 elementary loop.

As a simple illustration of all the terms defined above, let us consider the signal flow graph shown in Figure 3-17. The segments and the corresponding segment gains of the signal flow graph are

$$\sigma_1 = \sigma(1, Y_1) = (1, E, Y_1), \qquad S_1 = R^*$$

$$\sigma_2 = \sigma(Y_1, C) = (Y_1, C), \qquad S_2 = G$$

There are no type-1 elementary paths between the input node 1 and the output node C.

There is one type-2 elementary path between node 1 and node C.

$$u_1^{(2)} = (1, E, Y_1, C)$$

and the path gain is

$$P_1^{(2)} = R^*G$$

There are no type-1 elementary loops in the signal flow graph.

There is one type-2 elementary loop in the signal flow graph,

$$v_1^{(2)} = (Y_1, Y_2, E, Y_1)$$

and the loop gain is

$$L_1^{(2)} = -(GH)^*$$

Now let us consider a somewhat more elaborate example with reference to the signal flow graph shown in Figure 3-18.

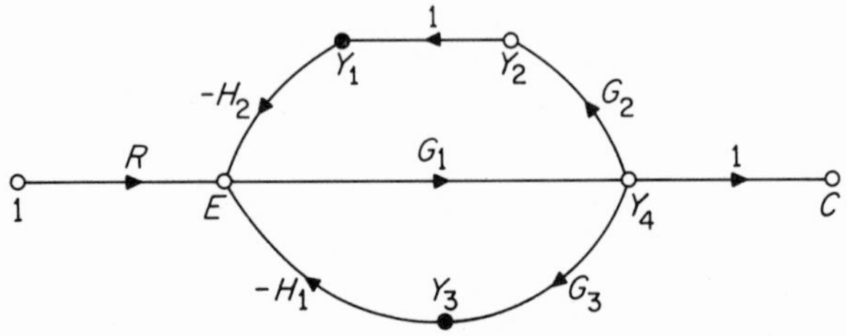

FIG. 3-18. Signal flow graph of a digital system.

The following elementary paths exist between the input node 1 and the output node C:

Type 1: $u_1^{(1)} = (1, E, Y_4, C), \quad P_1^{(1)} = RG_1$

Type-2: $u_1^{(2)} = (1, E, Y_4, Y_2, Y_1, E, Y_4, C)$

$$P_1^{(2)} = -(RG_1G_2)^*H_2G_1$$

$$u_2^{(2)} = (1, E, Y_4, Y_3, E, Y_4, C)$$

$$P_2^{(2)} = -(RG_1G_3)^*H_1G_1$$

$$u_3^{(2)} = (1, E, Y_4, Y_3, E, Y_4, Y_2, Y_1, E, Y_4, C)$$

$$P_3^{(2)} = (RG_1G_3)^*(H_1G_1G_2)^*H_2G_1$$

$$u_4^{(2)} = (1, E, Y_4, Y_2, Y_1, E, Y_4, Y_3, E, Y_4, C)$$

$$P_4^{(2)} = (RG_1G_2)^*(H_2G_1G_3)^*H_1G_1$$

Notice that in selecting the type-2 elementary path, white nodes may be traversed more than once so long as each time a white node is passed it belongs to a different segment. For instance, for the path $u_2^{(2)}$, the first time the node Y_4 is traversed, it belongs to the segment $(1, E, Y_4, Y_3)$; the second time Y_4 is traversed, however, it belongs to the segment (Y_3, E, Y_4, C). Similarly, the type-2 elementary path $u_3^{(2)}$ contains the following distinct segments:

$$\sigma_1(1, Y_3) = (1, E, Y_4, Y_3)$$

$$\sigma_2(Y_3, Y_1) = (Y_3, E, Y_4, Y_2, Y_1)$$

$$\sigma_3(Y_1, C) = (Y_1, E, Y_4, C)$$

Therefore, although the nodes E and Y_4 are passed three times in path $u_3^{(2)}$, each time they belong to a different segment.

There are no type-1 elementary loops in the signal flow graph.

The following type-2 loops are observed:

$$v_1^{(2)} = (Y_3, E, Y_4, Y_3), \quad L_1^{(2)} = -(G_1G_3H_1)^*$$
$$v_2^{(2)} = (Y_1, E, Y_4, Y_1), \quad L_2^{(2)} = -(G_1G_2H_2)^*$$
$$v_3^{(2)} = (Y_3, E, Y_4, Y_2, Y_1, E, Y_4, Y_3), \quad L_3^{(2)} = (G_1G_2H_1)^*(G_1G_3H_2)^*$$

The gain formula for the determination of input-output relationships on the new signal flow graph is given by

$$C = \frac{\Delta_j^{(1)}}{\Delta^{(1)}} \otimes \frac{\sum_{i=1}^{N} P_i \Delta_i^{(2)}}{\Delta^{(2)}} \tag{3-158}$$

where

$C =$ Output variable of the signal flow graph (the values of input are normalized to unity).

$P_i =$ Path gain of the ith elementary forward path (type 1 or type 2). The total number of elementary forward paths (type 1 and type 2) is N.

$\Delta^{(1)} =$ First determinant of the signal flow graph

$$= 1 - \sum L_i^{(1)} + \sum L_i^{(1)}L_j^{(1)} - \sum L_i^{(1)}L_j^{(1)}L_k^{(1)} + \dots \tag{3-159}$$

where

$\sum L_i^{(1)} =$ Sum of loop gains of all type-1 elementary loops. (3-160)

$\sum L_i^{(1)}L_j^{(1)} =$ Sum of the products of loop gains of all nonconnected (nontouching) type-1 elementary loops taken two at a time. (3-161)

$\sum L_i^{(1)}L_j^{(1)}L_k^{(1)} =$ Sum of the products of loop gains of all nonconnected (nontouching) type-1 elementary loops taken three at a time. (3-162)

$\Delta^{(2)} =$ Second determinant of the signal flow graph.

$$= 1 - \sum L_i^{(2)} + \sum L_i^{(2)}L_j^{(2)} - \sum L_i^{(2)}L_j^{(2)}L_k^{(2)} + \dots \tag{3-163}$$

The summation terms in Eq. (3-163) are interpreted the same way as above except that type-2 elementary loops are involved.

$\Delta_i^{(2)} =$ The second determinant, $\Delta^{(2)}$, of that part of the signal flow graph which is not *type-2 connected* (sharing at least one black node) with the ith forward path. Notice that if the ith forward path is type 1, $\Delta_i^{(2)} = \Delta^{(2)}$.

The symbol $\otimes$ represents the operation of multiplying $\Delta_j^{(1)}/\Delta^{(1)}$ by the gain of the jth segment for all the segments found in $\sum_{i=1}^{N} \frac{P_i \Delta_i^{(2)}}{\Delta^{(2)}}$. If any of the segment gain appears in sampled form, the multiplication must be performed on the corresponding continuous quantities, and the product sampled.

$\Delta_j^{(1)}$ = The first determinant, $\Delta^{(1)}$, of that part of the signal flow graph which is not connected with the jth segment σ_j.

An illustrative example at this point should help explain the steps described thus far for the direct signal flow graph method.

EXAMPLE 3-13

Consider the digital system whose signal flow graph is drawn as shown in Figure 3-17. The following information concerning the signal flow graph has been obtained earlier:

Type-1 forward elementary paths: none.
Type-2 forward elementary paths: $u_1^{(2)} = (1, E, Y_1, C)$
path gain, $P_1^{(2)} = R^*G$

Type-1 elementary loops: none
Therefore, $\Delta^{(1)} = 1$

Type-2 elementary loops: $v_1^{(2)} = (Y_1, C, E, Y_1)$
loop gain, $L_1^{(2)} = -(GH)^*$
Therefore, $\Delta^{(2)} = 1 - L_1^{(2)} = 1 + (GH)^*$

Now using the gain formula in Eq. (3-158), we have

$$C = \frac{\Delta_j{}^{(1)}}{\Delta^{(1)}} \otimes \frac{P_1^{(2)}\Delta_1^{(2)}}{\Delta^{(2)}} \tag{3-164}$$

Since $u_1^{(2)}$ is type-2 connected to $v_1^{(2)}$, $\Delta_1^{(2)} = 1$; Eq. (3-164) becomes

$$C = \frac{\Delta_j{}^{(1)}}{\Delta^{(1)}} \otimes \frac{R^*G}{1 + (GH)^*} \tag{3-165}$$

In the expression $R^*G/[1 + (GH)^*]$ of Eq. (3-165) we recognize that R^* and G are gains of individual segments, and the $\otimes$ operation must be performed on these quantities. However, since $\Delta^{(1)} = 1$, $\Delta_j^{(1)} = 1$ for all values of j, we have

$$C = \frac{R^*G}{1 + (GH)^*} \tag{3-166}$$

which agrees with the result obtained earlier in Eq. (3-143). The z-transform of the output is obtained by taking the z-transformation on both sides of Eq. (3-166).

EXAMPLE 3-14

Let us determine the input-output relation of the digital system shown in Figure 3-14 by means of the direct signal flow graph method.

The signal flow graph of the system with black nodes representing the digital operations is shown in Figure 3-19.

The following segments, forward paths, and loops are identified from the signal flow graph:

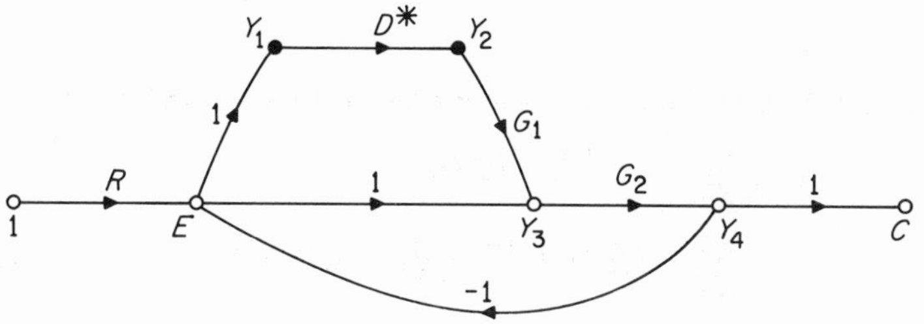

FIG. 3-19. Signal flow graph of the digital system shown in Fig. 3-14.

Segments:

$$\sigma_1 = (1, E, Y_3, Y_4, C) \qquad S_1 = RG_2$$
$$\sigma_2 = (1, E, Y_1) \qquad S_2 = R^*$$
$$\sigma_3 = (Y_1, Y_2) \qquad S_3 = D^*$$
$$\sigma_4 = (Y_2, Y_3, Y_4, C) \qquad S_4 = G_1 G_2$$
$$\sigma_5 = (Y_2, Y_3, Y_4, E, Y_1) \qquad S_5 = -(G_1 G_2)^*$$

Elementary Forward Paths Between Nodes 1 *and C:*

Type 1:

$$u_1^{(1)} = (1, E, Y_3, Y_4, C) \quad P_1^{(1)} = RG_2$$

Type 2:

$$u_1^{(2)} = (1, E, Y_1, Y_2, Y_3, Y_4, C) \quad P_1^{(2)} = R^*D^*G_1G_2$$

This is the only type-2 elementary path between nodes 1 and C on the signal flow graph. The path, $(1, E, Y_1, Y_2, Y_3, Y_4, E, Y_3, Y_4, C)$, is not a type-2 elementary path although no black nodes are traversed more than once. The path violates the condition that it must contain distinct segments.

Type-1 Elementary Loop:

$$v_1^{(1)} = (E, Y_3, Y_4, E) \quad L_1^{(1)} = -G_2$$

Type-2 Elementary Loop:

$$v_1^{(2)} = (Y_1, Y_2, Y_3, Y_4, E, Y_1) \quad L_1^{(2)} = -D^*(G_1G_2)^*$$

First Determinant:

$$\Delta^{(1)} = 1 - L_1^{(1)} = 1 + G_2 \tag{3-167}$$

Second Determinant:

$$\Delta^{(2)} = 1 - L_1^{(2)} = 1 + D^*(G_1G_2)^* \tag{3-168}$$

For $u_1^{(1)}$,

$$\Delta_1^{(2)} = 1 + D^*(G_1G_2)^*$$

and for $u_1^{(2)}$,

$$\Delta_2^{(2)} = 1$$

From Eq. (3-158), the input-output transfer relation of the system is written

$$C = \frac{\Delta_j^{(1)}}{\Delta^{(1)}} \otimes \frac{P_1^{(1)}\Delta_1^{(2)} + P_2^{(2)}\Delta_2^{(2)}}{\Delta^{(2)}}$$

$$= \frac{\Delta_j^{(1)}}{\Delta^{(1)}} \otimes \frac{RG_2[1 + D^*(G_1G_2)^*] + R^*D^*G_1G_2}{1 + D^*(G_1G_2)^*} \tag{3-169}$$

The second term on the right-hand side of Eq. (3-169) is identified to be

$$\frac{RG_2[1 + D^*(G_1G_2)^*] + R^*D^*G_1G_2}{1 + D^*(G_1G_2)^*} = \frac{S_1(1 - S_3S_5) + S_2S_3S_4}{1 - S_3S_5} \tag{3-170}$$

where S_j denotes the gain of the jth segment. The determinant $\Delta_j^{(1)}$ for $j = 1, 2, 3, 4, 5$ is found by determining the $\Delta^{(1)}$ of the part of the signal flow graph which is not connected to the jth segment. Therefore,

$$\sigma_1\text{:}\quad \Delta_1^{(1)} = 1$$

$$\sigma_2\text{:}\quad \Delta_2^{(1)} = 1$$

$$\sigma_3\text{:}\quad \Delta_3^{(1)} = \Delta^{(1)} = 1 + G_2$$

$$\sigma_4\text{:}\quad \Delta_4^{(1)} = 1$$

$$\sigma_5\text{:}\quad \Delta_5^{(1)} = 1$$

Now multiplying the segment gains by the corresponding $\Delta_j^{(1)}/\Delta^{(1)}$ in Eq. (3-170) (if the segment gain appears in sampled form multiplication is performed on the corresponding continuous quantities, and then the sampled operation is applied to the product), the output of the system is written

$$C = \frac{RG_2}{1 + G_2} + \frac{(R/1 + G_2)^*D^*(G_1G_2/1 + G_2)}{1 + D^*(G_1G_2/1 + G_2)^*} \tag{3-171}$$

which agrees with the result obtained earlier in Eq. (3-156).

EXAMPLE 3-15

Consider the digital control system shown in Figure 3-20.

The following segments, forward paths, and loops are identified from the signal flow graph of Figure 3-20(b):

Segments:

$$\sigma_1 = (1, E, Y_1) \qquad S_1 = R^*$$

$$\sigma_2 = (Y_1, Y_2, Y_4, C) \qquad S_2 = G_1G_2$$

$$\sigma_3 = (Y_3, Y_2, Y_4, C) \qquad S_3 = -G_2H$$

$$\sigma_4 = (Y_1, Y_2, Y_4, Y_3) \qquad S_4 = (G_1G_2)^*$$

$$\sigma_5 = (Y_3, Y_2, Y_4, E, Y_1) \qquad S_5 = (G_2H)^*$$

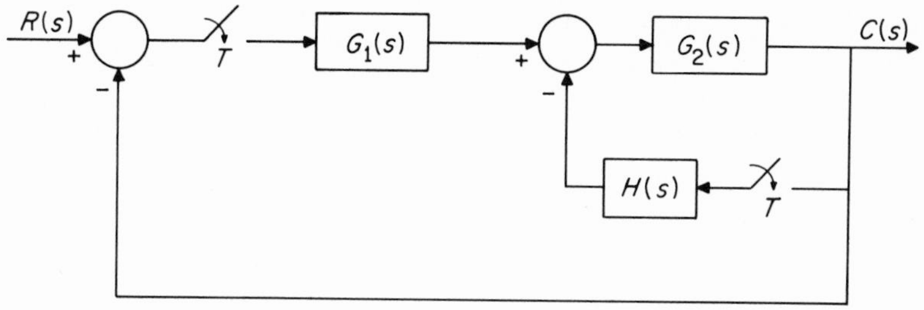

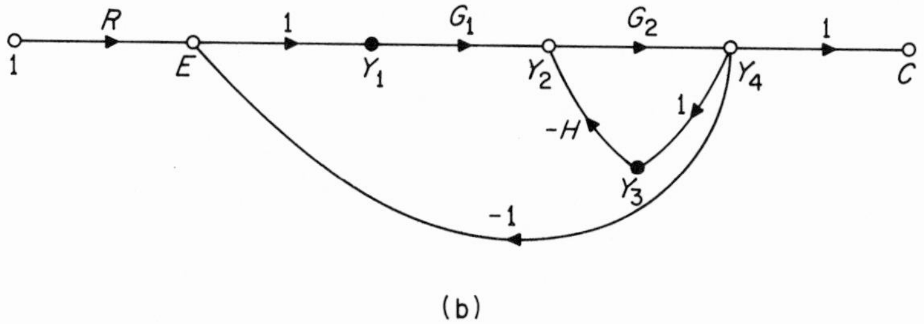

FIG. 3-20. (a) Block diagram of a digital control system; (b) signal flow graph.

Elementary Forward Paths Between Nodes 1 and C:

Type 1: none

$P_i^{(1)} = 0$ for all i

Type 2:

$u_1^{(2)} = (1, E, Y_1, Y_2, Y_4, C) \qquad P_1^{(2)} = R^*G_1G_2$

$u_2^{(2)} = (1, E, Y_1, Y_2, Y_4, Y_3, Y_2, Y_4, C) \qquad P_2^{(2)} = R^*(G_1G_2)^*(-G_2H)$

Elementary Loops:

Type 1: none

$v_i^{(1)} = 0$ for all i

Type 2:

$v_1^{(2)} = (E, Y_1, Y_2, Y_4, E), \qquad L_1^{(2)} = -(G_1G_2)^*$

$v_2^{(2)} = (Y_2, Y_4, Y_3, Y_2), \qquad L_2^{(2)} = -(G_2H)^*$

$v_3^{(2)} = (Y_1, Y_2, Y_4, Y_3, Y_2, Y_4, E, Y_1), \qquad L_3^{(2)} = (G_1G_2)^*(G_2H)^*$

First Determinant:

$\Delta^{(1)} = 1$ since there are no type-1 loops.

Second Determinant:

$$\Delta^{(2)} = 1 - (L_1^{(2)} + L_2^{(2)} + L_3^{(2)}) + L_1^{(2)}L_2^{(2)}$$

The last term on the right side of the last equation denotes the product of the loop gains of $v_1^{(2)}$ and $v_2^{(2)}$ which are type-2 nontouching. Since $\Delta^{(1)} = 1, \Delta_j^{(1)} = 1$ for all values of j.

The input-output relation for the system is now obtained by using the gain formula of Eq. (3-158).

$$C = \frac{\Delta_j^{(1)}}{\Delta^{(1)}} \otimes \frac{P_1^{(2)}\Delta_1^{(2)} + P_2^{(2)}\Delta_2^{(2)}}{\Delta^{(2)}}$$

$$= \frac{R^*G_1G_2[1 + (G_2H)^*] - R^*(G_1G_2)^*G_2H}{1 + (G_1G_2)^* + (G_2H)^*} \tag{3-172}$$

The z-transform of the output is obtained by taking the z-transformation on both sides of Eq. (3-172).

$$C(z) = \frac{G_1G_2(z)[1 + G_2H(z)] - G_1G_2(z)G_2H(z)}{1 + G_1G_2(z) + G_2H(z)} R(z) \tag{3-173}$$

From the illustrative examples worked out in this section, it appears that the direct signal flow graph method offers a simpler way of determining the input-output relations for digital systems than the sampled signal flow graph method. A further comparison of the two methods may show that the sampled signal flow graph method is perhaps easier to understand, whereas the direct signal flow graph method requires an understanding of the topological interpretation and definitions of the flow graph. However, we may find use in both methods for the purpose of checking the result obtained by each.

REFERENCES

1. Barker, R.H., "The Pulse Transfer Function and Its Application to Servo Systems," *Proc. I.E.E.*, London **99**, Part 4, December 1952, pp. 302–317.
2. Helm, H.A., "The Z-Transformation," *Bell System Technical Journal*, **38**, January 1959, pp. 177–196.
3. Jury, E.I., "Synthesis and Critical Study of Sampled-Data Control System," "*Trans. A.I.E.E.*, **75**, Part 2, 1956, pp. 141–151.
4. Jury, E.I., "Additions to the Modified z-Transform Method," *I.R.E. WESCON Convention Record*, Part 4, 1957, pp. 136–156.
5. Jury, E.I., and Farmanfarma, "Table of z-Transforms and Modified z-Transforms of Various Sampled-Data Systems Configurations." Univ. of California (Berkeley), Electronics Research Lab., Report 136A, Ser. 60, 1955.
6. Lago, G.V., "Additions to z-Transformation Theory for Sampled-Data System," *Trans. A.I.E.E.*, **74**, Part 2, January 1955, pp. 403–407.
7. Papoulis, A., "A New Method of Analysis of Sampled-Data Systems," *I.R.E. Trans. on Automatic Control*, Vol. AC-4, November 1959, pp. 67–73.
8. Lago, G.V., "Additions to Sampled-Data Theory," *Proc. National Electronics Conference*, **10**, 1954, pp. 758–766.

9. Pastel, M. P., and Thaler, G. J., "Sampled-Data Design by Log Gain Diagrams," *I. R. E. Trans. on Automatic Control*, Vol. AC-4, November 1959, pp. 192-197.

10. Ragazzini, J.R., and Zadeh, L.H., "The Analysis of Sampled-Data Systems," *Trans. A.I.E.E.*, **71**, Part 2, 1952, pp. 225-234.

11. Truxal, J.G., *Automatic Feedback Control System Synthesis*, Chapter 9. McGraw-Hill, New York, 1955.

12. Kuo, B.C., *Analysis and Synthesis of Sampled-Data Control Systems*. Prentice-Hall, Inc., Englewood Cliffs, N.J., 1963.

13. Ash, R., Kim, W.E., and Kranc, G.M., "A General Flow Graph Technique for the Solution of Multiloop Sampled Systems," *Trans. of A.S.M.E., Journal of Basic Engineering*, June 1960, pp. 360-370.

14. Kuo, B.C., "Composite Flow Graph Technique for the Solution of Multiloop, Multisampler, Sampled Systems," *I.R.E. Trans. on Automatic Control*, Vol. AC-6, 1961, pp. 343-344.

15. Lendaris, G. G., and Jury, E. I., "Input-Output Relationships for Multiple Sampled-Loop Systems," *Trans. A. I. E. E.*, **79**, Part 2, January 1960, pp. 375-385.

16. Salzer, J.M., "Signal Flow Techniques for Digital Compensation," *Proc. Computers in Control Systems Conference*, October 1957.

17. Salzer, J.M., "Signal Flow Reductions in Sampled-Data Systems," *I.R.E., WESCON Convention Record*, Part 4, 1957, pp. 166-170.

18. Tou, J.T., "Simplified Technique for the Determination of Output Transforms of Multiloop, Multisampler, Variable-Rate Discrete-Data Systems," *Proc. of the I.R.E.*, **49**, March 1961, pp. 646-647.

19. Mason, S.J., "Feedback Theory—Some Properties of Flow Graphs," *Proc. I.R.E.*, **41**, September 1953, pp. 1144-1156.

20. Mason, S.J., "Feedback Theory—Further Properties of Signal Flow Graphs," *Proc. I.R.E.*, **44**, July 1956, pp. 960-966.

21. Sedlar, M., and Bekey, G.A. "Signal Flow Graphs of Sampled-Data Systems: A New Formulation," *IEEE Trans. on Automatic Control*, AC-12, **2**, April 1967, pp. 154-161.

22. Jury, E.I., "A General z-Transform Formula for Sampled-Data Systems," *IEEE Trans. on Automatic Control*, AC-12, **5**, October 1967, pp. 606-608.

23. Jury, E.I., *Theory and Application of the z-Transform Method*. John Wiley & Sons, New York, 1964.

24. Kliger, I., and Lipinski, W.C., "The z-Transform of a Product of Two Functions," *IEEE Trans. on Automatic Control*, AC-9, October 1964, pp. 582-583.

25. Mesa, W., and Phillips, C.L., "A Theorem on the Modified z-Transform," *IEEE Trans. on Automatic Control*, AC-10, October 1965, p. 489.

26. Tsypkin, Y.Z., *Theory of Pulse Systems.* State Press for Physics and Math. USSR, 1958.

27. Lindorff, D.P., *Theory of Sampled-Data Control Systems.* John Wiley & Sons, New York, 1964.

28. Kuo, B.C., *Linear Networks and Systems.* McGraw-Hill, New York, 1967.

29. Jury, E.I., and Pai, M.A., "Convolution Z-Transform Method Applied to Certain Nonlinear Discrete Systems," *IRE Trans. on Automatic Control*, Vol. AC-7, January 1962, pp. 57-64.

4

THE STATE VARIABLE TECHNIQUE

4.1 INTRODUCTION

In general, the analysis and design of linear systems may be carried out by one of two major approaches. One approach relies on the use of Laplace and z-transforms, transfer functions, block diagrams or signal flow graphs. The other method, which has gained significant importance in control system theory and engineering is the state variable technique. The popularity of the state variable technique is also evidenced by the fact that a great majority of modern control system design techniques are based on the state variables formulation.

In a broad sense the state variable method has at least the following important advantages over the conventional transfer function method:

1. The state variable formulation is natural and convenient for computer solutions.
2. The state variable approach allows a unified representation of digital systems with various types of sampling schemes.
3. The state variable method allows a unified representation of single variable and multivariable systems.
4. The state variable method can be applied to certain types of nonlinear and time-varying systems.

In the state variable method a continuous-data system is represented by a set of first-order differential equations, called state equations. For a digital system with discrete-data components the state equations are first-order difference equations. Normally a pulsed-data control system contains both continuous-data and discrete-data components and the state equations of the system consist of both first-order differential and difference equations.

4.2 STATE EQUATIONS AND STATE TRANSITION EQUATIONS OF CONTINUOUS-DATA SYSTEMS[1,2,3]

Consider a continuous-data system with p inputs and q outputs as shown in Figure 4-1 characterized by the following set of n first-order differential equations, called state equations:

$$\frac{dx_i(t)}{dt} = f_i[x_1(t), x_2(t), \ldots, x_n(t), m_1(t), m_2(t), \ldots, m_p(t), t] \qquad (4\text{-}1)$$

$$(i = 1, 2, \ldots, n)$$

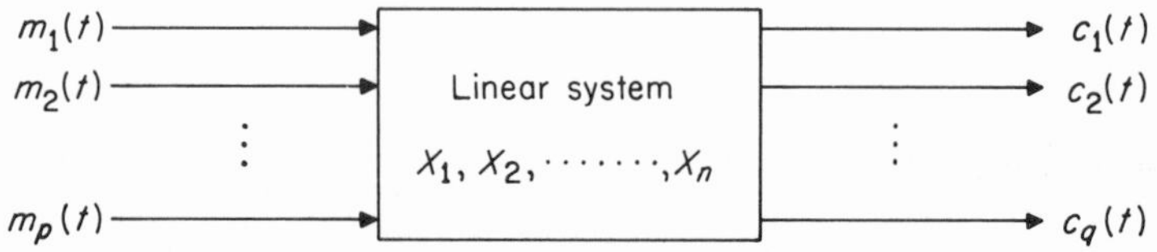

FIG. 4-1. A linear system with p inputs, q outputs, and n state variables.

where $x_1(t), x_2(t), \ldots, x_n(t)$ are the state variables, $m_1(t), m_2(t), \ldots, m_p(t)$ are the input variables, and f_i denotes the ith functional relationship. In general, f_i can be linear or nonlinear.

The q outputs of the system are related to the state variables and the inputs through the *output equations* which are of the form

$$c_k(t) = g_k[x_1(t), x_2(t), \ldots, x_n(t), m_1(t), m_2(t), \ldots, m_p(t), t] \qquad (4\text{-}2)$$

$$(k = 1, 2, \ldots, q)$$

Similar remarks can be made for g_k as for f_i.

The combined set of state equations and output equations are called the *dynamic equations* of the system.

It is customary to write the dynamic equations in vector-matrix form:

State equation $$\frac{d\mathbf{x}(t)}{dt} = \mathbf{f}[\mathbf{x}(t), \mathbf{m}(t), t] \qquad (4\text{-}3)$$

Output equation $$\mathbf{c}(t) = \mathbf{g}[\mathbf{x}(t), \mathbf{m}(t), t] \qquad (4\text{-}4)$$

where $\mathbf{x}(t)$ is an $n \times 1$ column matrix and is called the *state vector*; that is,

$$\mathbf{x}(t) = \begin{bmatrix} x_1(t) \\ x_2(t) \\ \cdot \\ \cdot \\ \cdot \\ x_n(t) \end{bmatrix} \tag{4-5}$$

The input vector, $\mathbf{m}(t)$, is a $p \times 1$ column matrix, and

$$\mathbf{m}(t) = \begin{bmatrix} m_1(t) \\ m_2(t) \\ \cdot \\ \cdot \\ \cdot \\ m_p(t) \end{bmatrix} \tag{4-6}$$

The output vector, $\mathbf{c}(t)$, is defined as

$$\mathbf{c}(t) = \begin{bmatrix} c_1(t) \\ c_2(t) \\ \cdot \\ \cdot \\ \cdot \\ c_q(t) \end{bmatrix} \tag{4-7}$$

and is a $q \times 1$ column matrix.

If a system is linear but time-varying, the dynamic equations of Eqs. (4-3) and (4-4) are written as

$$\frac{d\mathbf{x}(t)}{dt} = A(t)\mathbf{x}(t) + B(t)\mathbf{m}(t) \tag{4-8}$$

$$\mathbf{c}(t) = D(t)\mathbf{x}(t) + E(t)\mathbf{m}(t) \tag{4-9}$$

where $A(t)$ is an $n \times n$ square matrix with time-varying coefficients, and $B(t)$ is an $n \times p$ matrix. The dimensions of the matrices $D(t)$ and $E(t)$ are $q \times n$ and $q \times p$, respectively.

If the system is linear and time-invariant, Eqs. (4-8) and (4-9) are reduced to

$$\frac{d\mathbf{x}(t)}{dt} = A\mathbf{x}(t) + B\mathbf{m}(t) \tag{4-10}$$

and

$$\mathbf{c}(\mathrm{t}) = D\mathbf{x}(\mathrm{t}) + E\mathbf{m}(\mathrm{t}) \tag{4-11}$$

respectively. The matrices A, B, D, and E now all contain constant elements.

The solution of the state equation is called the *state transition equation.* First, let us consider Eq. (4-10), the time-invariant situation. The state equation can be solved by means of the Laplace transform. Taking the Laplace transform on both sides of Eq. (4-10), we have

$$s\mathbf{X}(s) - \mathbf{x}(0^+) = A\mathbf{X}(s) + B\mathbf{M}(s) \tag{4-12}$$

where $\mathbf{x}(0^+)$ is the initial state vector evaluated at $t = 0^+$. Solving for $\mathbf{X}(s)$ in Eq. (4-12) yields

$$\mathbf{X}(s) = (sI - A)^{-1}\mathbf{x}(0^+) + (sI - A)^{-1}B\mathbf{M}(s) \tag{4-13}$$

where I denotes the unity matrix and $(sI - A)^{-1}$ is the matrix inverse of $(sI - A)$. It is assumed that $(sI - A)$ is nonsingular so that its matrix inverse exists.

Now taking the inverse Laplace transform on both sides of Eq. (4-13), we get

$$\mathbf{x}(t) = \Phi(t)\mathbf{x}(0^+) + \int_0^t \Phi(t - \tau)B\mathbf{m}(\tau)\,d\tau \quad t \geq 0 \tag{4-14}$$

where

$$\Phi(t) = \mathscr{L}^{-1}[(sI - A)^{-1}] = e^{At} = \sum_{k=0}^{\infty} A^k \frac{t^k}{k!} \tag{4-15}$$

and $\Phi(t)$ is called the *state transition matrix* of A.

Several important properties of the state transition matrix $\Phi(t)$ are listed below:

1. $\Phi(0) = I$ (unity matrix) (4-16)
2. $\Phi(t_2 - t_1)\Phi(t_1 - t_0) = \Phi(t_2 - t_0)$ for any t_0, t_1, t_2 (4-17)
3. $\Phi(-t) = \Phi^{-1}(t)$ (4-18)
4. $\underbrace{\Phi(t)\Phi(t)\cdots\Phi(t)}_{k} = \Phi^k(t) = \Phi(kt)$ k = positive integer (4-19)

The state transition equation of Eq. (4-14) is useful only when the initial time is taken at $t = 0$. In the study of digital control systems it is desirable to use a more general time reference, t_0. Thus, the initial state vector is denoted by $\mathbf{x}(t_0)$, and it is assumed that the input vector $\mathbf{m}(t)$ is defined for $t \geq t_0$.

Letting $t = t_0$ in Eq. (4-14), and solving for $\mathbf{x}(0^+)$, we get

$$\mathbf{x}(0^+) = \Phi(-t_0)\mathbf{x}(t_0) - \Phi(-t_0)\int_0^{t_0} \Phi(t_0 - \tau)B\mathbf{m}(\tau)\,d\tau \tag{4-20}$$

where, in arriving at the final form of this equation, the property of $\Phi(t)$ in Eq. (4-18) has been used.

Substituting Eq. (4-20) into Eq. (4-14) and utilizing the properties of $\Phi(t)$ in Eqs. (4-16) through (4-18), we get

$$\mathbf{x}(t) = \Phi(t - t_0)\mathbf{x}(t_0) + \int_{t_0}^t \Phi(t - \tau)B\mathbf{m}(\tau)\,d\tau \quad t \geq t_0 \tag{4-21}$$

Equation (4-21) is referred to as the *state transition equation* of the system described by Eq. (4-10).

Ordinarily, Eq. (4-21) cannot be simplified any further without further knowledge of the system and input variables.

4.3 STATE TRANSITION EQUATIONS OF DIGITAL SYSTEMS WITH SAMPLE-AND-HOLD[4,5]

An open-loop pulse-data control system is obtained when sample-and-hold operation is applied to the inputs of the system of Fig. 4-1. The resulting system is shown in Fig. 4-2.

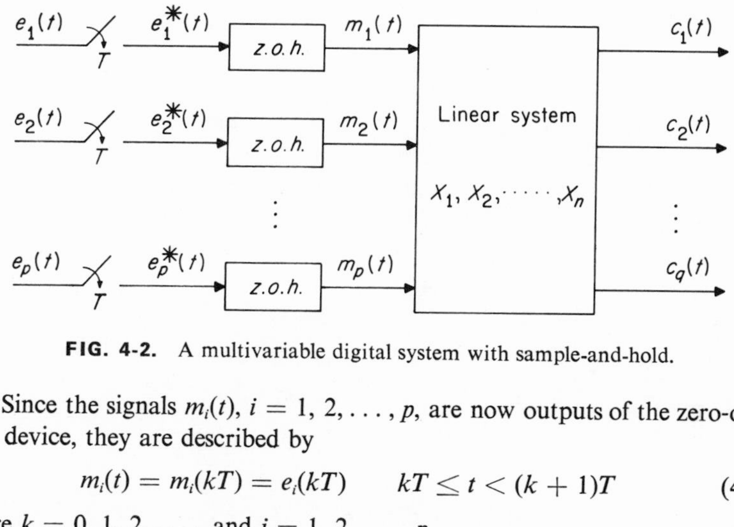

FIG. 4-2. A multivariable digital system with sample-and-hold.

Since the signals $m_i(t)$, $i = 1, 2, \ldots, p$, are now outputs of the zero-order hold device, they are described by

$$m_i(t) = m_i(kT) = e_i(kT) \qquad kT \leq t < (k+1)T \tag{4-22}$$

where $k = 0, 1, 2, \ldots,$ and $i = 1, 2, \ldots, p$.

Now letting $t_0 = kT$, and since $\mathbf{m}(\tau) = \mathbf{m}(kT) =$ constant vector, Eq. (4-21) is written

$$\mathbf{x}(t) = \Phi(t - kT)\mathbf{x}(kT) + \left[\int_{kT}^{t} \Phi(t - \tau)B\,d\tau\right]\mathbf{m}(kT) \tag{4-23}$$

for $kT \leq t \leq (k+1)T$.

Equation (4-23) is valid only for one sampling period since the input vector $\mathbf{m}(kT)$ is constant only for that duration. It is possible to express the integral on the right side of Eq. (4-23) as

$$\Theta(t - kT) = \int_{kT}^{t} \Phi(t - \tau)B\,d\tau \tag{4-24}$$

Then, Eq. (4-23) is simplified to

$$\mathbf{x}(t) = \boldsymbol{\Phi}(t - kT)\mathbf{x}(kT) + \boldsymbol{\Theta}(t - kT)\mathbf{m}(kT) \tag{4-25}$$

for $kT \leq t \leq (k + 1)T$.

If we are interested only in the response at the sampling instants, just as in the case of the z-transform solution, we set $t = (k + 1)T$, then Eq. (4-25) becomes

$$\mathbf{x}[(k + 1)T] = \boldsymbol{\Phi}(T)\mathbf{x}(kT) + \boldsymbol{\Theta}(T)\mathbf{m}(kT) \tag{4-26}$$

where

$$\boldsymbol{\Phi}(T) = e^{AT} \quad (n \times n) \tag{4-27}$$

and

$$\boldsymbol{\Theta}(T) = \int_{kT}^{(k+1)T} \boldsymbol{\Phi}[(k + 1)T - \tau]B\, d\tau$$

$$= \int_{0}^{T} \boldsymbol{\Phi}(T - \lambda)B\, d\lambda \quad (n \times p) \tag{4-28}$$

Equation (4-26) is of the form of a linear difference equation in vector-matrix form. Since the equation represents a set of first-order difference equations describing the state variables at discrete instants of time, it may be referred to as the *discrete state equation* of the system in Fig. 4-2.

The discrete state equation of Eq. (4-26) can be solved by means of a simple recursive procedure. Setting $k = 0, 1, 2, \ldots,$ in turn in Eq. (4-26), the following equations result:

$$k = 0 \qquad \mathbf{x}(T) = \boldsymbol{\Phi}(T)\mathbf{x}(0) + \boldsymbol{\Theta}(T)\mathbf{m}(0) \tag{4-29}$$

$$k = 1 \qquad \mathbf{x}(2T) = \boldsymbol{\Phi}(T)\mathbf{x}(T) + \boldsymbol{\Theta}(T)\mathbf{m}(T) \tag{4-30}$$

$$k = 2 \qquad \mathbf{x}(3T) = \boldsymbol{\Phi}(T)\mathbf{x}(2T) + \boldsymbol{\Theta}(T)\mathbf{m}(2T) \tag{4-31}$$

$$\vdots$$

$$k = N - 1 \quad \mathbf{x}(NT) = \boldsymbol{\Phi}(T)\mathbf{x}[(N - 1)T] + \boldsymbol{\Theta}(T)\mathbf{m}[(N - 1)T] \tag{4-32}$$

Now substituting Eq. (4-29) into Eq. (4-30), and then Eq. (4-30) into Eq. (4-31), . . . , and so on, the solution of Eq. (4-26) is obtained as

$$\mathbf{x}(NT) = \boldsymbol{\Phi}(NT)\mathbf{x}(0) + \sum_{k=0}^{N-1} \boldsymbol{\Phi}[(N - k - 1)T]\boldsymbol{\Theta}(T)\mathbf{m}(kT) \tag{4-33}$$

where N is a positive integer.

Equation (4-33) is defined as the *discrete state transition equation* of the digital system in Fig. 4-2. It is interesting to note that Eq. (4-33) is analogous to its continuous counterpart in Eq. (4-14). The state transition equation of Eq. (4-14) describes the state of the system of Fig. 4-1 for $t \geq t_0$ for any input defined over the same interval. *The discrete state transition equation of Eq.* (4-33) *is more restrictive in that it describes the state only at* $t = NT$ ($N = 0$, 1, 2, . . .) *and only for a system with sample-and-hold devices at the inputs.*

An expression analogous to that of Eq. (4-21) can be obtained from Eq. (4-33) by using $t = NT$ as the initial time. Starting with the expression

$$\mathbf{x}[(N+1)T] = \Phi(T)\mathbf{x}(NT) + \Theta(T)\mathbf{m}(NT) \tag{4-34}$$

and by recursion and continued substitution, we easily get

$$\mathbf{x}[(N+M)T] = \Phi(MT)\mathbf{x}(NT) + \sum_{k=0}^{M-1} \{\Phi[(M-k-1)T]\Theta(T)\mathbf{m}[(N+k)T]\} \tag{4-35}$$

where M is a positive integer.

The discrete state equation

$$\mathbf{x}[(k+1)T] = \Phi(T)\mathbf{x}(kT) + \Theta(T)\mathbf{m}(kT) \tag{4-36}$$

can be solved by means of the z-transform method. Taking the z-transform on both sides of Eq. (4-36) yields

$$z\mathbf{X}(z) - z\mathbf{x}(0) = \Phi(T)\mathbf{X}(z) + \Theta(T)\mathbf{M}(z) \tag{4-37}$$

Solving for $\mathbf{X}(z)$ from the last equation, we get

$$\mathbf{X}(z) = [zI - \Phi(T)]^{-1}\, z\mathbf{x}(0) + [zI - \Phi(T)]^{-1}\, \Theta(T)\mathbf{M}(z) \tag{4-38}$$

The inverse z-transform of the last equation is written as

$$\mathbf{x}(kT) = \mathscr{Z}^{-1}\{[zI - \Phi(T)]^{-1} z\mathbf{x}(0)\} + \mathscr{Z}^{-1}\{[zI - \Phi(T)]^{-1}\Theta(T)\mathbf{M}(z)\} \tag{4-39}$$

We shall show that the inverse z-transform of $[zI - \Phi(T)]^{-1}z$ is $\Phi(kT)$. The z-transform of $\Phi(kT)$is, by definition,

$$\mathscr{Z}[\Phi(kT)] = \sum_{k=0}^{\infty} \Phi(kT)z^{-k} = I + \Phi(T)z^{-1} + \Phi(2T)z^{-2} + \dots \tag{4-40}$$

Premultiplying both sides of Eq. (4-40) by $\Phi(T)z^{-1}$ and subtracting the result from Eq. (4-40), we have

$$[I - \Phi(T)z^{-1}]\mathscr{Z}[\Phi(kT)] = I \tag{4-41}$$

Therefore,

$$\mathscr{Z}[\Phi(kT)] = [I - \Phi(T)z^{-1}]^{-1} = [zI - \Phi(T)]^{-1}z \tag{4-42}$$

or

$$\Phi(kT) = \mathscr{Z}^{-1}\{[zI - \Phi(T)]^{-1}z\} \tag{4-43}$$

Equation (4-43) also represents a way of finding $\Phi(kT)$ using the z-transform method.

The inverse z-transform of $[zI - \Phi(T)]^{-1}\Theta(T)\mathbf{M}(z)$ is found from the real convolution theorem of Eq. (3-100). Therefore. Eq. (4-39) is written

$$\mathbf{x}(kT) = \Phi(kT)\mathbf{x}(0) + \sum_{i=0}^{k-1} \Phi[(k-i-1)T]\Theta(T)\mathbf{m}(iT) \tag{4-44}$$

which is of the same form as Eq. (4-33).

EXAMPLE 4-1

In this example we shall illustrate the analysis of an open-loop digital system by the state variable method presented above. The block diagram of the system under consideration is shown in Fig. 4-3. The dynamic equations that describe the linear process are

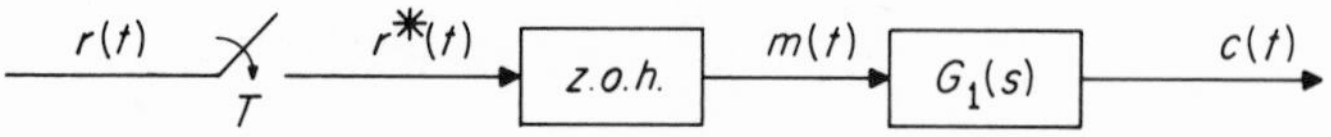

FIG. 4-3. An open-loop digital system.

$$\begin{bmatrix} \dfrac{dx_1(t)}{dt} \\ \dfrac{dx_2(t)}{dt} \end{bmatrix} = \begin{bmatrix} 0 & 1 \\ -2 & -3 \end{bmatrix} \begin{bmatrix} x_1(t) \\ x_2(t) \end{bmatrix} + \begin{bmatrix} 0 \\ 1 \end{bmatrix} m(t) \tag{4-45}$$

$$c(t) = x_1(t) \tag{4-46}$$

where $x_1(t)$ and $x_2(t)$ are the state variables, $c(t)$ is the scalar output, and $m(t)$ is the scalar input. Also, since $m(t)$ is the output of the zero-order hold,

$$m(t) = m(kT) = r(kT)$$

for $kT \leq t < (k+1)T$.

Comparing Eq. (4-45) with the standard state equation form of Eq. (4-10), we have

$$A = \begin{bmatrix} 0 & 1 \\ -2 & -3 \end{bmatrix} \tag{4-47}$$

and

$$B = \begin{bmatrix} 0 \\ 1 \end{bmatrix} \tag{4-48}$$

The following matrix is formed:

$$(sI - A) = \begin{bmatrix} s & -1 \\ 2 & s+3 \end{bmatrix} \tag{4-49}$$

Therefore,

$$(sI - A)^{-1} = \frac{1}{(s^2 + 3s + 2)} \begin{bmatrix} s+3 & 1 \\ -2 & s \end{bmatrix} \tag{4-50}$$

The state transition matrix of A is obtained by taking the inverse Laplace transform of $(sI - A)^{-1}$. Therefore,

$$\Phi(t) = \mathscr{L}^{-1}[(sI - A)^{-1}] = \begin{bmatrix} 2e^{-t} - e^{-2t} & e^{-t} - e^{-2t} \\ -2e^{-t} + 2e^{-2t} & -e^{-t} + 2e^{-2t} \end{bmatrix} \tag{4-51}$$

Substitution of Eqs. (4-48) and (4-51) into Eq. (4-24) with $k = 0$ yields

$$\Theta(t) = \int_0^t \Phi(t-\tau)B\,d\tau$$

$$= \int_0^t \begin{bmatrix} e^{-\tau} - e^{-2\tau} \\ -e^{-\tau} + 2e^{-2\tau} \end{bmatrix} d\tau = \begin{bmatrix} \frac{1}{2} - e^{-t} + \frac{1}{2}e^{-2t} \\ e^{-t} - e^{-2t} \end{bmatrix} \tag{4-52}$$

Now substituting Eqs. (4-51) and (4-52) into Eq. (4-26), the discrete state equation of the system is written

$$\begin{bmatrix} x_1[(k+1)T] \\ x_2[(k+1)T] \end{bmatrix} = \begin{bmatrix} 2e^{-T} - e^{-2T} & e^{-T} - e^{-2T} \\ -2e^{-T} + 2e^{-2T} & -e^{-T} + 2e^{-2T} \end{bmatrix} \begin{bmatrix} x_1(kT) \\ x_2(kT) \end{bmatrix} + \begin{bmatrix} \frac{1}{2} - e^{-T} + \frac{1}{2}e^{-2T} \\ e^{-T} - e^{-2T} \end{bmatrix} m(kT) \tag{4-53}$$

For $T = 1$ second, Eq. (4-53) is simplified to

$$\begin{bmatrix} x_1(k+1) \\ x_2(k+1) \end{bmatrix} = \begin{bmatrix} 0.6 & 0.233 \\ -0.466 & -0.098 \end{bmatrix} \begin{bmatrix} x_1(k) \\ x_2(k) \end{bmatrix} + \begin{bmatrix} 0.2 \\ 0.233 \end{bmatrix} m(k) \tag{4-54}$$

The solution of Eq. (4-54) is given by Eq. (4-33). Therefore, for any integer N,

$$\begin{bmatrix} x_1(N) \\ x_2(N) \end{bmatrix} = \begin{bmatrix} 0.6 & 0.233 \\ -0.466 & -0.098 \end{bmatrix}^N \begin{bmatrix} x_1(0) \\ x_2(0) \end{bmatrix} + \sum_{k=0}^{N-1} \begin{bmatrix} 0.6 & 0.233 \\ -0.466 & -0.098 \end{bmatrix}^{N-k-1} \times \begin{bmatrix} 0.2 \\ 0.233 \end{bmatrix} m(k) \tag{4-55}$$

or

$$\begin{bmatrix} x_1(N) \\ x_2(N) \end{bmatrix} = \begin{bmatrix} 2e^{-N} - e^{-2N} & e^{-N} - e^{-2N} \\ -2e^{-N} + 2e^{-2N} & -e^{-N} + 2e^{-2N} \end{bmatrix} \begin{bmatrix} x_1(0) \\ x_2(0) \end{bmatrix} + \sum_{k=0}^{N-1} \begin{bmatrix} 0.633e^{-(N-k-1)} - 0.433e^{-(N-k-1)} \\ -0.633e^{-(N-k-1)} + 0.866e^{-2(N-k-1)} \end{bmatrix} m(k) \tag{4-56}$$

Equation (4-55) or Eq. (4-56) gives the values of the state variables at any sampling instant $t = NT$, once the initial state $\mathbf{x}(0)$ and the input $r(t)$ for $t \geq 0$ are specified.

4.4 STATE TRANSITION EQUATIONS OF DIGITAL SYSTEMS WITH ALL DISCRETE ELEMENTS

When a pulsed-data system is composed of all discrete elements, that is, the input and the output of the system are all discrete-data, the system may be described by the following discrete state equation:

$$\mathbf{x}(k+1) = A\mathbf{x}(k) + B\mathbf{m}(k) \tag{4-57}$$

where A and B are constant matrices.

Comparing Eq. (4-57) with Eq. (4-26), the solution of the former is written

$$\mathbf{x}(N) = A^N\mathbf{x}(0) + \sum_{k=0}^{N-1} A^{N-k-1}B\mathbf{m}(k) \tag{4-58}$$

Notice that Eq. (4-26) is obtained from the continuous state equations of a system with sample-and-hold by characterizing the variables at the sampling instants. However, Eq. (4-57) is regarded as a starting point for the description of a pulsed-data system with all discrete-data elements.

4.5 RELATION BETWEEN STATE EQUATION AND TRANSFER FUNCTION

It is of interest to investigate the relationship between the transfer function and the state variable methods.

Consider that a digital system with multiple inputs and outputs is described by the transfer relation:

$$\mathbf{C}(z) = G(z)\mathbf{R}(z) \tag{4-59}$$

where

$$\mathbf{C}(z) = \begin{bmatrix} C_1(z) \\ C_2(z) \\ \vdots \\ C_q(z) \end{bmatrix} \tag{4-60}$$

is the $q \times 1$ output transform vector;

$$\mathbf{R}(z) = \begin{bmatrix} R_1(z) \\ R_2(z) \\ \vdots \\ R_p(z) \end{bmatrix} \tag{4-61}$$

is the $p \times 1$ input transform vector;

$$G(z) = \begin{bmatrix} G_{11}(z) & G_{12}(z) \dots G_{1p}(z) \\ G_{21}(z) & G_{22}(z) \dots G_{2p}(z) \\ \vdots & \\ G_{q1}(z) & G_{q2}(z) \dots G_{qp}(z) \end{bmatrix} \tag{4-62}$$

is the $q \times p$ z-transfer function matrix.

The elements of $G(z)$ may be described by

$$G_{ij}(z) = \frac{a_0 + a_1 z^{-1} + \dots + a_n z^{-n}}{b_0 + b_1 z^{-1} + \dots + b_m z^{-m}} \tag{4-63}$$

Now if a pulsed-data control system is described by the dynamic equations

$$\mathbf{x}[(k+1)T] = A\mathbf{x}(kT) + B\mathbf{r}(kT) \tag{4-64}$$

$$\mathbf{c}(kT) = D\mathbf{x}(kT) + E\mathbf{r}(kT) \tag{4-65}$$

Taking the z-transform on both sides of Eq. (4-64) and solving for $\mathbf{X}(z)$, we get

$$\mathbf{X}(z) = (zI - A)^{-1}z\mathbf{x}(0) + (zI - A)^{-1}B\mathbf{R}(z) \tag{4-66}$$

Substitution of Eq. (4-66) into the z-transform of Eq. (4-65) yields

$$\mathbf{C}(z) = D(zI - A)^{-1}z\mathbf{x}(0) + D(zI - A)^{-1}B\mathbf{R}(z) + E\mathbf{R}(z) \tag{4-67}$$

For the transfer function, we assume that the initial state $\mathbf{x}(0)$ is zero; therefore, Eq. (4-67) becomes

$$\mathbf{C}(z) = [D(zI - A)^{-1}B + E]\mathbf{R}(z) \tag{4-68}$$

Comparing Eq. (4-68) with Eq. (4-59) we see that the z-transfer function matrix of the system may be written as

$$G(z) = D(zI - A)^{-1}B + E \tag{4-69}$$

If a pulsed-data system has sample-and-hold operations and is described by the dynamic equations in Eqs. (4-10) and (4-11), we only have to replace A and B in Eq. (4-69) by $\Phi(T)$ and $\Theta(T)$, respectively. Of course, this is performed with the assumption that a z-transfer function can be written for the system with sample-and-hold. It was illustrated in Sec. 3.8 that sometimes transfer functions in the form of Eq. (4-59) cannot be defined for a system with sample-and-hold, and only input-output relations exist.

The inverse transform of the transfer function matrix is called the *impulse response matrix*. Taking the inverse z-transform on both sides of Eq. (4-69), we get

$$\mathbf{g}(kT) = D\Phi[(k-1)T]B + E \tag{4-70}$$

Since $\Phi[(k-1)T] = \mathbf{0}$ for $k<1$, $\mathbf{g}(kT)$in Eq. (4-70) can be written as

$$\mathbf{g}(kT) = E \qquad k = 0 \tag{4-71}$$

$$\mathbf{g}(kT) = D\Phi[(k-1)T]B + E \qquad k \geq 1 \tag{4-72}$$

EXAMPLE 4-2

As an illustrative example, let us derive the transfer function of the open-loop system shown in Fig. 4-3, and with the data of Example 4-1.

First, we shall use the state variable method described in the preceding paragraphs. From Example 4-1, we have

$$\Phi(1) = \begin{bmatrix} 0.6 & 0.233 \\ -0.466 & -0.098 \end{bmatrix} \tag{4-73}$$

$$\Theta(1) = \begin{bmatrix} 0.2 \\ 0.233 \end{bmatrix} \tag{4-74}$$

$$D = [1 \quad 0] \text{ and } E = 0 \tag{4-75}$$

Substituting Eqs. (4-73), (4-74), and (4-75) into Eq. (4-69), we have the transfer function of the system

$$G(z) = [1 \quad 0] \begin{bmatrix} z - 0.6 & -0.233 \\ 0.466 & z + 0.098 \end{bmatrix}^{-1} \begin{bmatrix} 0.2 \\ 0.233 \end{bmatrix}$$

$$= \frac{0.2z + 0.074}{(z - 0.135)(z - 0.368)} \tag{4-76}$$

To use the z-transform method we must first determine the transfer function of the linear process $G_1(s)$ (see Fig. 4-3). From the dynamic equations which describe the process, Eqs. (4-45) and (4-46), the differential equation for the relation between $m(t)$ and $c(t)$ is written

$$\frac{d^2c(t)}{dt^2} + 3\frac{dc(t)}{dt} + 2c(t) = m(t) \tag{4-77}$$

where $m(t)$ is the output of the zero-order hold. Therefore, the transfer function $G_1(s)$ is

$$G_1(s) = \frac{C(s)}{M(s)} = \frac{1}{s^2 + 3s + 2} \tag{4-78}$$

With reference to Fig. 4-3, the z-transfer function of the overall system is

$$G(z) = \mathscr{Z}\left[\frac{1 - e^{-Ts}}{s} G_1(s)\right] = \frac{C(z)}{R(z)} \tag{4-79}$$

or

$$G(z) = (1 - z^{-1})\, \mathscr{Z}\left[\frac{1}{s(s^2 + 3s + 2)}\right] \tag{4-80}$$

Evaluating the z-transform of Eq. (4-80), we have

$$G(z) = \frac{0.2z + 0.074}{(z - 0.135)(z - 0.368)} \tag{4-81}$$

which agrees with the result of Eq. (4-76).

The impulse sequence of the system can be obtained by taking the inverse z-transform of $G(z)$. A partial fraction expansion of $G(z)$ gives

$$G(z) = \frac{0.633}{z - 0.368} - \frac{0.433}{z - 0.135}$$

Therefore,

$$g(k) = 0.633e^{-(k-1)} - 0.433e^{-2(k-1)} \tag{4-82}$$

for $k > 0$. For $k = 0$, $g(0) = 0$. An alternate method of determining $g(k)$ calls for the use of Eqs. (4-71) and (4-72). Therefore,

$$g(0) = 0$$

and $$g(kT) = [1 \quad 0]\begin{bmatrix} 2e^{-(k-1)} - e^{-2(k-1)} & e^{-(k-1)} - e^{-2(k-1)} \\ -2e^{-(k-1)} + 2e^{-2(k-1)} & -e^{-(k-1)} + 2e^{-2(k-1)} \end{bmatrix}\begin{bmatrix} 0.2 \\ 0.233 \end{bmatrix}$$

$$= 0.633e^{-(k-1)} - 0.433e^{-2(k-1)} \qquad (4\text{-}83)$$

for $k = 0$.

4.6 THE CHARACTERISTIC EQUATION

The characteristic equation of a linear system can be defined with respect to the system's difference equation, transfer function, or dynamic equations. Consider that a linear single variable digital system is described by the difference equation

$$\begin{aligned} b_0c(k + n) + b_1c(k + n - 1) + b_2c(k + n - 2) + \ldots + b_{n-1}c(k + 1) \\ + b_nc(k) = a_0r(k + n) + a_1r(k + n - 1) + \ldots \\ + a_{n-1}r(k + 1) + a_nr(k) \end{aligned} \qquad (4\text{-}84)$$

The transfer function of the system is

$$G(z) = \frac{C(z)}{R(z)} = \frac{a_0z^n + a_1z^{n-1} + \ldots + a_{n-1}z + a_n}{b_0z^n + b_1z^{n-1} + \ldots + b_{n-1}z + b_n} \qquad (4\text{-}85)$$

The characteristic equation is defined as the equation obtained by equating the denominator of the transfer function to zero; that is

$$b_0z^n + b_1z^{n-1} + \ldots + b_{n-1}z + b_n = 0 \qquad (4\text{-}86)$$

Notice that the characteristic equation is a polynomial which contains only the coefficients of the homogeneous part of the difference equation of the system.

The characteristic equation can also be defined in terms of elements of the dynamic equations. Referring to Eq. (4-69), we can write the equation as

$$\begin{aligned} G(z) &= D\frac{[\Delta_{ij}]'}{|zI - A|}B + E \\ &= \frac{D[\Delta_{ij}]'B + |zI - A|E}{|zI - A|} \end{aligned} \qquad (4\text{-}87)$$

where Δ_{ij} represents the cofactor of the ijth element of the matrix $zI - A$, and $[\Delta_{ij}]'$ is the matrix transpose of $[\Delta_{ij}]$; $|zI - A|$ denotes the determinant of $zI - A$. Setting the denominator of $G(z)$ to zero gives

$$|zI - A| = 0 \qquad (4\text{-}88)$$

which is the characteristic equation.

The roots of the characteristic equation are referred to as the eigenvalues of the matrix A.

For the system described in Example 4-2 it is easy to see that the characteristic equation of the system is

$$z^2 - 0.502z + 0.0497 = 0 \tag{4-89}$$

and the eigenvalues are $z = 0.135$ and $z = 0.368$.

4.7 TIME-VARYING SYSTEMS

When a linear continuous-data system has time-varying elements, the dynamic equations of the system are written as

$$\frac{d\mathbf{x}(t)}{dt} = A(t)\mathbf{x}(t) + B(t)\mathbf{m}(t) \tag{4-90}$$

$$\mathbf{c}(t) = D(t)\mathbf{x}(t) + E(t)\mathbf{m}(t) \tag{4-91}$$

where $\mathbf{x}(t)$ is the $n \times 1$ state vector, $\mathbf{m}(t)$ is the $p \times 1$ input vector, and $\mathbf{c}(t)$ is the $q \times 1$ output vector; $A(t)$, $B(t)$, $D(t)$, and $E(t)$ are matrices with time-varying elements.

It can be shown that the solution of Eq. (4-90) is

$$\mathbf{x}(t) = \Phi(t, t_0)\mathbf{x}(t_0) + \int_{t_0}^{t} \Phi(t, \tau)B(\tau)\mathbf{m}(\tau)\, d\tau \tag{4-92}$$

which is similar to the solution of the time-invariant state equation, Eq. (4-21). In Eq. (4-92), $\Phi(t, t_0)$ is the state transition matrix of $A(t)$ and is related to $A(t)$ by

$$\Phi(t, t_0) = e^{\int_{t_0}^{t} A(\tau)\, d\tau} \tag{4-93}$$

However, Eq. (4-93) is valid only if $A(t)$ and $\int_{t_0}^{t} A(\tau)\, d\tau$ commute for all t, that is

$$A(t)\int_{t_0}^{t} A(\tau)\, d\tau = \int_{t_0}^{t} A(\tau)\, d\tau \cdot A(t) \tag{4-94}$$

If the input vector $\mathbf{m}(t)$ represents the outputs of sample-and-hold devices, the time-varying system may be characterized by the following difference equation for the time interval $kT \leq t \leq (k + 1)T$:

$$\mathbf{x}[(k + 1)T] = \Phi[(k + 1)T, kT]\mathbf{x}(kT + \Theta[(k + 1)T, kT]\mathbf{m}(kT) \tag{4-95}$$

where

$$\Theta[(k + 1)T, kT] = \int_{kT}^{(k+1)T} \Phi[(k + 1)T, \tau]B(\tau)\, d\tau \tag{4-96}$$

When a time-varying system is composed of all-discrete elements, the dynamic equations of the system may be written as

$$\mathbf{x}(k+1) = A(k)\mathbf{x}(k) + B(k)\mathbf{r}(k) \tag{4-97}$$

$$\mathbf{c}(k) = D(k)\mathbf{x}(k) + E(k)\mathbf{r}(k) \tag{4-98}$$

where it is assumed that the values of the elements of $A(k)$, $B(k)$, $D(k)$, and $E(k)$ can vary only at the discrete instants $k = 0, 1, 2, \ldots$.

The solution of the state equation in Eq. (4-97) is obtained in the same way as that described in Eqs. (4-29) through (4-32). The recursive procedure leads to

$$k = 0 \qquad \mathbf{x}(1) = A(0)\mathbf{x}(0) + B(0)\mathbf{r}(0) \tag{4-99}$$

$$\begin{aligned} k = 1 \qquad \mathbf{x}(2) &= A(1)\mathbf{x}(1) + B(1)\mathbf{r}(1) \\ &= A(1)A(0)\mathbf{x}(0) + A(1)B(0)\mathbf{r}(0) + B(1)\mathbf{r}(1) \\ &\;\;\vdots \end{aligned} \tag{4-100}$$

$$\begin{aligned} k = N-1 \qquad \mathbf{x}(N) &= A(N-1)\mathbf{x}(N-1) + B(N-1)\mathbf{r}(N-1) \\ &= \Phi(N, 0)\mathbf{x}(0) + \sum_{k=0}^{N-1} \Phi(N, k+1)B(k)\mathbf{r}(k) \end{aligned} \tag{4-101}$$

where

$$\Phi(N, k+1) = A(N-1)A(N-2)\ldots A(k+2)A(k+1) \tag{4-102}$$

Therefore, the discrete state transition equation for the time-varying state equation of Eq. (4-97) is

$$\mathbf{x}(N) = \Phi(N, 0)\mathbf{x}(0) + \sum_{k=0}^{N-1} \Phi(N, k+1)B(k)\mathbf{r}(k) \tag{4-103}$$

Similarly, the solution to Eq. (4-95) is written

$$\begin{aligned} \mathbf{x}(NT) = \Phi(NT, 0)\mathbf{x}(0) + \sum_{k=0}^{N-1} \Phi[NT, (N-k)T]\Theta[(N-k)T, \\ (N-k-1)T]\mathbf{m}[(N-k-1)T] \end{aligned} \tag{4-104}$$

In summary of the discussions on the state variable analysis given in the preceding sections, a tabulation of results is given in Table 4-1.

4.8 STATE DIAGRAMS[1,6]

The signal flow graph methods presented in Chapter 3 apply only to algebraic equations. Therefore, these conventional signal flow graphs can be used only for the derivation of the input-output relations of a linear system in the trans-

Table 4-1

TABULATION OF RESULTS OF STATE VARIABLE ANALYSIS OF LINEAR SYSTEMS

	Continuous-data System	*Digital System with Sample-and-hold*	*Digital System*
State Variables	$\mathbf{x}(t)$	$\mathbf{x}(kT)$	$\mathbf{x}(k)$
State Equation (Time-invariant)	$\dot{\mathbf{x}}(t) = A\mathbf{x}(t) + B\mathbf{r}(t)$	$\mathbf{x}[(k+1)T] = \Phi(T)\mathbf{x}(kT) + \Theta(T)\mathbf{m}(kT)$	$\mathbf{x}(k+1) = A\mathbf{x}(k) + B\mathbf{r}(k)$
State Equation (Time-varying)	$\dot{\mathbf{x}}(t) = A(t)\mathbf{x}(t) + B(t)\mathbf{r}(t)$	$\mathbf{x}[(k+1)T] = \Phi[(k+1)T, kT]\,\mathbf{x}(kT) + \Theta[(k+1)T, kT]\mathbf{m}(kT)$	$\mathbf{x}(k+1) = A(k)\mathbf{x}(k) + B(k)\mathbf{r}(k)$
State Transition Matrix (Time-invariant)	$\Phi(t) = e^{At}$	$\Phi(kT) = \Phi^k(T) = e^{AkT}$	$\Phi(kT) = A^k$
State Transition Matrix (Time-varying)	$\Phi(t, t_0) = e^{\int_{t_0}^{t} A(\tau)\,d\tau}$	$\Phi[(k+1)T, kT] = e^{\int_{kT}^{(k+1)T} A(\tau)\,d\tau}$	$\Phi(N, k) = A(N-1)A(N-2)\ldots A(k+1)A(k)$
Transform of State Transition Matrix	$\Phi(s) = (sI - A)^{-1}$	$\Phi(z) = [zI - \Phi(\mathrm{T})]^{-1} z$	$\Phi(z) = (zI - A)^{-1} z$
Impulse Response Matrix	$\mathbf{g}(t) = D\Phi(t)B + E$	$\mathbf{g}(kT) = D\Phi[(k-1)T]B \quad k \geq 1$ $= E \quad k = 0$	$g(k) = D\Phi(k-1)B \quad k \geq 1$ $= E \quad k = 0$
Transfer Matrix	$G(s) = D(sI - A)^{-1}B + E$	$G(z) = D[zI - \Phi(\mathrm{T})]^{-1}B + E$	$G(z) = D(zI - A)^{-1}B + E$
State Transition Equation (Time-invariant)	$\mathbf{x}(t) = \Phi(t - t_0)\mathbf{x}(t_0) + \int_{t_0}^{t} \Phi(t - \tau)B\mathbf{r}(\tau)\,d\tau$	$\mathbf{x}(NT) = \Phi(NT)\mathbf{x}(0) + \sum_{k=0}^{N-1} \Phi[(N-k-1)T]\Theta(\mathrm{T})\mathbf{m}(kT)$	$\mathbf{x}(N) = A^N\mathbf{x}(0) + \sum_{k=0}^{N-1} A^{N-k-1}B\mathbf{r}(k)$
State Transition Equation (Time-varying)	$\mathbf{x}(t) = \Phi(t, t_0)\mathbf{x}(t_0) + \int_{t_0}^{t} \Phi(t, \tau)B(\tau)\mathbf{r}(\tau)\,d\tau$	$\mathbf{x}(NT) = \Phi(NT, 0)\mathbf{x}(0) + \sum_{k=0}^{N-1} \Phi[NT, (N-k)T]\Theta[(N-k)T, (N-k-1)T]\mathbf{m}[(N-k-1)T]$	$\mathbf{x}(N) = \Phi(N, 0)\mathbf{x}(0) + \sum_{k=0}^{N-1} \Phi(N, k+1)B(k)\mathbf{r}(k)$
Characteristic Equation	$\lvert sI - A \rvert = 0$	$\lvert zI - \Phi(T) \rvert = 0$	$\lvert zI - A \rvert = 0$

form domain. In this section we shall apply the method of the state transition signal flow graph[1] to the representation of difference state equations. For simplicity, the *state transition signal flow graph* is called the *state diagram* in this text.

For a continuous-data system, the state diagram resembles the block diagram of the analog computer program. Therefore, once the state diagram is drawn, the problem can be solved either by an analog computer or by pencil and paper. For a pulsed-data system, the state diagram describes the operations on a digital computer, so that the problem can again be solved either by machine or analytical methods.

State Diagram of Continous-Data Systems

The fundamental linear operations that can be performed on an analog computer are *multiplication by a constant*, *addition*, *polarity inversion*, and *integration*. We shall now show that these computer operations are closely related to the elements of a state diagram.

1. *Multiplication by a constant.* Multiplication of a machine variable by a constant on an analog computer is done by potentiometers and amplifiers. Consider the operation

$$x_2(t) = ax_1(t) \tag{4-105}$$

where a is a constant. If a lies between zero and unity, a potentiometer is used for the purpose. If a is a negative integer less than -1 an operational amplifier is used to realize Eq. (4-105). The analog computer block diagram symbols of the potentiometer and the operational amplifier are shown in Figures 4-4(a) and 4-4(b), respectively. Since Eq. (4-105) is an algebraic equation it can be represented by a signal flow graph as shown in Figure 4-4(c). This signal flow graph can be regarded as a state diagram or an element of a state diagram if the variables $x_1(t)$ and $x_2(t)$ are state variables or linear combinations of state variables. Since the Laplace transform of Eq. (4-105) is

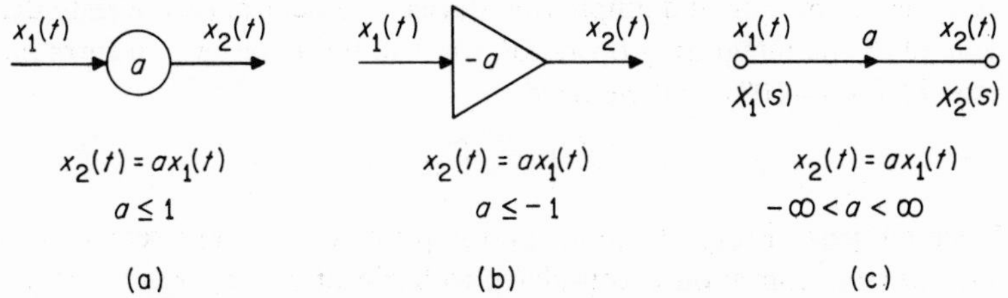

FIG. 4-4. (a) Block diagram symbol of a potentiometer; (b) block diagram symbol of an operational amplifier; (c) state diagram of the operation of multiplying a variable by a constant.

$$X_2(s) = aX_2(s) \qquad (4\text{-}106)$$

the state diagram symbol of Figure 4-4(c) also portrays the relation between the transformed variables.

2. *Algebraic sum of two or more machine variables.* The algebraic sum of two or more machine variables of an analog computer is obtained by means of an operational amplifier as shown in Figure 4-5(a). The illustrated case portrays the algebraic equation

$$x_4(t) = a_1x_1(t) + a_2x_2(t) + a_3x_3(t) \qquad (4\text{-}107)$$

where all the coefficients are assumed to be less than or equal to 0. The state diagram representation of Eq. (4-107) as well as its Laplace transform is shown in Figure 4-5(b).

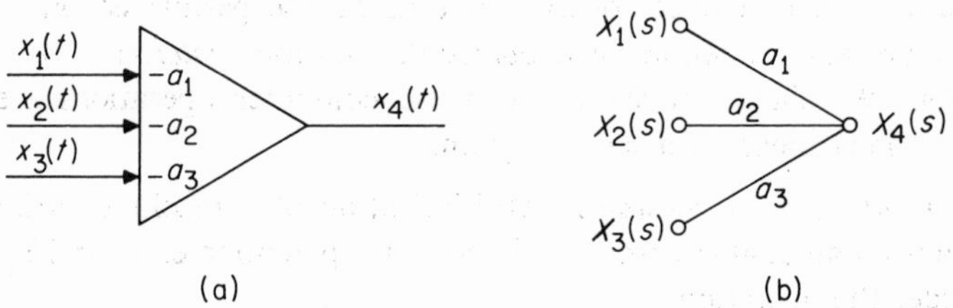

FIG. 4-5. (a) Block diagram symbol of the operational amplifier used as a summing device; (b) state diagram representation of the summing operation.

3. *Integration.* The integration of a machine variable on an analog computer is achieved by means of a computer element called the *integrator*. If $x_1(t)$ is the output of the integrator with the initial condition $x_1(t_0)$ given at $t = t_0$, and $x_2(t)$ is the input, the integrator performs the following operation:

$$x_1(t) = \int_{t_0}^{t} ax_2(\tau)d\tau + x_1(t_0) \qquad (4\text{-}108)$$

where a is an integer ≤ 0. The block diagram symbol of the integrator is shown in Figure 4-6(a). In general, the integrator can be used simultaneously as a summing and amplifying device. To determine the state diagram symbol of the integrating operation we take the Laplace transform on both sides of Eq. (4-108), and we have

$$X_1(s) = a\frac{X_2(s)}{s} + a\int_{t_0}^{0} x_2(\tau)d\tau + \frac{x_1(t_0)}{s} \qquad (4\text{-}109)$$

Since the past history of the integrator prior to t_0 is represented by $x_1(t_0)$, and the state transition is considered to begin at $t = t_0$, $x_2(\tau) = 0$ for $0 < \tau < t_0$. Thus, Eq. (4-109) becomes

$$X_1(s) = \frac{aX_2(s)}{s} + \frac{x_1(t_0)}{s} \qquad (t \geq t_0) \qquad (4\text{-}110)$$

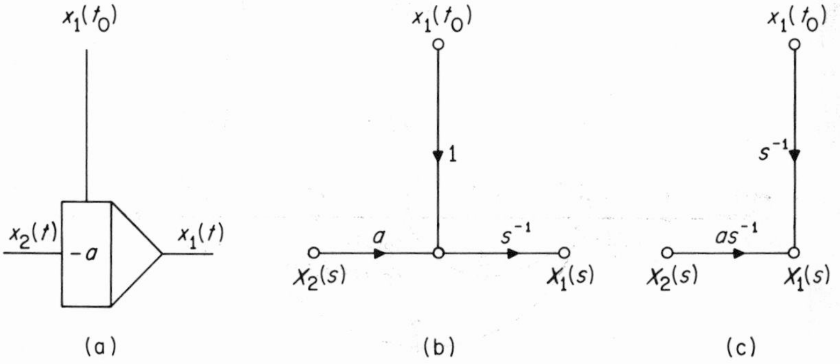

FIG. 4-6. (a) Block diagram representation of an integrator; (b), (c) state diagram symbols of an integrator.

It should be pointed out here that the transformed equation of Eq. (4-110) is defined only for the period $t \geq t_0$. Therefore, the inverse Laplace transform of $X_1(s)$ should read

$$x_1(t) = \mathscr{L}^{-1}[X_1(s)] = a\int_{t_0}^{t} x_2(\tau)d\tau + x_1(t_0) \tag{4-111}$$

A state diagram representation of the integration operation is obtained from the signal flow graph description of Eq. (4-110). Two parallel versions of the state diagram symbols of an integrator are shown in Figure 4-6(b), (c).

Besides leading to computer solutions, a state diagram can also provide the following information on a linear dynamic system:

1. Dynamic equations; that is, state equations and output equations.
2. State transition equation.
3. Transfer function.
4. State variables.

We shall use the following numerical example to illustrate the use of the state diagram for analyzing linear continuous-data systems.

EXAMPLE 4-3

Consider that a linear dynamic system is described by the differential equation

$$\frac{d^2c(t)}{dt^2} + 3\frac{dc(t)}{dt} + 2c(t) = r(t) \tag{4-112}$$

where $c(t)$ is the output and the input $r(t)$ is a unit-step function. The initial conditions of the system are represented by $c(t_0)$ and $dc(t_0)/dt$ at $t = t_0$.

A state diagram of the system is drawn by first equating the highest-order deriv-

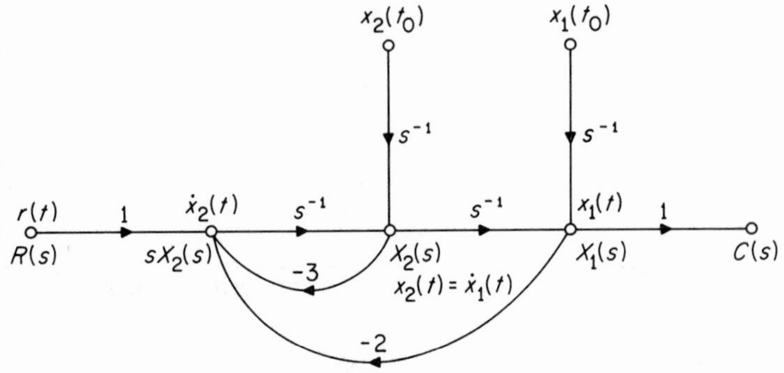

FIG. 4-7. A state diagram of the system described by Eq. (4-112).

ative term to the rest of the terms in Eq. (4-112), as shown in Figure 4-7. The state variables of the system, $x_1(t)$ and $x_2(t)$, are defined as output variables of the integrators. The state transition equations of the system in transformed form are written directly from the state diagram using Mason's gain formula.

$$\begin{bmatrix} X_1(s) \\ X_2(s) \end{bmatrix} = \frac{1}{(s+1)(s+2)} \begin{bmatrix} s+3 & 1 \\ -2 & s \end{bmatrix} \begin{bmatrix} x_1(t_0) \\ x_2(t_0) \end{bmatrix} + \begin{bmatrix} \dfrac{1}{(s+1)(s+2)} \\ \dfrac{s}{(s+1)(s+2)} \end{bmatrix} R(s) \tag{4-113}$$

For a unit-step input, $R(s) = 1/s$, the inverse Laplace transform of Eq. (4-113) is

$$\begin{bmatrix} x_1(t) \\ x_2(t) \end{bmatrix} = \begin{bmatrix} 2e^{-(t-t_0)} - e^{-2(t-t_0)} & e^{-(t-t_0)} - e^{-2(t-t_0)} \\ -2e^{-(t-t_0)} + 2e^{-2(t-t_0)} & -e^{-(t-t_0)} + 2e^{-2(t-t_0)} \end{bmatrix} \begin{bmatrix} x_1(t_0) \\ x_2(t_0) \end{bmatrix}$$
$$+ \begin{bmatrix} \frac{1}{2} - e^{-(t-t_0)} + \frac{1}{2}e^{-2(t-t_0)} \\ e^{-(t-t_0)} - e^{-2(t-t_0)} \end{bmatrix} \qquad t \geq t_0 \tag{4-114}$$

which is the state transition equation of the system.

Notice that in deriving the last equation the time origin is taken to be at $t = t_0$, so that the following inverse Laplace transform relation has been used:

$$\mathscr{L}^{-1}\left(\frac{1}{s+a}\right) = e^{-a(t-t_0)} \qquad t \geq t_0 \tag{4-115}$$

The interested reader may derive the state transition equation using the analytic expression in Eq. (4-21).

The state equations of the system are obtained from the state diagram by applying Mason's gain formula to the nodes $\dot{x}_1(t)$ and $\dot{x}_2(t)$ with $r(t)$, $x_1(t)$, and $x_2(t)$ as inputs. The branches with the gain s^{-1} are deleted from the state diagram in this operation. Therefore the dynamic equations of the system are

$$\begin{bmatrix} \dfrac{dx_1(t)}{dt} \\ \dfrac{dx_2(t)}{dt} \end{bmatrix} = \begin{bmatrix} 0 & 1 \\ -2 & -3 \end{bmatrix} \begin{bmatrix} x_1(t) \\ x_2(t) \end{bmatrix} + \begin{bmatrix} 0 \\ 1 \end{bmatrix} r(t) \tag{4-116}$$

$$c(t) = [1 \quad 0] \begin{bmatrix} x_1(t) \\ x_2(t) \end{bmatrix} \tag{4-117}$$

The transfer function of the system is ordinarily obtained by taking the Laplace transform of Eq. (4-112). However, applying Mason's gain formula to the state diagram of Figure 4-7 between $R(s)$ and $C(s)$ and setting the initial states to zero, we get

$$\frac{C(s)}{R(s)} = \frac{1}{s^2 + 3s + 2} \tag{4-118}$$

State Diagram of Pulsed-Data Systems

When a digital system is described by difference equations or discrete state equations, a state diagram may be drawn to represent the relations between the discrete state variables. In contrast to the similarity between an analog computer diagram and a continuous-data state diagram, the digital state diagram portrays operations on a digital computer.

Some of the linear operations of a digital computer are: multiplication by a constant, addition of several variables, and time delay or storage. The mathematical descriptions of these basic digital operations and their corresponding z-transform expressions are given below:

1. Multiplication by a constant.

$$x_2(kT) = ax_1(kT) \tag{4-119}$$

$$X_2(z) = aX_1(z) \tag{4-120}$$

2. Summing.

$$x_3(kT) = x_1(kT) + x_2(kT) \tag{4-121}$$

$$X_3(z) = X_1(z) + X_2(z) \tag{4-122}$$

3. Time delay or storage.

$$x_2(kT) = x_1[(k+1)T] \tag{4-123}$$

$$X_2(z) = zX_1(z) - zx_1(0) \tag{4-124}$$

or

$$X_1(z) = z^{-1}X_2(z) + x_1(0) \tag{4-125}$$

The state diagram representations and the corresponding digital computer

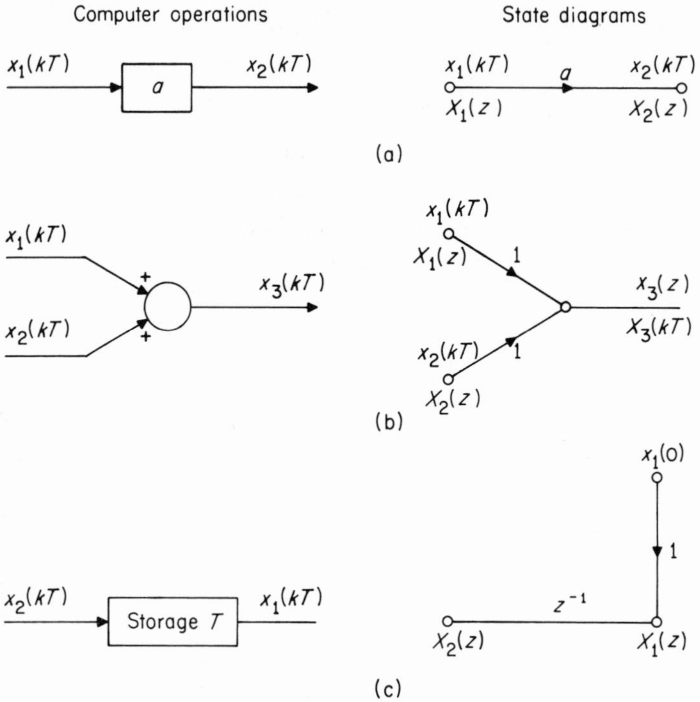

FIG. 4-8. Basic elements of a digital state diagram and the corresponding digital computer operations.

diagrams of these operations are shown in Figure 4-8. The following example serves to illustrate the construction and applications of the state diagram of a digital system.

EXAMPLE 4-4

Consider that a digital system is described by the difference equation

$$c(k+2) + 2c(k+1) + 3c(k) = r(k) \tag{4-126}$$

The state diagram of the system is drawn as shown in Figure 4-9 by first equating the highest-order term to the rest of the terms in Eq. (4-126).

The state variables of the system are designated as the output variables of the time delay units of the state diagram. Now neglecting the initial states and the branches with the gain z^{-1} on the state diagram, the state equations of the system are written

$$\begin{bmatrix} x_1(k+1) \\ x_2(k+1) \end{bmatrix} = \begin{bmatrix} 0 & 1 \\ -3 & -2 \end{bmatrix} \begin{bmatrix} x_1(k) \\ x_2(k) \end{bmatrix} + \begin{bmatrix} 0 \\ 1 \end{bmatrix} r(k) \tag{4-127}$$

and the output equation is

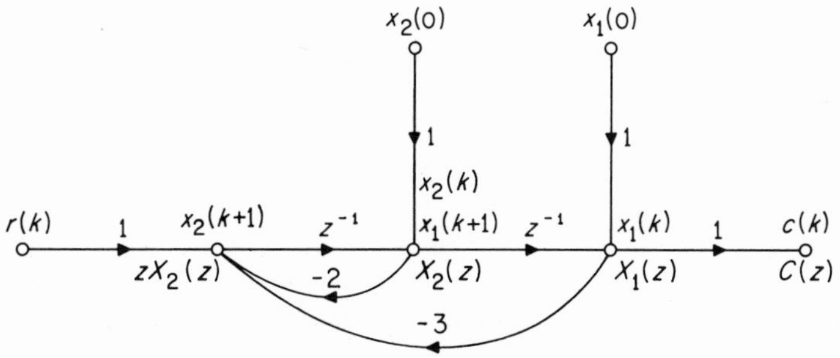

FIG. 4-9. A state diagram of the digital system described by Eq. (4-126).

$$c(k) = [1 \quad 0]\begin{bmatrix} x_1(k) \\ x_2(k) \end{bmatrix} \tag{4-128}$$

The state transition equation of the system, which is the solution of the state equation in Eq. (4-127), is obtained from the state diagram using Mason's gain formula. Using $X_1(z)$ and $X_2(z)$ as output nodes, and $R(z)$, $x_1(0)$ and $x_2(0)$ as input nodes, we have,

$$\begin{bmatrix} X_1(z) \\ X_2(z) \end{bmatrix} = \frac{1}{\Delta}\begin{bmatrix} 1 + 2z^{-1} & z^{-1} \\ -\ 3z^{-1} & 1 \end{bmatrix}\begin{bmatrix} x_1(0) \\ x_2(0) \end{bmatrix} + \frac{1}{\Delta}\begin{bmatrix} z^{-2} \\ z^{-1} \end{bmatrix} R(z) \tag{4-129}$$

where

$$\Delta = 1 + 2z^{-1} + 3z^{-2} \tag{4-130}$$

Equation (4-129) is the state transition equation in the z-transform domain. The general form of the equation is given by Eq. (4-66); that is

$$\mathbf{X}(z) = (zI - A)^{-1}z\mathbf{x}(0) + (zI - A)^{-1}B\mathbf{R}(z) \tag{4-131}$$

One advantage of using the state diagram is that Eq. (4-129) is obtained simply by use of Mason's gain formula. This saves the effort of performing the matrix inverse of $(zI - A)$ which is required if Eq. (4-131) is used.

The state transition equation in the time domain is obtained from Eq. (4-129) by taking the inverse z-transform.

The transfer function between the output and the input of the system is determined from the state diagram.

$$\frac{C(z)}{R(z)} = \frac{X_1(z)}{R(z)} = \frac{z^{-2}}{1 + 2z^{-1} + 3z^{-2}} = \frac{1}{z^2 + 2z + 3} \tag{4-132}$$

4.9 DECOMPOSITION OF DIGITAL SYSTEMS

In general, the transfer function of a digital controller or system, $D(z)$, may be realized by pulsed-data RC networks,[5,7] a general-purpose digital computer, or a special-purpose digital computer. Because of the advantages in computing speed, storage capacity, and flexibility, the use of digital computers in control systems has become increasingly important. Furthermore, due to the discrete nature of the signals received and processed by a digital controller, the realization of a discrete-data transfer function by a digital computer is quite natural.

The a priori requirement in the transfer function $D(z)$ before attempting to implement it is that it must be physically realizable. The condition of physical realizability of any linear system implies that no output signal of the system can appear before an input signal is applied. Now consider that the transfer function $D(z)$ of a digital controller is expressed as a ratio of two polynomials in z

$$D(z) = \frac{a_0 z^n + a_1 z^{n-1} + \ldots + a_n}{b_0 z^m + b_1 z^{m-1} + \ldots + b_m} \tag{4-133}$$

It is known that by expanding $D(z)$ into a power series in z^{-1}, the coefficients of the series represent the values of the weighting sequence of the digital system. The coefficient of the z^{-k} term corresponds to the value of the weighting sequence at $t = kT$. Clearly, for the digital system to be physically realizable, the power series expansion of $D(z)$ must not contain any positive power in z. Any positive power in z in the series simply indicates "prediction" or that the output precedes the input. Therefore, *for the $D(z)$ in Eq. (4-133) to be a physically realizable transfer function, the highest power of the denominator must be equal to or greater than that of the numerator, or simply $m \geq n$.*

Quite often $D(z)$ has equal number of poles and zeros and is written

$$D(z) = \frac{a_0 + a_1 z^{-1} + \ldots + a_n z^{-n}}{b_0 + b_1 z^{-1} + \ldots + b_m z^{-m}} \tag{4-134}$$

where n and m are any positive integers. In this case, the denominator of $D(z)$ must not contain a common factor of z^{-1}, or, in Eq. (4-134), $b_0 \neq 0$.

The implementation of a discrete-data transfer function by a digital computer can generally be effected in three different ways: direct decomposition, cascade decomposition, and parallel decomposition. These three methods of decomposition are illustrated in terms of state diagrams in the following.

1. Direct Decomposition

Consider that the transfer function of a digital controller is

$$D(z) = \frac{C(z)}{R(z)} = \frac{a_0 + a_1 z^{-1} + a_2 z^{-2} + \ldots + a_n z^{-n}}{b_0 + b_1 z^{-1} + b_2 z^{-2} + \ldots + b_m z^{-m}} \tag{4-135}$$

where $b_0 \neq 0$, m and n are positive integers, and $C(z)$ and $R(z)$ are the z-transforms of the output and input of the controller, respectively. We shall find a state diagram for the system since it also represents the digital computer realization of the system.

Let us multiply the numerator and the denominator of the right-hand side of Eq. (4-135) by a variable $X(z)$; we have

$$\frac{C(z)}{R(z)} = \frac{(a_0 + a_1 z^{-1} + a_2 z^{-2} + \ldots + a_n z^{-n})X(z)}{(b_0 + b_1 z^{-1} + b_2 z^{-2} + \ldots + b_m z^{-m})X(z)} \tag{4-136}$$

Now equating the numerators on both sides of the last equation gives

$$C(z) = (a_0 + a_1 z^{-1} + a_2 z^{-2} + \ldots + a_n z^{-n})X(z) \tag{4-137}$$

The same operation on the denominator brings

$$R(z) = (b_0 + b_1 z^{-1} + b_2 z^{-2} + \ldots + b_m z^{-m})X(z) \tag{4-138}$$

In order to construct a state diagram, Eq. (4-138) must first be written in a cause-and-effect relation. Therefore, solving for $X(z)$ in Eq. (4-138) gives

$$X(z) = \frac{1}{b_0} R(z) - \frac{b_1}{b_0} z^{-1} X(z) - \frac{b_2}{b_0} z^{-2} X(z) - \ldots - \frac{b_m}{b_0} z^{-m} X(z) \tag{4-139}$$

The state diagram portraying Eqs. (4-137) and (4-139) is now drawn as shown in Figure 4-10 for $m = n = 3$. For simplicity, the initial states are excluded from the state diagram. A digital computer program can easily be derived from the state diagram of Figure 4-10, where the branches with the gain z^{-1} are realized by a time delay or storage of T seconds.

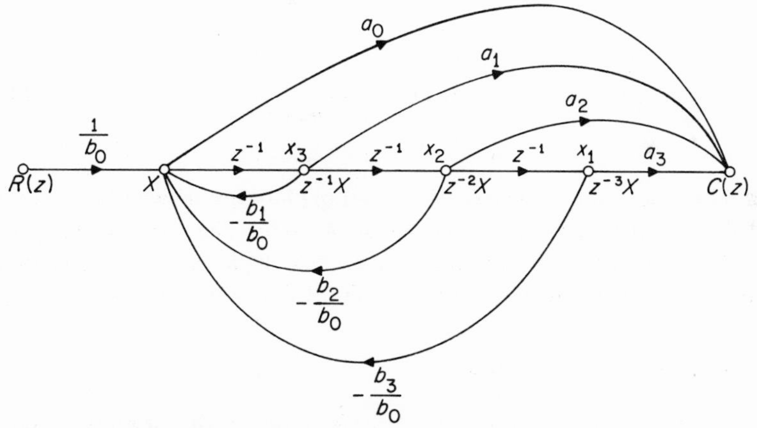

FIG. 4-10. State diagram representation of the transfer function of Eq. (4-135) with $m = n = 3$ by direction decomposition.

The state diagram of Figure 4-10 can of course be used for analytical purposes. By defining the variables at the output nodes of all the time-delay units as state variables, the dynamic equations and the state transition equation can all be determined directly from the state diagram.

In general, direct decomposition is preferred when the numerator and the denominator of the transfer function are not in factored form.

2. Cascade Decomposition

If a transfer function $D(z)$ is given in factored form, it may be written as the product of a number of first-order transfer functions each realizable by a simple digital program. The complete digital program for $D(z)$ is then represented by the series or cascade connections of the digital programs of the first-order transfer functions.

Now consider that Eq. (4-135) can be written

$$D(z) = \frac{C(z)}{R(z)} = \frac{a_0(1 + c_1 z^{-1})(1 + c_2 z^{-1}) \dots (1 + c_n z^{-1})}{b_0(1 + d_1 z^{-1})(1 + d_2 z^{-1}) \dots (1 + d_m z^{-1})} \tag{4-140}$$

where $-c_k$ and $-d_k$ are the zeros and poles of $D(z)$, respectively. For $m > n$, $D(z)$ is written as the product of a constant and m first-order transfer functions

$$D(z) = KD_1(z)D_2(z) \dots D_m(z) \tag{4-141}$$

where

$$K = \frac{a_0}{b_0} \tag{4-142}$$

and

$$D_k(z) = \frac{1 + c_k z^{-1}}{1 + d_k z^{-1}} \tag{4-143}$$

$$k = 1, 2, \dots, n$$

$$D_k(z) = \frac{1}{1 + d_k z^{-1}} \tag{4-144}$$

$$k = n + 1, n + 2, \dots, m$$

The state diagram representation of Eq. (4-143) is shown in Figure 4-11(a) and that of Eq. (4-144) is shown in Figure 4-11(b). The overall digital program for $D(z)$ iz obtained by connecting an element with gain $K = a_0/b_0$ in series with the first-order digital programs of $D_k(z)$ shown in Figure 4-11.

3. Parallel Decomposition

If only the denominator of the transfer function $D(z)$ is given in factored form, it is advantageous to represent $D(z)$ as a sum of partial fraction terms. The parallel decomposition involves taking the partial fraction expansion of $D(z)$; that is, Eq. (4-135) becomes

$$D(z) = \sum_{k=1}^{m} \frac{A_k}{(1 + d_k z^{-1})} \tag{4-145}$$

where it is assumed that $m > n$ and $D(z)$ has only simple poles denoted by $-d_k$; and A_k is a real constant.

Equation (4-145) can be written

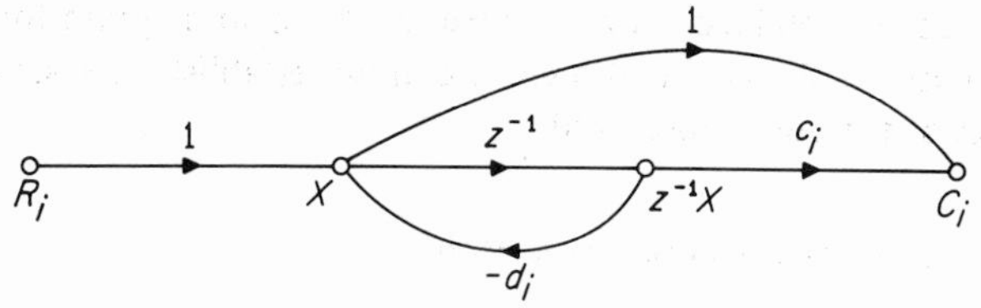

(a)

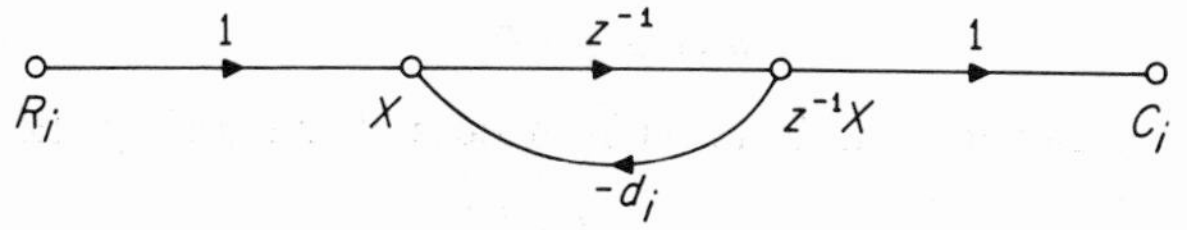

(b)

FIG. 4-11. State diagram representations of Eqs. (4-143) and (4-144).

$$D(z) = \sum_{k=1}^{m} D_k(z) \tag{4-146}$$

where

$$D_k(z) = \frac{A_k}{1 + d_k z^{-1}} \tag{4-147}$$

Then the transfer function $D(z)$ is realized by the m first-order digital programs for $D_1(z), D_2(z), \ldots, D_m(z)$, operating in parallel.

4.10 STATE DIAGRAMS OF SAMPLED-DATA CONTROL SYSTEMS

A sampled-data control system usually contains digital as well as analog elements. The two types of elements are coupled together by sample-and-hold devices. As an illustration the block diagram of a sampled-data control system is shown in Figure 4-12. The system is composed of a digital controller, and zero-order hold, and a continuous-data process which is to be controlled. We shall show how state diagrams and a state variable analysis of digital systems

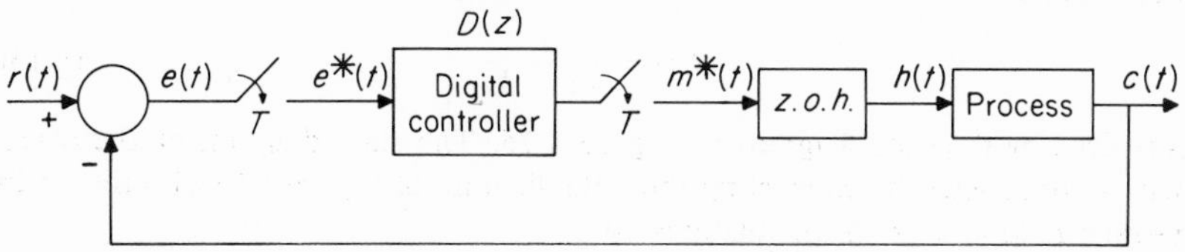

FIG. 4-12. A sampled-data control system.

of this type can be obtained. However before the state diagram for the entire system in Figure 4-12 can be drawn, we must establish the state diagram representation of the zero-order hold.

State Diagram of the Zero-Order Hold

Let the input and the output of the zero-order hold be denoted by $e^*(t)$ and $h(t)$, respectively. Then, for the time interval $kT \leq t < (k+1)T$,

$$h(t) = e(kT) \tag{4-148}$$

Taking the Laplace transform on both sides of the last equation, we get

$$H(s) = \frac{e(kT)}{s} \tag{4-149}$$

for $kT \leq t < (k+1)T$. Therefore, the state diagram of the zero-order hold consists of a single branch connecting between the nodes $e(kT)$ and $H(s)$ as shown in Figure 4-13. The gain of the branch is s^{-1}.

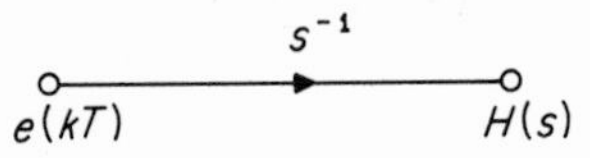

FIG. 4-13. State diagram of the zero-order hold for $kT \leq t < (k+1)T$.

EXAMPLE 4-5

Consider the sampled-data control system shown in Figure 4-14. It is desired to construct a state diagram and write the state equations for the system.

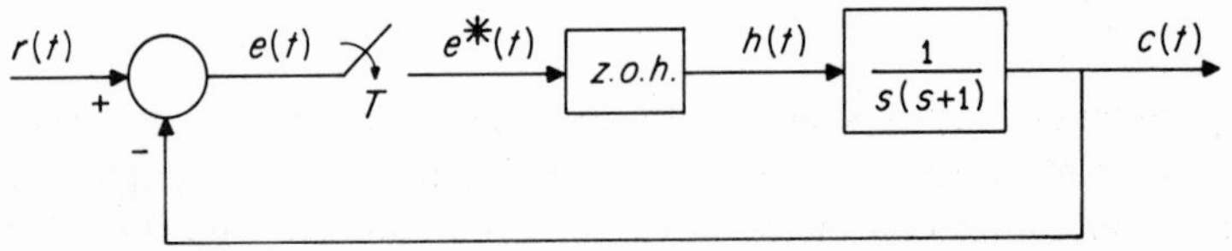

FIG. 4-14. A sampled-data control system.

Using the method of direct decomposition, the transfer function of the controlled process

$$G(s) = \frac{1}{s(s+1)} \tag{4-150}$$

is portrayed by the state diagram of Figure 4-15. The state diagram of the overall system is constructed by connecting the state diagram of Figure 4-15 with that of the zero-order hold and using the relationship

$$\begin{aligned} e(kT) &= r(kT) - c(kT) \\ &= r(kT) - x_1(kT) \end{aligned} \tag{4-151}$$

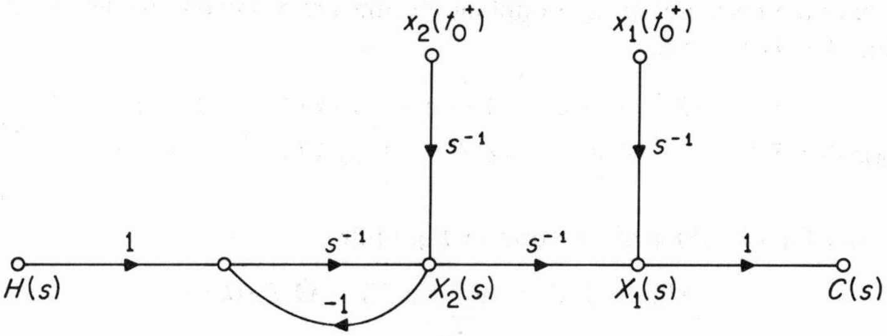

FIG. 4-15. A state diagram of $G(s) = 1/s(s + 1)$.

Also, letting $t_0 = kT$, and

$$h(kT^+) = h(t) = e(kT) \quad kT \leq t < (k + 1)T \tag{4-152}$$

the complete state diagram of the system is shown in Figure 4-16.

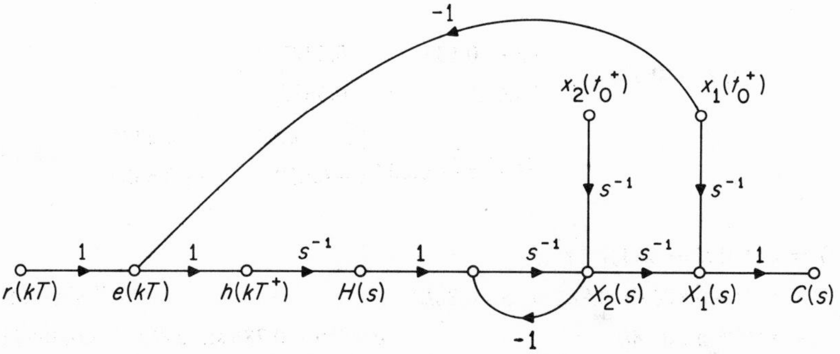

FIG. 4-16. A state diagram of the sampled-data control system shown in Fig. 4-14.

The state transition equations in transformed vector-matrix form are written directly by inspection from the state diagram using Mason's gain formula.

$$\begin{bmatrix} X_1(s) \\ X_2(s) \end{bmatrix} = \begin{bmatrix} \frac{1}{s} - \frac{1}{s^2(s+1)} & \frac{1}{s(s+1)} \\ -\frac{1}{s(s+1)} & \frac{1}{s+1} \end{bmatrix} \begin{bmatrix} x_1(kT) \\ x_2(kT) \end{bmatrix} + \begin{bmatrix} \frac{1}{s^2(s+1)} \\ \frac{1}{s(s+1)} \end{bmatrix} r(kT) \tag{4-153}$$

for $kT \leq t \leq (k + 1)T$.

Taking the inverse Laplace transform on both sides of the last equation, we get

$$\begin{bmatrix} x_1(t) \\ x_2(t) \end{bmatrix} = \begin{bmatrix} 2 - (t - kT) - e^{-(t-kT)} & 1 - e^{-(t-kT)} \\ -1 + e^{-(t-kT)} & e^{-(t-kT)} \end{bmatrix} \begin{bmatrix} x_1(kT) \\ x_2(kT) \end{bmatrix} + \begin{bmatrix} (t - kT) - 1 + e^{-(t-kT)} \\ 1 - e^{-(t-kT)} \end{bmatrix} r(kT) \tag{4-154}$$

for $kT \leq t \leq (k + 1)T$.

If only the responses at the sampling instants are of interest, we let $t = (k+1)T$; then Eq. (4-154) becomes

$$\begin{bmatrix} x_1(k+1)T \\ x_2(k+1)T \end{bmatrix} = \begin{bmatrix} 2 - T - e^{-T} & 1 - e^{-T} \\ -1 + e^{-T} & e^{-T} \end{bmatrix} \begin{bmatrix} x_1(kT) \\ x_2(kT) \end{bmatrix} + \begin{bmatrix} T - 1 + e^{-T} \\ 1 - e^{-T} \end{bmatrix} r(kT) \tag{4-155}$$

Notice that Eq. (4-155) is of the form of Eq. (4-26):

$$\mathbf{x}[(k+1)T] = \mathbf{\Phi}(T)\mathbf{x}(kT) + \mathbf{\Theta}(T)\mathbf{r}(kT) \tag{4-156}$$

For a sampling period of one second, $T = 1$ sec and a unit-step input, $r(t) = u(t)$, we have

$$\mathbf{\Phi}(1) = \begin{bmatrix} 0.632 & 0.632 \\ -0.632 & 0.368 \end{bmatrix} \tag{4-157}$$

and

$$\mathbf{\Theta}(1) = \begin{bmatrix} 0.368 \\ 0.632 \end{bmatrix} \tag{4-158}$$

Then,

$$\begin{aligned}[zI - \mathbf{\Phi}(1)]^{-1}z &= \begin{bmatrix} z - 0.632 & -0.632 \\ 0.632 & z - 0.368 \end{bmatrix}^{-1} z \\ &= \frac{z}{z^2 - z + 0.632} \begin{bmatrix} z - 0.368 & 0.632 \\ -0.632 & z - 0.632 \end{bmatrix} \end{aligned} \tag{4-159}$$

and

$$\begin{aligned}\mathbf{\Phi}(k) &= \mathscr{Z}^{-1}\{[zI - \mathbf{\Phi}(1)]^{-1}z\} \\ &= \begin{bmatrix} e^{-0.23k}(-0.378 \sin 0.88k + \cos 0.88k) & e^{-0.23k} \sin 0.88k \\ -e^{-0.23k} \sin 0.88k & e^{-0.23k}(-0.786 \sin 0.88k + \cos 0.88k) \end{bmatrix}\end{aligned} \tag{4-160}$$

Also,

$$\begin{aligned}&\mathbf{\Phi}[(N-k-1)]\mathbf{\Theta}(1) \\ &\quad = \begin{bmatrix} e^{-0.23(N-k-1)}[0.493 \sin 0.88(N-k-1) + 0.368 \cos 0.88(N-k-1) \\ e^{-0.23(N-k-1)}[-0.865 \sin 0.88(N-k-1) + 0.632 \cos 0.88(N-k-1) \end{bmatrix}\end{aligned} \tag{4-161}$$

Therefore, the discrete state transition equation of the system is

$$\begin{aligned}\mathbf{x}(N) &= \begin{bmatrix} e^{-0.23N}(-0.378 \sin 0.88N + \cos 0.88N) & e^{-0.23N} \sin 0.88N \\ -e^{-0.23N} \sin 0.88N & e^{-0.23N}(-0.786 \sin 0.88N + \cos 0.88N) \end{bmatrix} \\ &\times \begin{bmatrix} x_1(0) \\ x_2(0) \end{bmatrix} + \sum_{k=0}^{N-1} \begin{bmatrix} e^{-0.23(N-k-1)}[0.493 \sin 0.88(N-k-1) + 0.368 \cos 0.88(N-k-1)] \\ e^{-0.23(N-k-1)}[-0.865 \sin 0.88(N-k-1) + 0.632 \cos 0.88(N-k-1)] \end{bmatrix}\end{aligned} \tag{4-162}$$

for $N = 1, 2, 3, \ldots$.

EXAMPLE 4-6

In this example we shall conduct a state analysis of a sampled-data control system which has a digital controller. Let us consider the block diagram shown in Figure 4-12. The digital controller, which actually may be a digital computer, is described by

$$D(z) = \frac{a_0 + a_1 z^{-1}}{1 + b_1 z^{-1}} \tag{4-163}$$

and $G(s)$ is as given in Eq. (4-150).

It is desired to draw a state diagram and to obtain the state transition equations for the system.

Applying the direct decomposition scheme to $D(z)$, we have

$$D(z) = \frac{M(z)}{E(z)} = \frac{(a_0 + a_1 z^{-1})X(z)}{(1 + b_1 z^{-1})X(z)} \tag{4-164}$$

Let

$$M(z) = (a_0 + a_1 z^{-1})X(z) \tag{4-165}$$

and

$$E(z) = (1 + b_1 z^{-1})X(z) \tag{4-166}$$

From Eq. (4-166), we have

$$X(z) = E(z) - b_1 z^{-1} X(z) \tag{4-167}$$

The state diagram for the digital controller is constructed as shown in Figure 4-17 using Eqs. (4-165) and (4-167).

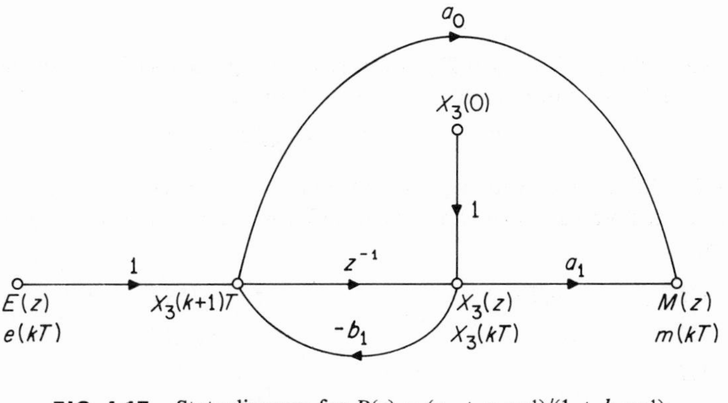

FIG. 4-17. State diagram for $D(z) = (a_0 + a_1 z^{-1})/(1 + b_1 z^{-1})$.

From Figure 4-17, the dynamic equations which characterize $D(z)$ are written

$$x_3(k + 1)T = e(kT) - b_1 x_3(kT) \quad \text{(State Equation)} \tag{4-168}$$

$$m(kT) = a_0 e(kT) + (a_1 - a_0 b_1)x_3(kT) \quad \text{(Output Equation)} \tag{4-169}$$

The state diagram for $G(s)$ and the zero-order hold were established earlier in

Example 4-5. The state diagram for the entire system is now drawn as shown in Figure 4-18.

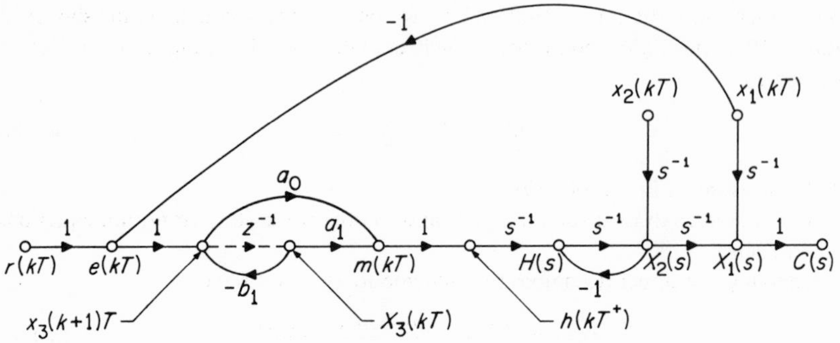

FIG. 4-18. State diagram for the sampled-data control system shown in Fig. 4-17.

Applying Mason's gain formula to Figure 4-18 with $X_1(s)$ and $X_2(s)$ as output nodes, we get

$$X_1(s) = \left[\frac{1}{s} - \frac{a_0}{s^2(s+1)}\right]x_1(kT) + \frac{1}{s(s+1)}x_2(kT) + \frac{a_1 - a_0 b_1}{s^2(s+1)}x_3(kT) + \frac{a_0}{s^2(s+1)}r(kT) \tag{4-170}$$

$$X_2(s) = \frac{-a_0}{s(s+1)}x_1(kT) + \frac{1}{s+1}x_2(kT) + \frac{a_1 - a_0 b_1}{s(s+1)}x_3(kT) + \frac{a_0}{s(s+1)}r(kT) \tag{4-171}$$

Notice that $x_1(kT)$, $x_2(kT)$, $x_3(kT)$, and $r(kT)$ are regarded as inputs in this case. Also applying Mason's formula to $x_3(k+1)T$, we have

$$x_3(k+1)T = -x_1(kT) - b_1 x_3(kT) + r(kT) \tag{4-172}$$

Equation (4-172) together with Eqs. (4-170) and (4-171) when the latter equations are inverse-transformed and with $t = (k+1)T$, form the discrete state equations of the system.

$$\begin{bmatrix} x_1(k+1)T \\ x_2(k+1)T \\ x_3(k+1)T \end{bmatrix} = \begin{bmatrix} 1 - a_0(T-1+e^{-T}) & 1-e^{-T} & (a_1 - a_0 b_1)(T-1+e^{-T}) \\ -a_0(1-e^{-T}) & e^{-T} & (a_1 - a_0 b_1)(1-e^{-T}) \\ -1 & 0 & -b_1 \end{bmatrix} \begin{bmatrix} x_1(kT) \\ x_2(kT) \\ x_3(kT) \end{bmatrix} + \begin{bmatrix} a_0(T-1+e^{-T}) \\ a_0(1-e^{-T}) \\ 1 \end{bmatrix} r(kT) \tag{4-173}$$

The output equation is $c(kT) = x_1(kT)$.

Equation (4-173) can now be solved by means of the same method as described in the last example.

4.11 RESPONSE OF SAMPLED-DATA SYSTEMS BETWEEN SAMPLING INSTANTS USING THE STATE APPROACH

The state-variable method may be applied to the evaluation of systems response between sampling instants. The method represents an alternative to the modified z-transform method.[5,8]

From Eq. (4-21), the state vector $\mathbf{x}(t)$ for any $t > t_0$ is represented by

$$\mathbf{x}(t) = \mathbf{\Phi}(t - t_0)\mathbf{x}(t_0) + \int_{t_0}^{t} \mathbf{\Phi}(t - \tau)\mathbf{B}\mathbf{r}(\tau)\, d\tau \tag{4-174}$$

when $\mathbf{x}(t_0)$ and $\mathbf{r}(\tau)$ for $t \geq t_0$ are specified. When $\mathbf{r}(t)$ is constant for $t_0 \leq \tau < t$, Eq. (4-174) is written as

$$\mathbf{x}(t) = \mathbf{\Phi}(t - t_0)\mathbf{x}(t_0) + \mathbf{\Theta}(t - t_0)\mathbf{r}(t_0) \tag{4-175}$$

where

$$\Theta(t - t_0) = \int_{t_0}^{t} \mathbf{\Phi}(t - \tau)\mathbf{B}\, d\tau$$

and

$$\mathbf{r}(t_0) = \mathbf{r}(\tau) \qquad t_0 \leq \tau < t$$

Now if the response in between the sampling instants is desired, we let

$$t = kT + \Delta T = (k + \Delta)T \tag{4-176}$$

where $k = 0, 1, 2, \ldots,$ and $0 \leq \Delta \leq 1$. Then, Eq. (4-175) gives

$$\mathbf{x}(kT + \Delta T) = \mathbf{\Phi}(\Delta T)\mathbf{x}(kT) + \mathbf{\Theta}(\Delta T)\mathbf{r}(kT) \tag{4-177}$$

where the initial time has been taken as $t_0 = kT$.

Thus, by varying the value of Δ between 0 and 1, essentially all information on $\mathbf{x}(t)$ for all t can be obtained.

For example, for the system described in Example 4-5, the state transition equation in Eq. (4-155) gives values of the state variables at the sampling instants only. However, letting $t = (k + \Delta)T$, Eq. (4-155) becomes

$$\begin{bmatrix} x_1(kT + \Delta T) \\ x_2(kT + \Delta T) \end{bmatrix} = \begin{bmatrix} 2 - \Delta T - e^{-\Delta T} & 1 - e^{-\Delta T} \\ -1 + e^{-\Delta T} & e^{-\Delta T} \end{bmatrix} \begin{bmatrix} x_1(kT) \\ x_2(kT) \end{bmatrix} + \begin{bmatrix} \Delta T - 1 + e^{-\Delta T} \\ 1 - e^{-\Delta T} \end{bmatrix} r(kT) \tag{4-178}$$

To complete the iteration procedure, it is necessary to let $t = (k + 1)T$ and $t_0 = (k + \Delta)T$, and Eq. (4-155) becomes

$$\mathbf{x}[(k + 1)T] = \mathbf{\Phi}(T - \Delta T)\mathbf{x}[(k + \Delta)T] + \mathbf{\Theta}(T - \Delta T)\mathbf{r}[(k + \Delta)T] \tag{4-179}$$

The state transition equation for the system between $t_0 = (k + \Delta)T$ and $t = (k + 1)T$ is

$$\begin{bmatrix} x_1[(k+1)T] \\ x_2[(k+1)T] \end{bmatrix} = \begin{bmatrix} 2-(1-\Delta)T-e^{-(1-\Delta)T} & 1-e^{-(1-\Delta)T} \\ -1+e^{-(1-\Delta)T} & e^{-(1-\Delta)T} \end{bmatrix} \begin{bmatrix} x_1[(k+\Delta)T] \\ x_2[(k+\Delta)T] \end{bmatrix} + \begin{bmatrix} (1-\Delta)T-1+e^{-(1-\Delta)T} \\ 1-e^{-(1-\Delta)T} \end{bmatrix} r[(k+\Delta)T] \tag{4-180}$$

4.12 ANALYSIS OF DISCRETE-TIME SYSTEMS WITH MULTIRATE, SKIP-RATE, AND NONSYNCHRONOUS SAMPLINGS

The state variable method discussed in the preceding sections can readily be applied to digital systems with nonuniform sampling schemes. The analysis of digital systems with nonuniform sampling by means of the z-transform method usually encounters difficulty and special equivalent models with uniform sampling must first be made. However, the state variable method offers a unified approach to the analysis of a wide class of digital systems.

The sampling scheme shown in Figure 4-19 is often referred to as a cyclic

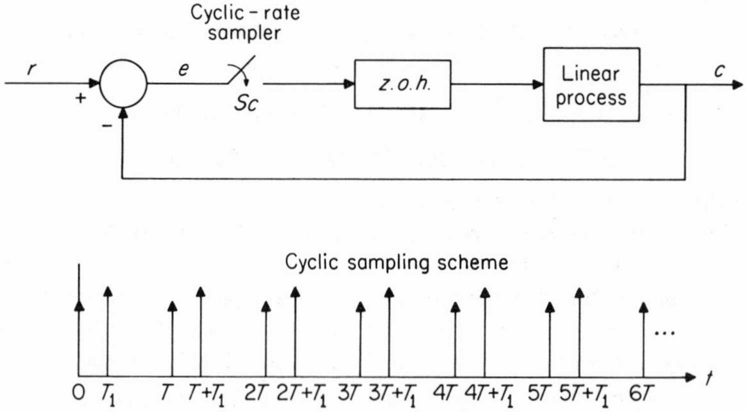

FIG. 4-19. A sampled-data system with cyclic-rate sampling.

sampling operation. In general, the transition of state of a digital system may be considered to take place from t_0 to $t_0 + \tau$, where $0 < \tau \leq T$, and then from $t = \tau$ to T. For the cyclic sampling, we let $\tau = T_1$; the state transition equation of the system at $t = kT + T_1$ is written as

$$\mathbf{x}[(kT+T_1)] = \mathbf{\Phi}(T_1)\mathbf{x}(kT) + \mathbf{\Theta}(T_1)\mathbf{r}(kT) \tag{4-181}$$

Then, at $t = (k+1)T$ the state transition equation is

$$\mathbf{x}[(k+1)T] = \mathbf{\Phi}(T-T_1)\mathbf{x}[(kT+T_1)] + \mathbf{\Theta}(T-T_1)\mathbf{r}[(kT+T_1)] \tag{4-182}$$

Substituting Eq. (4-181) into Eq. (4-182) yields

$$\mathbf{x}[(k+1)T] = \Phi(T - T_1)\Phi(T_1)\mathbf{x}(kT) + \Phi(T - T_1)\Theta(T_1)\mathbf{r}(kT) + \Theta(T - T_1)\mathbf{r}(kT + T_1) \tag{4-183}$$

In general, if the cyclic sampler samples at $t = kT, kT + T_1, kT + T_1 + T_2, \ldots, (k+1)T, (k+1)T + T_1, (k+1)T + T_1 + T_2, \ldots, (k+1)T + T_1 + T_2 + \ldots + T_q, \ldots$ so that $T = T_1 + T_2 + \ldots + T_q$, then,

$$\mathbf{x}(kT + T_1) = \Phi(T_1)\mathbf{x}(kT) + \Theta(T_1)\mathbf{r}(kT) \tag{4-184}$$

$$\mathbf{x}(kT + T_1 + T_2) = \Phi(T_2)\mathbf{x}(kT + T_1) + \Theta(T_2)\mathbf{r}(kT + T_1) \tag{4-185}$$

$$\vdots \qquad\qquad \vdots \qquad\qquad \vdots$$

$$\mathbf{x}(kT + T_1 + T_2 + \ldots + T_{q-1}) = \Phi(T_{q-1})\mathbf{x}(kT + T_1 + T_2 + \ldots + T_{q-2}) + \Theta(T_{q-1})\mathbf{r}(kT + T_1 + T_2 + \ldots + T_{q-2}) \tag{4-186}$$

At $t = (k+1)T$,

$$\mathbf{x}[(k+1)T] = \Phi(T_q)\mathbf{x}(kT + T_1 + T_2 + \ldots + T_{q-1}) + \Theta(T_q)\mathbf{r}(kT + T_1 + T_2 + \ldots + T_{q-1}) \tag{4-187}$$

Substituting Eq. (4-184) into Eq. (4-185) and so on, we see that Eq. (4-187) can be written as

$$\begin{aligned}\mathbf{x}[(k+1)T] = {} & \Phi(T_q)\Phi(T_{q-1})\ldots\Phi(T_2)\Phi(T_1)\mathbf{x}(kT) \\ & + \Phi(T_q)\Phi(T_{q-1})\ldots\Phi(T_2)\Theta(T_1)\mathbf{r}(kT) \\ & + \Phi(T_q)\Phi(T_{q-1})\ldots\Phi(T_3)\Theta(T_2)\mathbf{r}[kT + T_1] \\ & + \ldots + \Phi(T_q)\Theta(T_{q-1})\mathbf{r}[kT + T_1 + T_2 + \ldots + T_{q-2}] \\ & + \Theta(T_q)\mathbf{r}(kT + T_1 + T_2 + \ldots + T_{q-1})\end{aligned} \tag{4-188}$$

EXAMPLE 4-7

Consider that in the system of Figure 4-19, the linear process is described by

$$G(s) = \frac{4}{s+1} \tag{4-189}$$

and $T = 1$ sec, $T_1 = 0.25$ sec. The input signal is a unit-step function applied at $t = 0$. The system is initially at rest and $c(0) = 0$. The output response of the system for $t > 0$ is desired.

The periods of state transition in the case are $0 \leq t \leq T_1$ and $T_1 \leq t \leq T$, and in general, $kT \leq t \leq kT + T_1$ and $kT + T_1 \leq t \lesssim (k+1)T$. The state diagram of the system for $t \geq t_0$, where t_0 can be kT or $kT + T_1$, is drawn as shown in Figure 4-20. The state transition equation in the Laplace domain is written as

$$X_1(s) = \frac{1}{s+1}x_1(t_0^+) - \frac{4}{s(s+1)}x_1(t_0^+) + \frac{4}{s(s+1)}r(t_0^+) \tag{4-190}$$

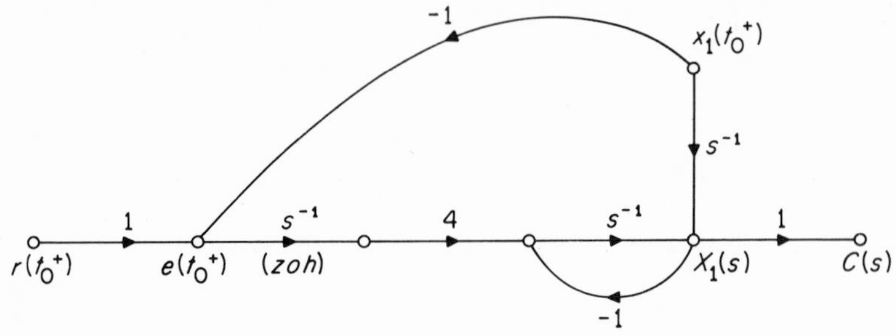

FIG. 4-20. State diagram of the system in Fig. 4-19.

The inverse Laplace transform of $X_1(s)$ is

$$x_1(t) = e^{-(t-t_0)}x_1(t_0^+) - 4[1 - e^{-(t-t_0)}]x_1(t_0^+) + 4[1 - e^{-(t-t_0)}]r(t_0^+) \qquad (4\text{-}191)$$

Therefore,

$$\Phi(t - t_0) = -4 + 5e^{-(t-t_0)} \qquad (4\text{-}192)$$

$$\Theta(t - t_0) = 4(1 - e^{-(t-t_0)}) \qquad (4\text{-}193)$$

From Eq. (4-191), with $t = kT + T_1$ and $t_0 = kT$, we have

$$\begin{aligned} x_1(kT + T_1) &= \Phi(T_1)x_1(kT) + \Theta(T_1)r(kT) \\ &= [-4 + 5e^{-T_1}]x_1(kT) + 4(1 - e^{-T_1})r(kT) \end{aligned} \qquad (4\text{-}194)$$

In the same equation with $t = (k + 1)T$ and $t_0 = kT + T_1$, we get

$$x_1[(k + 1)T] = [-4 + 5e^{-(T-T_1)}]x_1(kT + T_1) + 4[1 - e^{-(T-T_1)}]r(kT + T_1) \qquad (4\text{-}195)$$

For $T_1 = 0.25$ sec and $T = 1$ sec, the last two equations become

$$\begin{aligned} x_1(k + 0.25) &= \Phi(0.25)x_1(kT) + \Theta(0.25)r(kT) \\ &= -0.1x_1(kT) + 0.884r(kT) \end{aligned} \qquad (4\text{-}196)$$

and

$$\begin{aligned} x_1(k + 1) &= \Phi(0.75)x_1(k + 0.25) + \Theta(0.75)r(kT + 0.25) \\ &= -1.64x_1(k + 0.25) + 2.1r(k + 0.25) \end{aligned} \qquad (4\text{-}197)$$

Equations (4-196) and (4-197) are two recursion relations from which the response of the system at $t = kT + T_1$ and $(k + 1)T$ for $k = 0, 1, 2, \ldots$ can be computed. Note that the solutions are valid for any input $r(t)$ whose values at $t = kT$ and $kT + T_1$ are defined. For $r(t) = u(t)$ the following results are obtained.

$x_1(0)$	0	$x_1(2.25)$	0.81
$x_1(0.25)$	0.884	$x_1(3)$	0.77
$x_1(1)$	0.65	$x_1(3.25)$	0.807
$x_1(1.25)$	0.82	$x_1(4)$	0.78
$x_1(2)$	0.75	$x_1(4.25)$	0.806

The unit-step response of the system is sketched as shown in Figure 4-21.

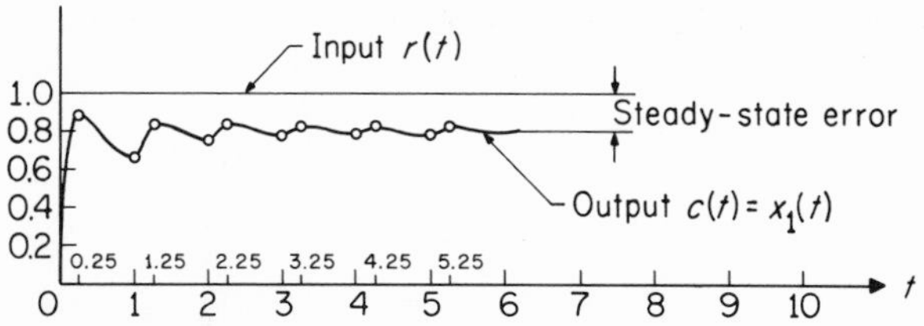

FIG. 4-21. Unit-step response of the digital control system of Example 4-7, Fig. 4-19.

In this case, it is shown that the system has a steady-state error of 0.2, or 20 per cent.

EXAMPLE 4-8

A digital control system with nonsynchronous sampling is shown in Figure 4-22. The sampling instants of the sampler S_2 lag behind that of S_1 by T_1 second.

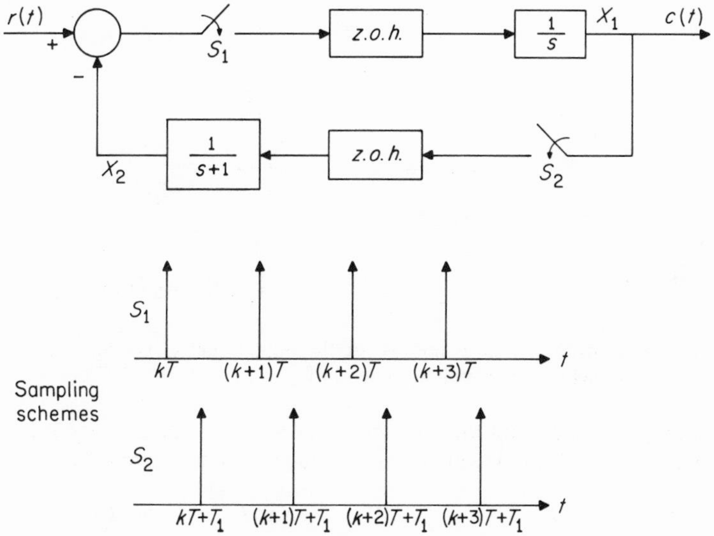

FIG. 4-22. A digital control system with nonsynchronous sampling.

We shall write the state transition equations for the system using the same principles as discussed in this section.

Since S_1 and S_2 close at different sampling instants, the forward path of the system has a transition interval from $t = kT$ to $t = (k + 1)T$, and the feedback path has a transition interval from $t = kT + T_1$ to $t = (k + 1)T + T_1$. The state diagrams for the two paths with the corresponding transition intervals are shown in Figure 4-23.

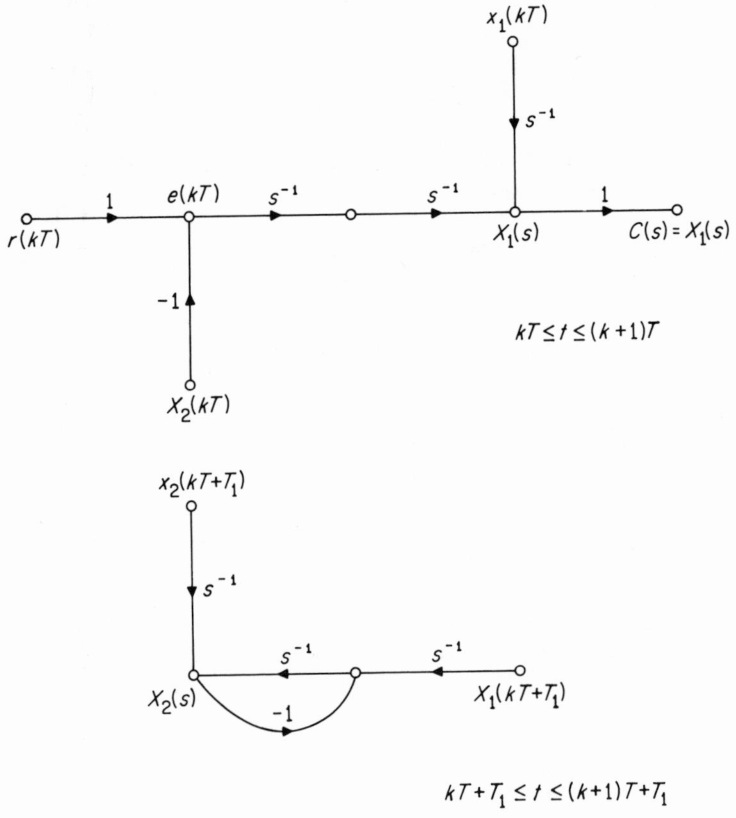

FIG. 4-23. State diagrams for the digital control system with nonsynchronous sampling shown in Fig. 4-22.

In the Laplace domain the state transition equations for the forward path and the feedback path of the system are written directly from Figure 4-23.

$$kT \le t \le (k+1)T \quad : \quad X_1(s) = \frac{1}{s}x_1(kT) - \frac{1}{s^2}x_2(kT) + \frac{1}{s^2}r(kT) \tag{4-198}$$

$$kT + T_1 \le t \le kT + T + T_1: \quad X_2(s) = \frac{1}{s(s+1)}x_1(kT+T_1) + \frac{1}{s+1}x_2(kT+T_1) \tag{4-199}$$

Taking the inverse Laplace transform of the last two equations yields

$$kT \leq t \leq (k+1)T \qquad : \quad x_1(t) = x_1(kT) - (t - kT)x_2(kT) + (t - kT)r(kT) \tag{4-200}$$

$$kT + T_1 \leq t \leq kT + T + T_1: \quad x_2(t) = [1 - e^{-(t-kT-T_1)}]x_1(kT + T_1) + e^{-(t-kT-T_1)}x_2(kT + T_1) \tag{4-201}$$

Since the initial states are $x_1(kT)$, $x_1(kT + T_1)$, $x_2(kT)$ and $x_2(kT + T_1)$, it is necessary to compute these quantities from the two transition equations. Therefore, setting $t = kT + T_1$ and $t = (k + 1)T$ in Eq. (4-200), we get,

$$x_1(kT + T_1) = x_1(kT) - T_1x_2(kT) + T_1r(kT) \tag{4-202}$$

$$x_1(k + 1)T = x_1(kT) - Tx_2(kT) + Tr(kT) \tag{4-203}$$

Setting $t = (k + 1)T$ and $t = (k + 1)T + T_1$ in Eq. (4-201), we get

$$x_2(k + 1)T = [1 - e^{-(T-T_1)}]x_1(kT + T_1) + e^{-(T-T_1)}x_2(kT + T_1) \tag{4-204}$$

$$x_2[(k + 1)T + T_1] = (1 - e^{-T})x_1(kT + T_1) + e^{-T}x_2(kT + T_1) \tag{4-205}$$

Now by setting $k = 0, 1, 2, 3, \ldots$ into these four transition equations, the states and the output of the system can be computed at the sampling instants of both samplers.

4.13 STATE PLANE ANALYSIS

The phase plane method[9] is a well-known graphical technique of studying linear and nonlinear control systems. However, the method has been applied predominantly to the studies of nonlinear second-order control systems for the following reasons:

1. Nonlinear systems are difficult to solve analytically, and very few nonlinear differential equations have closed form solutions.
2. The phase plane diagram can be used only for systems with an order less than or equal to two. For higher-order systems, the multidimensional "phase space" diagram is very difficult to visualize and construct.

If a second-order system is characterized by the dependent variable $c(t)$ and its first derivative with respect to time $dc(t)/dt$, the phase plane diagram of the system is a plot of $dc(t)/dt = \dot{c}$ against $c(t)$ with time t as a variable parameter. At any time $t_1 \geq 0$, the coordinate $[c(t_1), \dot{c}(t_1)]$ in the phase plane describes the "state" of the system at that time. As time is varied from zero to infinity, the representing point $(c, \dot{c})$ will trace out a continuous path which represents the complete history of the system in regard to c and $\dot{c}$ at all times. The path itself is called the *phase plane trajectory*. For instance, for a second-order control system, the phase plane diagram is usually a plot of the output velocity $\dot{c}$ against the output displacement c. Sometimes it may be more convenient to use the error function e and its derivative for the phase plane plot.

In the state variable analysis, a second-order system is characterized by its two state variables $x_1(t)$ and $x_2(t)$. A state plane diagram is obtained when a trajectory of $x_2(t)$ versus $x_1(t)$ is plotted for $t \geq 0$. The trajectory is called the *state plane trajectory*. It is apparent that the phase plane and the state plane diagrams are of the same nature; only that the coordinate variables of the latter are state variables.

Phase Plane Trajectories of a Second-Order System

In this section we shall demonstrate the derivation of the phase plane trajectories of a second-order open-loop control system. The input to the system is assumed to be connected to the output of a sample-and-hold device. The block diagram of the system is shown in Figure 4-24.

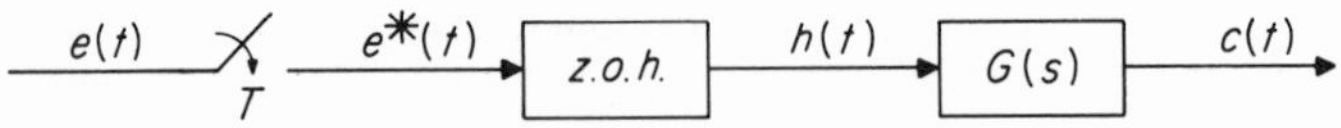

FIG. 4-24. Block diagram of an open-loop digital control system.

Let the linear process of the control system be described by

$$G(s) = \frac{C(s)}{H(s)} = \frac{K}{s(1 + \tau s)} \tag{4-206}$$

where K and τ are real constants. A state diagram of the process is drawn as shown in Figure 4-25 for $kT \leq t \leq (kT + 1)T$. The input of the state diagram $h(t_0^+)$ is defined as

$$h(t_0^+) = e(kT)$$

for $kT \leq t < (k + 1)T$.

The state transition equation of the process is written

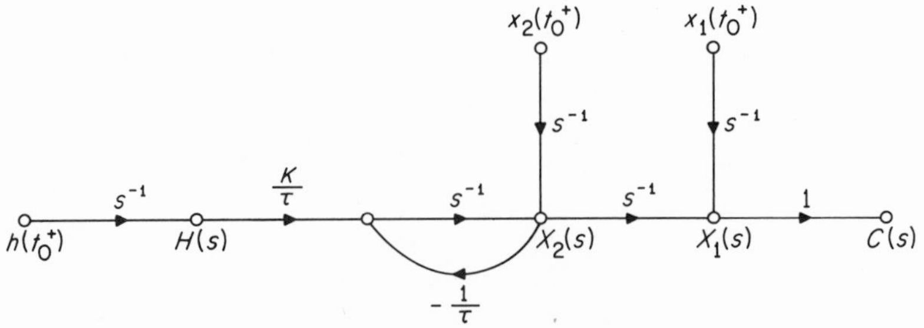

FIG. 4-25. A state diagram of the system shown in Fig. 4-24.

$$\begin{bmatrix} x_1(t) \\ x_2(t) \end{bmatrix} = \begin{bmatrix} 1 & \tau(1 - e^{-(t-t_0)/\tau}) \\ 0 & e^{-(t-t_0)/\tau} \end{bmatrix} \begin{bmatrix} x_1(t_0^+) \\ x_2(t_0^+) \end{bmatrix} + \begin{bmatrix} K(t - t_0 - \tau + \tau e^{-(t-t_0)/\tau}) \\ K(1 - e^{-(t-t_0)/\tau}) \end{bmatrix} h(t_0^+) \tag{4-207}$$

for $kT \leq t \leq (k + 1)T$; where $kT = t_0$.

It is convenient to normalize the state variables by letting

$$x_{1n}(t) = x_1(t)$$
$$x_{2n}(t) = \tau x_2(t)$$
$$t_n = t/\tau$$
$$t_{0n} = t_0/\tau$$
$$h_n(t) = \tau K h(t)$$

Substituting these relationships into Eq. (4-207), the normalized state transition equation is

$$\begin{bmatrix} x_{1n}(t_n) \\ x_{2n}(t_n) \end{bmatrix} = \begin{bmatrix} 1 & 1 - e^{-(t_n - t_{0n})} \\ 0 & e^{-(t_n - t_{0n})} \end{bmatrix} \begin{bmatrix} x_{1n}(t_{0n}^+) \\ x_{2n}(t_{0n}^+) \end{bmatrix} + \begin{bmatrix} t_n - t_{0n} - 1 + e^{-(t_n - t_{0n})} \\ 1 - e^{-(t_n - t_{0n})} \end{bmatrix} h_n(t_{0n}^+) \tag{4-208}$$

The equation of the state plane trajectory in the x_{2n} versus x_{1n} plane is obtained by eliminating t_n from Eq. (4-208). Substituting the second equation in the first equation in Eq. (4-208), we get

$$x_{1n}(t_n) = x_{1n}(t_{0n}^+) + x_{2n}(t_{0n}^+) - x_{2n}(t_n) + (t_n - t_{0n})h_n(t_{0n}^+) \tag{4-209}$$

Also, solving for $t_n - t_{0n}$ from the second equation in Eq. (4-208), we have

$$t_n - t_{0n} = -\ln\left[\frac{h_n(t_{0n}^+) - x_{2n}(t_n)}{h_n(t_{0n}^+) - x_{2n}(t_{0n}^+)}\right] \tag{4-210}$$

Now substitution of Eq. (4-210) in Eq. (4-209) gives

$$x_{1n}(t_n) = x_{1n}(t_{0n}^+) + x_{2n}(t_{0n}^+) - x_{2n}(t_n) - h_n(t_{0n}^+) \times \ln\left[\frac{h_n(t_{0n}^+) - x_{2n}(t_n)}{h_n(t_{0n}^+) - x_{2n}(t_{0n}^+)}\right] \tag{4-211}$$

or

$$x_{1n}(t_n) - x_{1n}(t_{0n}^+) = -[x_{2n}(t_n) - x_{2n}(t_{0n}^+)] - h_n(t_{0n}^+) \times \ln\left[\frac{h_n(t_{0n}^+) - x_{2n}(t_n)}{h_n(t_{0n}^+) - x_{2n}(t_{0n}^+)}\right] \tag{4-212}$$

The last equation describes the normalized state plane trajectory for Eq. (4-206) during the time interval $t_0 = kT \leq t \leq (k+1)T$, and when the input is obtained from a zero-order hold. Since the normalized input $h_n(t_{0n}^+)$ is equal to the output of the zero-order hold at $t = t_0$, and it is held constant for one sampling duration, we need to consider only three values of $h_n(t_{0n}^+)$, namely, $+1$, -1, and 0. The state plane trajectories described by Eq. (4-212) are plotted in Figure 4-26 for $h_n(t_{0n}^+) = +1$ and -1 for various initial states. These are subsequenly referred to as *the* $+1$*-trajectories* and *the* -1*-trajectories*.

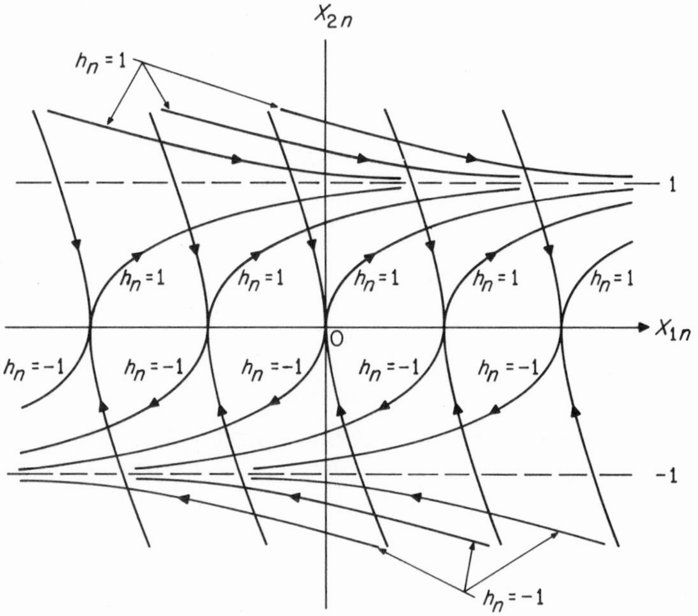

FIG. 4-26. State plane trajectories of a second-order system $G(s) = K/s(1 + \tau s)$ with input normalized to $h_n = +1$ and -1.

The data of the zero-initial state trajectory with $h_n(t_{0n}^+) = 1$ are tabulated in Table 4-2. These data correspond to the $h_n(t_{0n}^+)$ trajectory which passes through the origin in Figure 4-26. It is shown that for nonzero initial states, the trajectory is simply shifted horizontally until it passes through the point $[x_{1n}(t_{0n}), x_{2n}(t_{0n})]$ in the normalized state plane. The state plane trajectory for $h_n(t_{0n}^+) = -1$ has the same shape as that of the trajectory of $h_n(t_{0n}^+) = +1$ except that it is inverted.

When $h_n(t_{0n}^+) = 0$, Eq. (4-212) is reduced to

$$x_{1n}(t_n) - x_{1n}(t_{0n}^+) = -[x_{2n}(t_n) - x_{2n}(t_{0n}^+)] \tag{4-213}$$

which describes a straight line in the state plane (Figure 4-27). These lines are subsequently referred to as the 0*-trajectories*.

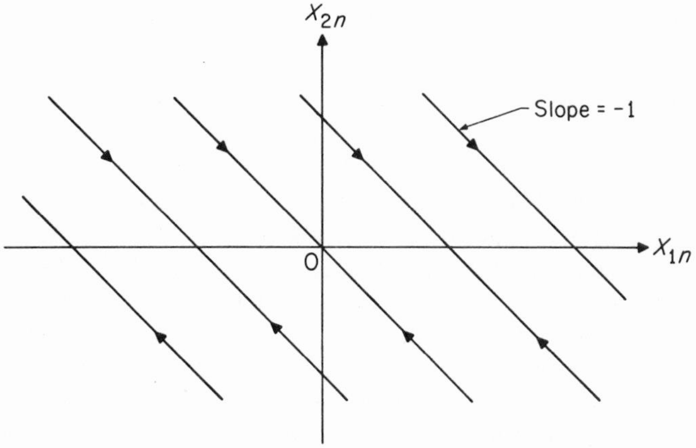

FIG. 4-27. State plane trajectories of a second-order system $G(s) = K/s(1 + \tau s)$ with $h_n = 0$.

For digital control systems, if we are interested only in the response at the sampling instants, we let $t_n = (k + 1)T_n$, $t_{0n}^+ = kT_n$, where $T_n = T/\tau$, Eqs. (4-212) and (4-213) become

Table 4-2

x_{2n}	x_{1n}	x_{2n}	x_{1n}	x_{2n}	x_{1n}
0.1	0.0054	1.05	1.95	−0.2	0.018
0.2	0.023	1.10	1.20	−0.4	0.064
0.3	0.057	1.20	0.41	−0.5	0.095
0.4	0.112	1.28	0	−0.7	0.170
0.5	0.194	1.30	−0.10	−0.8	0.212
0.6	0.317	1.50	−0.80	−0.9	0.258
0.7	0.503			−1.0	0.307
0.8	0.810			−1.2	0.412
0.84	1.00			−1.4	0.525
0.90	1.40			−1.6	0.644
0.95	2.05			−1.8	0.770
0.98	2.93			−2.0	0.90
1.00	∞			−3.0	1.61

$$x_{1n}[(k + 1)T_n] - x_{1n}(kT_n) = -x_{2n}[(k + 1)T_n] + x_{2n}(kT_n) - h_n(kT_n)\ln\left[\frac{h_n(kT_n) - x_{2n}[(k + 1)T_n]}{h_n(kT_n) - x_{2n}(kT_n)}\right] \tag{4-214}$$

and

$$x_{1n}[(k+1)T_n] - x_{1n}(kT_n) = -x_{2n}[(k+1)T_n] + x_{2n}(kT_n) \tag{4-215}$$

respectively.

Thus far we have investigated only the state plane trajectories of a second-order open-loop system when the input is constant. The state plane trajectories in this case are shown to be those illustrated in Figures 4-26 and 4-27 when normalized input of $h_n = +1, -1$, or 0 is applied. The state plane analysis does not pose any advantage for the study of linear closed-loop digital systems since in a closed-loop system the magnitude of the input to the zero-order hold varies with time, and therefore, the magnitude of $h_n(t_{0n}^+)$ cannot be uniformly normalized to just the three values, $+1, -1$, or 0. However, the state plane method is convenient for the analysis and design of digital systems with relay-type or bang-bang control, since the output of a relay can assume values of only $+M$, $-M$, or 0 for any given input. Later we shall show that optimal digital control systems usually require a bang-bang type of control.

Now we shall demonstrate how the entire state plane trajectory of a digital system with relay-type control can be constructed using the normalized trajectories of Figures 4-26 and 4-27.

Consider the system shown in Figure 4-28. The linear process is assumed

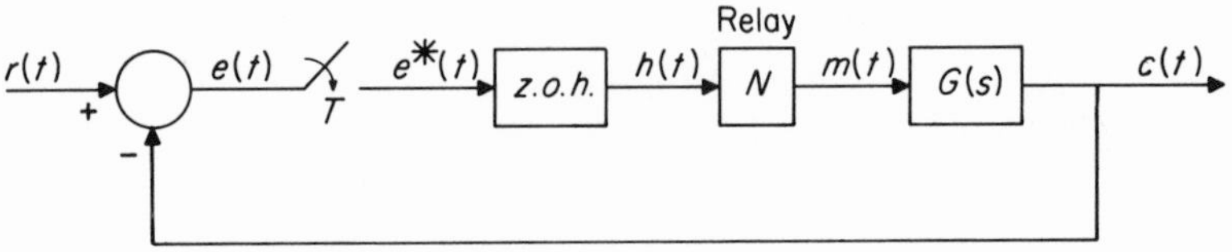

FIG. 4-28. A discrete-data system with relay control.

to be of the second-order and is described by Eq. (4-206). The nonlinearity is a relay with dead zone and its input-output characteristics are defined by

$$m(t) = \begin{cases} +M & h(t) \geq D \\ -M & h(t) \leq -D \\ 0 & -D < h(t) < D \end{cases} \tag{4-216}$$

where M is a positive constant, and D denotes the magnitude of the dead zone. A pictorial representation of the relay characteristic is shown in Figure 4-29. Since the input to the linear second-order process is a constant or zero, the state plane trajectories described by Eqs. (4-214) and (4-215) can be applied.

It is well-known that for a continuous-data system with relay-type control, the switching of the relay from one mode to another depends on the magnitude of the error signal $e(t)$. However, for a digital system, because of the sampling operation, the switching operation, the switching of the relay mode depends not only upon the magnitude of $e(t)$ at the sampling instants, but can occur only at the sampling instants. The construction of the entire state plane

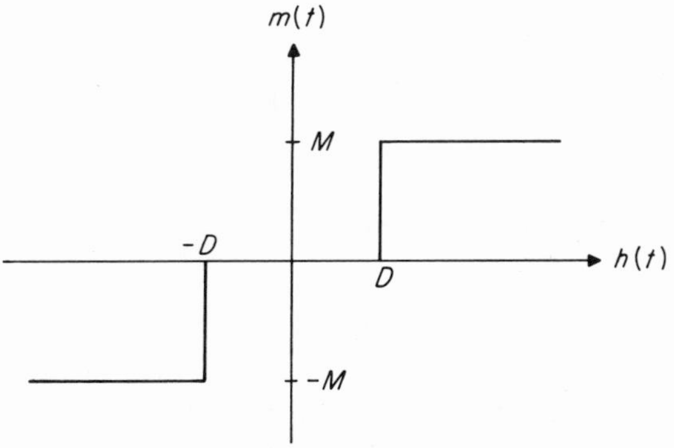

FIG. 4-29. Characteristic of relay with dead zone.

trajectory of the system shown in Figure 4-28 may appear to be difficult, since time does not appear explicitly on the trajectory. However, for the second-order system under consideration a set of switching lines can be derived in the state plane. The intersections of the state plane trajectory and the switching lines indicate the possibility of relay switchings. The details of the method are described as follows:

With reference to Figure 4-28, the system's variables are normalized as

$$x_{1n}(t) = x_1(t) = c(t)$$

$$x_{2n}(t) = \tau x_2(t) = \frac{dx_1}{dt}$$

$$t_n = t/\tau$$

$$t_{0n} = t_0/\tau = kT/\tau$$

$$T_n = [(k+1)T - t_0]/\tau$$

$$m_n(t) = \tau KMm(t)$$

where

$$m_n(t) = \begin{cases} +1 & h(t) \geq D \\ -1 & h(t) \leq -D \\ 0 & -D < h(t) < D \end{cases} \tag{4-217}$$

Then, the state plane trajectories of the second-order system are described by

$$x_{1n}[(k+1)T_n] - x_{1n}(kT_n) = -x_{2n}[(k+1)T_n] + x_{2n}(kT_n) - m_n(kT_n) \ln\left[\frac{m_n(kT_n) - x_{2n}[(k+1)T_n]}{m_n(kT_n) - x_{2n}(kT_n)}\right] \tag{4-218}$$

where $m_n(kT_n)$ can be $+1$, -1, or 0 only, and can change value only at the sampling instants.

At discrete sampling instants Eq. (4-209) is written

$$x_{1n}[(k+1)T_n] + x_{2n}[(k+1)T_n] = x_{1n}(kT_n) + x_{2n}(kT_n) + T_n m_n(kT_n) \tag{4-219}$$

The last equation gives the sum of the coordinates of the switching points as a function of the initial state $[x_{1n}(kT_n), x_{2n}(kT_n)]$, the input signal $m_n(kT_n)$, and the normalized sampling period T_n. In general, $m_n(kT_n)$ and T_n are known so that given the values of $x_{1n}(kT_n)$ and $x_{2n}(kT_n)$, the value of $x_{1n}[(k+1)T_n] + x_{2n}[(k+1)T_n]$ is determined from Eq. (4-219). However, to find the individual values of $x_{1n}[(k+1)T_n]$ and $x_{2n}[(k+1)T_n]$ another independent relation is needed. Since the state represented by $\{x_{1n}[(k+1)T_n], x_{2n}[(k+1)T_n]\}$ must also lie on the state plane trajectory, it must satisfy Eq. (4-218).

Let the sum of $x_{1n}(kT_n)$ and $x_{2n}(kT_n)$ be designated by a constant A. Then, from Eq. (4-219),

$$x_{1n}[(k+1)T_n] + x_{2n}[(k+1)T_n] = A + T_n \quad \text{for } m_n(kT_n) = 1 \tag{4-220}$$

$$x_{1n}[(k+1)T_n] + x_{2n}[(k+1)T_n] = A - T_n \quad \text{for } m_n(kT_n) = -1 \tag{4-221}$$

$$x_{1n}[(k+1)T_n] + x_{2n}[(k+1)T_n] = A \quad \text{for } m_n(kT_n) = 0 \tag{4-222}$$

which means that given the state $[x_{1n}(kT_n), x_{2n}(kT_n)]$, and $m_n(kT_n)$, the state $\{x_{1n}[(k+1)T_n], x_{2n}[(k+1)T_n]\}$ is determined from the intersection of the straight-line equations in (4-220), (4-221), or (4-222), and the corresponding state plane trajectories. Figure 4-30 illustrates that once the initial state $[x_{1n}(kT_n), x_{2n}(kT_n)]$ is given, if $m_n(kT_n) = 0$, the state at t_n must satisfy Eqs. (4-218) and (4-222), and the latter is a straight line with a slope of -1 and passing through the initial point in the state plane. When $m_n(kT_n) = 1$, the state point in the state plane moves from the initial state along the $+1$-trajectory shown in Figure 4-26 until it reaches a point where the trajectory intersects the line described by Eq. (4-220); this intersection gives the point $\{x_{1n}[(k+1)T_n], x_{2n}[(k+1)T_n]\}$. At this instant, $t_n = (k+1)T_n$, and if $m_n[(k+1)T_n]$ is again equal to $+1$, the state point continues to move on the same trajectory until it reaches the line described by the equation $x_{1n} + x_{2n} = A + 2T_n$. However, at $t_n = (k+1)T_n$, if $m_n[(k+1)T_n] = -1$, the operating point will move from $\{x_{1n}[(k+1)T_n], x_{2n}[(k+1)T_n]\}$ along the -1-trajectory until it intersects the line described by Eq. (4-222).

For $m_n(kT_n) = 0$, indicating that the input signal lies inside the relay dead zone, Eqs. (4-218) and (4-219) both become

$$x_{1n}[(k+1)T_n] + x_{2n}[(k+1)T_n] = x_{1n}(kT_n) + x_{2n}(kT_n) \tag{4-223}$$

Therefore, when $m_n(kT_n) = 0$, the swiching lines and the state plane trajectories (0-trajectories) are identical and are straight lines with slopes equal to -1. In Figure 4-29 it is shown that when $m_n(kT_n) = 0$ the state point moves from the initial state along the line $x_{1n}[(k+1)T_n] + x_{2n}[(k+1)T_n] = A$.

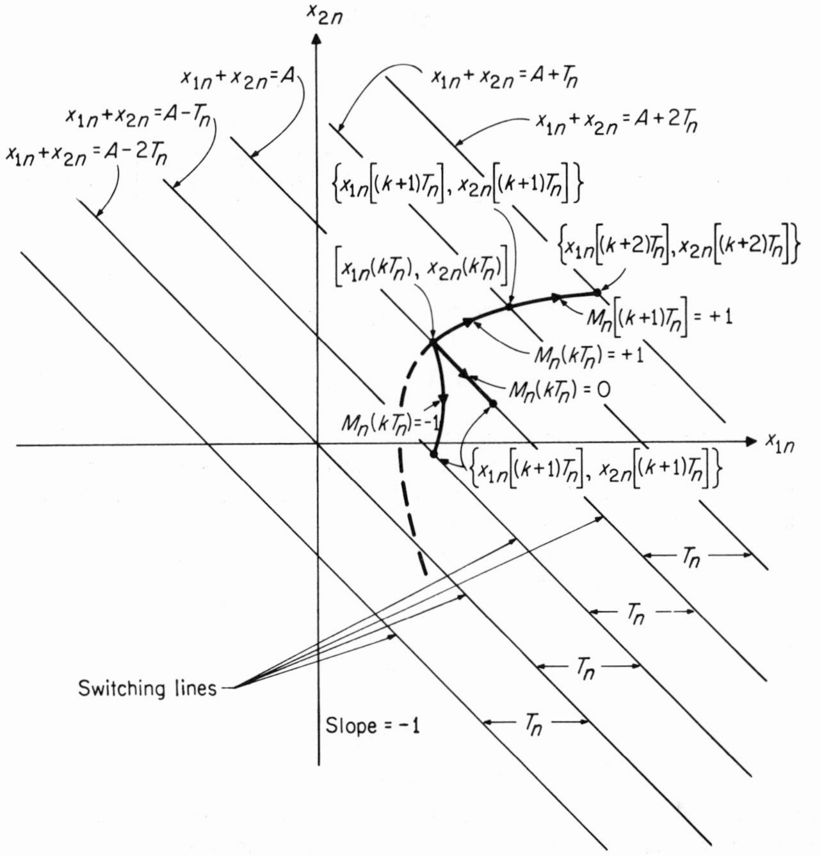

FIG. 4-30. Switching lines and state plane trajectories of a second-order digital system with relay control.

However, since time does not appear explicitly on the state plane trajectory, the location of the $\{x_{1n}[(k+1)T_n], x_{2n}[(k+1)T_n]\}$ point on the switching line is uncertain. To find the switching point, we must turn to the second equation in Eq. (4-208).

$$x_{2n}[(k+1)T_n] = e^{-T_n}x_{2n}(kT_n) + m_n(kT_n)(1 - e^{-T_n}) \qquad (4\text{-}224)$$

When $m_n(kT_n) = 0$, the last equation gives

$$x_{2n}[(k+1)T_n] = e^{-T_n}x_{2n}(kT_n) \qquad (4\text{-}225)$$

Since T_n and $x_{2n}(kT_n)$ are known, Eq. (4-225) gives the switching point on the switching line described by Eq. (4-222) at $t_n = (k+1)T_n$.

The state plane method of analyzing digital systems with relay control is illustrated by the following numerical examples.

EXAMPLE 4-9

Consider that for the system shown in Figure 4-31,

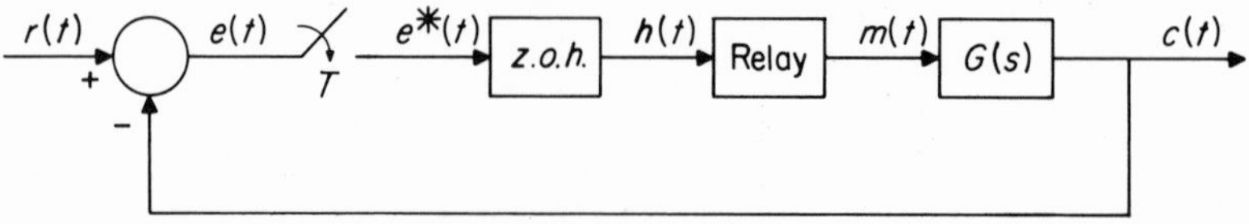

FIG. 4-31. A closed-loop digital system with relay control.

$$G(s) = \frac{1}{s(s+1)} \tag{4-226}$$

$T = 1$ second, and the relay characteristic is that shown in Figure 4-29 with $M = 1$ and $D = 0$ (ideal relay). The input to the system is a unit-step function, $r(t) = u(t)$. The output of the system is desired with the initial conditions of the system given as $c(0) = 0$ and $\dot{c}(0) = 0$.

The first step of the state plane analysis is to define the normalized state variables and time variables. These are:

$$x_{1n}(t) = x_1(t) = c(t)$$

$$x_{2n}(t) = x_2(t) = \dot{c}(t)$$

$$t_n = t$$

$$T_n = T = 1 \text{ second}$$

$$e_n = r_n - x_{1n} = 1 - x_{1n}$$

The switching lines for $T_n = 1$ are drawn in the normalized state plane as shown in Figure 4-32. The reference switching line, which is described by the equation $x_{1n}[(k+1)T_n] + x_{2n}[(k+1)T_n] = 0$, has a slope of -1, and passes through the initial state $x_{1n}(0) = x_{2n}(0) = 0$ in the state plane. All other switching lines have the same slope and are spaced at a horizontal distance of integral multiples of T_n from the reference switching line.

At $t_n = 0$, the error signal e_n is equal to $+1.0$; hence $m_n(0) = +1$. The operating point will move from the origin along the $+1$-trajectory until it intersects the switching line $x_{1n}[(k+1)T_n] + x_{2n}[(k+1)T_n] = T_n$ and the intersection point is the $[x_{1n}(T_n), x_{2n}(T_n)]$ point. At this instant, $t_n = T_n$, since $x_{1n}(T_n)$ is less than unity, $e_n(T_n)$ is positive, and therefore $m_n(T_n) = +1$. The operating point continues to move along the same $+1$-trajectory until it intersects the next switching line $x_{1n}[(k+1)T_n] + x_{2n}[(k+1)T_n] = 2T_n$ at $t_n = 2T_n$, and the intersection represents $[x_{1n}(2T_n), x_{2n}(2T_n)]$. Now since $x_{1n}(2T_n)$ is greater than unity, the magnitude of the reference input, $e_n(2T_n) < 0$, and $m_n(2T_n) = -1$. The operating point is now switched to and moves along the -1-trajectory which passes through the $[x_{1n}(2T_n), x_{2n}(2T_n)]$ point, until the trajectory intersects the switching line $x_{1n}[(k+1)T_n] + x_{2n}[(k+1)T_n] = T_n$ and the intersect gives the coordinates for $[x_{1n}(3T_n), x_{2n}(3T_n)]$.

As is apparent, $x_{3n}(3T_n)$ is still greater than unity; therefore, $m_n(3T_n) = -1$, and the operating point continues to move along the -1-trajectory until it intersects the switching line $x_{1n}[(k+1)T_n] + x_{2n}[(k+1)T_n] = 0$ at $t_n = 4T_n$. At this instant,

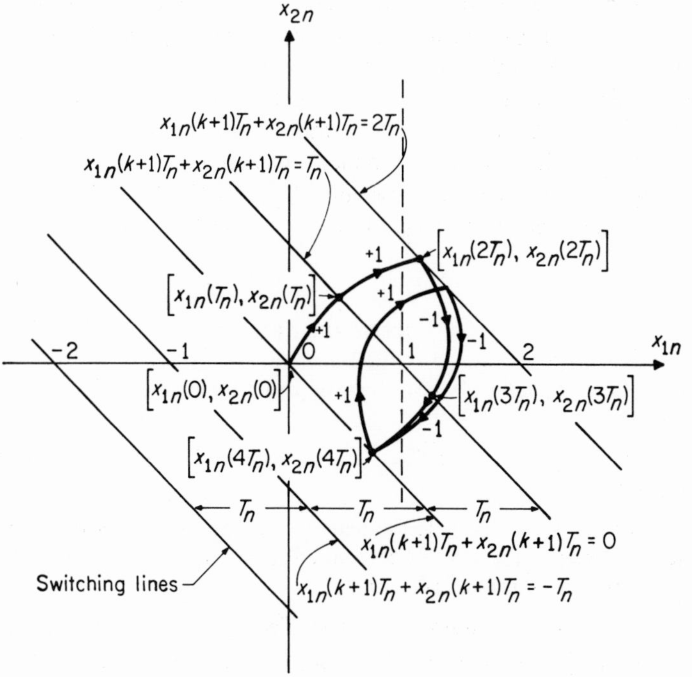

FIG. 4-32. State plane diagram of the control system with relay control in Example 4-9.

$x_{1n}(4T_n)$ is less than one, which gives $m_n(4T_n) = +1$ and the operating point is switched to a $+1$-trajectory. Figure 4-32 shows that as the switching process continues, the system eventually enters into sustained oscillation or a limit cycle. The period of the limit cycle as shown in Figure 4-32 is 4 seconds, and the amplitude is approximately 0.85. The output response and the relay switching function $m_n(t_n)$ of the system are sketched as shown in Figure 4-33.

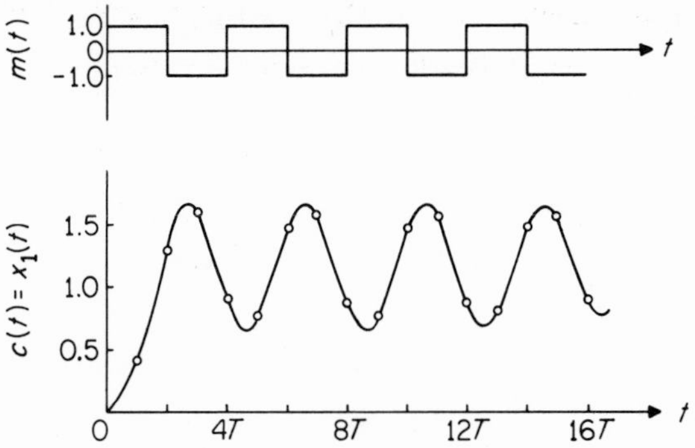

FIG. 4-33. Unit-step response of the relay control system in Example 4-9.

In this section we have discussed and illustrated the state plane method of analyzing digital control systems. The discussion is not intended to be exhaustive since the graphical method is not emphasized here. In general, the method can be applied to second-order digital systems which have inputs defined at the sampling instants. Example 4-9 illustrates a case when the input is a step function. The method is readily applied to sinusoidal or any periodic inputs.[5] For a ramp input, the state plane should be defined in terms of $e(t)$ and $\dot{e}(t)$. Therefore, a new set of state variables $x_1(t) = e(t)$ and $x_2(t) = \dot{e}(t)$ should be defined.

REFERENCES

1. Kuo, B.C., *Automatic Control Systems*, 2nd edition. Prentice-Hall, Englewood Cliffs, N. J., 1967.
2. Zadeh, L.A., and Desoer, C.A., *Linear System Theory*. McGraw-Hill, New York, 1963.
3. Zadeh, L.A., "An Introduction to State Space Techniques," Workshop on State Space Techniques for Control Systems, Proceedings, Joint Automatic Control Conference, Boulder, Colorado, 1962.
4. Kalman, R.E., and Bertram, J.E., "A Unified Approach to the Theory of Sampling Systems," *Journal of Franklin Inst.*, **267**, May 1959, pp. 405–436.
5. Kuo, B.C., *Analysis and Synthesis of Sampled-Data Control Systems*. Prentice-Hall, Englewood Cliffs, N.J., 1963.
6. Kuo, B.C., *Linear Networks and Systems*. McGraw-Hill, New York, 1967.
7. Tou, J.T., *Digital and Sampled-Data Control Systems*. McGraw-Hill, New York, 1959.
8. Jury, E.I., *Theory and Application of the z-Transform Method*. John Wiley & Sons, New York, 1964.
9. Gibson, J.E., *Nonlinear Automatic Control*. McGraw-Hill, New York, 1963.
10. Tou, J.T., *Modern Control Theory*. McGraw-Hill, New York, 1964.
11. Lindorff, D.P., *Theory of Sampled-Data Control Systems*. John Wiley & Sons, New York, 1964.

5

STABILITY OF DISCRETE-DATA SYSTEMS

5.1 INTRODUCTION

One of the most important requirements in the performance of control systems is stability. This is true whether the system has continuous-data, discrete-data, or a combination of the two kinds of signals. In this chapter we shall investigate the methods of testing for the stability condition of discrete-data systems. These methods are also applicable to continuous-data systems whose outputs are measured at discrete time intervals

The following methods of stability analysis are well known for continuous-data systems: (1) the Routh-Hurwitz criterion;[1] (2) the Nyquist criterion;[1] (3) the root locus plot;[1] (4) the Bode plot;[1] (5) Liapunov's direct method.[2,3] Most of these methods can be applied directly to the stability analysis of discrete-data systems with slight modifications. However, since the stability boundary in the z-plane is the unit circle $|z| = 1$, in contrast to the imaginary axis of the s-plane, the Routh-Hurwitz criterion cannot be applied for the discrete case. Therefore, the Schur-Cohn criterion and Jury's criterion are devised as replacements of the Routh-Hurwitz criterion for discrete-data systems. Finally, the Liapunov's direct method may be used to study the stability of linear as well as nonlinear systems with discrete data.

Before embarking on the discussions of the aforementioned methods of stability analysis, let us collect some essential definitions and terminologies on stability of discrete-data systems.

The discrete-data system under consideration can be described by the state equation

$$\mathbf{x}(t_{k+1}) = \mathbf{f}[\mathbf{x}(t_k)] \tag{5-1}$$

where $\mathbf{x}(t_k)$ is the $n \times 1$ state vector and $\mathbf{f}[\mathbf{x}(t_k)]$ is an $n \times 1$ vector whose elements are $f_1[\mathbf{x}(t_k)], f_2[\mathbf{x}(t_k)], \ldots, f_n[\mathbf{x}(t_k)]$; t_k denotes discrete instants of time with $k = 0, 1, 2, \ldots$. In general, $\mathbf{f}[\mathbf{x}(t_k)]$ may be a linear or a nonlinear function of $\mathbf{x}(t_k)$. The solution of Eq. (5-1) at $t_k = t_N$, given the initial state $\mathbf{x}(t_0)$, is

$$\mathbf{x}(t_N) = \Phi[t_N, \mathbf{x}(t_0)] \tag{5-2}$$

The initial state is also denoted by

$$\mathbf{x}(t_0) = \Phi[t_0, \mathbf{x}(t_0)] \tag{5-3}$$

Without loss of generality, we define the state

$$\mathbf{f}[\mathbf{x}_e(t_k)] = 0$$

as the equilibrium state. For linear time-invariant systems, Eq. (5-1) is written

$$\mathbf{x}(t_{k+1}) = A\mathbf{x}(t_k) \tag{5-4}$$

where A is an $n \times n$ constant coefficient matrix. If A is nonsingular, there is only one equilibrium state. If the state equation is nonlinear, then there may exist more than one equilibrium state. The concept of stability is centered on the study of the system's behavior in the neighborhood of the equilibrium state.

5.2 DEFINITION OF STABILITY

Let $\mathbf{x}_e(t_N)$ denote an equilibrium state of the system of Eq. (5-1). Then $\mathbf{x}_e(t_N)$ is said to be stable if given any real number $\epsilon > 0$, there exists a real number $\delta(\epsilon) > 0$ such that for all initial states $\mathbf{x}(t_0)$ in the sphere of radius δ

$$||\mathbf{x}(t_0) - \mathbf{x}_e|| \leq \delta \qquad (5\text{-}5)\dagger$$

the solution for all k,

$$\mathbf{x}(t_k) = \Phi[t_k, \mathbf{x}(t_0)] \tag{5-6}$$

remain within a sphere of radius; that is,

$$||\mathbf{x}(t_k) - \mathbf{x}_e|| < \epsilon \tag{5-7}$$

This definition of stability is also referred to as *stability in the sense of Liapunov.*

† $||x||$ denotes the Euclidean norm $(\mathbf{x}'\mathbf{x})^{1/2}$.

A graphical interpretation of the definition of stability given above is shown in Figure 5-1 for a second-order case.

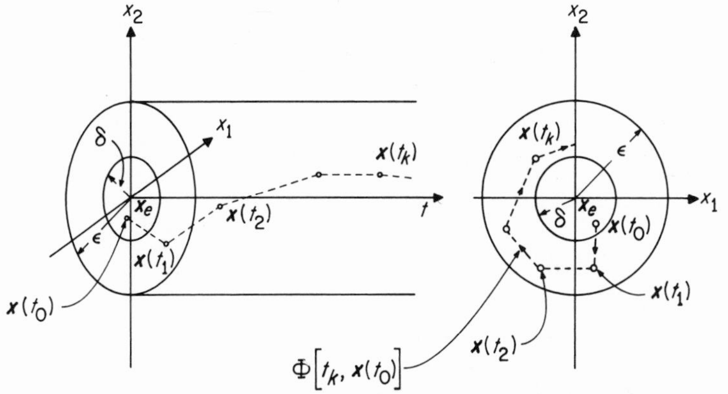

FIG. 5-1. Definition of stability.

Asymptotic Stability

An important type of stability in linear time-invariant systems is *asymptotic stability*.

An equilibrium state $\mathbf{x}_e$ of the system described by Eq. (5-1) is said to be asymptotically stable if

1. It is stable in the sense of Liapunov.
2. For any initial time t_0 and any $\mathbf{x}(t_0)$ sufficiently close to $\mathbf{x}_e$, $\Phi[t_k, \mathbf{x}(t_0)]$ converges to $\mathbf{x}_e$ as t_k approaches infinity.

Mathematically, we can state that given $\delta > 0$ and $\mu > 0$ there are real numbers $\epsilon > 0$ and $t_f(\mu, \delta, t_0)$ such that

$$||\mathbf{x}(t_0) - \mathbf{x}_e|| \leq \delta \tag{5-8}$$

implies

$$||\Phi[t_k, \mathbf{x}(t_0)] - \mathbf{x}_e|| \leq \epsilon \quad \text{for } k \geq 0 \tag{5-9}$$

and

$$||\Phi[t_k, \mathbf{x}(t_0)] - \mathbf{x}_e|| \leq \epsilon \tag{5-10}$$

for all $t_k \geq t_0 + t_f(\mu, \delta, t_0)$. The interpretation of asymptotic stability is illustrated as shown in Figure 5-2.

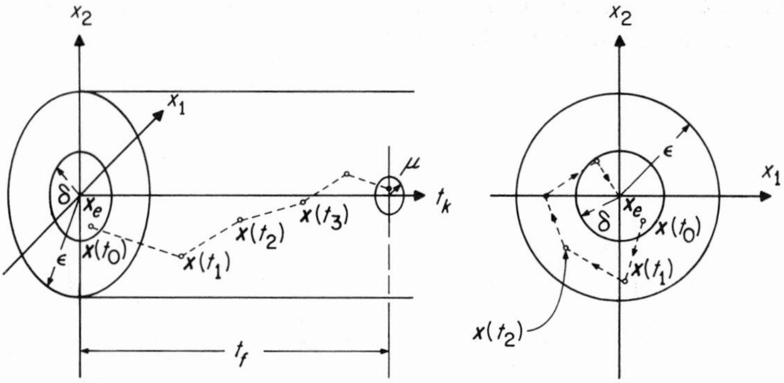

FIG. 5-2. Definition of asymptotic stability.

Asymptotic Stability in the Large

If the solution of Eq. (5-1) is asymptotically stable and if δ, the radius of the initial disturbance sphere (circle for the second-order case) can be arbitrarily large, we say that the equilibrium state is asymptotically stable in the large, or simply, *globally stable*.

Instability

An equilibrium state $\mathbf{x}_e$ is said to be unstable if it is neither stable nor asymptotically stable. With reference to Figure 5-3, the definition of instability can be stated, as for some real $\epsilon > 0$ and any real $\delta > 0$, no matter how small, there is always a state $\mathbf{x}(t_0)$ in the neighborhood of $\mathbf{x}_e$ such that the motion starting from this state will reach the boundary of $|\mathbf{x}| = \epsilon$.

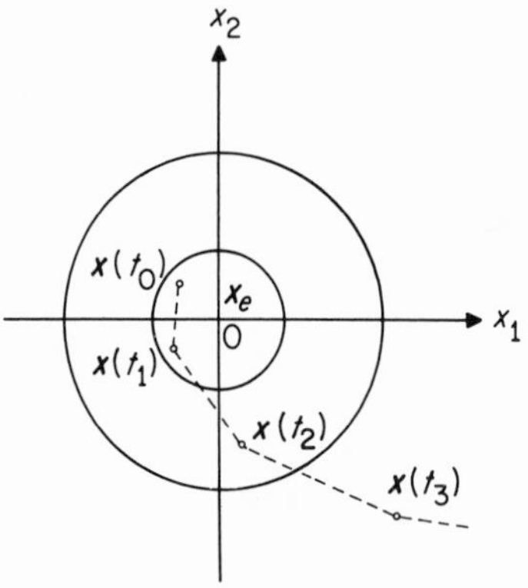

FIG. 5-3. Definition of instability.

We shall now derive the conditions of stability of linear time-invariant discrete-data systems with respect to the roots of the characteristic equation.

The state transition equation of the system described by Eq. (5-4) is written

$$\mathbf{x}(t_k) = \mathbf{\Phi}(t_k - t_0)\mathbf{x}(t_0) \quad (5\text{-}11)$$

where

$$\mathbf{\Phi}(t_k - t_0) = e^{A(t_k - t_0)} \quad (5\text{-}12)$$

is known as the state transition matrix. Taking the norm on both sides of Eq. (5-11) gives

$$||\mathbf{x}(t_k)|| = ||\Phi(t_k - t_0)\mathbf{x}(t_0)|| \tag{5-13}$$

An important property of the norm of a vector leads to

$$||\mathbf{x}(t_k)|| \leq ||\Phi(t_k - t_0)||\;||\mathbf{x}(t_0)|| \tag{5-14}$$

Then, with the equilibrium state set at $\mathbf{x}_e = \mathbf{0}$, the condition on asymptotic stability implies that

$$\lim_{t_k \to \infty} ||\Phi(t_k - t_0)|| = \mathbf{0} \tag{5-15}$$

The state transition matrix can be written

$$\Phi(t_k - t_0) = \mathscr{Z}^{-1}[(zI - A)^{-1}z] \quad t \geq t_0 \tag{5-16}$$

or

$$\Phi(t_k - t_0) = \mathscr{Z}^{-1}\left[\frac{\text{Adj}\,(zI - A)}{|zI - A|}\,z\right] \tag{5-17†}$$

Since $|zI - A| = 0$ is the characteristic equation of the system, Eq. (5-17) leads to the conclusion that in order for Eq. (5-15) to hold true, the roots of the characteristic equation must all lie inside the unit circle in the z-plane.

The following examples illustrate the ideas on stability, asymptotic stability, and instability of linear discrete-data systems.

EXAMPLE 5-1

Consider that a second-order discrete-data system is described by the state equation

$$\mathbf{x}(k + 1) = A\mathbf{x}(k) \tag{5-18}$$

where

$$A = \begin{bmatrix} 0 & 1 \\ -1 & 0 \end{bmatrix}$$

Taking the z-transform on both sides of Eq. (5-18) and solving for $\mathbf{X}(z)$, we get

$$\mathbf{X}(z) = (zI - A)^{-1}z\mathbf{x}(0) \tag{5-19}$$

or

$$\begin{bmatrix} X_1(z) \\ X_2(z) \end{bmatrix} = \frac{1}{z^2 + 1}\begin{bmatrix} z^2 & z \\ -z & z^2 \end{bmatrix}\begin{bmatrix} x_1(0) \\ x_2(0) \end{bmatrix} \tag{5-20}$$

The responses of $x_1(k)$ and $x_2(k)$ for two arbitrary initial states $\mathbf{x}(0)$ are computed by carrying out the synthetic division on the right-hand side of Eq. (5-20). The state plane trajectories are shown in Figure 5-4. The states are presented as circles, and

†Adj$(zI - A)$ denotes the adjoint matrix of $(zI - A)$.

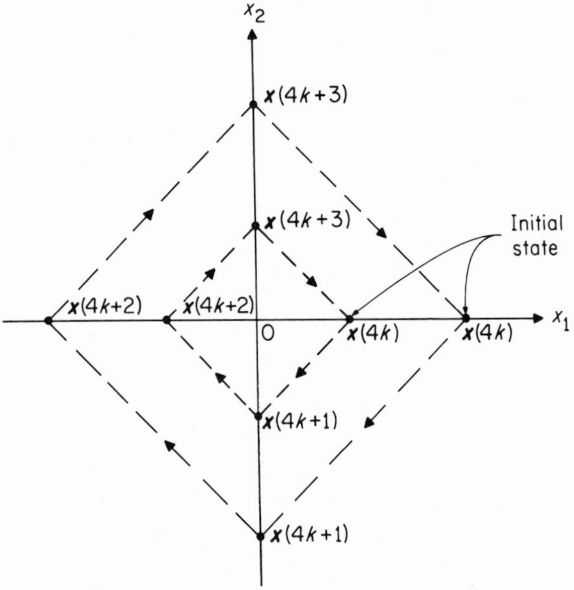

FIG. 5-4. State plane trajectories of system in Example 5-1.

these circles are joined by dotted straight lines to indicate the transition of state from one instant to another. In this case, the responses are in the form of limit cycles. The time domain response is oscillatory. Therefore, the system is stable but not asymptotically stable. The characteristic equation of the system is

$$|zI - A| = z^2 + 1 = 0 \tag{5-21}$$

Therefore, the two roots of the characteristic equation are on the unit circle in the z-plane.

EXAMPLE 5-2

Consider that the A matrix of the system described in the last example is given by

$$A = \begin{bmatrix} -0.5 & 0 \\ 0 & -0.5 \end{bmatrix} \tag{5-22}$$

Then the solution of Eq. (5-18) for $\mathbf{X}(z)$ gives

$$\mathbf{X}(z) = (zI - A)^{-1}\mathbf{x}(0) = \begin{bmatrix} \dfrac{z}{z+0.5} & 0 \\ 0 & \dfrac{z}{z+0.5} \end{bmatrix} \begin{bmatrix} x_1(0) \\ x_2(0) \end{bmatrix} \tag{5-23}$$

For arbitrary initial states of $\mathbf{x}(0) = [1 \quad 1]'$ and $[-1 \quad -1]'$, the responses of $\mathbf{x}(k)$ are shown in the state plane in Figure 5-5. In this case, the responses converge to

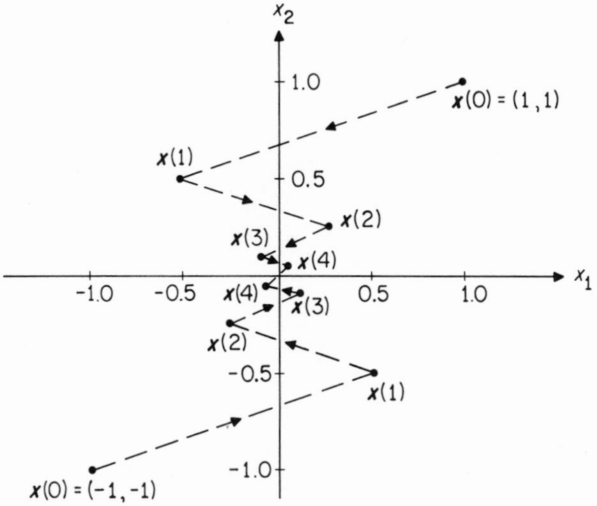

FIG. 5-5. State plane trajectories of system in Example 5-2.

$\mathbf{x} = 0$ as k approaches infinity. The system is asymptotically stable in the large. Notice that the characteristic equation is

$$|zI - A| = (z + 0.5)^2 = 0 \tag{5-24}$$

whose two roots are inside the unit circle at $z = -0.5$.

EXAMPLE 5-3

Now consider that the A matrix of the system of Eq. (5-18) is

$$A = \begin{bmatrix} -1.5 & 0 \\ 0 & 0.5 \end{bmatrix} \tag{5-25}$$

Then

$$\mathbf{X}(z) = \begin{bmatrix} \dfrac{z + 1.5}{z} & 0 \\ 0 & \dfrac{z}{z - 0.5} \end{bmatrix} \mathbf{x}(0) \tag{5-26}$$

For an initial state $\mathbf{x}(0) = [1 \quad 1]'$, the state plane trajectory of Figure 5-6 shows that the response of $x_1(k)$ diverges as k approaches infinity. Therefore, the system is unstable, that is, neither stable nor asymptotically stable.

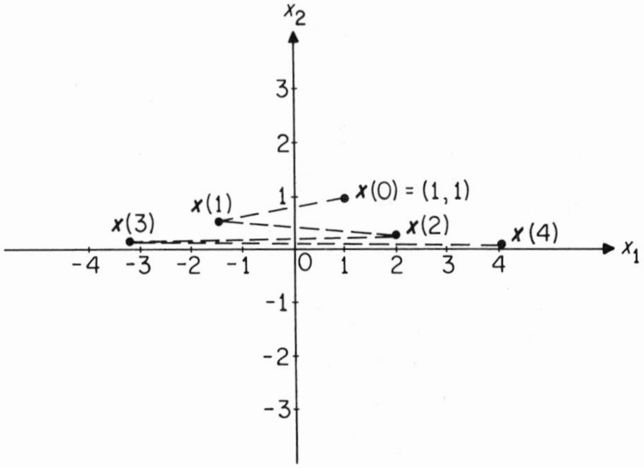

FIG. 5-6. State plane trajectories of system in Example 5-3.

5.3 THE SCHUR-COHN STABILITY CRITERION[4,5]

It has been established in the preceding section that a discrete-data system is asymptotically stable if all the roots of the characteristic equation lie inside the unit circle in the z-plane. The Routh-Hurwitz criterion which is normally useful for continuous-data systems cannot be applied directly to the z-domain since the stability boundary is now different. A test for the location of the roots of a polynomial in z with respect to the unit circle can be made by use of the Schur-Cohn criterion.

In general, if the characteristic equation of a linear discrete-data system is

$$F(z) = a_n z^n + a_{n-1} z^{n-1} + \dots + a_2 z^2 + a_1 z + a_0 = 0 \qquad (5\text{-}27)$$

where $a_0, a_1, \dots, a_n$ are either real or complex coefficients, the Schur-Cohn criterion states that

The roots of the characteristic equation lie inside the unit circle $z = 1$ if and only if the sequence of "Schur-Cohn determinants," $1, \Delta_1, \Delta_2, \dots, \Delta_k, \dots, \Delta_n$, has n variations in signs. The kth Schur-Cohn determinant is defined as

$$\Delta_k = \left|\begin{array}{ccccc|cccc}
a_0 & 0 & 0 & \cdots & 0 & a_n & a_{n-1} & \cdots & a_{n-k+1} \\
a_1 & a_0 & 0 & \cdots & 0 & 0 & a_n & \cdots & a_{n-k+2} \\
a_2 & a_1 & a_0 & \cdots & \cdot & \cdot & \cdot & \cdots & \cdot \\
\cdot & \cdot & \cdot & \cdots & \cdot & \cdot & \cdot & \cdots & \cdot \\
a_{k-1} & a_{k-2} & a_{k-3} & \cdots & a_0 & 0 & 0 & \cdots & a_n \\
\hline
\bar{a}_n & 0 & 0 & \cdots & 0 & \bar{a}_0 & \bar{a}_1 & \cdots & \bar{a}_{k-1} \\
\bar{a}_{n-1} & \bar{a}_n & 0 & \cdots & 0 & 0 & \bar{a}_0 & \cdots & \bar{a}_{k-2} \\
\bar{a}_{n-2} & \bar{a}_{n-1} & \bar{a}_n & \cdots & 0 & 0 & 0 & \cdots & \bar{a}_{k-3} \\
\cdot & \cdot & \cdot & \cdots & \cdot & \cdot & \cdot & \cdots & \cdot \\
\bar{a}_{n-k+1} & \bar{a}_{n-k+2} & \cdot & \cdots & \bar{a}_n & 0 & 0 & \cdots & \bar{a}_0
\end{array}\right| \qquad (5\text{-}28)$$

where $k = 1, 2, 3, \ldots, n$, and $\bar{a}_k$ is the conjugate of a_k. Note that when all the coefficients of the polynomial are real, the determinant Δ_k is symmetrical with respect to its principal diagonal. The stability criterion can also be stated as to require that

$$\begin{aligned} \Delta_k &< 0 \quad \text{for } k \text{ odd} \\ \Delta_k &> 0 \quad \text{for } k \text{ even} \end{aligned} \qquad (5\text{-}29)$$

However, if the conditions stated in Eq. (5-29) are not satisfied, then the characteristic equation has at least one root that falls outside the unit circle. Unlike the Routh-Hurwitz criterion, the Schur-Cohn test does not indicate in any definite manner how many roots are outside the unit circle. Nor does it tell if any of the roots are located on the unit circle.

EXAMPLE 5-4

Consider the characteristic equation of the system described in Example 5-2, Eq. (5-24),

$$(z + 0.5)^2 = z^2 + z + 0.25 = 0 \qquad (5\text{-}30)$$

Comparing the equation with Eq. (5-27), we have $\bar{a}_2 = a_2 = 1$, $\bar{a}_1 = a_1 = 1$, $\bar{a}_0 = a_0 = 0.25$. The Schur-Cohn determinants are determined as follows:

$$\Delta_1 = \begin{vmatrix} a_0 & a_2 \\ \bar{a}_2 & a_0 \end{vmatrix} = \begin{vmatrix} 0.25 & 1 \\ 1 & 0.25 \end{vmatrix} = -0.9375 \qquad (5\text{-}31)$$

$$\Delta_2 = \begin{vmatrix} a_0 & 0 & a_2 & a_1 \\ a_1 & a_0 & 0 & a_2 \\ \bar{a}_2 & 0 & \bar{a}_0 & \bar{a}_1 \\ \bar{a}_1 & \bar{a}_2 & 0 & \bar{a}_0 \end{vmatrix} = \begin{vmatrix} 0.25 & 0 & 1 & 1 \\ 1 & 0.25 & 0 & 1 \\ 1 & 0 & 0.25 & 1 \\ 1 & 1 & 1 & 0.25 \end{vmatrix} = 0.121 \qquad (5\text{-}32)$$

Thus, the sequence of Schur-Cohn determinants reads 1, -0.9375, 0.121, which has two changes in sign, and the system is asymptotically stable as determined earlier.

EXAMPLE 5-5

Consider the characteristic equation of the system described in Example 5-3,

$$(z - 1.5)(z + 0.5) = z^2 - z - 0.75 = 0 \tag{5-33}$$

The coefficients are $a_0 = \bar{a}_0 = -0.75$, $a_1 = \bar{a}_1 = -1$, $a_2 = \bar{a}_2 = 1$. The Schur-Cohn determinants are determined as follows:

$$\Delta_1 = \begin{vmatrix} a_0 & a_2 \\ \bar{a}_2 & a_0 \end{vmatrix} = \begin{vmatrix} -0.75 & 1 \\ 1 & -0.75 \end{vmatrix} = -0.4375 \tag{5-34}$$

$$\Delta_2 = \begin{vmatrix} a_0 & 0 & a_2 & a_1 \\ a_1 & a_0 & 0 & a_2 \\ \bar{a}_2 & 0 & \bar{a}_0 & \bar{a}_1 \\ \bar{a}_1 & \bar{a}_2 & 0 & \bar{a}_0 \end{vmatrix} = \begin{vmatrix} -0.75 & 0 & 1 & -1 \\ -1 & -0.75 & 0 & 1 \\ 1 & 0 & -0.75 & -1 \\ -1 & 1 & 0 & -0.75 \end{vmatrix} = -1.56 \tag{5-35}$$

Thus, the sequence of Schur-Cohn determinants reads 1, -0.4375, -1.56, which has only one change in sign. Therefore, the system is not asymptotically stable.

From these two examples, we see that the Schur-Cohn stability criterion is not nearly as convenient to apply as the Routh-Hurwitz criterion ordinarily used for continuous-data systems. For higher-order systems, the use of the Schur-Cohn test relies on the evaluation of higher-order determinants, thus making the task laborious. Unfortunately, there does not seem to be any simple way of making a tabulation of the equation coefficients so that the stability condition can be determined without evaluating the Schur-Cohn determinants. Certainly, the attractiveness of the Routh-Hurwitz criterion is due in part to the fact that the equation coefficients can be tabulated into the form of the Routh tabulation, and then the stability condition is determined simply by manipulation of the coefficients of the tabulation. Furthermore, the Routh's test will give information on asymptotic stability, stability, or instability, whereas the Schur-Cohn test tells only whether the system is asymptotically stable or not.

5.4 JURY'S STABILITY TEST[6,7,8]

A stability test which has some of the advantages of the Routh's test, and is simpler to apply than the Schur-Cohn criterion, is devised by Jury and Blanchard. Referring to the polynomial of Eq. (5-27), with $a_n > 0$, the first step in Jury's test is the formulation of the following table using the coefficients of the polynomial:

Row	z^0	z^1	z^2	$\cdots$	z^{n-k}	$\cdots$	z^{n-1}	z^n
1	a_0	a_1	a_2	$\cdots$	a_{n-k}	$\cdots$	a_{n-1}	a_n
2	a_n	a_{n-1}	a_{n-2}	$\cdots$	a_k	$\cdots$	a_1	a_0
3	b_0	b_1	b_2	$\cdots$	b_{n-k}	$\cdots$	b_{n-1}	
4	b_{n-1}	b_{n-2}	b_{n-3}	$\cdots$	b_k	$\cdots$	b_0	
5	c_0	c_1	c_2	$\cdots$		c_{n-2}		
6	c_{n-2}	c_{n-3}	c_{n-4}	$\cdots$		c_0		
.	.	.	.					
.	.	.	.	$\cdots$				
.	.	.	.					
$2n-5$	p_0	p_1	p_2	p_3				
$2n-4$	p_3	p_2	p_1	p_0				
$2n-3$	q_0	q_1	q_2					

Note that the elements of the $(2k+2)$th row $(k = 0, 1, 2, \ldots)$ consist of the coefficients of the $(2k+1)$th row written in the reverse order. The elements in the table are defined as

$$b_k = \begin{vmatrix} a_0 & a_{n-k} \\ a_n & a_k \end{vmatrix} \qquad c_k = \begin{vmatrix} b_0 & b_{n-1-k} \\ b_{n-1} & b_k \end{vmatrix}$$

$$d_k = \begin{vmatrix} c_0 & c_{n-2-k} \\ c_{n-2} & c_k \end{vmatrix} \qquad \cdots$$

$$\cdots \qquad q_0 = \begin{vmatrix} p_0 & p_3 \\ p_3 & p_0 \end{vmatrix} \qquad q_2 = \begin{vmatrix} p_0 & p_1 \\ p_3 & p_2 \end{vmatrix}$$

The necessary and sufficient conditions for the polynomial $F(z)$ to have no roots on and inside the unit circle in the z-plane (asymptotic stability) are:

$$F(1) > 0$$

$$F(-1) \begin{matrix} > 0 & n \text{ even} \\ < 0 & n \text{ odd} \end{matrix}$$

$$\left.\begin{matrix} |a_0| < a_n \\ |b_0| > |b_{n-1}| \\ |c_0| > |c_{n-2}| \\ |d_0| > |d_{n-3}| \\ \cdot \quad \cdot \\ \cdot \quad \cdot \\ |q_0| > |q_2| \end{matrix}\right\} \quad (n-1) \text{ constraints} \tag{5-36}$$

For a second-order system, $n = 2$, Jury's tabulation contains only one row.

Therefore, the requirement listed in Eq. (5-36) are reduced to $F(1) > 0$, $F(-1) > 0$ for $n =$ even, $F(-1) < 0$ for $n =$ odd, and $|a_0| < a_n$.

EXAMPLE 5-6

Consider the same system treated in Example 5-4. The characteristic equation is

$$F(z) = z^2 + z + 0.25 = 0 \tag{5-37}$$

The first step of Jury's test find that

$$F(1) = 2.25 > 0 \quad \text{and} \quad F(-1) = 0.25 > 0$$

Since $n = 2$ is even, these conditions satisfy the $F(1) > 0$ and $F(-1) > 0$ requirements. Next, we tabulate the coefficients of Eq. (5-37) as follows:

Row	z^0	z^1	z^2
1	0.25	1	1

Since $2n - 3 = 1$, the Jury's tabulation consists of only one row. Thus,

$$|a_0| = 0.25 < a_2 = 1$$

and the test shows that the system is asymptotically stable.

EXAMPLE 5-7

Consider the polynomial

$$F(z) = z^2 - z - 0.75 = 0 \tag{5-38}$$

which is the characteristic equation of the system in Examples 5-3 and 5-5.

Applying Jury's test to the last equation, we have

$$F(1) = -0.75 \qquad F(-1) = 1.25$$

and

$$|a_0| = 0.75 < |a_2| = 1$$

Since $F(1)$ is not positive, the system is not asymptotically stable.

5.5 EXTENSION OF ROUTH-HURWITZ CRITERION TO DISCRETE-DATA SYSTEMS[5]

As pointed out earlier, since the stability boundary in the z-plane is the unit circle $|z| = 1$, the Routh-Hurwitz criterion cannot be applied directly to the characteristic equation $F(z) = 0$. However, if we map the interior of the unit circle in the z-plane onto the left-half of a complex variable plane by a bilinear transformation, the Routh-Hurwitz criterion may again be applied directly to the equation in the new variable.

Let us consider the bilinear transformation

$$z = \frac{aw + b}{cw + d} \tag{5-39}$$

where w is a complex variable. We can show that the transformations

$$z = \frac{1 + w}{1 - w} \tag{5-40}$$

and

$$z = -\frac{1 + w}{1 - w} \tag{5-41}$$

will transform the interior of the unit circle onto the left-half of the w plane.

5.6 THE SECOND METHOD OF LIAPUNOV[2,3,8]

A powerful method of determining the stability of linear and nonlinear systems with continuous-data is the second method of Liapunov. The subject is well-published and readily extended to the study of stability of discrete-data systems. The second method of Liapunov is based on the determination of a V function called the *Liapunov function.* From the properties of the V function one is able to show stability or instability of the system. However, the main disadvantage of Liapunov's stability criterion is that it gives only the sufficient conditions but not the necessary conditions for stability. Furthermore, there are no unique methods of determining the V function for a wide class of systems.

The Liapunov Function

Any function $V = V[\mathbf{x}(k)]$ *of definite sign (positive definite or negative definite) is called a Liapunov function if* $V(\mathbf{0}) = 0$ *and* $\mathbf{x}(k)$ *is the solution of the state equation*

$$\mathbf{x}(k + 1) = \mathbf{f}[\mathbf{x}(k)] \tag{5-42}$$

We define a difference operator Δ so that $\Delta V[\mathbf{x}(k)]$ is defined as

$$\Delta V[\mathbf{x}(k)] = V[\mathbf{x}(k + 1)] - V[\mathbf{x}(k)] \tag{5-43}$$

Stability Theorem of Liapunov

Consider that a discrete-data system is described by

$$\mathbf{x}(k + 1) = \mathbf{f}[\mathbf{x}(k)] \tag{5-44}$$

where $\mathbf{x}(k)$ *is the* $n \times 1$ *state vector,* $\mathbf{f}[\mathbf{x}(k)]$ *is an* $n \times 1$ *function vector with the property that*

$$\mathbf{f}[\mathbf{x}(k) = \mathbf{0}] = \mathbf{0} \quad \text{for all } k \tag{5-45}$$

Suppose that there exists a scalar function $V[\mathbf{x}(k)]$ *continuous in* $\mathbf{x}(k)$, *such that*

$$\begin{aligned} &1. \quad V[\mathbf{x}(k) = 0] = V(\mathbf{0}) = \mathbf{0} \\ &2. \quad V[\mathbf{x}(k)] = V(\mathbf{x}) > 0 \quad \text{for } \mathbf{x} \neq \mathbf{0} \\ &3. \quad V(\mathbf{x}) \textit{ approaches infinity as } \|\mathbf{x}\| \to \infty \\ &4. \quad \Delta V(\mathbf{x}) < 0 \qquad\qquad \text{for } \mathbf{x} \neq \mathbf{0} \end{aligned} \tag{5-46}$$

Then the equilibrium state $\mathbf{x}(k) = \mathbf{0}$ *(for all* k*) is asymptotically stable in the large and* $V(\mathbf{x})$ *is a Liapunov function.*

For a proof of the theorem the reader may refer to the literature.

Instability Theorem of Liapunov

For the system described by Eq. (5-44), if there exists a scalar function $V(\mathbf{x})$ *continuous in* $\mathbf{x}(k)$, *such that*

$$\Delta V(\mathbf{x}) < 0, \quad \textit{then:}$$

1. *The system is unstable in the finite region for which V is not positive semidefinite* (≥ 0).
2. *The response is unbounded as k approaches infinity if V is not positive semidefinite for all* $\mathbf{x}(k)$.

EXAMPLE 5-8

Consider the discrete-data system which is described by

$$x_1(k+1) = -0.5x_1(k) \tag{5-47}$$

$$x_2(k+1) = -0.5x_2(k) \tag{5-48}$$

Let us assign the Liapunov function to be

$$V(\mathbf{x}) = x_1^2(k) + x_2^2(k) \tag{5-49}$$

which is positive for all values of $x_1(k)$ and $x_2(k)$ not equal to zero. Then, the function $\Delta V(\mathbf{x})$ is evaluated using Eq. (5-43),

$$\begin{aligned} \Delta V(\mathbf{x}) &= V[\mathbf{x}(k+1)] - V[\mathbf{x}(k)] \\ &= x_1^2(k+1) + x_2^2(k+1) - x_1^2(k) - x_2^2(k) \\ &= -0.75x_1^2(k) - 0.75x_2^2(k) \end{aligned} \tag{5-50}$$

Since $\Delta V(\mathbf{x})$ is negative for all $\mathbf{x}(k) \neq 0$, the system is asymptotically stable.

EXAMPLE 5-9

Consider the discrete-data system which is described by

$$x_1(k + 1) = -1.5x_1(k) \tag{5-51}$$

$$x_2(k + 1) = -0.5x_2(k) \tag{5-52}$$

We know that the system is unstable. However, assuming, that without prior knowledge, the stability theorem of Liapunov is applied. Let us again try the Liapunov function

$$V(\mathbf{x}) = x_1^2(k) + x_2^2(k) \tag{5-53}$$

Then,

$$\Delta V(\mathbf{x}) = V[\mathbf{x}(k + 1)] - V[\mathbf{x}(k)]$$
$$= 1.25x_1^2(k) - 0.75x_2^2(k) \tag{5-54}$$

However, since $\Delta V(\mathbf{x})$ is indefinite in sign, the Liapunov test using the V function of Eq. (5-53) failed.

Now let us turn to the instability theorem of Liapunov. Let the Liapunov function be defined as

$$V(\mathbf{x}) = a_1x_1^2(k) + 2a_2x_1(k)x_2(k) + a_3x_2^2(k) \tag{5-55}$$

and let the function $\Delta V(\mathbf{x})$ be of the form

$$\Delta V(\mathbf{x}) = -x_1^2(k) - x_2^2(k) \tag{5-56}$$

so that it is negative definite for all $x_1(k)$ and $x_2(k) \neq 0$.

Forming $\Delta V(\mathbf{x})$ according to Eq. (5-43), we get

$$\Delta V(\mathbf{x}) = V[\mathbf{x}(k + 1)] - V[\mathbf{x}(k)]$$
$$= 1.25a_1x_1^2(k) + 1.25a_3x_2^2(k) \tag{5-57}$$

Therefore,

$$a_1 = a_3 = -0.8 \qquad a_2 = 0$$

and

$$V(\mathbf{x}) = -0.8x_1^2(k) - 0.8x_2^2(k) \tag{5-58}$$

Since $V(\mathbf{x})$ is negative definite for all $x_1(k)$ and $x_2(k) \neq 0$, the system is unstable.

Liapunov Stability Theorem for Linear Discrete-Data Systems

The stability and instability theorems of Liapunov discussed above are valid for both linear and nonlinear systems. However, if a discrete-data system is linear, a simpler test procedure is available.

Consider that a linear time-invariant discrete-data system is described by the difference equation

$$\mathbf{x}(k + 1) = A\mathbf{x}(k) \tag{5-59}$$

where $\mathbf{x}(k)$ *is the* $n \times 1$ *state vector and* A *is an* $n \times n$ *constant matrix. The equilibrium state* $\mathbf{x}(0) = \mathbf{0}$ *is asymptotically stable if and only if given any positive definite real symmetric matrix* Q, *there exists a positive definite real symmetric matrix* P *such that*

$$A'PA - P = -Q \tag{5-60}$$

Then

$$V(\mathbf{x}) = \mathbf{x}'(k)P\mathbf{x}(k) \tag{5-61}$$

is a Liapunov function for the system, and further,

$$\Delta V(\mathbf{x}) = -\mathbf{x}'(k)Q\mathbf{x}(k) \tag{5-62}$$

where $\Delta V(\mathbf{x})$ *is as defined in Eq. (5-43).*

EXAMPLE 5-10

Consider the discrete-data system

$$x_1(k+1) = -0.5x_1(k) \tag{5-63}$$

$$x_2(k+1) = -0.5x_2(k) \tag{5-64}$$

where $\mathbf{x}(0) = \mathbf{0}$ is the equilibrium state.

Let

$$Q = \begin{bmatrix} 1 & 0 \\ 0 & 1 \end{bmatrix} = I$$

and

$$P = \begin{bmatrix} p_{11} & p_{12} \\ p_{21} & p_{22} \end{bmatrix}$$

Then Eq. (5-60) gives

$$\begin{bmatrix} -0.5 & 0 \\ 0 & -0.5 \end{bmatrix}\begin{bmatrix} p_{11} & p_{12} \\ p_{21} & p_{22} \end{bmatrix}\begin{bmatrix} -0.5 & 0 \\ 0 & -0.5 \end{bmatrix} - \begin{bmatrix} p_{11} & p_{12} \\ p_{21} & p_{22} \end{bmatrix} = -\begin{bmatrix} 1 & 0 \\ 0 & 1 \end{bmatrix}$$

Solving for the P matrix from the last equation yields

$$P = \begin{bmatrix} 1.33 & 0 \\ 0 & 1.33 \end{bmatrix}$$

which is positive definite. Therefore,

$$V(\mathbf{x}) = \mathbf{x}'(k)P\mathbf{x}(k)$$

which is the Liapunov function and is positive definite.

Therefore, $V(\mathbf{x})$ is given by Eq. (5-62), is negative definite, and the equilibrium state $\mathbf{x}(0) = \mathbf{0}$ is asymptotically stable.

REFERENCES

1. Kuo, B. C., *Automatic Control Systems*, 2nd edition. Prentice-Hall, Englewood Cliffs, N. J., 1967.
2. Kalman, R. E., and Bertram, J. E., "Control Systems Analysis and Design Via the Second Method of Liapunov: II, Discrete Time Systems," *Trans. ASME, J. Basic Eng.*, Series D, No. 3, June 1960, pp. 371-400.
3. Hahn, W., *Theory and Application of Liapunov's Direct Method.* Prentice-Hall, Englewood Cliffs, N. J., 1963.
4. Marden, M., *The Geometry of the Zeros of a Polynomial in a Complex Variable*, Chapter 10. American Math. Society, New York, 1949.
5. Kuo, B. C., *Analysis and Synthesis of Sampled-Data Control Systems.* Prentice-Hall, Englewood Cliffs, N. J., 1963.
6. Jury, E. I., and Bharucha, B. H., "Notes on the Stability Criterion for Linear Discrete Systems," *IRE Trans. on Automatic Control*, AC-6, February 1961, pp. 88-90.
7. Jury, E. I,. and Blanchard, J., "A Stability Test for Linear Discrete Systems in Table Form," *IRE Proc.*, **49**, No. 12, December 1961, pp. 1947-1948.
8. Ogata, K., *State Space Analysis of Control Systems.* Prentice-Hall, Englewood Cliffs, N. J., 1967.
9. Jury, E. I., "The Number of Roots of a Real Polynomial Inside (or Outside) the Unit Circle Using the Determinant Method," *IEEE Trans. on Automatic Control*, Vol. AC-10, July 1965, pp. 371-372.
10. Cohen, M. L., "A Set of Stability Constraints on the Denominator of a Sampled-Data Filter," *IEEE Trans. on Automatic Control*, Vol. AC-11, April 1966, pp. 327-328.
11. Pearson, J. B., Jr., "A Note on the Stability of a Class of Optimum Sampled-Data Systems," *IEEE Trans. on Automatic Control*, Vol. AC-10, January 1965, pp. 117-118.
12. Szego, G. P., and Pearson, J. B., Jr., "On the Absolute Stability of Sampled-Data Systems: the Indirect Control Case," *IEEE Trans. on Automatic Control*, Vol. AC-9, April 1964, pp, 160-163.
13. O'Shea, R. P. O., and Younis, M. I., "A Frequency-Time Domain Stability Criterion for Sampled-Data Systems," *IEEE Trans. on Automatic Control*, Vol. AC-12, December 1967, pp. 719-724.
14. Wu, S. H., "A Circle Stability Criterion for a Class of Discrete Systems," *IEEE Trans. on Automatic Control*, Vol. AC-12, February 1967, pp. 114-115.
15. Brockett, R. W., "The State of Stability Theory for Deterministic Systems," *IEEE Trans. on Automatic Control*, Vol. AC-11, July 1966, pp. 596-606.

16. Kodama, S., "Stability of a Class of Discrete Control Systems Containing a Nonlinear Gain Element," *IRE Trans. on Automatic Control*, Vol. AC-7, October 1962, pp. 102–109.

17. Kodama, S., "Stability on Nonlinear Sampled-Data Control Systems," *IRE Trans. on Automatic Control*, Vol. AC-7, January 1962, pp. 15–23.

18. Tsypkin, Y. S., "On the Stability in the Large of Nonlinear Sampled-Data Systems," *Dokl. Akad. Nank.*, Vol. 145, July 1962, pp. 52–55.

19. Tsypkin, Y. S., "Absolute Stability of Equilibrium Positions and of Responses in Nonlinear Sampled-Data, Automatic Systems," *Avtomat, i Telemekhan*, Vol. 24, December 1963, pp. 1601–1615.

20. Tsypkin, Y. S., "Frequency Criteria for the Absolute Stability of Nonlinear Sampled-Data Systems," *Avtomat. i Telemekhan*, Vol. 25, March 1964, pp. 281–290.

21. Iwens, R. P., and Bergen, A. R., "Frequency Criteria for Bounded-Input-Bounded-Output Stability of Nonlinear Sampled-Data Systems," *IEEE Trans. on Automatic Control*, Vol. AC-12, February 1967, pp. 46–53.

22. Iwens, R. P., and Bergen, A. R., "On Bounded-Input-Bounded-Output Stability of a Certain Class of Nonlinear Sampled-Data Systems," *J. Franklin Inst.*, Vol. 282, October 1966, pp. 193–205.

23. O'Shea, R. P., "Approximation of the Asymptotic Stability Boundary of Discrete-time Control Using an Inverse Transformation Approach," *IEEE Trans. on Automatic Control*, Vol. AC-9, October 1964, pp. 441–448.

24. Jury, E. I., and Lee, B. W., "On the Absolute Stability of Nonlinear Sampled-Data Systems," *IEEE Trans. on Automatic Control*, Vol. AC-9, October 1964, pp. 551–554.

25. Murphy, G. J., and Wu, S. H., "A Stability Criterion for Pulse-Width Modulated Feedback Control Systems," *IEEE Trans. on Automatic Control*, Vol. AC-9, October 1964, pp. 434–441.

26. Pearson, J. B., and Gibson, J. E., "On the Asymptotic Stability of a Class of Saturating Sampled-Data Systems," *IEEE Trans. on Application and Industry*, No. 71, March 1964, pp. 81–86.

27. Chen, C. T., "On the Stability of Nonlinear Sampled-Data Feedback Systems," *J. Franklin Inst.*, Vol. 280, October 1965, pp. 316–324.

6

TIME-OPTIMAL CONTROL OF DISCRETE-TIME SYSTEMS

6.1 INTRODUCTION

The design problem in control systems can generally be described as: given a *plant* or *controlled process* as shown in the block diagram of Figure 6-1,

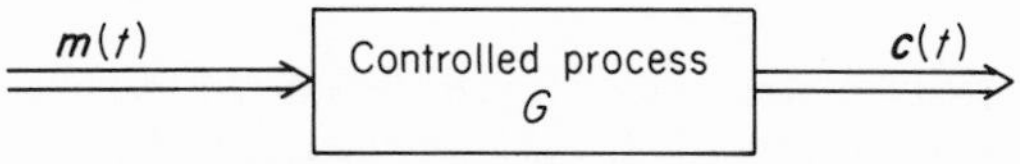

FIG. 6-1. Block diagram of a controlled process.

determine the control or actuating signal $\mathbf{m}(t)$ so that the prescribed *design specifications* or *performance criteria* are satisfied. In general, the control vector $\mathbf{m}(t)$ is a $p \times 1$ vector and the output vector $c(t)$ is $q \times 1$. In the continuous-data case, the control vector $\mathbf{m}(t)$ is a continuous function of time and acts on the process at all times. In the digital case, the elements of $\mathbf{m}(t)$ may be pulse signals or a sequence of numbers. Most often, they are outputs of zero-order hold devices.

Although the design problem stated above seems to be that of the open-loop nature, however, the designed overall system is closed loop since $\mathbf{m}(t)$ is usually obtained as the output of a controller which performs specific operations on the reference input $\mathbf{r}(t)$ and the state vector $\mathbf{x}(t)$ of the process. Therefore, a complete system is usually a feedback system as illustrated by the block diagram in Figure 6-2.

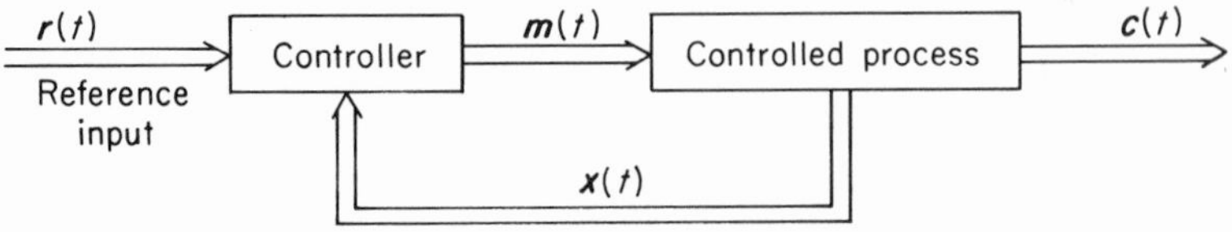

FIG. 6-2. Block diagram of a feedback control system.

When it comes to the design techniques, we can generally divide them into two categories: (1) conventional design and synthesis, and (2) modern control theory. The conventional design principle, which is also referred to as the compensation of control systems, was developed during World War II and the postwar years immediately following. The method is characterized by first selecting a proper configuration for the overall system, and then the compensator or controller is designed so that the control system performs according to prescribed specifications. In carrying out the design the designer usually has to rely to a great extent on his past experience in choosing a satisfactory compensator. The design procedure also depends heavily on trial and error. For deterministic inputs, the performance criteria used in the conventional design often include the following terms:

1. Frequency-domain specifications: bandwidth, peak resonance, gain margin, phase margin, etc.
2. Time-domain specifications: overshoot, rise time, settling time, steady-state error, etc.

A great majority of the conventional design methods rely on graphical techniques. These include the Bode diagram, Nyquist plot, Nichols chart, root locus diagram,[1] etc. These graphical techniques which were originally devised for continuous-data systems can all be extended to the design of digital control systems.

The synthesis of linear continuous-data control systems was first introduced by Aaron and Truxal[2] in early 1950. The philosophy of synthesis is that the overall system transfer function, $C(s)/R(s)$, is first determined to satisfy simultaneously the design criterion and the physical realizability of the controller. Once the desired $C(s)/R(s)$ is determined, the system configuration is then selected and the transfer function of the controller is finally determined. In contrast to the conventional design technique which starts out with the controller and then works toward the closed-loop configuration, the synthesis method starts with the closed-loop transfer function of the controller, and then works toward the transfer function. Furthermore, the synthesis method is more of an analytic approach and it reduces the uncertainties and chores of trial and error inherently found in conventional designs. It has been shown that a synthesis method similar to that proposed by Truxal can be applied to linear digital control systems.

One of the disadvantages of conventional designs and synthesis of control systems is that these methods can lead to a system which merely satisfies the specified design criteria. There is no guarantee that the designed system is the best under certain criteria.

Another disadvantage of the conventional design is that the method does not guarantee that a solution exists at the outset of the design. It is entirely possible that the specifications include contradictory requirements which are impossible to meet by physical elements.

Perhaps the most important limitation of the conventional design and synthesis is that these methods were not devised to take advantage and make use of high-speed digital computers. With the advent of space age technology, modern control systems often are quite complex and with multiple number of inputs and outputs. Because of the many interdependent relations that exist between inputs and outputs of a multivariable system, it is virtually impossible to carry out a design of this type of system using a trial-and-error or graphical method.

Since the fifties the modern control theory has come into vogue and has proved to be invaluable for the design of complex control systems and systems with more stringent performance requirements. Modern control theory places more emphasis on the analytic approach and the use of applied mathematics and computers. The product of modern design principles is the optimal control system which can be claimed to be the "best" according to some prescribed performance criterion. Just as the design specifications in the frequency-domain design which are selected because of convenience for interpretation graphically, the performance criteria used in modern control systems design are often selected based on mathematical convenience. Therefore, the optimal control is judged the best not entirely from the true performance of the system, but rather based on a certain performance criterion which is chosen because it makes possible the mathematical design of the system.

Among the many different types of optimal control design problems, the time-optimal control and the minimum-integral control are the most general. The first type deals with the system which is able to bring the states of the system from one state to another in the shortest possible time, and is the subject of this chapter. The second type is the minimization (or maximization) of a certain functional of the system variable. For instance, the minimum mean-square error design concerns the design of an optimum system which minimizes the average value of the square of the error between the actual and the desired outputs.

6.2 CONTROLLABILITY AND OBSERVABILITY OF LINEAR CONTROL SYSTEMS[3,4,5]

The concepts of controllability and observability introduced first by Kalman play an important role in the design of time-optimal control systems. The significance of controllability can be stated with reference to the block diagram of Figure 6-1. The controlled process G is said to be completely controllable if every state variable of G can be affected or controlled in finite time by some unconstrained control signal $\mathbf{m}(t)$. Intuitively, we understand that if any one of the state variables is independent of the control signal $\mathbf{m}(t)$, then there is no way of driving this particular state variable to a desired state in finite time by means of some control effort. Therefore, this particular state variable is said to be uncontrollable, and the system is said to be not completely controllable or simply uncontrollable.

The concept of observability is the dual of controllability. Essentially, the process G is said to be observable if every state variable of the process eventually affects some of the outputs of the process. In other words, it is often desirable to obtain information on the state variables from measurements of the outputs and the inputs. If any one of the states cannot be observed from the measurements of the outputs, then the state is said to be unobservable and the system is not completely observable or is simply unobservable.

The descriptions of controllability and observability given above are only general in nature. Before giving formal definitions of the subjects, let us first consider some simple illustrative examples on the ideas. Figure 6-3 shows the

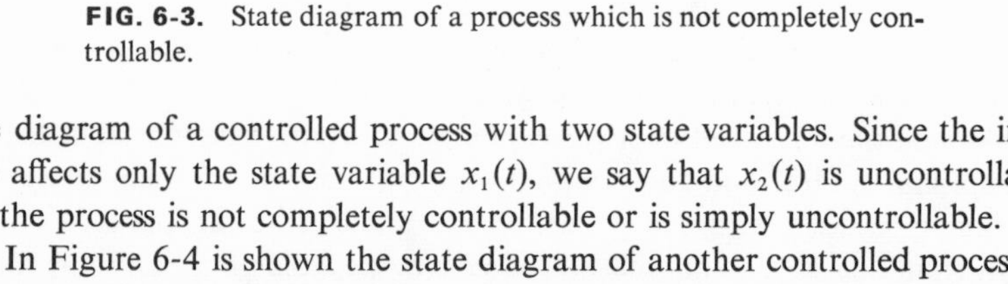

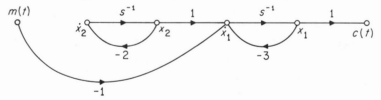

FIG. 6-3. State diagram of a process which is not completely controllable.

state diagram of a controlled process with two state variables. Since the input $m(t)$ affects only the state variable $x_1(t)$, we say that $x_2(t)$ is uncontrollable, and the process is not completely controllable or is simply uncontrollable.

In Figure 6-4 is shown the state diagram of another controlled process. It is observed that the state variable $x_2(t)$ is not connected to the output $y(t)$ in any way. Therefore, if once we have measured $y(t)$, we can observe from it the state variable $x_1(t)$, since $y(t) = x_1(t)$. However, the state variable $x_2(t)$ cannot be observed from the information on $y(t)$. Thus the process is described as not completely observable or simply unobservable.

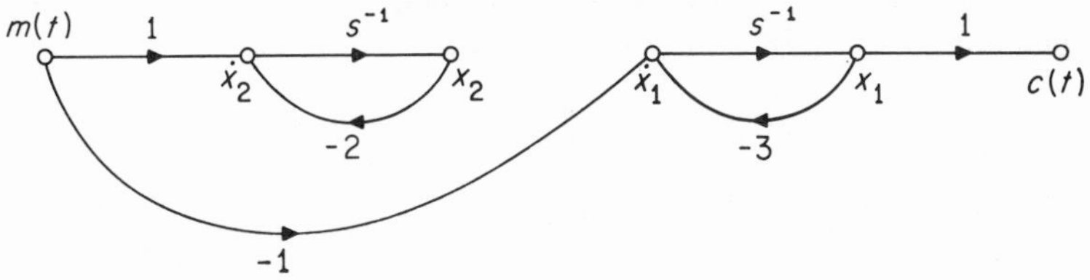

FIG. 6-4. State diagram of a process with is not completely observable.

Definitions of Controllability

Consider that the system shown in Fig. 6-1 is an nth order linear multivariable time-invariant system and is described by the following dynamic equations:
State equation:

$$\frac{d\mathbf{x}(t)}{dt} = A\mathbf{x}(t) + B\mathbf{m}(t) \tag{6-1}$$

Output equation:

$$\mathbf{c}(t) = D\mathbf{x}(t) + E\mathbf{m}(t) \tag{6-2}$$

where

$$\begin{aligned}
\mathbf{x}(t) &= n \times 1 \quad \text{state vector} \\
\mathbf{m}(t) &= r \times 1 \quad \text{input vector} \\
\mathbf{c}(t) &= p \times 1 \quad \text{output vector} \\
A &= n \times n \quad \text{constant matrix} \\
B &= n \times r \quad \text{constant matrix} \\
D &= p \times r \quad \text{constant matrix} \\
E &= p \times n \quad \text{constant matrix}
\end{aligned}$$

The state $x_i(t)$ $(i = 1, 2, \ldots, n)$ of the process G is said to be controllable at $t = t_0$ if there exists an unconstrained piecewise continuous input $\mathbf{m}(t)$ which will drive the state to any final state $\mathbf{x}_i(t_f)$ for the finite time interval $(t_f - t_0) \geq 0$. If every state of the process is controllable in some finite time interval, then the process G is said to be completely controllable or simply controllable.

We shall show that the condition of controllability depends on the matrices A and B of the process.

The solution of Eq. (6-1) for $t \geq t_0$ is

$$\mathbf{x}(t) = \Phi(t - t_0)\mathbf{x}_0(t) + \int_{t_0}^{t} \Phi(t - \tau)B\mathbf{m}(\tau)\, d\tau \tag{6-3}$$

Without losing any generality we can assume that the desired final state for some finite $t_f > t_0$ is $\mathbf{x}(t_f) = \mathbf{0}$. Then, Eq. (6-3) gives

$$\mathbf{x}(t_0) = -\int_{t_0}^{t} \Phi(t_0 - \tau)B\mathbf{m}(\tau)\,d\tau \tag{6-4}$$

The state transition matrix is written as

$$\Phi(t) = e^{At} = \sum_{k=0}^{n-1} \alpha_k(t)A^k \tag{6-5}$$

[1]

where $\alpha_k(t)$ is a scalar function of t. Therefore, Eq. (6-4) becomes

$$\begin{aligned}\mathbf{x}(t_0) &= -\int_{t_0}^{t_f} \sum_{k=0}^{n-1} \alpha_k(t_0 - \tau)A^k B\mathbf{m}(\tau)\,d\tau \\ &= -\sum_{k=0}^{n-1} \int_{t_0}^{t_f} \alpha_k(t_0 - \tau)\mathbf{m}(\tau)\,d\tau \end{aligned} \tag{6-6}$$

The last equation is written

$$\mathbf{x}(t_0) = -\,[B\ AB\ A^2B \ldots A^{n-1}B]\begin{bmatrix} U_0 \\ U_1 \\ \cdot \\ \cdot \\ U_{n-1} \end{bmatrix} = -\,SU \tag{6-7}$$

where

$$U_k = \int_{t_0}^{t_f} \alpha_k(t_0 - \tau)\mathbf{m}(\tau)\,d\tau \qquad (r \times 1 \text{ matrix}) \tag{6-8}$$

$$S = [B\ AB\ A^2B \ldots A^{n-1}B) \qquad (n \times nr) \tag{6-9}$$

$$U = [U_0\, U_1 \ldots U_{n-1}]' \qquad (nr \times 1) \tag{6-10}$$

Equation (6-7) can be interpreted as: Given any initial state $\mathbf{x}(t_0)$, find the control vector $\mathbf{m}(t)$ so that the final state is $\mathbf{x}(t_f) = \mathbf{0}$ for finite $t_f - t_0$. Therefore, a unique solution exists if and only if there is a set of n linearly independent column vectors in the matrix

$$S = [B\ AB\ A^2B \ldots A^{n-1}B] \tag{6-11}$$

In particular, if the process has only one input, $r = 1$, S is an $n \times n$ square matrix. Then Eq. (6-7) represents a set of n linear independent equations if S in Eq. (6-11) is nonsingular, or, the determinant of S is nonzero.

Based on the foregoing discussion, we state that the system of Eq. (6-1) is completely controllable if S in Eq. (6-11) contains n linearly independent column vectors, or, alternately, S has a rank n.

An alternate definition and test of controllability is given below.

First, we will consider that the eigenvalues of A are distinct and are denoted by λ_i, $i = 1, 2, \ldots, n$, then there exists an nth order nonsingular matrix P which transforms A into a diagonal matrix Λ, such that

$$P^{-1}AP = \Lambda = \begin{bmatrix} \lambda_1 & 0 & 0 & \cdots & 0 \\ 0 & \lambda_2 & 0 & \cdots & 0 \\ 0 & 0 & \lambda_3 & \cdots & 0 \\ \cdot & \cdot & \cdot & \cdots & \cdot \\ \cdot & \cdot & \cdot & \cdots & \cdot \\ 0 & 0 & 0 & \cdots & \lambda_n \end{bmatrix} \tag{6-12}$$

The matrix P is formed from the eigenvectors of A; that is, if $X_i (i = 1, 2, \ldots, n)$ is the eigenvector associated with the eigenvalue λ_i, then

$$\mathbf{P} = [\mathbf{X}_1, \mathbf{X}_2, \ldots \mathbf{X}_i \ldots \mathbf{X}_n] \tag{6-13}$$

The eigenvector of the eigenvalue λ_i is a vector X_i which satisfies the matrix equation

$$(\lambda I - A)\mathbf{X}_i = \mathbf{0} \tag{6-14}$$

In determining the eigenvectors, in general, we can always select the first element of any $\mathbf{X}_i$ to be unity. Therefore, it can be shown that if A has distinct eigenvalues, a general form for $\mathbf{P}$ is

$$\mathbf{P} = \begin{bmatrix} 1 & 1 & 1 & \cdots & 1 \\ \lambda_1 & \lambda_2 & \lambda_3 & \cdots & \lambda_n \\ \lambda_1^2 & \lambda_2^2 & \lambda_3^2 & \cdots & \lambda_n^2 \\ \cdot & \cdot & \cdot & \cdots & \cdot \\ \cdot & \cdot & \cdot & \cdots & \cdot \\ \lambda_1^{n-1} & \lambda_2^{n-1} & \lambda_3^{n-1} & \cdots & \lambda_n^{n-1} \end{bmatrix} \tag{6-15}$$

which is the well-known Vandermonde matrix.

Once the matrix P is found, the new state vector may be defined as

$$\mathbf{y}(t) = P^{-1}\mathbf{x}(t) \tag{6-16}$$

Substituting Eq. (6-16) into Eqs. (6-1) and (6-2), the new state equation in vector-matrix form is

$$\dot{\mathbf{y}}(t) = \Lambda\mathbf{y}(t) + \mathbf{\Gamma}\mathbf{m}(t) \tag{6-17}$$

and the new output equation is

$$\mathbf{v}(t) = F\mathbf{y}(t) + E\mathbf{m}(t) \tag{6-18}$$

where

$$\Gamma = P^{-1}B \tag{6-19}$$

$$F = DP \tag{6-20}$$

Equations (6-17) and (6-18) are often referred to as the normalized or canonical-form dynamic equations.

Since Λ is a diagonal matrix, the matrix equation in Eq. (6-17) represents a set of first-order differential equations which are of the form

$$\dot{y}_i(t) = \lambda_i y_i(t) + \sum_{k=1}^{r} \gamma_{ik} m_k(t) \qquad (i = 1, 2, \ldots, n) \tag{6-21}$$

where γ_{ik} denotes the ikth element of the matrix Γ. It is easy to see that if any one row of Γ contains all zeros, the corresponding state will not be controlled by any of the inputs, and the state is uncontrollable. Therefore, an alternative definition of controllability for a system with distinct eigenvalues (or with multiple eigenvalues but with A which can be diagonalized) is given as follows:

The process G is said to be completely controllable if the matrix Γ has no rows which are zero.

It is apparent that this alternate definition of controllability is not as simple to apply as the condition given by Eq. (6-11), since it involves the diagonalization of A and the determination of the matrix P.

The condition on controllability by examining the rows of $\Gamma = P^{-1}B$ applies only to a system with matrix A which can be diagonalized. However, when A has multiple-order eigenvalues, sometimes it cannot be diagonalized. Under this condition, we can transform A into the Jordan canonical form. For instance, if A has four eigenvalues, $\lambda_1, \lambda_1, \lambda_1, \lambda_2$, three of which are equal, then there is a nonsingular 4×4 matrix P which transforms A into the Jordan canonical form

$$\Lambda = P^{-1}AP = \left[\begin{array}{ccc:c} \lambda_1 & 1 & 0 & 0 \\ 0 & \lambda_1 & 1 & 0 \\ 0 & 0 & \lambda_1 & 0 \\ \hdashline 0 & 0 & 0 & \lambda_2 \end{array}\right] \tag{6-22}$$

Then the dynamic equations in the Jordan canonical form are

$$\text{State equation:} \qquad \dot{\mathbf{y}}(t) = \Lambda\mathbf{y}(t) + \Gamma\mathbf{m}(t) \tag{6-23}$$

$$\text{Output equation:} \qquad \mathbf{v}(t) = F\mathbf{y}(t) + E\mathbf{m}(t) \tag{6-24}$$

where

$$\Gamma = P^{-1}B \tag{6-25}$$

$$F = DP \tag{6-26}$$

The matrix Λ in Eq. (6-22) is shown to be partitioned into a 2×2 diagonal matrix. The submatrices on the main diagonal of the partitioned matrix are called the *Jordan blocks*. Then the condition of controllability of the system of Eq. (6-23) is that *all the elements of Γ that correspond to the last row of each Jordan block are nonzero*. The reason for this condition is that the last row of each Jordan block will always correspond to a state equation which is similar to Eq. (6-21). The elements in the other rows of Γ need not all be nonzero, because the corresponding state variables are not uncoupled.

EXAMPLE 6-1

Consider the state diagram shown in Figure 6-3. It was reasoned earlier that the system is uncontrollable. Let us try to arrive at the same answer using the methods discussed above.

The state equation of the system is written

$$\begin{bmatrix} \dfrac{dx_1(t)}{dt} \\ \dfrac{dx_2(t)}{dt} \end{bmatrix} = \begin{bmatrix} -3 & 1 \\ 0 & -2 \end{bmatrix} \begin{bmatrix} x_1(t) \\ x_2(t) \end{bmatrix} + \begin{bmatrix} 1 \\ 0 \end{bmatrix} m(t) \tag{6-27}$$

Therefore, from Eq. (6-11),

$$S = [B\ AB] = \begin{bmatrix} 1 & -3 \\ 0 & 0 \end{bmatrix} \tag{6-28}$$

which is singular, and the system is uncontrollable.

The eigenvalues of A are $\lambda_1 = -3$ and $\lambda_2 = -2$. Substituting $\lambda_1 = -3$ into Eq. (6-14), we have

$$\begin{bmatrix} \lambda_1 + 3 & -1 \\ 0 & \lambda_1 + 2 \end{bmatrix} \begin{bmatrix} x_1 \\ x_2 \end{bmatrix} = \begin{bmatrix} 0 & -1 \\ 0 & -1 \end{bmatrix} \begin{bmatrix} x_1 \\ x_2 \end{bmatrix} = \mathbf{0} \tag{6-29}$$

which corresponds to $x_2 = 0$. Since x_1 is not specified by Eq. (6-29), we can choose x_1 to be unity, and thus the eigenvector associated with $\lambda_1 = -3$ is

$$\mathbf{x}_1 = \begin{bmatrix} 1 \\ 0 \end{bmatrix} \tag{6-30}$$

For $\lambda = \lambda_2 = -2$, Eq. (6-14) gives

$$\begin{bmatrix} \lambda_2 + 3 & -1 \\ 0 & \lambda_2 + 2 \end{bmatrix} \begin{bmatrix} x_1 \\ x_2 \end{bmatrix} = \begin{bmatrix} +1 & -1 \\ 0 & 0 \end{bmatrix} \begin{bmatrix} x_1 \\ x_2 \end{bmatrix} = \mathbf{0} \tag{6-31}$$

which corresponds to

$$x_1 - x_2 = 0 \tag{6-32}$$

or $x_1 = x_2$. For simplicity, we can choose $x_1 = x_2 = 1$. Therefore, the eigenvector associated with $\lambda_2 = -2$ is

$$\mathbf{x}_2 = \begin{bmatrix} 1 \\ 1 \end{bmatrix} \tag{6-33}$$

Now using Eq. (6-13) we obtain the matrix P as

$$P = [\mathbf{x}_1\ \ \mathbf{x}_2] = \begin{bmatrix} 1 & 1 \\ 0 & 1 \end{bmatrix} \tag{6-34}$$

From Eq. (6-25), we have

$$\Gamma = P^{-1}B = \begin{bmatrix} 1 & 1 \\ 0 & 1 \end{bmatrix}^{-1} \begin{bmatrix} 1 \\ 0 \end{bmatrix} = \begin{bmatrix} 1 \\ 0 \end{bmatrix} \tag{6-35}$$

Since the second row of $\mathbf{\Gamma}$ contains a zero, the process is uncontrollable, the same result as obtained earlier.

Definitions of Observability

With reference to the system described by Eqs. (6-1), and (6-2), *the state $x_i(t)$ $(i = 1, 2, \ldots, n)$ is said to be observable if given any input $\mathbf{m}(t)$ for $t_0 \leq t < t_f$, the matrices A, B, D, and E and the output $\mathbf{c}(t)$ for $t_0 \leq t < t_f$ is sufficient to determine $x_i(t_0)$. If every state of the system is observable for some finite t_f, then we say that the system is completely observable, or simply observable.*

We shall show that the condition of observability depends on the matrices A and D of the process.

Substituting Eq. (6-3) into Eq. (6-2), we get

$$\mathbf{c}(t) = D\Phi(t - t_0)\mathbf{x}(t_0) + D\int_{t_0}^{t} \Phi(t - \tau)B\mathbf{m}(\tau)\,d\tau + E\mathbf{m}(t) \qquad (6\text{-}36)$$

Based on the definition of observability, it is apparent that the observability of the state variable $\mathbf{x}(t_0)$ depends essentially on the first term on the right side of Eq. (6-36).

Therefore, when $\mathbf{m}(t) = \mathbf{0}$ the output of the system is

$$\mathbf{c}(t) = D\Phi(t - t_0)\mathbf{x}(t_0) \qquad (6\text{-}37)$$

and, using Eq. (6-6), we have

$$\mathbf{c}(t) = \sum_{k=0}^{n-1} \alpha_k(t - t_0)DA^k\mathbf{x}(t_0) \qquad (6\text{-}38)$$

Based on the last expression we can state that knowing the output $\mathbf{c}(t)$ over the interval $t_0 \leq t < t_f$, $\mathbf{x}(t_0)$ is uniquely determined if and only if $\mathbf{x}(t_0)$ is a linear combination of the vector $(D_jA^k)'$, $k = 0, 1, 2, \ldots, n - 1$, $j = 1, 2, \ldots, P$, where D_j denotes the $1 \times n$ matrix formed from elements in the jth row of D. Since $(D_jA^k)' = (A')^k D'_j$, if we let $\mathbf{T}$ be the $n \times np$ matrix defined by

$$\mathbf{T} = [D' \quad A'D' \quad (A')^2D' \quad \ldots \quad (A')^{n-1}D'] \qquad (6\text{-}39)$$

then the system is completely observable if and only if there is a set of n linearly independent column vectors in $\mathbf{T}$. In particular, if the system has only one output, D is a $1 \times n$ matrix, $\mathbf{T}$ is an $n \times n$ square matrix. Then the system is completely observable if $\mathbf{T}$ is nonsingular.

An alternative definition of observability is stated as: if A is diagonalized as in Eq. (6-12) with distinct entries, then the process is completely observable if the matrix F in Eq. (6-26) has no column containing all zeros. The reason behind this is that if the jth column of F contains all zeros $(j = 1, 2, \ldots, n)$ the state variable y_j will not appear in Eq. (6-18) and is not related to the output $\mathbf{v}(t)$. The states that correspond to zero columns of F are said to be unobservable and those corresponding to nonzero columns are called observable.

EXAMPLE 6-2

Consider the state diagram shown in Figure 6-4. It was reasoned earlier that the process is unobservable. The dynamic equations of the process are

$$\begin{bmatrix} \dfrac{dx_1(t)}{dt} \\ \dfrac{dx_2(t)}{dt} \end{bmatrix} = \begin{bmatrix} -3 & 0 \\ 0 & -2 \end{bmatrix} \begin{bmatrix} x_1(t) \\ x_2(t) \end{bmatrix} + \begin{bmatrix} 1 \\ 1 \end{bmatrix} m(t) \tag{6-40}$$

$$c(t) = [1 \quad 0] \begin{bmatrix} x_1(t) \\ x_2(t) \end{bmatrix}$$

Therefore,

$$D = [1 \quad 0] \qquad D' = \begin{bmatrix} 1 \\ 0 \end{bmatrix}$$

$$A'D' = \begin{bmatrix} -3 & 0 \\ 0 & -2 \end{bmatrix} \begin{bmatrix} 1 \\ 0 \end{bmatrix} = \begin{bmatrix} -3 \\ 0 \end{bmatrix} \tag{6-41}$$

and from Eq. (6-39),

$$\mathbf{T} = [D' \quad A'D'] = \begin{bmatrix} 1 & -3 \\ 0 & 0 \end{bmatrix} \tag{6-42}$$

Since $\mathbf{T}$ is singular, the system is unobservable. Now let us apply the alternative test of observability to the system of Figure 6-4. Since A is already a diagonal matrix, the condition of observability requires that the matrix D must not contain zero columns. However, in this case the second column of D is zero. Therefore $x_2(t)$ is not observable.

Relationship Between Controllability, Observability, and Transfer Functions

The concepts of controllability and observability can be related to transfer functions, which are often used in linear systems.

Let us focus our attention to the system in Figure 6-3. The transfer function of the system is

$$\frac{C(s)}{M(s)} = \frac{s+2}{s^2+5s+6} = \frac{s+2}{(s+3)(s+2)} \tag{6-43}$$

Since this transfer function has a common pole and zero at $s = -2$, it is simplified to

$$\frac{C(s)}{M(s)} = \frac{1}{s+3} \tag{6-44}$$

after the pole-zero cancellation. We shall prove in the following that a system is either uncontrollable or unobservable when its transfer function has identical poles and zeros.

Consider that an nth order system with a single input and single output and distinct characteristic roots is described by the state equation

$$\frac{dx(t)}{dt} = A\mathbf{x}(t) + Bm(t) \tag{6-45}$$

The output equation is

$$c(t) = D\mathbf{x}(t) \tag{6-46}$$

Let the A matrix be diagonalized by an $n \times n$ Vandermonde matrix P of Eq. (6-15), so that the state equation in canonical form is

$$\dot{\mathbf{z}}(t) + \Lambda\mathbf{z}(t) + \Gamma m(t) \tag{6-47}$$

where $\Lambda = P^{-1}AP$. The output equation is transformed into

$$c(t) = F\mathbf{z}(t) \tag{6-48}$$

where $F = DP$. The state variables $\mathbf{x}(t)$ and $\mathbf{z}(t)$ are related by

$$\mathbf{x}(t) = P\mathbf{z}(t) \tag{6-49}$$

Since Λ is a diagonal matrix, the ith equation of Eq. (6-47) is

$$\dot{z}_i(t) = \lambda_i z_i(t) + \gamma_i m(t) \tag{6-50}$$

where λ_i is the ith characteristic root of A, and γ_i is the ith element of Γ which is $n \times 1$ in this case. Taking the Laplace transform on both sides of Eq. (6-50) and assuming zero initial conditions, we obtain the transfer function relation between $Z_i(s)$ and $M(s)$

$$Z_i(s) = \frac{\gamma_i}{s - \lambda_i} M(s) \tag{6-51}$$

The transformed version of the output equation in Eq. (6-48) is

$$C(s) = F\mathbf{Z}(s) = DP\mathbf{Z}(s) \tag{6-52}$$

Now if we assume that

$$D = [d_1 \quad d_2 \quad \dots \quad d_n] \tag{6-53}$$

then

$$F = DP = [f_1 \quad f_2 \quad \dots \quad f_n] \tag{6-54}$$

where

$$f_i = d_1 + d_2\lambda_i + \dots + d_n\lambda_i^{n-1} \tag{6-55}$$

for $i = 1, 2, \dots, n$. Therefore, Eq. (6-52) can be written as

$$C(s) = [f_1 \quad f_2 \ldots f_n]\mathbf{Z}(s)$$

$$= [f_1 \quad f_2 \quad \ldots \quad f_n]\begin{bmatrix} \dfrac{\gamma_1}{s-\lambda_1} \\ \dfrac{\gamma_2}{s-\lambda_2} \\ \vdots \\ \dfrac{\gamma_n}{s-\lambda_n} \end{bmatrix} M(s) \tag{6-56}$$

For an nth order system, the input-output transfer function can be written as

$$\frac{C(s)}{M(s)} = \frac{K(s-a_1)(s-a_2)\ldots(s-a_m)}{(s-\lambda_1)(s-\lambda_2)\ldots(s-\lambda_n)} \quad (n > m) \tag{6-57}$$

For distinct characteristic roots, Eq. (6-57) is expanded by partial fraction expansion into

$$\frac{C(s)}{M(s)} = \sum_{i=1}^{n} \frac{\sigma_i}{s-\lambda_i} \tag{6-58}$$

where σ_i denotes the residue of $C(s)/M(s)$ at $s = \lambda_i$.

It was established earlier that for the system described by Eq. (6-47) to be controllable, all the rows of Γ must be nonzero; i.e., $\gamma_i \neq 0$ for $i = 1, 2, \ldots, n$. If $C(s)/M(s)$ has at least one pair of identical pole and zero, for instance in Eq. (6-57), $a_1 = \lambda_1$, then in Eq. (6-58), $\sigma_1 = 0$. Now comparing Eq. (6-56) with Eq. (6-58), we see that in general

$$\sigma = f_i\gamma_i \tag{6-59}$$

Therefore, when $\sigma_i = 0$, γ_i will be zero if $f_i \neq 0$, and the state $z_i(t)$ is uncontrollable.

For observability, it was established earlier that F must not have columns containing zeros. Or, in the present case, $f_i \neq 0$ for $i = 1, 2, \ldots, n$. However, from Eq. (6-59),

$$f_i = \frac{\sigma_i}{\gamma_i} \tag{6-60}$$

when the transfer function has an identical pair of pole and zero at $a_i = \lambda_i$, $\sigma_i = 0$. Therefore, $f_i = 0$ if $\gamma_i \neq 0$.

The discussions given above lead to the following conclusion on the relationship between the transfer function and controllability and observability of a linear system.

If the input-output function of a system has pole-zero cancellation, the system will either be uncontrollable or unobservable, depending on how the state variables are defined.

If the input-output transfer function of a system does not have pole-zero cancellation, the system can always be represented by dynamic equations as a completely controllable and observable system.

6.3 CONTROLLABILITY AND OBSERVABILITY OF DIGITAL SYSTEMS

Controllability of Digital Systems

The concept of controllability of a digital system is similar to that of a continuous-data system except that the state equations are difference equations.

Consider that a digital system is described by the state equation

$$\mathbf{x}[(k+1)T] = \Phi(T)\mathbf{x}(kT) + \Theta(T)\mathbf{m}(kT) \tag{6-61}$$

where $\mathbf{m}(kT)$ is an $r \times 1$ input vector, and $m_i(kT)$, $i = 1, 2, \ldots, r$ is a constant for $kT \leq t < (k+1)T$. $\Phi(T)$ is an $n \times n$ coefficient matrix, and $\Theta(T)$ is an $n \times r$ input matrix.

The condition of controllability of a digital system is

Given $\Phi(T)$ *and* $\Theta(T)$, *there exists a set of unconstrained input vectors* $\mathbf{m}(kT)$, $k = 0, 1, \ldots, (N-1)$, *which will bring the states from* $\mathbf{x}(0)$ *to* $\mathbf{x}(NT)$ *in N sampling periods, where N is a finite positive integer.*

The solution of Eq. (6-61) is

$$\mathbf{x}(kT) = \Phi(kT)\mathbf{x}(0) + \sum_{i=0}^{k-1} \Phi[(k-i-1)T]\Theta(T)\mathbf{m}(iT) \tag{6-62}$$

Assuming that the state of the process is to reach the equilibrium state $\mathbf{x}(NT) = \mathbf{0}$ in N sampling periods, and solving for $\mathbf{x}(0)$ in Eq. (6-62), we get

$$\mathbf{x}(0) = -\Phi(-NT)\sum_{i=0}^{N-1} \Phi[(N-i-1)T]\Theta(T)\mathbf{m}(iT) \tag{6-63}$$

or

$$\mathbf{x}(0) = -\sum_{i=0}^{N-1} \Phi[(-i-1)T]\Theta(T)\mathbf{m}(iT) \tag{6-64}$$

If we set

$$S_i(T) = \Phi[(-i-1)T]\Theta(T) \tag{6-65}$$

Eq. (6-64) becomes

$$\mathbf{x}(0) = -\sum_{i=0}^{N-1} S_i(T)\mathbf{m}(iT) \tag{6-66}$$

which represents a set of n simultaneous equations. Given any arbitrary initial state $\mathbf{x}(0)$, the solution for the control vector $\mathbf{m}(iT)$, $i = 0, 1, 2, \ldots, (N-1)$

to bring $\mathbf{x}(0)$ to $\mathbf{x}(NT)$ exists only if the n equations represented by Eq. (6-66) are all linearly independent. In other words, the necessary and sufficient condition for $\mathbf{x}(0)$ to $\mathbf{x}(NT)$ for a finite N is that all the vectors

$$\begin{aligned} S_0(T) &= \Phi(-T)\Theta(T) \\ S_1(T) &= \Phi(-2T)\Theta(T) \\ &\vdots \\ S_{N-1}(T) &= \Phi(-NT)\Theta(T) \end{aligned} \tag{6-67}$$

are linearly independent.

Observability of Digital Systems[6,14]

Observability was defined earlier as a condition of determining the state variables from measurements of the outputs. This is an important step in the implementation of an optimal control system since the optimal control signal $\mathbf{m}(kT)$ usually depends on the state variables.

The observability of a digital system is defined with respect to the dynamic equations

$$\mathbf{x}[(k+1)T] = \Phi(T)\mathbf{x}(kT) + \Theta(T)\mathbf{m}(kT) \tag{6-68}$$

$$\mathbf{c}(kT) = D\mathbf{x}(kT) + E\mathbf{m}(kT) \tag{6-69}$$

A practical way of stating observability for the digital system is that $\mathbf{x}(kT)$ can be determined from information on $\Phi(T)$, $\Theta(T)$, D, and E, and the past inputs $\mathbf{m}[(k-j)T]$ and outputs $\mathbf{c}[(k-j)T]$, $j = 1, 2, \ldots, n$; $k \geq n$.

From Eq. (6-68), we have

$$\mathbf{x}(kT) = \Phi(T)\mathbf{x}[(k-1)T] + \Theta(T)\mathbf{m}[(k-1)T] \quad (k \geq 1) \tag{6-70}$$

Solving for $\mathbf{x}[(k-1)T]$ from the last equation gives

$$\mathbf{x}[(k-1)T] = \Phi(-T)\mathbf{x}(kT) - \Phi(-T)\Theta(T)\mathbf{m}[(k-1)T] \quad (k \geq 1) \tag{6-71}$$

In general, Eq. (6-71) leads to

$$\mathbf{x}[(k-n)T] = \Phi(-T)\mathbf{x}[(k-n+1)T] - \Phi(-T)\Theta(T)\mathbf{m}[(k-n)T] \tag{6-72}$$

where n is a positive integer and $k \geq n$.

From Eq. (6-69), we have

$$\mathbf{c}[(k-1)T] = D\mathbf{x}[(k-1)T] + E\mathbf{m}[(k-1)T] \tag{6-73}$$

Substitution of Eq. (6-71) into the last equation yields

$$\begin{aligned} \mathbf{c}[(k-1)T] = {} & D\Phi(-T)\mathbf{x}(kT) \\ & - [D\Phi(-T)\Theta(T) - E]\mathbf{m}[(k-1)T] \quad (k \geq 1) \end{aligned} \tag{6-74}$$

Similarly,

$$\mathbf{c}[(k-2)T] = D\Phi(-T)\mathbf{x}[(k-1)T] - [D\Phi(-T)\Theta(T) - E]\mathbf{m}[(k-1)T] \quad (k \geq 2) \tag{6-75}$$

Now substituting Eq. (6-71) into the last equation for $\mathbf{x}[(k-1)T]$, we get

$$\mathbf{c}[(k-2)T] = D\Phi(-2T)\mathbf{x}(kT) - D\Phi(-2T)\Theta(T)\mathbf{m}[(k-1)T] - [D\Phi(-T)\Theta(T) - E]\mathbf{m}[(k-2)T] \quad (k \geq 2) \tag{6-76}$$

Continuing the process described above, we have

$$\mathbf{c}[(k-3)T] = D\Phi(-3T)\mathbf{x}(kT) - D\Phi(-3T)\Theta(T)\mathbf{m}[(k-1)T] - D\Phi(-2T)\Theta(T)\mathbf{m}[(k-2)T] - [D\Phi(-T)\Theta(T) - E]\mathbf{m}[(k-3)T] \quad (k \geq 3) \tag{6-77}$$

In general,

$$\mathbf{c}[(k-n)T] = D\Phi(-nT)\mathbf{x}(kT) - \sum_{i=1}^{n} D\Phi[(-n+i-1)T]\Theta(T)\mathbf{m}[(k-i)T] + E\mathbf{m}[(k-\mathrm{n})T] \quad (k \geq n) \tag{6-78}$$

Rearranging and writing in matrix form, Eqs. (6-74) through (6-78) become

$$\begin{bmatrix} D\Phi(-T) \\ D\Phi(-2T) \\ \vdots \\ D\Phi(-nT) \end{bmatrix} \mathbf{x}(kT) = \begin{bmatrix} \mathbf{c}[(k-1)T] \\ \mathbf{c}[(k-2)T] \\ \vdots \\ \mathbf{c}[(k-n)T] \end{bmatrix}$$
$$+ \begin{bmatrix} D\Phi(-T)\Theta(T) - E & 0 & 0 & \cdots & 0 \\ D\Phi(-2T)\Theta(T) & D\Phi(-T)\Theta(T) - E & 0 & \cdots & 0 \\ \vdots & \vdots & \vdots & \cdots & \vdots \\ D\Phi(-nT)\Theta(T) & D\Phi[(-n+1)T]\Theta(T) & \cdots & & D\Phi(-T)\Theta(T) - E \end{bmatrix}$$
$$\times \begin{bmatrix} \mathbf{m}[(k-1)T] \\ \mathbf{m}[(k-2)T] \\ \vdots \\ \mathbf{m}[(k-n)T] \end{bmatrix} \tag{6-79}$$

If the inverse of the matrix

$$[D\Phi(-T) \quad D\Phi(-2T) \quad \ldots \quad D\Phi(-nT)] \tag{6-80}$$

exists, Eq. (6-79) is written

$$\mathbf{x}(kT) = \begin{bmatrix} D\Phi(-T) \\ D\Phi(-2T) \\ \cdot \\ \cdot \\ \cdot \\ D\Phi(-nT) \end{bmatrix}^{-1} \left\{ \begin{bmatrix} \mathbf{c}[(k-1)T] \\ \mathbf{c}[(k-2)T] \\ \cdot \\ \cdot \\ \cdot \\ \mathbf{c}[(k-n)T] \end{bmatrix} \right.$$

$$+ \begin{bmatrix} D\Phi(-T)\Theta(T) - E & 0 & 0 & \cdots & 0 \\ D\Phi(-2T)\Theta(T) & D\Phi(-T)\Theta(T) - E & 0 & \cdots & 0 \\ \cdot & \cdot & \cdot & \cdots & \cdot \\ \cdot & \cdot & \cdot & & \cdot \\ \cdot & \cdot & \cdot & & \cdot \\ D\Phi(-nT)\Theta(T) & D\Phi[(-n+1)T]\Theta(T) & \cdots & & D\Phi(-T)\Theta(T) - E \end{bmatrix}$$

$$\left. \times \begin{bmatrix} \mathbf{m}[(k-1)T] \\ \mathbf{m}[(k-2)T] \\ \cdot \\ \cdot \\ \cdot \\ \mathbf{m}[(k-n)T] \end{bmatrix} \right\} \quad (k \geq n) \tag{6-81}$$

Equation (6-81) implies that the state variables at any sampling instant kT can be determined as a linear combination of n past output measurements and n past input measurements, taken at the sampling instants. Therefore, the condition of observability of the system is that the inverse of the matrix in Eq. (6-80) exists.

It should be pointed out, however, that the method described above would allow the computation of $\mathbf{x}(kT)$ only for $k \geq n$, where n is the dimension of $\mathbf{x}(kT)$. For $k < n$, measurements of the output vector and the input vector for negative times would be required. For instance, if $n = 2$, that is, a second-order system, Eq. (6-81) gives only the values of $\mathbf{x}(kT)$ for $k = 2, 3, \ldots$, once the input $\mathbf{m}$ and the output $\mathbf{c}$ at $t = 0$ and $t = T$ are known. The equation does not provide an answer to $\mathbf{x}(T)$ from $\mathbf{c}(0)$, $\mathbf{c}(T)$ $\mathbf{m}(0)$, and $\mathbf{m}(T)$. Truly, $\mathbf{x}(T)$ must be computed from $\mathbf{c}(-T)$, $\mathbf{c}(0)$, $\mathbf{m}(-T)$, and $\mathbf{m}(0)$, using Eq. (6-81). However, in practice there may be no way of measuring $\mathbf{c}(-T)$ and $\mathbf{m}(-T)$.

6.4 CONTROLLABILITY AND OBSERVABILITY VERSUS SAMPLING PERIOD[5]

Investigation of Eqs. (6-67) and (6-80) reveals that when

$$\Phi(-T) = \Phi(-2T) = \ldots = \Phi(-nT) \tag{6-82}$$

the digital system will be neither controllable nor observable. It is interesting to discover that the condition of Eq. (6-82) depends on the sampling period

T. Let us consider that the transfer function of a linear process is described by

$$G(s) = \frac{\omega}{s^2 + \omega^2} \tag{6-83}$$

Ordinarily, in a continuous-data system, $G(s)$ represents a completely controllable and observable system, if it did not have common poles and zeros. With sampled data, the control signal to the process is $m(t) = m(kT)$ for $kT \leq t < (k + 1)T$, where T is the sampling period. The state transition matrix of $G(s)$ is

$$\Phi(t) = \begin{bmatrix} \cos \omega t & \frac{1}{\omega} \sin \omega t \\ -\omega \sin \omega t & \cos \omega t \end{bmatrix} \tag{6-84}$$

Therefore, when $t = T = 2\pi n/\omega$, $\Phi(T) = \Phi(2T) = \ldots = \Phi(nT)$ for $n =$ positive integer, and the digital system is uncontrollable and unobservable.

In terms of the transfer function approach, the z-transform of $G(s)$ is

$$G(z) = \frac{z \sin \omega T}{z^2 - 2z \cos \omega T + 1} \tag{6-85}$$

When $T = 2n\pi/\omega$,

$$G(z) = 0 \tag{6-86}$$

which indicates that the sampled output of the process is zero, but the true output of the system is not.

6.5 TIME-OPTIMAL CONTROL OF SINGLE-VARIABLE REGULATOR SYSTEMS[7,8]

In this section we shall discuss the time-optimal control of digital systems with single input and output. Our discussion will be centered on the regulator-type systems for which the reference input is zero.

The block diagram of a typical system considered here is shown in Figure 6-5. In general, of course, the digital controller may be located elsewhere in the system. The saturation nonlinearity shown on the diagram is used to represent the amplitude constraint on the control signal $m(t)$. The time-optimal design problem is stated as:

> Given the linear process $G(s)$ and an arbitrary initial state $\mathbf{x}(0)$, find an optimum control vector $\mathbf{m}(kT)$, $k = 0, 1, 2, \ldots, (N - 1)$, under the constraint $|m(kT)| \leq M$, which will bring the process to the equilibrium state $\mathbf{x}(NT)$ in minimum time $t = NT$, where N is a positive integer.

The state equation of the linear process is given as

$$\dot{\mathbf{x}}(t) = A\mathbf{x}(t) + Bm(t) \tag{6-87}$$

where $\mathbf{x}(t)$ is an $n \times 1$ state vector, A is an $n \times n$ constant matrix, B is an $n \times 1$ constant matrix, and $m(t)$ is a scalar control input.

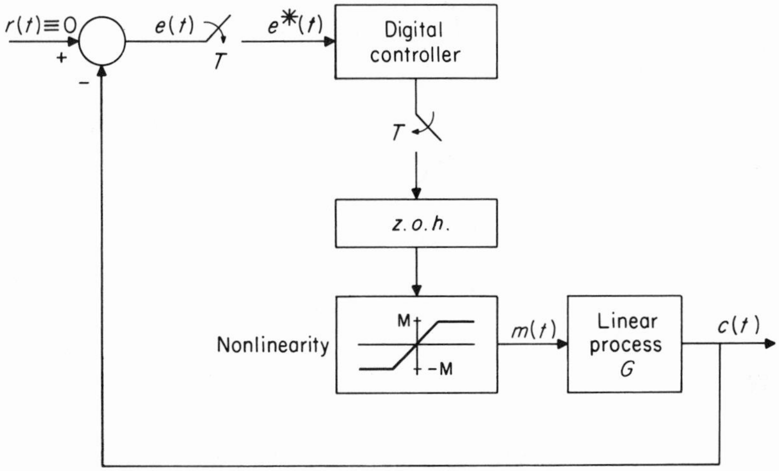

FIG. 6-5. Block diagram of a single-variable digital control system.

The discrete state transition equation of the system is

$$\mathbf{x}[(k+1)T] = \mathbf{\Phi}(T)\mathbf{x}(kT) + \mathbf{\Theta}(T)m(kT) \tag{6-88}$$

where

$$\mathbf{\Phi}(T) = e^{AT} \tag{6-89}$$

and

$$\mathbf{\Theta}(T) = \int_0^T \mathbf{\Phi}(T-\tau)B\,d\tau \tag{6-90}$$

The solution of Eq. (6-88) is

$$\mathbf{x}(kT) = \mathbf{\Phi}(kT)\mathbf{x}(0) + \sum_{i=0}^{k-1} \mathbf{\Phi}[(k-i-1)T]\mathbf{\Theta}(T)m(iT) \tag{6-91}$$

For $k = N$, we let the equilibrium state be $\mathbf{x}(NT) = \mathbf{0}$, and Eq. (6-91) becomes

$$\mathbf{x}(NT) = \mathbf{0} = \mathbf{\Phi}(NT)\mathbf{x}(0) + \sum_{i=0}^{N-1} \mathbf{\Phi}[(N-i-1)T]\mathbf{\Theta}(T)m(iT) \tag{6-92}$$

Solving for $\mathbf{x}(0)$ from the last equation, we get

$$\mathbf{x}(0) = -\sum_{i=0}^{N-1} \mathbf{\Phi}[(i-1)T]\mathbf{\Theta}(T)m(iT) \tag{6-93}$$

where $|m(iT)| \leq M$.

Equation (6-93) can be written as

$$\mathbf{x}(0) = -[S_0(T) \quad S_1(T) \quad \ldots \quad S_{N-1}(T)] \begin{bmatrix} m(0) \\ m(T) \\ \cdot \\ \cdot \\ \cdot \\ m[(N-1)T] \end{bmatrix} \tag{6-94}$$

or

$$\mathbf{x}(0) = -S \begin{bmatrix} m(0) \\ m(T) \\ \cdot \\ \cdot \\ \cdot \\ m[(N-1)T] \end{bmatrix} \tag{6-95}$$

where

$$S = [S_0(T) \quad S_1(T) \quad \ldots \quad S_{N-1}(T)] \tag{6-96}$$

and

$$S_i(T) = \Phi[(-i-1)T]\Theta(T) \qquad i = 0, 1, 2, \ldots, N-1 \tag{6-97}$$

Solving for the control signals in Eq. (6-95), we get

$$\begin{bmatrix} m(0) \\ m(T) \\ \cdot \\ \cdot \\ \cdot \\ m[(N-1)T] \end{bmatrix} = -S^{-1}\mathbf{x}(0) \tag{6-98}$$

if the inverse of the matrix S exists. If the magnitude of $m(kT)$ is unconstrained, then any arbitrary initial state $\mathbf{x}(0)$ can be brought to the equilibrium state $\mathbf{x}(NT) = \mathbf{0}$ in $N = n$ sampling period or less. The value of N may increase if amplitude constraint is imposed on $m(kT)$. Earlier we have defined that if for any unconstrained control signal the process can be brought from an arbitrary initial state $\mathbf{x}(0)$ to $\mathbf{x}(NT) = \mathbf{0}$ in N finite sampling periods, the process is said to be controllable.

Region of Controllable States

When the control signal is subject to amplitude constraints, not all the initial states can be brought to equilibrium in a specific number of sampling periods. We shall now investigate the points in the state space which can be brought to equilibrium under amplitude constraint on $m(kT)$. Because of the difficulty in portraying high-order state space, we will consider only second-order systems in this discussion. Also, for simplicity, the constraint on the control input is $|m(kT)| \leq M = 1$.

Let R'_N represent the set of initial states that can be brought to equilibrium in N sampling periods *or less*. Let R_N be the initial states that can be brought to equilibrium in N sampling periods and *no less*. It is apparent that

$$R_N = R'_N - R'_{N-1} \tag{6-99}$$

since R'_{N-1} denotes the set of initial states that can be brought to equilibrium in $N - 1$ sampling periods or less.

We shall show in the following that R'_N forms a convex polygon in the state plane $x_1 - x_2$.

Theorem 6-1

The region R'_N is convex. This implies that if two initial states represented by the points P_k and P_{k+1} can be brought to equilibrium in N sampling periods or less, the same is true for any initial state that lies on the line segment $P_k P_{k+1}$.

Proof. With reference to Figure 6-6, let OP_k represent the vector $\mathbf{x}(0)$. Since P_k lies in R'_N, $\mathbf{x}(0)$ must satisfy Eq. (6-93); that is,

$$\mathbf{x}(0) = -\sum_{i=0}^{N-1} S_i(T)m(iT) \tag{6-100}$$

where $S_i(T)$ is given by Eq. (6-97), and $m(iT)$ is the optimal control required to bring $\mathbf{x}(0)$ to $\mathbf{x}(NT) = \mathbf{0}$ in N sampling periods or less, and $|m(iT)| \leq 1$ for $i = 0, 1, 2, \ldots, N - 1$. Similarly, we let $\mathbf{x}'(0)$ be represented by OP_{k+1}. Since P_{k+1} also lies in R'_N,

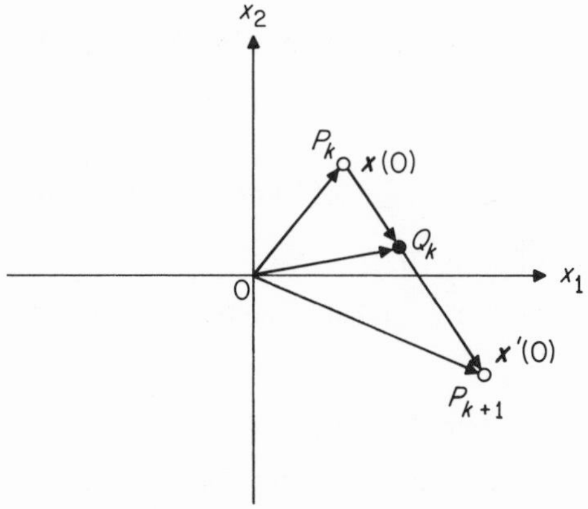

FIG. 6-6. Proof that the region R'_N is convex.

$$\mathbf{x}'(0) = -\sum_{i=1}^{N} S_i(T)m'(iT) \tag{6-101}$$

where $m'(iT)$ denotes the optimal control required to bring $\mathbf{x}'(0)$ to $\mathbf{x}(NT) = \mathbf{0}$ in N sampling periods or less, and $|m'(iT)| \leq 1$ for $i = 0, 1, 2, \ldots, N-1$. Let us write

$$m'(iT) = m(iT) + q(iT) \tag{6-102}$$

and substitute it into Eq. (6-101). We have

$$\begin{aligned}\mathbf{x}'(0) &= -\sum_{i=0}^{N-1} S_i(T)m(iT) - \sum_{i=0}^{N-1} S_i(T)q(iT) \\ &= \mathbf{x}(0) - \sum_{i=0}^{N-1} S_i(T)q(iT)\end{aligned} \tag{6-103}$$

From Figure 6-6 we have

$$\overline{OP_k} - \overline{OP_{k+1}} = \overline{P_k P_{k+1}} \tag{6-104}$$

Therefore,

$$\mathbf{x}(0) - \mathbf{x}'(0) = \overline{P_k P_{k+1}} \tag{6-105}$$

and comparing with Eq. (6-103) gives

$$\overline{P_k P_{k+1}} = -\sum_{i=0}^{N-1} S_i(T)q(iT) \tag{6-106}$$

Let us designate an arbitrary point on the line segment $\overline{P_k P_{k+1}}$ as Q_k, and let

$$\overline{OQ_k} = \mathbf{x}''(0) \tag{6-107}$$

Then Figure 6-6 shows that

$$\overline{OQ_k} = \overline{OP_k} + \overline{P_k Q_k} \tag{6-108}$$

Since $\overline{OP_k} = \mathbf{x}(0)$ and $\overline{P_k Q_k} = \delta \overline{P_k P_{k+1}}$, where $0 \leq \delta \leq 1$, (6-108) is written

$$\overline{OQ_k} = \mathbf{x}''(0) = \mathbf{x}(0) - \delta \sum_{i=1}^{N-1} S_i(T)q(iT) \tag{6-109}$$

or

$$\overline{OQ_k} = -\sum_{i=0}^{N-1} S_i(T)[m(iT) + \delta q(iT)] \tag{6-110}$$

However, since from Eq. (6-102),

$$|m(iT) + q(iT)| = |m'(iT)| \leq M = 1 \tag{6-111}$$

it is simple to see that

$$|m(iT) + \delta q(iT)| \leq M = 1 \tag{6-112}$$

Therefore, Eq. (6-110) has the same form as Eqs. (6-100) and (6-101), and it is possible to bring $\mathbf{x}''(0)$ to equilibrium in N sampling periods or less. In other words, Q_k is a point in R'_N.

Theorem 6-2

R'_N is the closed set whose boundary is a the convex polygon $\prod_N$ which has the following $2N$ vertices:

$$OP_1 = -S_0 + S_1 + S_2 + \ldots + S_{N-1}$$

$$OP_2 = -S_0 - S_1 + S_2 + \ldots + S_{N-1}$$

$$\vdots$$

$$OP_k = -S_0 - S_1 - S_2 - \ldots - S_{k-1} + S_k + \ldots + S_{N-1} \qquad (6\text{-}113)$$

$$\vdots$$

$$OP_N = -S_0 - S_1 - S_2 - \ldots - S_{N-1}$$

$$OP_{-1} = S_0 - S_1 - S_2 - \ldots - S_{N-1}$$

$$OP_{-2} = S_0 + S_1 - S_2 - \ldots - S_{N-1}$$

$$\vdots$$

$$OP_{-k} = S_0 + S_1 + S_2 + \ldots + S_{k-1} - S_k - \ldots - S_{N-1} \qquad (6\text{-}114)$$

$$\vdots$$

$$OP_{-N} = S_0 + S_1 + S_2 + \ldots + S_{N-1}$$

where O is the origin of the state space, and for simplicity, S_i denotes $S_i(T)$. Notice that these $2N$ points are obtained by assigning the $2N$ combinations of $m(iT) = +M$ and $-M$ for $i = 0, 1, 2, \ldots, N-1$, to Eq. (6-100). It can be shown that any other combinations of $m(iT)$ will correspond to an initial state which lies inside the convex polygon.

Proof. We shall prove that any point on the boundary or inside the polygon $\prod_N$ belongs to R'_N, and no point outside of $\prod_N$ can belong to R'_N.

Let us refer to Figure 6-7 in which P_k, P_{k+1}, P_{-k-1}, and P_{-k} are shown to be four vertices of $\prod_N$. Notice that P_k is a symmetric point to P_{-k} and P_{k+1} is symmetric to P_{-k-1}, with respect to the origin O. Now let P be an arbitrary point inside R'_N. In Figure 6-7 P is shown to lie on the line segment $\overline{OQ_k}$. Then

$$\overline{OQ_k} = \overline{OP_k} + \overline{P_kQ_k} \qquad (6\text{-}115)$$

From Eq. (6-113), we have

$$\overline{P_kP_{k+1}} - \overline{OP_{k+1}} - \overline{OP_k} = -2S_k \qquad (6\text{-}116)$$

Thus

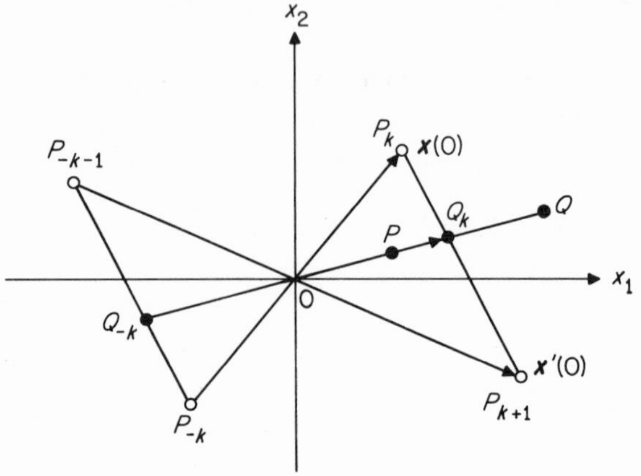

FIG. 6-7. Proof of Theorem 6-2.

$$\begin{aligned}\overline{OQ}_k &= \overline{OP}_k + \overline{P_kQ_k} = \overline{OP}_k + \delta\,\overline{P_kP_{k+1}} \\ &= -S_0 - S_1 - \ldots - S_{k-1} + S_k + \ldots + S_{N-1} - 2\delta S_k \\ &= -S_0 - S_1 - \ldots - S_{k-1} - (2\delta - 1)S_k + \ldots + S_{N-1} \\ &\qquad (0 \le \delta \le 1) \quad (6\text{-}117)\end{aligned}$$

Since $|2\delta - 1| \le 1$, Eq. (6-117) implies that $|m(kT) \le 1$, and Q_k is a point inside R'_N.

Similarly, Q_{-k} can be proved to be in R'_N. Then, using Theorem 6-1, any point P on $\overline{Q_kQ_{-k}}$ is also inside R'_N. Hence we have proved that any interior and any boundary point of the convex polygon $\prod_N$ belongs to R'_N.

Now let us consider an arbitrary point Q which lies outside of $\prod_N$. We must show that Q cannot be brought to the origin in N sampling periods or less. Suppose that Q can be brought to the origin in N sampling periods, then

$$\overline{OQ} = -\sum_{i=0}^{N-1} S_i(T)m(iT) \qquad (6\text{-}118)$$

where $|m(iT)| \le 1$. Since Q is outside $\prod_N$, from Figure 6-7, it is clear that

$$\overline{OQ} = K\,\overline{OQ}_k \qquad (6\text{-}119)$$

where $K > 1$.

Therefore, from Eqs. (6-117) and (6-118), we have

$$\begin{aligned}|m(iT)| &= K > 1 && \text{for } i = 0, 1, 2, \ldots, k \\ |m(iT)| &= (2\delta - 1)K && \text{for } i = k \\ |m(iT)| &= K > 1 && \text{for } i = k + 1, \ldots, N - 1\end{aligned}$$

Thus, all the control signals with the possible exception of $m(kT)$ have absolute values greater than unity. Therefore, these results contradict the condition that $|m(iT)| \leq 1$ for all i, and it is impossible to bring the state at point Q to equilibrium in N sampling periods or less.

Now let us investigate some further properties of the convex polygons, $\prod_{i+1}, i = 0, 1, 2, \ldots, N-1$.

For the initial state $\mathbf{x}(0)$ to reach equilibrium in $N-1$ sampling periods or less, Eq. (6-100) gives

$$\mathbf{x}(0) = -\sum_{i=0}^{N-2} S_i(T)m(iT) \tag{6-120}$$

and for an N-sampling-period-or-less process

$$\mathbf{x}(0) = -\sum_{i=0}^{N-1} S_i(T)m(iT) = -\sum_{i=0}^{N-2} S_i(T)m(iT) - S_{N-1}(T)m[(N-1)T] \tag{6-121}$$

From these last two equations it is clear that the convex polygon $\prod_N$ for N sampling periods or less is obtained by adding the vectors $-\Phi(-NT)\Theta(T)$ $m[(N-1)T]$, with $m[(N-1)T] = 1$ and -1, to the $2(N-1)$ vertices of the polygon $\prod_{N-1}$. Although this will result in a total of $4(N-1)$ points, in view of the conditions given in Eqs. (6-113) and (6-114) only $2N$ of these are vertices of the convex polygon $\prod_N$. The remaining $2N-4$ points belong to R'_N but are inside $\prod_N$. Stating in a different way, if the $2(N-1)$ vertices of the convex polygon $\prod_{N-1}$ are denoted by Q_k ($k = 1, 2, \ldots, N-1, -1, -2, \ldots, -N+1$), then P_k ($k = 1, 2, \ldots, N, -1, -2, \ldots, -N$), the vertices of $\prod_N$, are obtained from

$$\overline{OP_k} = \overline{OQ_k} - S_{N-1}(T)m[(N-1)T] \tag{6-122}$$

with $m[(N-1)T] = 1$ and/or -1.

In view of the preceding discussion, we conclude that if R'_{N+1} denotes the initial states which can be brought to equilibrium in $N+1$ sampling periods or less, its boundary is obtained by adding the vectors S_N and $-S_N$ to the vertices of R'_N in the following fashion:

If $Q_1, Q_2, \ldots, Q_N, Q_{-1}, Q_{-2}, \ldots, Q_{-N}$ are the $2N$ vertices of R'_N, then the vertices of R'_{N+1} are obtained as

$$\begin{aligned}
OP_k &= OQ_k + S_N \qquad & k = 1, 2, \ldots, N \\
OP_{N+1} &= OQ_N - S_N & \\
OP_{-k} &= OQ_{-k} - S_N \qquad & k = 1, 2, \ldots, N \\
OP_{-N-1} &= OQ_{-N} + S_N &
\end{aligned} \tag{6-123}$$

The equations in Eq. (6-123) represent a total of $2(N+1)$ vertices for R'_{N+1}.

EXAMPLE 6-3

Consider that a linear second-order process is characterized by the state equation

$$\mathbf{x}[(k+1)T] = \Phi(T)\mathbf{x}(kT) + \Theta(T)m(kT) \tag{6-124}$$

where

$$\Phi(T) = \begin{bmatrix} 1 & 1-e^{-T} \\ 0 & e^{-T} \end{bmatrix} \qquad \Theta(T) = \begin{bmatrix} T+e^{-T}-1 \\ 1-e^{-T} \end{bmatrix} \tag{6-125}$$

Therefore,

$$\begin{aligned} S = [S_0 \quad S_1] &= [\Phi(-T)\Theta(T) \qquad \Phi(-2T)\Theta(T)] \\ &= \begin{bmatrix} T-e^{T}+1 & T-e^{2T}+e^{T} \\ e^{T}-1 & -e^{T}+e^{2T} \end{bmatrix} \end{aligned} \tag{6-126}$$

Since S is nonsingular for all values of T, the process is controllable. Let us assume that $T = 1$ second and $|m(iT)| \leq 1$ for $i = 0, 1, 2, \ldots$.

Then, Eq. (6-125) becomes

$$\Phi(1) = \begin{bmatrix} 1 & 0.632 \\ 0 & 0.368 \end{bmatrix} \quad \Theta(1) = \begin{bmatrix} 0.368 \\ 0.632 \end{bmatrix} \tag{6-127}$$

We shall generate the convex polygons Π_N for $N = 1, 2,$ and 3 by the method described in the preceding sections.

1. $N = 1$:
   Equation (6-100) gives

$$\begin{aligned} \mathbf{x}(0) &= -\Phi(-T)\Theta(T)m(0) = -S_0(T)m(0) \\ &= \begin{bmatrix} 0.72 \\ -1.72 \end{bmatrix} m(0) \end{aligned} \tag{6-128}$$

The two vertices of the polygon Π_1 are obtained from

$$m(0) = 1, \qquad \mathbf{x}(0) = \begin{bmatrix} 0.72 \\ -1.72 \end{bmatrix} = -S_0 \tag{6-129}$$

$$m(0) = -1, \qquad \mathbf{x}(0) = \begin{bmatrix} -0.72 \\ 1.72 \end{bmatrix} = S_0 \tag{6-130}$$

Therefore, in this case the convex polygon which covers the initial states in the x_1, x_2 plane which can be brought to the origin in one sampling period is just a line segment, as shown in Figure 6-8.

2. $N = 2$:
   When $N = 2$, Eq. (6-100) gives

$$\begin{aligned} \begin{bmatrix} x_1(0) \\ x_2(0) \end{bmatrix} &= \begin{bmatrix} -T+e^{T}-1 \\ 1-e^{T} \end{bmatrix} m(0) + \begin{bmatrix} -T+e^{2T}-e^{T} \\ e^{T}-e^{2T} \end{bmatrix} m(T) \\ &= \begin{bmatrix} 0.72 \\ -1.72 \end{bmatrix} m(0) + \begin{bmatrix} 3.67 \\ -4.67 \end{bmatrix} m(1) \end{aligned} \tag{6-131}$$

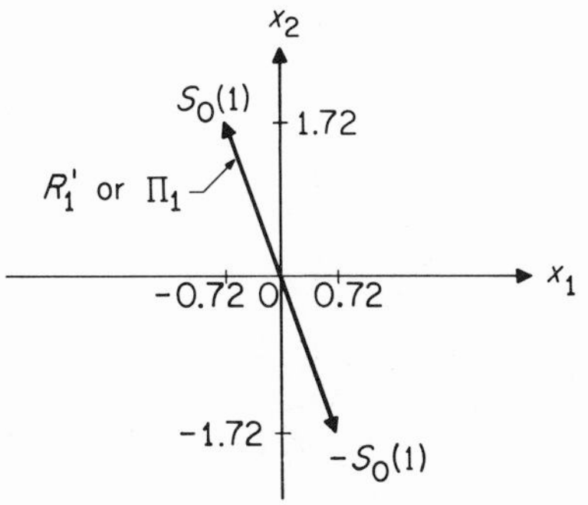

FIG. 6-8. Convex polygon Π_1 of a second-order process. R_1' represents the set of initial states which can be brought to the origin in one sampling period.

The four (2N) vertices of the convex polygon Π_2 are obtained as follows:

$$m(0) = 1,\ m(1) = 1, \qquad \begin{bmatrix} x_1(0) \\ x_2(0) \end{bmatrix} = \begin{bmatrix} 4.39 \\ -6.39 \end{bmatrix}$$

$$m(0) = 1,\ m(1) = -1, \qquad \begin{bmatrix} x_1(0) \\ x_2(0) \end{bmatrix} = \begin{bmatrix} -2.95 \\ 2.95 \end{bmatrix}$$

$$m(0) = -1,\ m(1) = 1, \qquad \begin{bmatrix} x_1(0) \\ x_2(0) \end{bmatrix} = \begin{bmatrix} 2.95 \\ -2.95 \end{bmatrix}$$

$$m(0) = -1,\ m(1) = -1, \qquad \begin{bmatrix} x_1(0) \\ x_2(0) \end{bmatrix} = \begin{bmatrix} -4.39 \\ 6.39 \end{bmatrix}$$

The construction procedure of the polygon Π_2 from the vertices of Π_1 using Eq. (6-123) is illustrated in Figure 6-9, and Π_2 is shown in Figure 6-10. The values of the control signal sequence [$m(0)$, $m(1)$] are shown at the vertices of the polygon.

3. $N = 3$:

When $N = 3$, Eq. (6-100) gives

$$\mathbf{x}(0) = -\Phi(-T)\Theta(T)m(0) - \Phi(-2T)\Theta(T)m(T) - \Phi(-3T)\Theta(T)m(2T) \tag{6-132}$$

or

$$\begin{bmatrix} x_1(0) \\ x_2(0) \end{bmatrix} = \begin{bmatrix} -T + e^T - 1 \\ 1 - e^T \end{bmatrix} m(0) + \begin{bmatrix} -T + e^{2T} - e^T \\ e^T - e^{2T} \end{bmatrix} m(T)$$

$$+ \begin{bmatrix} -T + e^{3T} - e^{2T} \\ e^{2T} - e^{3T} \end{bmatrix} m(2T)$$

$$= \begin{bmatrix} 0.72 \\ -1.72 \end{bmatrix} m(0) + \begin{bmatrix} 3.67 \\ -4.67 \end{bmatrix} m(1) + \begin{bmatrix} 11.72 \\ -12.72 \end{bmatrix} m(2) \tag{6-133}$$

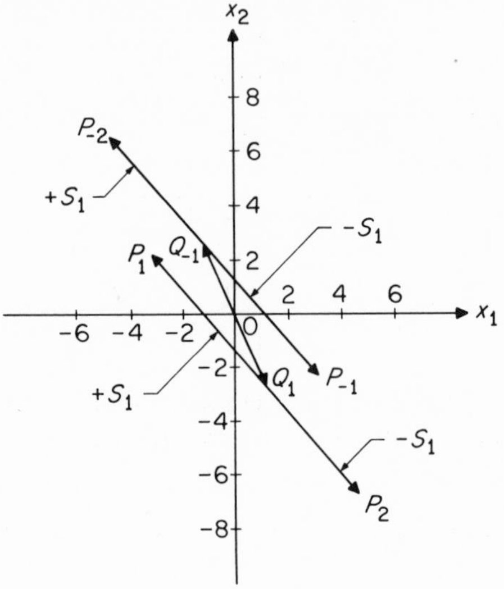

FIG 6-9. Construction procedure of Π_2 of a second-order process.

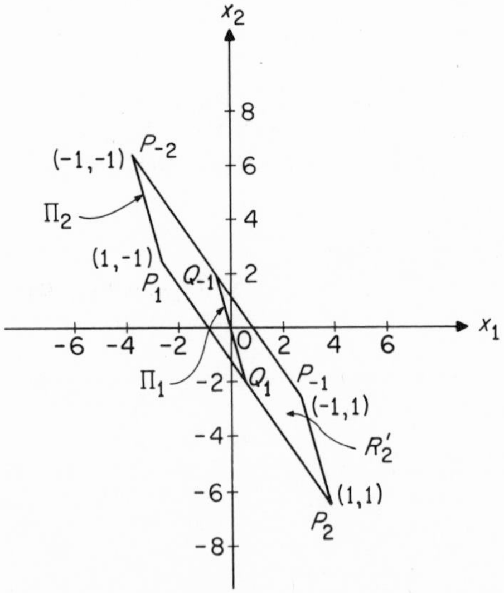

FIG. 6-10. Convex polygon Π_2 of a second-order process.

The vertices of the convex polygon Π_3 are computed as follows:

$$m(0) = 1,\ m(1) = -1,\ m(2) = -1, \qquad \begin{bmatrix} x_1(0) \\ x_2(0) \end{bmatrix} = \begin{bmatrix} -14.67 \\ 15.67 \end{bmatrix}$$

$$m(0) = 1,\ m(1) = 1,\ m(2) = -1, \qquad \begin{bmatrix} x_1(0) \\ x_2(0) \end{bmatrix} = \begin{bmatrix} -7.33 \\ 6.33 \end{bmatrix}$$

$$m(0) = 1,\ m(1) = 1,\ m(2) = 1, \qquad \begin{bmatrix} x_1(0) \\ x_2(0) \end{bmatrix} = \begin{bmatrix} 16.11 \\ -19.11 \end{bmatrix}$$

The other three vertices are symmetrical to the above three points. Notice that a total of eight points will result if Eq. (6-123) is used. However, two of the points,

$$\begin{bmatrix} x_1(0) \\ x_2(0) \end{bmatrix} = \begin{bmatrix} 8.77 \\ -9.77 \end{bmatrix} \qquad m(0) = 1,\ m(1) = -1,\ m(2) = 1$$

and

$$\begin{bmatrix} x_1(0) \\ x_2(0) \end{bmatrix} = \begin{bmatrix} -8.77 \\ 9.77 \end{bmatrix} \qquad m(0) = -1,\ m(1) = 1,\ m(2) = -1$$

are inside Π_3 and, therefore, are not the vertices of the convex polygon Π_3. The construction of Π_3 using the vertices of Π_2 is illustrated in Figure 6-11, and Π_3 is

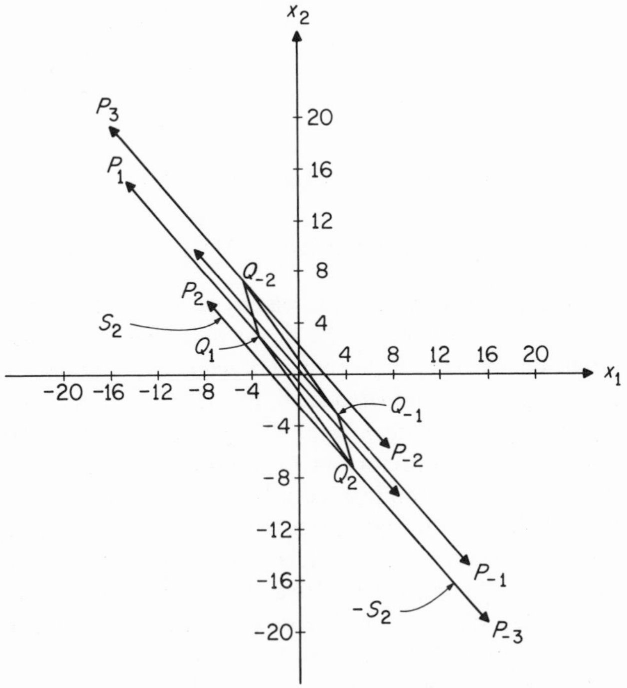

FIG. 6-11. Construction of Π_3 of a second-order process.

shown in Figure 6-12. The values of the control sequence $[m(0), m(1), m(2)]$ are shown at the vertices of the polygon.

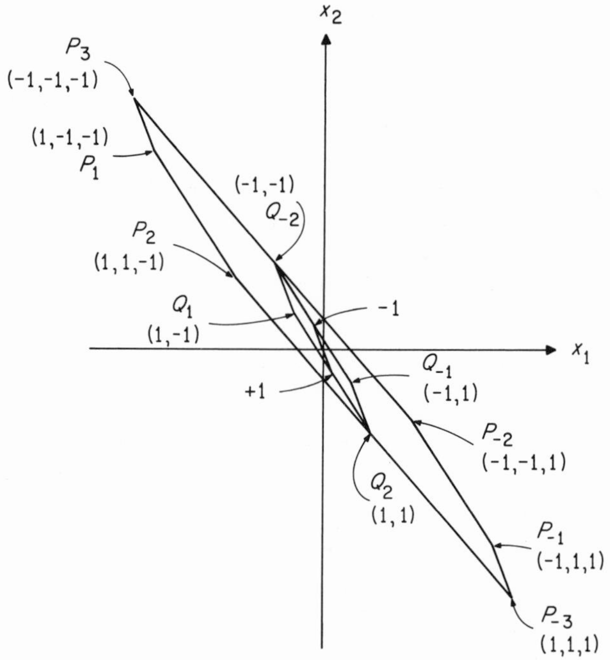

FIG. 6-12. Convex polygon Π_3 of a second-order process.

The Optimum Switching Trajectory (OST)

For continuous-data systems with relay-type control, the optimum switching trajectory (OST) is defined as the initial states which can be brought to the equilibrium state $\mathbf{0}$ by a single control of $m(t) = +1$ or $m(t) = -1$. For the second-order process considered here, the OST is formed by taking the lower half plane portion of the $h_n = +1$ trajectory and the upper half plane portion of the $h_n = -1$ trajectory in Figure 4-26. The OST with $\mathbf{x} = \mathbf{0}$ as the equilibrium state is shown in Figure 6-13. In the case of continuous-data systems, the OST divides the state plane into two regions. The region above the OST represents states which must be driven by a -1 control first to the OST, and then a $+1$ control will drive the state to the equilibrium state. Similarly, the region below the OST represents states which must be driven by a $+1$ control first to the OST, and then a -1 control will drive the state to the equilibrium state. Since the switch of control can occur at any instant for a continuous-data system, for a second-order system any state in the state plane can be brought to the equilibrium state with two switches or less.

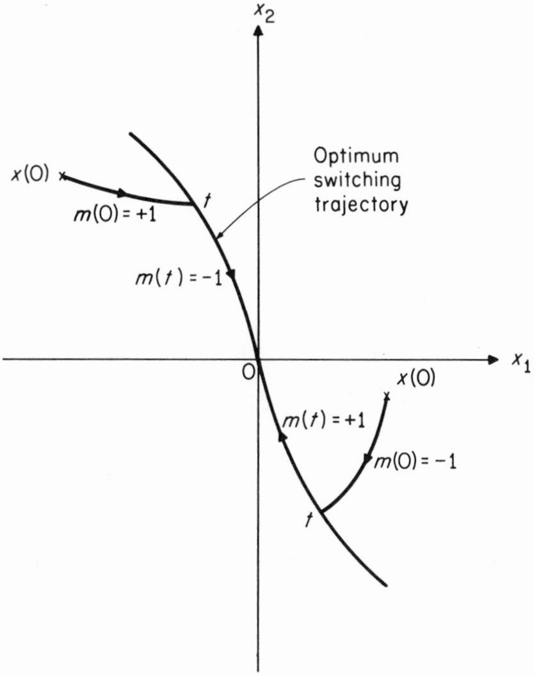

FIG. 6-13. Optimum switching trajectory of a continuous-data system with relay-control.

The optimum switching trajectory can be related to the vector $S_i(T)$ of the digital system. Let $s(t)$ denote the vector defining the OST of Figure 6-13 when $m = +1$, and $-s(t)$ be that for $m = -1$. In other words $s(t)$ represents a point in the state plane which can be brought to $\mathbf{x} = \mathbf{0}$ in t seconds with control $m = +1$. Then

$$s(t) = -\int_0^t \Phi(-\tau)B\,d\tau \tag{6-134}$$

and

$$-s(t) = \int_0^t \Phi(-\tau)B\,d\tau \tag{6-135}$$

However, from Eq. (6-90),

$$\Theta(T) = \int_0^T \Phi(T-\tau)B\,d\tau = \Phi(T)\int_0^T \Phi(-\tau)B\,d\tau \tag{6-136}$$

Comparing Eq. (6-136) with Eq. (6-135), we have

$$\Theta(T) = -\,\Phi(T)s(T) \tag{6-137}$$

Substituting Eq. (6-137) into Eq. (6-97) for $i = 0, 1$, we have

$$S_0(T) = \Phi(-T)\Theta(T) = -\Phi(-T)\Phi(T)s(T) = -s(T) \quad (6\text{-}138)$$

and

$$S_1(T) = \Phi(-2T)\Theta(T) = -\Phi(-2T)\Phi(T)s(T) = -\Phi(-T)s(T) \quad (6\text{-}139)$$

Also, we can write

$$s(kT) - s[(k-1)T] = -\int_0^{kT} \Phi(-\tau)B\,d\tau + \int_0^{(k-1)T} \Phi(-\tau)B\,d\tau$$
$$= -\int_{(k-1)T}^{kT} \Phi(-\tau)B\,d\tau \quad (6\text{-}140)$$

Let $\tau' = \tau - (k-1)T$, then Eq. (6-140) is written

$$s(kT) - s[(k-1)T] = -\int_0^T \Phi(-kT + T - \tau')B\,d\tau' = -\Phi(-kT)\Theta(T) \quad (6\text{-}141)$$

or

$$s(kT) - s[(k-1)T] = -S_{k-1}(T) \quad (6\text{-}142)$$

which indicates that the vector $S_k(T)$ sweeps out the optimum switching trajectory in the (x_1, x_2) state plane. Figure 6-14 illustrates the relationship between

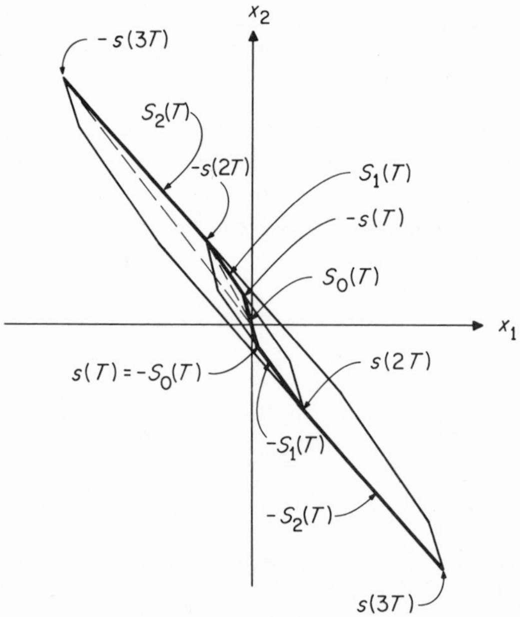

FIG. 6-14. Optimum switching trajectory (OST) of a second-order process.

$S_k(T)$ and $s(kT)$ for $k = 1, 2$, and 3, and the convex polygons $\prod_1, \prod_2, \prod_3$ of the second-order process. Notice that the tips of the vectors $\pm s(kT)$, $k = 1, 2, \ldots$ fall on the OST shown in Figure 6-13. The trajectory drawn in heavy lines in Figure 6-14 is also called the *polygonal curve*. It can be shown that this polygonal curve is a piecewise linear approximation to the OST of Figure 6-13, and as the sampling period T approaches zero, the two curves become identical.

The Optimal Control Strategy

The sequence of control signals $m(0), m(T), \ldots, m[(N-1)T]$, which can bring the linear process from an arbitrary initial state to equilibrium in a minimum number of N sampling periods is called the optimal control strategy or the optimal control law. The regions of controllable states investigated in the last section for a second-order system determine the value of N when the initial state is specified. In general, the optimal control strategy is not unique, except under the following two conditions:

1. The initial state $\mathbf{x}(0)$ falls on the vertices of the region $\prod_N$. Under this condition, the optimal control signals will be either $+1$ or -1, as specified by the particular vertex.
2. For a second-order process, when $N \leq 2$, the optimal control signals $m(0)$ and $m(T)$ are uniquely determined. In general, this is true for an nth order system if $N \leq n$.

The statement made in Case 1 is verified by referring to Figure 6-12. If $\mathbf{x}(0)$ falls on a vertex of $\prod_N$, a control signal $m(0) = +1$ or -1 will drive it to a vertex of $\prod_{N-1}$, and so on, until the equilibrium state is reached. The optimal control strategy is clearly marked by the sequence $[m(0), m(T)\ \mathrm{m}(2T)]$ at the vertices of the polygons in Figure 6-12.

For a second-order process, if $N \leq 2$, the optimal control strategy is also unique. For instance, if $\mathbf{x}(0)$ is in R'_2, $N \leq 2$, Eq. (6-94) reads

$$\mathbf{x}(0) = -[S_0(T) \quad S_1(T)] \begin{bmatrix} m(0) \\ m(T) \end{bmatrix} \tag{6-143}$$

Therefore, if the process is controllable, i.e., $[S_0(T) \quad S_1(T)]$ is nonsingular, the control law can be determined uniquely from

$$\begin{bmatrix} m(0) \\ m(T) \end{bmatrix} = -[S_0(T) \quad S_1(T)] \begin{bmatrix} x_1(0) \\ x_2(0) \end{bmatrix} \tag{6-144}$$

Similarly, if $\mathbf{x}(0)$ is in R'_1, $N = 1$, the control law is determined from

$$\mathbf{x}(0) = -S_0(T)m(0) \tag{6-145}$$

which represents two dependent equations with one unknown.

For the second-order process, if $N > 2$, the optimal control strategy will not be unique and there is an infinite number of combinations of control signals which will drive $\mathbf{x}(0)$ to equilibrium. If N is greater than two, this means that $N - 2$ control signals will be constrained and will take on values of $+1$ or -1. Once the state enters R_2', the control becomes unique. The signs of these constrained controls $m(0), m(1), \ldots, m(N - n - 1)$ can be determined by means of the "critical curve" in the state plane. The critical curve is obtained by joining successively the vertices defined by

$$\begin{aligned}
OC_1 &= S_1 \\
OC_2 &= S_1 + S_2 \\
&\cdot \\
&\cdot \\
&\cdot \\
OC_{N-2} &= \sum_{i=1}^{N-2} S_i \\
OC_{N-1} &= \sum_{i=1}^{N-1} S_i \\
OC_{-1} &= -S_1 \\
OC_{-2} &= -S_1 - S_2 \\
&\cdot \\
&\cdot \\
&\cdot \\
OC_{-N+2} &= -\sum_{i=1}^{N-2} S_i \\
QC_{-N+1} &= -\sum_{i=1}^{N-1} S_i
\end{aligned} \tag{6-146}$$

Notice that these vectors represent the condition when $m(0) = 0$, and for the second-order system under consideration, the critical curve is shown in Figure 6-15. This critical curve divides the state plane into two regions: I and II. Region (I) represents states which should be driven by $0 \leq m \leq 1$ controls and region (II) represents states that should be driven by $0 \geq m \geq -1$ controls. Therefore, once the initial state is given, its position in the state plane with respect to the critical curve determines the sign of the control signal $m(0)$.

The following numerical example illustrates the determination of the optimal control strategy and the applications of the regions of controllable states.

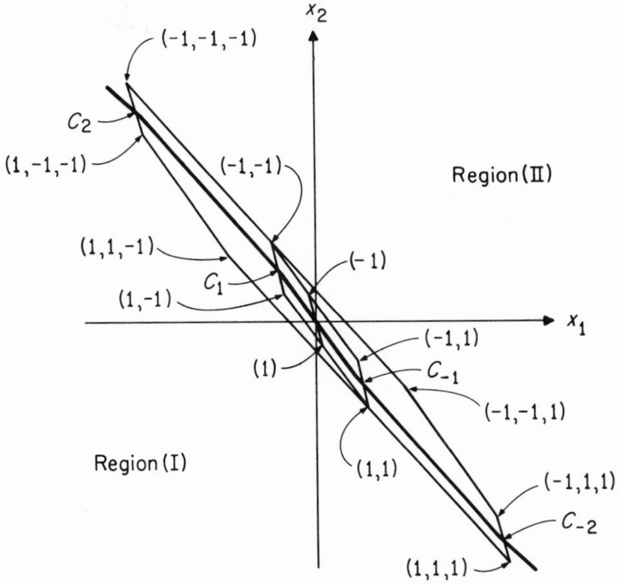

FIG. 6-15. The critical curve in the state plane.

EXAMPLE 6-4

Consider that a linear second-order process is described by the state equation

$$\mathbf{x}[(k+1)T] = \mathbf{\Phi}(T)\mathbf{x}(kT) + \mathbf{\Theta}(T)m(kT)$$

where

$$\mathbf{\Phi}(T) = \begin{bmatrix} 1 & 1-e^{-T} \\ 0 & e^{-T} \end{bmatrix} \qquad \mathbf{\Theta}(T) = \begin{bmatrix} T+e^{-T}-1 \\ 1-e^{-T} \end{bmatrix} \tag{6-147}$$

With $T = 1$ second,

$$\mathbf{\Phi}(1) = \begin{bmatrix} 1 & 0.632 \\ 0 & 0.368 \end{bmatrix} \qquad \mathbf{\Theta}(1) = \begin{bmatrix} 0.368 \\ 0.632 \end{bmatrix} \tag{6-148}$$

The regions of controllable states for $|m(kT)| \leq 1$ are shown in Figure 6-12. Let us assume that the initial state is

$$\mathbf{x}(0) = \begin{bmatrix} -7.33 \\ +6.33 \end{bmatrix} \tag{6-149}$$

which is at the vertex P_2 of Π_3. Therefore, $N = 3$; that is, it requires three seconds to bring the system to the equilibrium state $\mathbf{x}(3) = 0$.

Equation (6-94) gives

$$\mathbf{x}(0) = -S_0 m(0) - S_1 m(1) - S_2 m(2) \tag{6-150}$$

where

$$S_0 = \Phi(-1)\Theta(1) = \begin{bmatrix} -0.72 \\ 1.72 \end{bmatrix} \tag{6-151}$$

$$S_1 = \Phi(-2)\Theta(1) = \begin{bmatrix} -3.67 \\ 4.67 \end{bmatrix} \tag{6-152}$$

$$S_2 = \Phi(-3)\Theta(1) = \begin{bmatrix} -11.72 \\ 12.72 \end{bmatrix} \tag{6-153}$$

Equation (6-150) represents two equations with three unknowns. However, since $\mathbf{x}(0)$ is in region I, i.e., below the critical curve, we set $m(0) = 1$. Substituting $m(0) = 1$ into Eq. (6-146), with $k = 0$, we get

$$\mathbf{x}(T) = \Phi(1)\mathbf{x}(0) + \Theta(1)m(0) = \begin{bmatrix} 1 & 0.632 \\ 0 & 0.368 \end{bmatrix}\begin{bmatrix} -7.33 \\ 6.33 \end{bmatrix} + \begin{bmatrix} 0.368 \\ 0.632 \end{bmatrix}$$

$$= \begin{bmatrix} -2.95 \\ 2.95 \end{bmatrix} \tag{6-154}$$

which is recognized to be the vertex Q_1 of Π_2. The next two optimal control signals, $m(1) = 1$ and $m(2) = -1$, are determined by solving the following equation:

$$\begin{bmatrix} m(1) \\ m(2) \end{bmatrix} = -[S_0 \quad S_1]^{-1} \begin{bmatrix} -2.95 \\ 2.95 \end{bmatrix} \tag{6-155}$$

Therefore, the optimal control strategy is $m(0) = 1$, $m(1) = 1$, and $m(2) = -1$. Of course, once we have identified that $\mathbf{x}(0)$ is at the vertex P_2 of the polygon Π_3, the optimal control strategy is readily determined from Figure 6-12 or Figure 6-15.

Now let us select the initial state to be at

$$\mathbf{x}(0) = \begin{bmatrix} -4 \\ 4 \end{bmatrix} \tag{6-156}$$

Since $\mathbf{x}(0)$ is in R_3', $N = 3$. The optimal control strategy is not unique in this case since $\mathbf{x}(0)$ is not on Π_3. We observe that $\mathbf{x}(0)$ is in region II with respect to the critical curve, therefore, we set $m(0) = 1$. Then, Eq. (6-150) gives

$$\begin{bmatrix} m(1) \\ m(2) \end{bmatrix} = -[S_1 \quad S_2]^{-1}[\mathbf{x}(0) + S_0 m(0)]$$

$$= \frac{1}{8.05}\begin{bmatrix} -12.72 & -11.72 \\ 4.67 & 3.67 \end{bmatrix}\begin{bmatrix} -4.72 \\ 5.72 \end{bmatrix} = \begin{bmatrix} -0.869 \\ -0.13 \end{bmatrix} \tag{6-157}$$

Therefore, the system is brought from $\mathbf{x}(0)$ to $\mathbf{x}(3) = \mathbf{0}$ in three seconds, and the optimal control strategy is $m(0) = 1$, $m(1) = -0.869$, and $m(2) = -0.13$. The states for $t > 0$ are

$$\mathbf{x}(1) = \begin{bmatrix} -1.104 \\ 2.104 \end{bmatrix} \qquad \mathbf{x}(2) = \begin{bmatrix} -0.095 \\ 0.224 \end{bmatrix} \qquad \mathbf{x}(3) = \begin{bmatrix} 0 \\ 0 \end{bmatrix} \tag{6-158}$$

The optimal trajectory is shown in Figure 6-16.

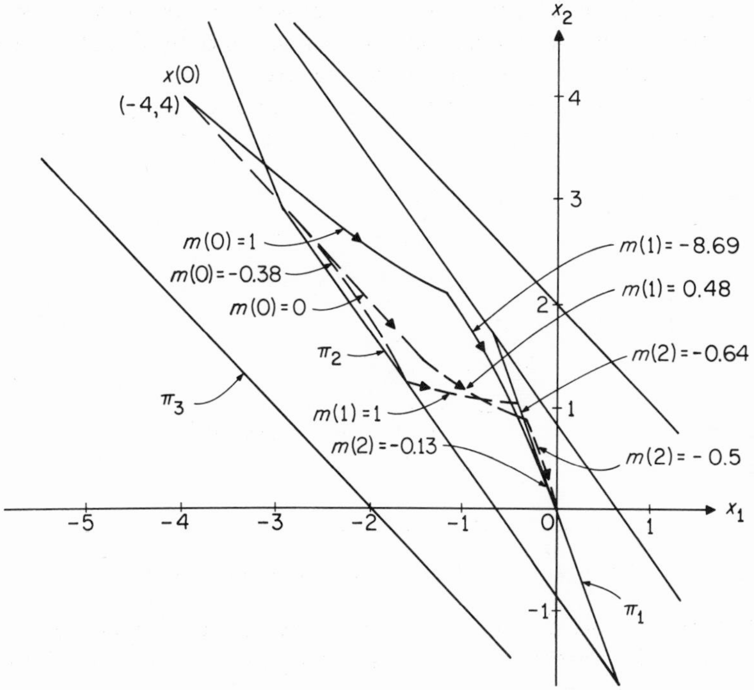

FIG. 6-16. Optimal trajectories of Example 6-4.

It is apparent that the optimal strategy determined above is not unique. Equation (6-150) is written

$$\begin{bmatrix} m(1) \\ m(2) \end{bmatrix} = -[S_1 \quad S_2]^{-1}[\mathbf{x}(0) + S_0 m(0)]$$

$$= \begin{bmatrix} -1.58 & -1.46 \\ 0.58 & 0.456 \end{bmatrix} [\mathbf{x}(0) + S_0 m(0)] \tag{6-159}$$

For the initial state of Eq. (6-156), the last equation becomes

$$\begin{bmatrix} m(1) \\ m(2) \end{bmatrix} = \begin{bmatrix} 0.48 - 1.37m(0) \\ -0.5 + 0.367m(0) \end{bmatrix} \tag{6-160}$$

It is apparent that $m(1)$ and $m(2)$ will both lie between -1 and 1 for all $0 \leq m(0) \leq 1$. If $m(0)$ is negative, it must lie between -0.38 and 0. Figure 6-16 illustrates several optimal trajectories that correspond to various optimal control strategies.

The Digital Controller

The time-optimal control design considered thus far in this chapter yields the optimum control strategy $m(t)$ for the system of Figure 6-5 when the linear process G and the initial state $\mathbf{x}(0)$ are prescribed. To complete the design problem, we must now determine the transfer function of the digital controller once the characteristics of the nonlinearity are given.

The block diagram of the overall system is repeated in Figure 6-17. The reference input $r(t)$ is considered to be zero for the present case.

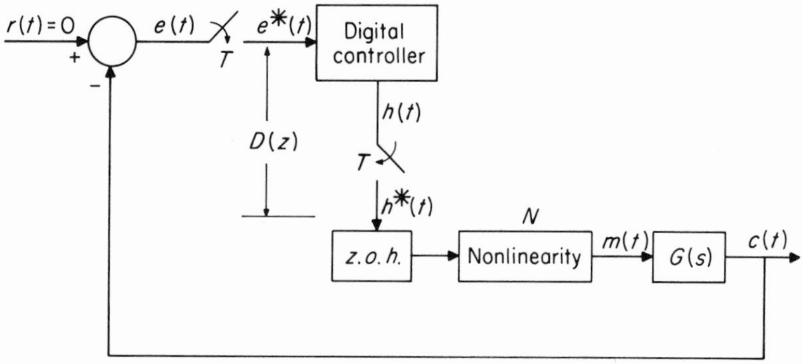

FIG. 6-17. Block diagram of a digital control system.

The design problem is to determine the transfer function of the digital controller, $D(z)$, so that the initial state $\mathbf{x}(0)$ can be brought to the equilibrium state $\mathbf{x}(NT) = \mathbf{0}$ in a minimum number of sampling periods.

The transfer function of the controller is written as

$$D(z) = \frac{H(z)}{E(z)} \tag{6-161}$$

where

$$H(z) = h(0) + h(T)z^{-1} + \ldots + h[(N-1)T]z^{-N+1} \tag{6-162}$$

$$E(z) = e(0) + e(T)z^{-1} + \ldots + e[(N-1)T]z^{-N+1} \tag{6-163}$$

Once the optimum control strategy $m(0), m(T), \ldots, m[(N-1)T]$ is determined, the corresponding sequence, $h(kT)$, $k = 0, 1, 2, \ldots, (N-1)$, is determined using the nonlinear characteristics.

If the output equation of the linear process G is $c(t) = x_1(t)$, the sequence $e(kT)$, $k = 0, 1, 2, \ldots, (N-1)$, is determined from

$$e(kT) = -x_1(kT) \tag{6-164}$$

It is apparent that the method is easily extended to systems with nonzero inputs; then,

$$e(kT) = r(kT) - x_1(kT) \tag{6-165}$$

The following example illustrates the design principle outlined above.

EXAMPLE 6-5

Consider that for the digital control system shown in Figure 6-17,

$$G(s) = \frac{1}{s(s+1)}, \quad T = 1 \text{ second}, \quad r(t) = 0$$

The nonlinearity N is described by (Figure 6-18)

$$\begin{aligned} m(t) &= 1 & 1 \leq h(t) < \infty \\ m(t) &= -1 & -\infty < h(t) \leq -1 \\ m(t) &= h^2(t)\,\text{sgn}\,h(t) & -1 \leq h(t) \leq 1 \end{aligned} \tag{6-166}$$

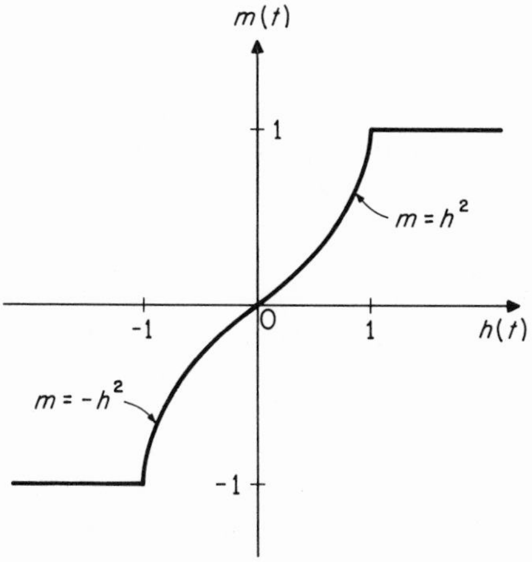

FIG. 6-18. Nonlinear characteristics of system of Example 6-5.

The initial state of the system is

$$\mathbf{x}(0) = \begin{bmatrix} -4 \\ 4 \end{bmatrix}$$

Find the transfer function $D(z)$ of the digital controller so that $\mathbf{x}(0)$ is brought to the equilibrium state $\mathbf{x}(NT) = \mathbf{0}$ in minimum time.

The design of the optimum control strategy has been carried out in Example 6-4, and it is found that $N = 3$, and one set of optimum $m(kT)$ is

$$m(0) = 1 \qquad m(1) = -0.869 \qquad m(2) = -0.13$$

The states for $t > 0$ are given by Eq. (6-158)

$$\mathbf{x}(1) = \begin{bmatrix} -1.104 \\ 2.104 \end{bmatrix} \qquad \mathbf{x}(2) = \begin{bmatrix} -0.095 \\ 0.224 \end{bmatrix} \qquad \mathbf{x}(3) = \begin{bmatrix} 0 \\ 0 \end{bmatrix} \tag{6-167}$$

Therefore, from Eq. (6-164), we get

$$e(0) = -x_1(0) = 4 \qquad e(1) = -x_1(1) = 1.104 \qquad e(2) = -x_1(2) = 0.095$$

For $m(0) = 1$, the value of $h(0)$ is not unique, since any value of $h(t)$ greater than unity will yield an $m(t)$ equal to unity. For practical reasons, let us choose $h(0) = 1$. For $m(1) = -0.869$, the nonlinear characteristics show that $h(1) = -0.932$. Similarly, $h(2)$ is found to be equal to -0.36.

Now using Eqs. (6-161), (6-162), and (6-163), the transfer function of the digital controller is determined as

$$D(z) = \frac{H(z)}{E(z)} = \frac{1 - 0.932z^{-1} - 0.36z^{-2}}{4 + 1.104z^{-1} + 0.095z^{-2}} \tag{6-168}$$

It must be pointed out that this digital controller is optimum only for the initial state $\mathbf{x}(0) = [-4 \quad 4]'$. For a different initial state, of course, a different digital controller must be used.

6.6 TIME-OPTIMAL CONTROL BY STATE VARIABLE FEEDBACK

The minimum-time control design discussed in the previous section is tuned to the initial state of the system when the system configuration of Figure 6-17 is adopted. In other words, the optimum control strategy depends on the initial state of the system. This places a severe limitation on the practical application of the method.

In this section we shall present a method which allows the computation of the optimum control strategy from information on the state variables of the system measured at the corresponding instants. The computation can usually be carried out by a special digital controller or an on-line digital computer.

Let us assume that the nth order linear process is described by

$$\mathbf{x}[(k+1)T] = \mathbf{\Phi}(T)\mathbf{x}(kT) + \mathbf{\Theta}(T)m(kT) \tag{6-169}$$

where $m(kT)$ is a scalar control signal.

By successive substitution, Eq. (6-169) leads to

$$\mathbf{x}[(k+N)T] = \mathbf{\Phi}(NT)\mathbf{x}(kT) + \sum_{i=0}^{N-1} \mathbf{\Phi}[(N-i-1)T]\mathbf{\Theta}(T)m[(k+i)T] \tag{6-170}$$

Setting $\mathbf{x}[(k+N)T] = \mathbf{0}$, Eq. (6-170) gives

$$\mathbf{x}(kT) = -\sum_{i=0}^{N-1} \mathbf{\Phi}[(-i-1)T]\mathbf{\Theta}(T)m[(k+i)T] \tag{6-171}$$

or

$$\mathbf{x}(kT) = -[S_0(T) \quad S_1(T) \quad \ldots \quad S_{N-1}(T)] \begin{bmatrix} m(kT) \\ m[(k+1)T] \\ \cdot \\ \cdot \\ \cdot \\ m[(k+N-1)T] \end{bmatrix} \tag{6-172}$$

where

$$S_i(T) = \Phi[(-i-1)T]\Theta(T) \qquad i = 0, 1, \ldots N-1 \tag{6-173}$$

Therefore, the optimum control strategy is written

$$\begin{bmatrix} m(kT) \\ m[(k+1)T] \\ \cdot \\ \cdot \\ \cdot \\ m[(k+N-1)T] \end{bmatrix} = -[S_0(T) \quad S_1(T) \quad \ldots \quad S_{N-1}(T)]^{-1}\mathbf{x}(kT)\ (n = N) \tag{6-174}$$

if the matrix inverse exists.

It is interesting to point out that Eq. (6-174) implies that the optimum control strategy $m(kT)$ for $k = 0, 1, 2, \ldots, N-1$ can be written in terms of $\mathbf{x}(kT)$ as

$$m(kT) = -[1 \quad 0 \quad 0 \ldots 0][S_0(T) \quad S_1(T) \quad \ldots \quad S_{N-1}(T)]^{-1}\mathbf{x}(kT) \tag{6-175}$$

or simply

$$m(kT) = C\mathbf{x}(kT) \tag{6-176}$$

where

$$C = -[1 \quad 0 \quad 0 \ldots 0][S_0(T) \quad S_1(T) \quad \ldots \quad S_{N-1}(T)]^{-1} \tag{6-177}$$

Equation (6-176) clearly points out the fact that the optimum control signal at $t = kT$ depends implicitly on the state vector $\mathbf{x}(kT)$ measured at the same instant. Physically, this is accomplished by feeding back the state variables $x_1(kT), x_2(kT), \ldots, x_n(kT)$ through a computer or controller which performs the operation of the matrix C, defined by Eq. (6-177), on $\mathbf{x}(kT)$. A block diagram illustrating the optimization scheme is shown in Figure 6-19.

One difficulty in the physical implementation of the optimal control with state variable feedback is that time will be required for the computer to compute $m(kT)$ from the state vector $\mathbf{x}(kT)$ so that time delay is usually inevitable. Another difficulty lies in the fact that in practice not all the state variables in

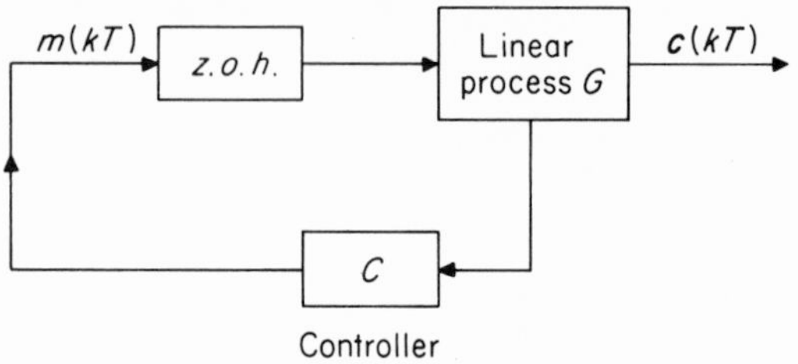

FIG. 6-19. Block diagram of time-optimal control system with state variable feedback.

a physical process are accessible. Usually, only the outputs of the process are directly measurable. Therefore, a computer or an "estimator" is needed to compute or estimate the state variables from the outputs. For instance, the method described in Sec. 6.3, Eq. (6-81), may be used to determine the state variables if past inputs and past outputs are known.

Figure 6-20 shows the block diagram of the time-optimum control system with a computer or estimator for determination of inaccessible state variables. The notation $\mathbf{x}(kT)$ denotes the estimated state vector.

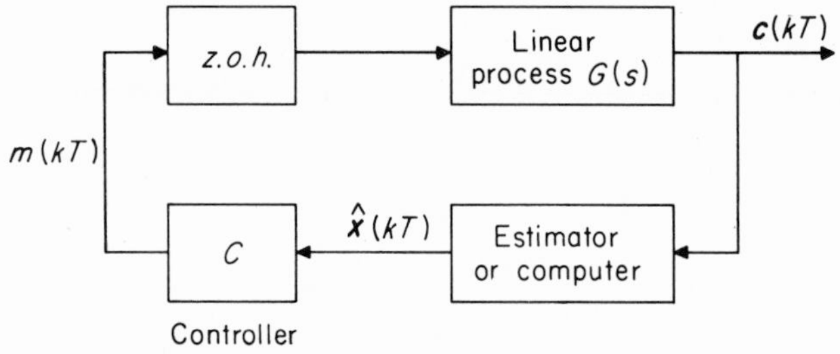

FIG. 6-20. Block diagram of time-optimal control system with inaccessible state variables.

EXAMPLE 6-6

Let us consider the numerical problem in Example 6-4, with

$$\mathbf{x}(0) = \begin{bmatrix} -4 \\ 4 \end{bmatrix}$$

We shall determine the optimal control strategy using the method of state variable feedback.

Substituting $S_1(T)$ and $S_2(T)$ from Eqs. (6-151) and (6-152) into Eq. (6-175), and with $N = n = 2$, we have

$$m(kT) = [1 \quad 0]\begin{bmatrix} 0.72 & 3.67 \\ -1.72 & -4.67 \end{bmatrix}^{-1}\begin{bmatrix} x_1(kT) \\ x_2(kT) \end{bmatrix}$$

$$= [-1.58 \quad -1.24]\begin{bmatrix} x_1(kT) \\ x_2(kT) \end{bmatrix} \tag{6-178}$$

When $k = 0$, Eq. (6-178) becomes

$$m(0) = [-1.58 \quad -1.24]\begin{bmatrix} -4 \\ 4 \end{bmatrix} = 1.36 \tag{6-179}$$

Since $m(0)$ exceeds the limiting value of unity, we set $m(0) = 1$. Next the state vector $\mathbf{x}(1)$ is computed

$$\begin{bmatrix} x_1(1) \\ x_2(1) \end{bmatrix} = \begin{bmatrix} 1 & 0.632 \\ 0 & 0.368 \end{bmatrix}\begin{bmatrix} -4 \\ 4 \end{bmatrix} + \begin{bmatrix} 0.368 \\ 0.632 \end{bmatrix} = \begin{bmatrix} -1.104 \\ 2.104 \end{bmatrix} \tag{6-180}$$

Substituting $\mathbf{x}(1)$ into Eq. (6-178), with $k = 1$, we get

$$m(1) = [-1.58 \quad -1.24]\begin{bmatrix} -1.104 \\ 2.104 \end{bmatrix} = -0.87 \tag{6-181}$$

Repeating the same procedure, $\mathbf{x}(2)$ is determined to be

$$\begin{bmatrix} x_1(2) \\ x_2(2) \end{bmatrix} = \begin{bmatrix} -0.095 \\ 0.224 \end{bmatrix}$$

and subsequently, $m(2) = -0.13$. Apparently, these results agree with those obtained earlier in Example 6-4.

6.7 TIME-OPTIMAL CONTROL OF SYSTEMS WITH STEP INPUTS

The time-optimal design considered in the preceding sections deals with systems which have zero reference inputs. In this section, the design is extended to step function inputs; that is, $r(t) = Ru(t)$, where R is a constant.

The block diagram of the system under consideration is shown in Figure 6-21. The dynamic equations of the linear process are

$$\frac{d\mathbf{x}(t)}{dt} = A\mathbf{x}(t) + Bm(t) \tag{6-182}$$

$$c(t) = x_1(t) \tag{6-183}$$

where $\mathbf{x}(t)$ is the $n \times 1$ state vector, A and B are constant matrices, and $m(t)$ and $c(t)$ are the scalar control and output variables, respectively.

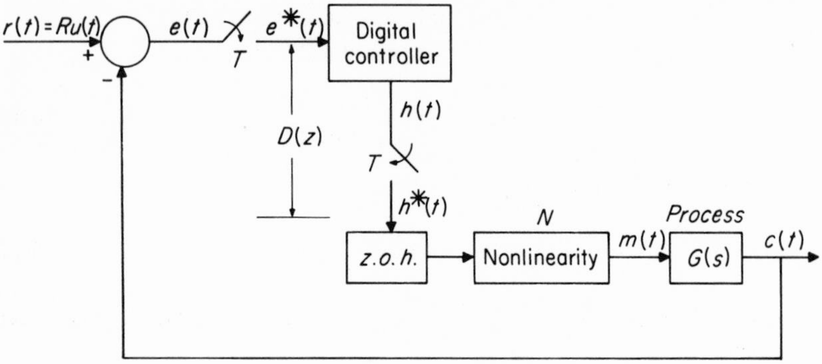

FIG. 6-21. Block diagram of a digital control system.

We define a new state vector $\mathbf{y}(t)$, such that

$$\mathbf{y}(t) = \mathbf{x}(t) - \mathbf{r}(t) \tag{6-184}$$

where

$$\mathbf{r}(t) = \begin{bmatrix} r(t) \\ 0 \\ \cdot \\ \cdot \\ \cdot \\ 0 \end{bmatrix} \text{(}n \times 1 \text{ input vector)} \tag{6-185}$$

Then, for $r(t) = Ru(t)$, the state equation of the process in terms of $\mathbf{y}(t)$ becomes

$$\frac{d\mathbf{y}(t)}{dt} = A\mathbf{y}(t) + A\mathbf{r}(t) = Bm(t) \tag{6-186}$$

It is assumed here that the derivative of the unit-step function is zero for $t \geq 0$.

Equation (6-186) is written

$$\frac{d\mathbf{y}(t)}{dt} = A\mathbf{y}(t) + [a_{j1}]r(t) + Bm(t) \tag{6-187}$$

where $[a_{j1}]$ is an $n \times 1$ column matrix which contains elements in the first column of A.

The state transition equation of Eq. (6-187) is

$$\mathbf{y}[(k+1)T] = \Phi(T)\mathbf{y}(kT) + \int_0^T \Phi(T-\tau)B\,d\tau\, m(kT) + \int_0^T \Phi(T-\tau)[a_{j1}]R\,d\tau \tag{6-188}$$

We define the following integrals:

$$\Theta(T) = \int_0^T \Phi(T - \tau)B\, d\tau \tag{6-189}$$

$$\Lambda(T) = \int_0^T \Phi(T - \tau)[a_{j1}]\, d\tau \tag{6-190}$$

Then, Eq. (6-188) becomes

$$\mathbf{y}[(k + 1)T] = \Phi(T)\mathbf{y}(kT) + \Theta(T)m(kT) + R\Lambda(T) \tag{6-191}$$

Solving Eq. (6-191) by recursion, we get

$$\mathbf{y}(NT) = \Phi(NT)\mathbf{y}(0) + \sum_{i=1}^{N} \Phi[(N - i)T]\Theta(T)m[(i - 1)T] + \sum_{i=0}^{N-1} R\Phi(iT)\Lambda(T) \tag{6-192}$$

To drive the system from $\mathbf{y}(0)$ to $\mathbf{y}(NT) = \mathbf{0}$ in N sampling periods, we set $\mathbf{y}(NT) = \mathbf{0}$ in Eq. (6-192), and solving for $\mathbf{y}(0)$, we get

$$\mathbf{y}(0) = -\sum_{i=1}^{N} \{\Phi[(-iT)]\Theta(T)m[(i - 1)T] - \Phi(-iT)R\Lambda(T)\} \tag{6-193}$$

Now comparing Eq. (6-193) with Eq. (6-93), it is simple to see that the region of controllable states R'_N in the $\mathbf{y}$ domain can be obtained by adding the vector $-\Phi(-NT)R\Lambda(T)$ to the vertices of R'_N in the $\mathbf{x}$ domain, and carrying out the linear transformation of Eq. (6-184).

For a second-order process, if the transfer function of the process is of the form:

$$G(s) = \frac{K}{s^2} \tag{6-194}$$

or

$$G(s) = \frac{K}{s(s + p_1)} \tag{6-195}$$

the A matrix contains all zeros in its first column; that is, $[a_{j1}] = \mathbf{0}$. This means that $\Lambda(T) = 0$, and Eq. (6-193) is of the same form as Eq. (6-93). Therefore, this means that the convex polygon for R'_N in the (x_1, x_2) state plane for $r(t) = Ru(t)$ is identical in shape to those for zero input, except that it is shifted along the x_1 axis by R units.

6.8 TIME-OPTIMAL CONTROL OF SYSTEMS WITH RAMP INPUTS

The method presented in the previous section can be extended to the time-optimal control of digital systems with ramp inputs.

Consider that the block diagram of the system is as shown in Figure 6-21. The reference input is

$$r(t) = Rt\,u(t) \tag{6-196}$$

where R is a constant. The dynamic equations of the linear process are given by Eqs. (6-182) and (6-183).

For the ramp input, the following new state vector is defined:

$$\mathbf{y}(t) = \mathbf{x}(t) - \mathbf{r}(t) \tag{6-197}$$

where

$$\mathbf{r}(t) = \begin{bmatrix} Rtu(t) \\ Ru(t) \\ 0 \\ \cdot \\ \cdot \\ \cdot \\ 0 \end{bmatrix} = \begin{bmatrix} r(t) \\ \dfrac{dr(t)}{dt} \\ 0 \\ \cdot \\ \cdot \\ \cdot \\ 0 \end{bmatrix} \quad (n \times 1 \text{ input vector}) \tag{6-198}$$

Then, the new state equation of the system is written

$$\frac{d\mathbf{y}(t)}{dt} = A\mathbf{y}(t) + A\mathbf{r}(t) - [1 \quad 0 \quad \ldots \quad 0]'R + Bm(t) \tag{6-199}$$

The state transition equation of the last equation is

$$\mathbf{y}[(k+1)T] = \Phi(T)\mathbf{y}(kT) + \Theta(T)m(kT) + \Delta(T)R \tag{6-200}$$

where $\Theta(T)$ is as given in Eq. (6-189), and

$$\Delta(T) = \int_0^T \Phi(T-\tau)\left\{a_{j1} + a_{j2} - \begin{bmatrix} 1 \\ 0 \\ \cdot \\ \cdot \\ \cdot \\ 0 \end{bmatrix}\right\} d\tau \tag{6-201}$$

where a_{j1} and a_{j2} are column matrices formed from the elements in the first and the second columns of the matrix A, respectively.

Solving Eq. (6-200) by recursion, we get

$$\mathbf{y}(NT) = \Phi(NT)\mathbf{y}(0) + \sum_{i=1}^{N} \Phi[(N-i)T]\Theta(T)m[(i-1)T] + \sum_{i=0}^{N-1} R\Phi(iT)\Delta(T) \tag{6-202}$$

Since Eq. (6-202) is of the same form as Eq. (6-192) which is for the step input case, the same conclusions can be drawn in regard to the regions of controllable states.

As a simple illustration of the procedure proposed above, let us consider that the linear process of the system shown in Figure 6-21 is described by the dynamic equations of Eqs. (6-182) and (6-183). The input to the system is a ramp input, $r(t) = Rtu(t)$. The A and B matrices are

$$A = \begin{bmatrix} 0 & 1 \\ 0 & -1 \end{bmatrix} \qquad B = \begin{bmatrix} 0 \\ 1 \end{bmatrix} \tag{6-203}$$

Defining the new state vector $\mathbf{y}(t)$ according to Eq. (6-197), the new state equation becomes

$$\frac{d\mathbf{y}(t)}{dt} = A\mathbf{y}(t) + A\mathbf{r}(t) - \begin{bmatrix} 1 \\ 0 \end{bmatrix} R + Bm(t) \tag{6-204}$$

Notice that $a_{j1} = \mathbf{0}$ in this case, and $\mathbf{r}(t)$ is given by Eq. (6-198). Equation (6-204) is simplified to

$$\frac{d\mathbf{y}(t)}{dt} = A\mathbf{y}(t) + B[m(t) - R] \tag{6-205}$$

For any initial state $\mathbf{y}(0)$ to be brought to the equilibrium $\mathbf{x}(NT) = \mathbf{0}$ in N sampling periods, we have

$$\mathbf{y}(0) = -\sum_{i=1}^{N} \Phi(-iT)\Theta(T)\{m[(i-1)T] - R\} \tag{6-206}$$

We notice that Eq. (6-206) is similar to the equation for controllability for zero input, except that the control signal is changed to $m[(i-1)T] - R$.

The vertices of the convex polygons for $R'_N(N = 1, 2, \ldots)$ are computed from Eq. (6-206) in the usual fashion.

For a sampling period of one second, the convex polygons for R'_N for several different cases are shown in Figure 6-22, and the vertices of R'_N are tabulated below.

$|m(kT)| \leq 1, N = 1$

$m(0)$	$y_1(0)$	$y_2(0)$
1	0	0
−1	−1.44	3.44

$|m(kT)| \leq 1, N = 2$

$m(0)$	$m(1)$	$y_1(0)$	$y_2(0)$
1	1	0	0
−1	1	−1.44	3.44
1	−1	−7.34	9.34
−1	−1	−8.78	12.78

$|m(kT)| \leq 2, N = 2$

$m(0)$	$m(1)$	$y_1(0)$	$y_2(0)$
2	2	4.39	−6.39
−2	2	1.51	0.49
2	−2	−10.29	12.39
−2	−2	−13.17	19.47

$|m(kT)| \leq 3, N = 2$

$m(0)$	$m(1)$	$y_1(0)$	$y_2(0)$
3	3	8.77	−12.77
−3	3	4.45	−2.45
3	−3	−13.23	15.23
−3	−3	−17.55	25.55

A word of caution must be given here concerning the use of the convex polygons for ramp input signals. One special nature of the ramp input is that when $t = 0$, $r(t) = 0$. Therefore, according to Eq. (6-197), if $x_1(0) = 0$, $y_1(0)$ also equals zero. Since $y_1(0) = e(0)$, and the output of the digital controller $h(kT)$ depends on $e(kT)$, we have $m(0) = h(0) = 0$. This means that if $y_1(0) = 0$, during the first period of state transition, $(0 \leq t \leq T)$,

$$\mathbf{r}(T) = \mathbf{\Phi}(T)\mathbf{y}(0) + \mathbf{\Theta}(T)[m(0) - R] \tag{6-207}$$

in which $m(0) = 0$. Therefore, with zero control signal applied during the first sampling interval, the convex polygon R'_N is described by

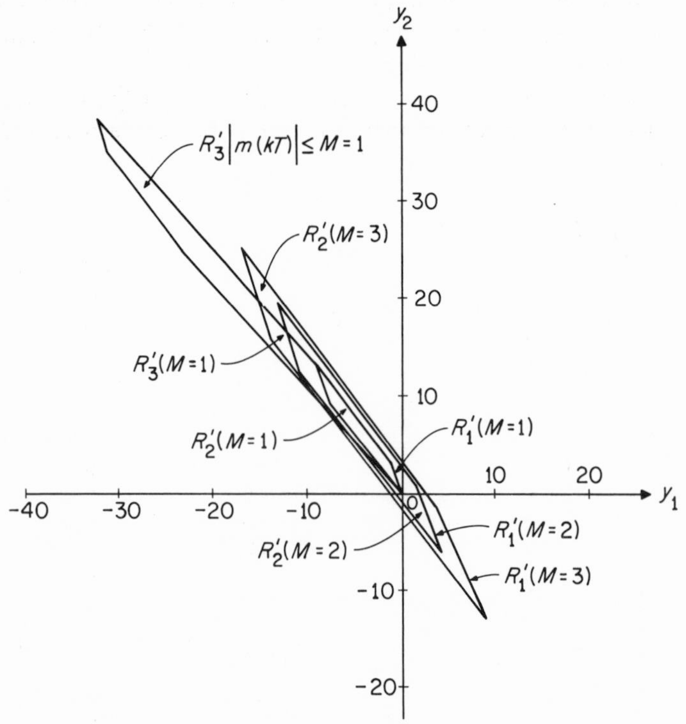

FIG. 6-22. Regions of controllable states of the second-order process, $G(s) = 1/s(s + 1)$, with $r(t) = tu(t)$.

$$\mathbf{y}(0) = \Theta(T)R - \sum_{i=2}^{N} \Phi(iT)\Theta(T)m[(i-1)T] \tag{6-208}$$

Therefore, when the initial state $\mathbf{y}(0)$ falls on the $y_1 = 0$ axis in the state plane, the polygons shown in Figure 6-22 are not valid, and Eq. (6-208) should be used.

EXAMPLE 6-7

Consider that the linear process of the system shown in Figure 6-21 is described by the following dynamic equations:

$$\begin{bmatrix} \dfrac{dx_1(t)}{dt} \\ \dfrac{dx_2(t)}{dt} \end{bmatrix} = \begin{bmatrix} 0 & 1 \\ 0 & -1 \end{bmatrix} \begin{bmatrix} x_1(t) \\ x_2(t) \end{bmatrix} + \begin{bmatrix} 0 \\ 1 \end{bmatrix} m(t) \tag{6-209}$$

$$c(t) = x_1(t)$$

The problem is to determine the optiomal control sequence $m(0)$, $m(T)$, . . . , so that the output of the system $c(t)$ will follow a unit ramp function input without error after a minimum number of sampling periods.

The nonlinearity of the system is defined as:

$$\begin{aligned} m(t) &= 2 & h(t) &\geq 2 \\ m(t) &= h(t) & 2 \geq h(t) &\geq -2 \\ m(t) &= -2 & -2 &\geq h(t) \end{aligned} \tag{6-210}$$

The sampling period is one second, and the initial state is

$$\begin{bmatrix} x_1(0) \\ x_2(0) \end{bmatrix} = \begin{bmatrix} 1 \\ 0 \end{bmatrix} \tag{6-211}$$

Using the state vector defined by Eq. (6-197), the new initial state vector is

$$\mathbf{y}(0) = \mathbf{x}(0) - \mathbf{r}(0) = \begin{bmatrix} 1 \\ 0 \end{bmatrix} - \begin{bmatrix} 0 \\ 1 \end{bmatrix} = \begin{bmatrix} 1 \\ -1 \end{bmatrix} \tag{6-212}$$

From the convex polygons shown in Figure 6-22, we observe that when $|m(kT)| \leq 2$, the initial state of Eq. (6-212) lies in R_2'; that is, $N = 2$. Therefore, for $N = 2$, Eq. (6-202) gives

$$\mathbf{y}(2) = \Phi(2)\mathbf{y}(0) + \sum_{i=1}^{2} \Phi(2-i)\Theta(1)[m(i-1) - 1] \tag{6-213}$$

For the system,

$$\Phi(1) = \begin{bmatrix} 1 & 0.632 \\ 0 & 0.368 \end{bmatrix}, \quad \Phi(2) = \begin{bmatrix} 1 & 0.865 \\ 0 & 0.135 \end{bmatrix} \tag{6-214}$$

and

$$\Theta(1) = \begin{bmatrix} 0.368 \\ 0.632 \end{bmatrix} \tag{6-215}$$

Therefore, solving for $m(0)$ and $m(1)$ from Eq. (6-213), we get

$$m(0) = 0.65 \tag{6-216}$$

$$m(1) = 1.34 \tag{6-217}$$

and $m(k) = 1.0$ for $k > 1$. The response of the optimum system is shown in Figure 6-23.

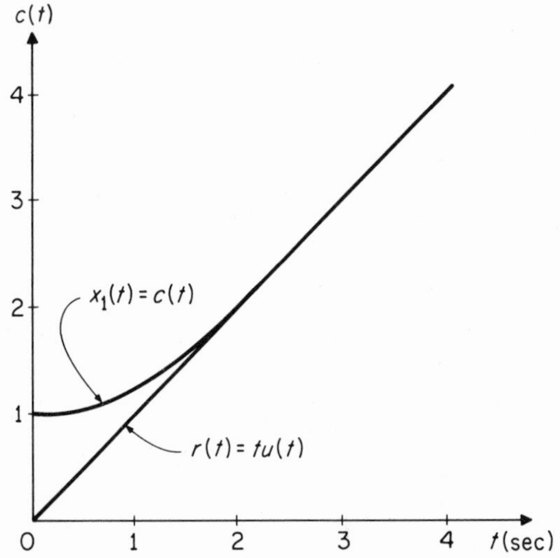

FIG. 6-23. Ramp response of time-optimal control system in Example 6-7.

6.9 TIME-OPTIMAL CONTROL OF MULTIVARIABLE SYSTEMS

In this section the minimum-time control of a linear process with a multiple number of controls and outputs will be considered. The process is described by the state equation

$$\frac{d\mathbf{x}(t)}{dt} = A\mathbf{x}(t) + B\mathbf{m}(t) \tag{6-218}$$

where

$$\mathbf{x}(t) = n \times 1 \text{ state vector}$$

$$A = n \times n \text{ constant matrix}$$

$$B = n \times r \text{ constant matrix}$$

$$\mathbf{m}(t) = r \times 1 \text{ control vector } (r \leq n)$$

and

$$\mathbf{m}(t) = \mathbf{m}(kT) \quad kT \leq t < (k+1)T$$

The design problem requires the determination of the optimal control strategy $\mathbf{m}(0), \mathbf{m}(T), \ldots, \mathbf{m}(N-1)T]$, so that the process is driven from $\mathbf{x}(0)$ to $\mathbf{x}(NT) = \mathbf{0}$ in minimum time NT.

The state transition equation of Eq. (6-218) at discrete time intervals is written

$$\mathbf{x}[(k+1)T] = \mathbf{\Phi}(T)\mathbf{x}(kT) + \mathbf{\Theta}(T)\mathbf{m}(kT) \tag{6-219}$$

The solution of the last equation is

$$\mathbf{x}(NT) = \mathbf{\Phi}(NT)\mathbf{x}(0) + \sum_{i=1}^{N-1} \mathbf{\Phi}[(N-i-1)T]\mathbf{\Theta}(T)\mathbf{m}(iT) \tag{6-220}$$

where N is a positive integer.

If the process is to be brought to the equilibrium $\mathbf{x}(NT) = \mathbf{0}$, Eq. (6-220) gives

$$\mathbf{x}(0) = -\sum_{i=0}^{N-1} \mathbf{\Phi}[(-i-1)T]\mathbf{\Theta}(T)\mathbf{m}(iT) \tag{6-221}$$

or

$$\mathbf{x}(0) = -\sum_{i=0}^{N=1} S_i(T)\mathbf{m}(iT) \tag{6-222}$$

where

$$S_i(T) = \mathbf{\Phi}[(-i-1)T]\mathbf{\Theta}(T) \tag{6-223}$$

Equation (6-222) can be written as

$$\mathbf{x}(0) = -\underbrace{[S_0(T) \quad S_1(T) \quad \ldots \quad S_{N-1}(T)]}_{\longleftarrow n \times rN \longrightarrow} \underbrace{\begin{bmatrix} \mathbf{m}(0) \\ \mathbf{m}(T) \\ \cdot \\ \cdot \\ \cdot \\ \mathbf{m}[(N-1)T] \end{bmatrix}}_{\leftarrow rN \times 1 \rightarrow} \tag{6-224}$$

or

$$\mathbf{x}(0) = -SM \tag{6-225}$$

Where S and M are implicitly defined by Eqs. (6-224) and (6-225). Since $S_i(T), i = 0, 1, \ldots, N-1$, is an $n \times r$ matrix, S is $n \times rN$. Unless $n = rN$, S will not be a square matrix, as is always the case when $m(kT)$ is a scalar control.

Given an arbitrary initial state $\mathbf{x}(0)$, the optimal control strategy is determined from Eq. (6-224) by forming nonsingular $n \times n$ matrix from n linearly independent columns of S. In other words, the necessary and sufficient condition for complete controllability of the states is that the matrix

$$S = [S_0(T) \quad S_1(T) \quad \dots \quad S_{N-1}(T)] \tag{6-226}$$

be of rank n. It is apparent that since S has n rows and rN columns, there is a certain degree of nonuniqueness in the selection of a nonsingular $n \times n$ matrix from S. We shall define a minimum-time control strategy by selecting the first n linearly independent columns from S.

Let Q be the $n \times n$ matrix which is formed by the first n linearly independent columns of S, and let M_t be the vector containing the rows of

$$M = \begin{bmatrix} \mathbf{m}(0) \\ \mathbf{m}(t) \\ \cdot \\ \cdot \\ \cdot \\ \mathbf{m}[(N-1)T] \end{bmatrix} \tag{6-227}$$

which correspond to the columns of Q taken from S. Then,

$$\mathbf{x}(0) = -QM_t \tag{6-228}$$

Since Q is a nonsingular matrix, the optimal control vector is determined from

$$M_t = -Q^{-1}\mathbf{x}(0) \tag{6-229}$$

The minimum time interval required for the optimal system to reach the desired equilibrium is determined as follows:

1. If the first n columns of S are linearly independent, the value of N is given by

$$\frac{n}{r} \leq N < \frac{n}{r} + 1 \tag{6-230}$$

In the special case when n/r is an integer, $N = n/r$, and the N optimum control vectors $\mathbf{m}(0), \mathbf{m}(T), \dots, \mathbf{m}[(N-1)T]$, are determined uniquely from

$$M = \begin{bmatrix} \mathbf{m}(0) \\ \mathbf{m}(T) \\ \cdot \\ \cdot \\ \cdot \\ \mathbf{m}[(N-1)T] \end{bmatrix} = -S^{-1}\mathbf{x}(0) \tag{6-231}$$

2. If the first L columns of S contain n linearly independent columns with $L \geq n$, then the number of sampling periods to reach equilibrium is given by

$$\frac{L}{r} \leq N < \frac{L}{r} + 1 \tag{6-232}$$

It should be noted that in either case, if n/r is not an integer, there may be $(Nr - n)$ degrees of freedom in selecting the optimal control strategy.

Optimal Control Strategy From State Variable Feedback

Since Eq. (6-231) is of the same form as Eq. (6-174), the optimal control in the multivariable case can also be obtained from state variable feedback. Therefore, the optimal control can be written

$$\mathbf{m}(kT) = -[\mathbf{I} \quad \mathbf{0}]S^{-1}\mathbf{x}(kT) \tag{6-233}$$

where I is an $r \times r$ unity matrix, $k = 0, 1, \ldots, N-1$. Or,

$$\mathbf{m}(kT) = C\mathbf{x}(kT) \tag{6-234}$$

where

$$C = -[\mathbf{I} \quad \mathbf{0}]S^{-1} \tag{6-235}$$

If S is not a square matrix Q^{-1} should replace S^{-1} in the last two equations.

EXAMPLE 6-8

Consider that a linear multivariable process is described by the following transfer function matrix.

$$\begin{bmatrix} C_1(s) \\ C_2(s) \end{bmatrix} = \begin{bmatrix} \frac{1}{s+2} & 0 \\ \frac{1}{s+1} & \frac{1}{s} \end{bmatrix} \begin{bmatrix} M_1(s) \\ M_2(s) \end{bmatrix} \tag{6-236}$$

The control signals $m_1(t)$ and $m_2(t)$ are outputs of zero-order hold devices. The sampling period is one second.

The dynamic equations of the system are written

$$\begin{bmatrix} \frac{dx_1}{dt} \\ \frac{dx_2}{dt} \\ \frac{dx_3}{dt} \end{bmatrix} = \begin{bmatrix} -2 & 0 & 0 \\ 0 & -1 & 0 \\ 0 & 0 & 0 \end{bmatrix} \begin{bmatrix} x_1 \\ x_2 \\ x_3 \end{bmatrix} + \begin{bmatrix} 1 & 0 \\ 1 & 0 \\ 0 & 1 \end{bmatrix} \begin{bmatrix} m_1 \\ m_2 \end{bmatrix} \tag{6-237}$$

$$\begin{bmatrix} C_1 \\ C_2 \end{bmatrix} = \begin{bmatrix} 1 & 0 & 0 \\ 0 & 1 & 1 \end{bmatrix} \begin{bmatrix} x_1 \\ x_2 \\ x_3 \end{bmatrix} \tag{6-238}$$

It is desired to determine the optimal control strategy so that the process is driven from $\mathbf{x}(0) = [1 \quad 1 \quad 1]'$ to $\mathbf{x}(NT) = \mathbf{0}$ in minimum time.

The discrete state transition equation of Eq. (6-237) is

$$\mathbf{x}(k+1) = \mathbf{\Phi}(1)\mathbf{x}(k) + \mathbf{\Theta}(1)\mathbf{m}(k) \tag{6-239}$$

where

$$\Phi(1) = \begin{bmatrix} e^{-2} & 0 & 0 \\ 0 & e^{-1} & 0 \\ 0 & 0 & 0 \end{bmatrix} \tag{6-240}$$

and

$$\Theta(1) = \begin{bmatrix} (1 - e^{-2})/2 & 0 \\ 1 - e^{-1} & 0 \\ 0 & 1 \end{bmatrix} \tag{6-241}$$

In the present case, $n = 3$ and $r = 2$, thus $n/r = 3/2$. Equation (6-230) suggests that if the process is controllable at all, $N = 2$. From Eq. (6-226),

$$S = [S_0(1) \quad S_1(1)] = [\Phi(-1)\Theta(1) \quad \Phi(-2)\Theta(1)]$$

$$= \begin{bmatrix} \frac{1}{2}(e^2 - 1) & 0 & \frac{1}{2}e^2(e^2 - 1) & 0 \\ (e^T - 1) & 0 & e^1(e^1 - 1) & 0 \\ 0 & 1 & 0 & 1 \end{bmatrix} \tag{6-242}$$

Substituting Eq. (6-242) into Eq. (6-224), we have

$$\begin{bmatrix} x_1(0) \\ x_2(0) \\ x_3(0) \end{bmatrix} = \begin{bmatrix} -\frac{1}{2}(e^2 - 1) & 0 & -\frac{1}{2}e^2(e^2 - 1) & 0 \\ -(e^1 - 1) & 0 & -e^1(e^1 - 1) & 0 \\ 0 & -1 & 0 & -1 \end{bmatrix} \begin{bmatrix} m_1(0) \\ m_2(0) \\ m_1(1) \\ m_2(1) \end{bmatrix} \tag{6-243}$$

It is apparent that the first three columns of S are linearly independent. Therefore, the optimal control strategy can be determined using these first three columns of S to form the nonsingular matrix Q in Eq. (6-229). Therefore, the result is

$$M_t = \begin{bmatrix} m_1(0) \\ m_2(0) \\ m_1(1) \end{bmatrix} = \begin{bmatrix} -\frac{1}{2}(e^2 - 1) & 0 & -\frac{1}{2}e^2(e^2 - 1) \\ -(e^1 - 1) & 0 & -e^1(e^1 - 1) \\ 0 & -1 & 0 \end{bmatrix}^{-1} \begin{bmatrix} x_1(0) \\ x_2(0) \\ x_3(0) \end{bmatrix}$$

$$= \begin{bmatrix} 0.182 & -0.922 & 0 \\ 0 & 0 & -1 \\ -0.067 & 0.124 & 0 \end{bmatrix} \begin{bmatrix} x_1(0) \\ x_2(0) \\ x_3(0) \end{bmatrix}$$

$$= \begin{bmatrix} -0.74 \\ -1 \\ 0.058 \end{bmatrix} \tag{6-244}$$

Since only three control signals are needed, by selecting the first three columns of S, the control $M_2(1)$ is set to zero.

Substituting the optimal controls into Eq. (6-239), the responses are determined.

$$\mathbf{x}(1) = \begin{bmatrix} -0.185 \\ -1 \\ 0 \end{bmatrix} \qquad \mathbf{x}(2) = \begin{bmatrix} 0 \\ 0 \\ 0 \end{bmatrix}$$

which proves that the system reaches the equilibrium state in two seconds.

As an alternative, the design is carried out using the state variable feedback method. Equation (6-233) gives

$$\mathbf{m}(kT) = \begin{bmatrix} m_1(kT) \\ m_2(kT) \end{bmatrix} = \begin{bmatrix} 1 & 0 & 0 \\ 0 & 1 & 0 \end{bmatrix} Q^{-1}\mathbf{x}(kT)$$

$$= \begin{bmatrix} 1 & 0 & 0 \\ 0 & 1 & 0 \end{bmatrix} \begin{bmatrix} 0.182 & -0.922 & 0 \\ 0 & 0 & -1 \\ -0.067 & 0.124 & 0 \end{bmatrix} \begin{bmatrix} x_1(kT) \\ x_2(kT) \\ x_3(kT) \end{bmatrix} \tag{6-245}$$

When $k = 0$,

$$\begin{bmatrix} m_1(0) \\ m_2(0) \end{bmatrix} = \begin{bmatrix} 0.182 & -0.922 & 0 \\ 0 & 0 & -1 \end{bmatrix} \begin{bmatrix} 1 \\ 1 \\ 1 \end{bmatrix} = \begin{bmatrix} -0.74 \\ -1 \end{bmatrix} \tag{6-246}$$

When $k = 1$,

$$\begin{bmatrix} m_1(1) \\ m_2(1) \end{bmatrix} = \begin{bmatrix} 0.182 & -0.922 & 0 \\ 0 & 0 & -1 \end{bmatrix} \begin{bmatrix} -0.185 \\ -1 \\ 0 \end{bmatrix} = \begin{bmatrix} 0.058 \\ 0 \end{bmatrix} \tag{6-247}$$

These results apparently agree with those obtained earlier.

6.10 TIME-OPTIMAL CONTROL OF SINGLE-VARIABLE SYSTEMS WITH INPUT DELAY[11,12]

Many industrial processes have signals transmitted through transmission lines or transducers with time delays. For a single input and output system, with time delay in the control signal, the block diagram of the system is as shown in Figure 6-24. For simplicity, we assume at present that the delay time is an

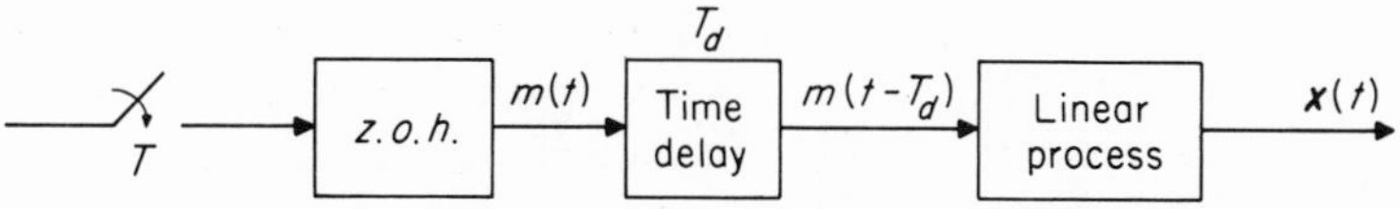

FIG. 6-24. A single-variable digital system with input delay.

integral multiple of the sampling period. In practice, if the sampling period. is small compared with the delay time, we can expect good results from this approximation. Thus, let T_d be the delay time and T the sampling period; then

$$T_d = pT \tag{6-248}$$

where p is a positive integer.

The state equation of the process with input delay is written

$$\frac{d\mathbf{x}(t)}{dt} = A\mathbf{x}(t) + Bm(t - pT) \tag{6-249}$$

where

$$\mathbf{x}(t) = n \times 1 \quad \text{state vector}$$
$$A = n \times n \quad \text{constant matrix}$$
$$B = n \times 1 \quad \text{constant matrix}$$

and

$$m(t - pT) = m(kT - pT) \tag{6-250}$$

$kT - pT \leq t < (k - p + 1)T$

The discrete state transition equation of Eq. (6-249) is

$$\mathbf{x}[(k + 1)T] = \Phi(T)\mathbf{x}(kT) + \Theta(T)m(kT - pT) \tag{6-251}$$

where

$$\Phi(T) = e^{AT}$$

and

$$\Theta(T) = \int_0^T \Phi(T - \tau)B\, d\tau \tag{6-252}$$

The solution of Eq. (6-251) is

$$\mathbf{x}(NT) = \Phi(NT)\mathbf{x}(0) + \sum_{i=0}^{N-1} \Phi[(N - i - 1)T]\Theta(T)m[(i - p)T] \tag{6-253}$$

The time-optimal design problem is to find the optimal control strategy, $m(0)$, $m(T), \ldots, m[(N - p - 1)T]$, given the initial state $\mathbf{x}(0)$ and the past inputs $m(-pT)$, $m[(-p + 1)T], \ldots, m(-T)$, so that the desired equilibrium state $\mathbf{0}$ is reached in a minimum number of sampling periods, N. Notice that in this case, in addition to the initial state, the past inputs to the process must also be specified.

Setting $\mathbf{x}(NT) = \mathbf{0}$ in Eq. (6-253), we get after rearranging

$$\mathbf{x}(0) = -\sum_{i=0}^{N-1} \Phi[(-i - 1)T]\Theta(T)m[(i - p)T] \tag{6-254}$$

The last equation is written as

$$\mathbf{x}(0) = -\underset{(n \times N)}{[S_0(T) \quad S_1(T) \quad \ldots \quad S_{N-1}(T)]} \underset{(N \times 1)}{\begin{bmatrix} m(-pT) \\ m[(-p+1)T] \\ \cdot \\ \cdot \\ \cdot \\ m(0) \\ m(T) \\ \cdot \\ \cdot \\ \cdot \\ m[(N-p-1)T] \end{bmatrix}} \tag{6-255}$$

where

$$S_i(T) = \Phi[(-i-1)T]\Theta(T) \tag{6-256}$$

Since the past inputs $m(-pT), m[(-p+1)T], \ldots, m(-T)$, are known, Eq. (6-255) can be written

$$\mathbf{x}(0) + \underset{(n \times p)}{[S_0(T) \quad S_1(T) \quad \ldots \quad S_{p-1}(T)]} \begin{bmatrix} m(-pT) \\ m[(-p+1)T] \\ \cdot \\ \cdot \\ \cdot \\ m(-T) \end{bmatrix}$$

$$= -\underset{n \times (N-p)}{[S_p(T) \quad S_{p+1}(T) \quad \ldots \quad S_{N-1}(T)]} \begin{bmatrix} m(0) \\ m(T) \\ \cdot \\ \cdot \\ \cdot \\ m[(N-p-1)T] \end{bmatrix} \tag{6-257}$$

The desired control strategy must be solved from the last equation. Since the matrix $[S_p(T) \quad S_{p+1}(T) \ldots S_{N-1}(T)]$ is n by $N - p$, in order that its inverse exists, $N - p$ must be at least equal to n; that is

$$N \geq n + p \tag{6-258}$$

This inequality establishes the minimum number of sampling periods for the process to reach the equilibrium in terms of the order of the system and the time delay, for arbitrary initial state and past inputs. Let

$$S = [S_p(T) \quad S_{p+1}(T) \quad \ldots \quad S_{N-1}(T)] \tag{6-259}$$

and if the first n columns of S are linearly independent, we can set $N = n + p$.

The optimal control strategy is obtained from Eq. (6-257):

$$\begin{bmatrix} m(0) \\ m(T) \\ \cdot \\ \cdot \\ \cdot \\ m[(N-p-1)T] \end{bmatrix} = -S^{-1}\left\{ \mathbf{x}(0) + [S_0(T) \quad S_1(T) \quad \dots \quad S_{p-1}(T)] \begin{bmatrix} m(-pT) \\ m[(-p+1)T] \\ \cdot \\ \cdot \\ \cdot \\ m(-T) \end{bmatrix} \right\} \tag{6-260}$$

If the n linearly independent columns are found from the first L columns of S, where $L > n$, the number of sampling periods to reach equilibrium is

$$N = L + p \tag{6-261}$$

Let the $n \times n$ matrix formed by the n linearly independent columns from the first L columns of S be Q. Then, the optimal control strategy is given by

$$\begin{bmatrix} m(0) \\ m(T) \\ \cdot \\ \cdot \\ \cdot \\ m[(N-p-1)T] \end{bmatrix}_Q = -Q^{-1}\left\{ \mathbf{x}(0) + [S_0(T) \quad S_1(T) \quad \dots \quad S_{p-1}(T)] \begin{bmatrix} m(-pT) \\ m[(-p+1)T] \\ \cdot \\ \cdot \\ \cdot \\ m(-T) \end{bmatrix} \right\} \tag{6-262}$$

where $[\ ]_Q$ denotes the $L \times 1$ column matrix which is formed by selecting the controls which correspond to the columns of Q.

For the single-variable system with input delay under consideration, the design of the optimal control strategy can also be effected by the state variable feedback approach. Let the delayed input be designated as

$$m(kT - pT) = w(kT) \tag{6-263}$$

Substitution of Eq. (6-263) into Eq. (6-251) gives

$$\mathbf{x}[(k+1)T] = \Phi(T)\mathbf{x}(kT) + \Theta(T)w(kT) \tag{6-264}$$

which is equivalent to the state transition equation of a system without delay. Now the problem is that of determing the optimal $w(0), w(T), \ldots, w(N-1)T$, so that $\mathbf{x}(NT) = \mathbf{0}$ for minimum N, a problem which was solved in Sec. 6.6. The optimal strategy for Eq. (6-264) is

$$w(kT) = C\mathbf{x}(kT) \tag{6-265}$$

where

$$C = -[1 \quad 0 \ldots 0][S_0(T) \quad S_1(T) \quad \ldots \quad S_{N-1}(T)]^{-1} \tag{6-266}$$

Once the optimal $w(kT)$ is determined, the optimal control $m(kT)$ is written

$$\begin{aligned} m(kT) &= w(kT + pT) \\ &= C\mathbf{x}[(k+p)T] \end{aligned} \tag{6-267}$$

The right-hand side of the last equation is expressed in terms of $\mathbf{x}(kT)$ and the past inputs by iterating Eq. (6-264). Therefore,

$$\mathbf{x}[(k+p)T] = \mathbf{\Phi}(pT)\mathbf{x}(kT) + \sum_{i=0}^{p-1} \mathbf{\Phi}[(-p-i-1)T]\mathbf{\Theta}(T)w[(k+i)T] \tag{6-268}$$

and

$$m(kT) = C\{\mathbf{\Phi}(pT)\mathbf{x}(kT) + \sum_{i=0}^{p-1} \mathbf{\Phi}[(p-i-1)T]\mathbf{\Theta}(T)m[(k+i-p)T]\} \tag{6-269}$$

The last expression gives the optimal control strategy in terms of the state vector $\mathbf{x}(kT)$ and the past controls $m[(k-p)T]$, $m[(k-p+1)T], \ldots,$ $m(-T)$.

EXAMPLE 6-9

Consider that the linear process with input delay shown in Figure 6-24 is described by the discrete state transition equation

$$\mathbf{x}[(k+1)T] = \mathbf{\Phi}(kT) + \mathbf{\Theta}(T)m[(k-1)T] \tag{6-270}$$

where

$$\mathbf{\Phi}(T) = \begin{bmatrix} 1 & 1-e^{-T} \\ 0 & e^{-T} \end{bmatrix} \qquad \mathbf{\Theta}(T) = \begin{bmatrix} T + e^{-T} - 1 \\ 1 - e^{-T} \end{bmatrix}$$

The sampling period is one second.

Given the initial state

$$\mathbf{x}(0) = \begin{bmatrix} -0.5 \\ 1.093 \end{bmatrix}$$

and the past input $m(-T) = 0$, it is desired to determine the optimal control strategy $m(kT)$, $k = 0, 1, \ldots, N-1$, so that the system is brought to the equilibrium state $\mathbf{x}(NT) = 0$ for minimum N.

Since $p = 1$ and $n = 2$, we must evaluate the matrices $S_0(T)$, $S_1(T)$, and $S_2(T)$. We have

$$S_0(T) = \Phi(-T)\Theta(T) = \begin{bmatrix} -0.72 \\ 1.72 \end{bmatrix}$$

$$S_1(T) = \Phi(-2T)\Theta(T) = \begin{bmatrix} -3.67 \\ 4.67 \end{bmatrix}$$

$$S_2(T) = \Phi(-3T)\Theta(T) = \begin{bmatrix} -11.72 \\ 12.72 \end{bmatrix}$$

Since

$$S = [S_1(T) \quad S_2(T)] = \begin{bmatrix} -3.67 & -11.72 \\ 4.67 & 12.72 \end{bmatrix}$$

is nonsingular, the optimal strategy is given by Eq. (6-260) with $N = 3$.

$$\begin{aligned} \begin{bmatrix} m(0) \\ m(1) \end{bmatrix} &= -\begin{bmatrix} -3.67 & -11.72 \\ 4.67 & 12.72 \end{bmatrix}^{-1} \left\{ \begin{bmatrix} -0.5 \\ 1.093 \end{bmatrix} + \begin{bmatrix} -0.72 \\ 1.72 \end{bmatrix} m(-1) \right\} \\ &= -\frac{1}{8}\begin{bmatrix} 12.72 & 11.72 \\ -4.67 & -3.67 \end{bmatrix}\begin{bmatrix} -0.5 \\ 1.093 \end{bmatrix} = \begin{bmatrix} -0.81 \\ 0.21 \end{bmatrix} \end{aligned} \tag{6-271}$$

Therefore, the process will be driven to the equilibrium state in three seconds with the control $m(0) = -0.81$ and $m(1) = 0.21$.

As an alternative, we may use the expression in Eq. (6-269). Equation (6-266) gives

$$\begin{aligned} C &= -[1 \quad 0][S_0(T) \quad S_1(T)]^{-1} \\ &= [-1.58 \quad -1.24] \end{aligned} \tag{6-272}$$

Substituting Eq. (6-272) into Eq. (6-269), we have

$$\begin{aligned} m(kT) &= [-1.58 \quad -1.24] \left\{ \begin{bmatrix} 1 & 0.632 \\ 0 & 0.368 \end{bmatrix} \begin{bmatrix} x_1(kT) \\ x_2(kT) \end{bmatrix} + \begin{bmatrix} 0.368 \\ 0.632 \end{bmatrix} m[(k-1)T] \right\} \\ &= [-1.58 \quad -1.46]\mathbf{x}(kT) - 1.36m[(k-1)T] \end{aligned} \tag{6-273}$$

For $k = 0$, and given that $m(-T) = 0$, the last equation gives

$$m(0) = [-1.58 \quad -1.46]\mathbf{x}(0) = -0.81 \tag{6-274}$$

Substituting this optimal control into Eq. (6-270), for $k = 0$, we get

$$\begin{bmatrix} x_1(1) \\ x_2(1) \end{bmatrix} = \begin{bmatrix} 1 & 0.632 \\ 0 & 0.368 \end{bmatrix} \begin{bmatrix} -0.5 \\ 1.093 \end{bmatrix} = \begin{bmatrix} 0.192 \\ 0.402 \end{bmatrix} \tag{6-275}$$

When $k = 1$, Eq. (6-273) gives

$$m(1) = [-1.58 \quad -1.46] \begin{bmatrix} 0.192 \\ 0.402 \end{bmatrix} - 1.36(-0.81) = 0.21 \tag{6-276}$$

which is the optimal control at $t = 1$ second. Substitution of $m(1)$ into Eq. (6-270), with $k = 1$, we obtain

$$\mathbf{x}(2) = \begin{bmatrix} 0.152 \\ -0.357 \end{bmatrix}$$

Substitution of $\mathbf{x}(2)$ and the value of $m(1)$ in Eq. (6-276) into Eq. (6-273), with $k = 2$, we obtain

$$m(2) = 0$$

as expected. Therefore, with $k = 3$, Eq. (6-270) brings the result

$$\mathbf{x}(3T) = \begin{bmatrix} 1 & 0.632 \\ 0 & 0.368 \end{bmatrix} \begin{bmatrix} 0.152 \\ -0.357 \end{bmatrix} + \begin{bmatrix} 0.368 \\ 0.632 \end{bmatrix} 0.21 = \begin{bmatrix} 0 \\ 0 \end{bmatrix} \tag{6-277}$$

which is the desired equilibrium state.

Therefore, we have obtained the same results using the two different methods. The second method using the state variable feedback is also a recursive method and is convenient for computer solution.

6.11 TIME-OPTIMAL CONTROL OF MULTIVARIABLE SYSTEMS WITH INPUT DELAYS

In this section we shall consider the design of minimum-time multivariable systems with input delays. The block diagram of a typical multivariable process with time delays in the inputs is shown in Figure 6-25. The delays are assumed to be integral multiples of the sampling periods, that is,

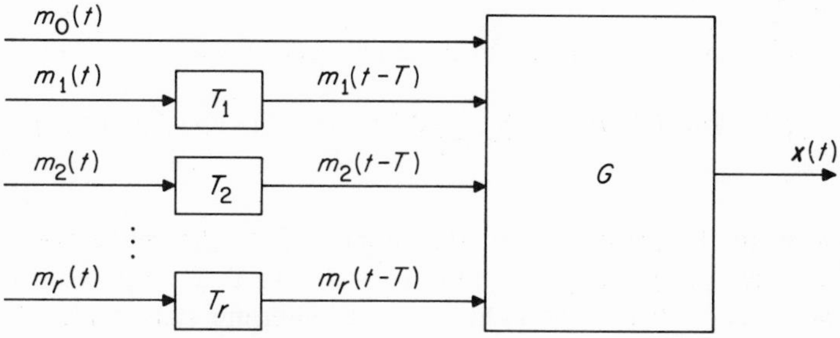

FIG. 6-25. Block diagram of a multivariable process with input delays.

$$T_j = p_j T \qquad j = 0, 1, 2, \ldots, r \tag{6-278}$$

where p_j is a positive integer, and T is the sampling period. For the convenience of representation and analysis, we assume that the inputs are numbered in order of increasing delay times so that

$$0 = p_0 < p_1 \leq p_2 \leq \ldots \leq p_r \tag{6-279}$$

The state equations of the process may be written as

$$\frac{d\mathbf{x}(t)}{dt} = A\mathbf{x}(t) + \sum_{j=0}^{r} B_j \mathbf{m}(t - p_j T) \tag{6-280}$$

where

$\mathbf{x}(t) = n \times 1$ state vector

$\mathbf{m}(t - p_j T) = (r + 1) \times 1$ input vector with the jth time delay

$B_j = n \times (r + 1)$ constant matrix

It is assumed that

$$\mathbf{m}(t - p_j T) = \mathbf{m}(kT - p_j T) \qquad kT \leq t < (k + 1)T \tag{6-281}$$

The solution of Eq. (6-280) for $t \geq t_0$ is

$$\mathbf{x}(t) = \Phi(t - t_0)\mathbf{x}(t_0) + \int_{t_0}^{t} \Phi(t - \tau) \sum_{j=0}^{r} B_j \mathbf{m}(\tau - p_j T)\, d\tau \tag{6-282}$$

and using the property of Eq. (6-281), with $t_0 = kT$, $t = (k + 1)T$,

$$\mathbf{x}[(k + 1)T] = \Phi(T)\mathbf{x}(kT) + \sum_{j=0}^{r} \Theta_j(T)\mathbf{m}[(k - p_j)T] \tag{6-283}$$

where

$$\Theta_j(T) = \int_0^T \Phi(T - \tau)B_j\, d\tau \tag{6-284}$$

Solving Eq. (6-283) by recursion in the usual fashion, we have

$$\mathbf{x}[(k+N)T] = \Phi(NT)\mathbf{x}(kT) + \sum_{i=0}^{N-1} \sum_{j=0}^{r} \Phi[(N-i-1)T]\Theta_j(T)\mathbf{m}[(k-p_j+i)T] \tag{6-285}$$

Similar to the previous case, the problem is to determine the optimal control strategy $\mathbf{m}(0)$, $\mathbf{m}(T), \ldots, \mathbf{m}[(N - 1)T]$, so that given $\mathbf{x}(0)$ and the past inputs, the system is brought to the equilibrium state $\mathbf{x}(NT) = \mathbf{0}$ for $N =$ minimum.

Setting $\mathbf{x}[(k + N)T] = \mathbf{0}$, Eq. (6-285) gives

$$\mathbf{x}(kT) = -\sum_{i=0}^{N-1} \sum_{j=0}^{r} \Phi[(-i - 1)T]\Theta_j(T)\mathbf{m}[(k - p_j + i)T] \tag{6-286}$$

Now expanding the right-hand side of the last equation, we get

$$\begin{aligned}
\mathbf{x}(kT) = &- \sum_{i=0}^{N-1} \Phi[(-i-1)T]\{\Theta_0(T)\mathbf{m}[(k-p_0+i)T] \\
&+ \Theta_1(T)\mathbf{m}[(k-p_1+i)T] + \ldots + \Theta_r(T)\mathbf{m}[(k-p_r+i)T]\} \\
= &- \Phi(-T)\{\Theta_0(T)\mathbf{m}[(k-p_0)T] + \Theta_1(T)\mathbf{m}[(k-p_1)T] \\
&+ \ldots + \Theta_r(T)\mathbf{m}[(k-p_r)T]\} \\
&- \Phi(-2T)\{\Theta_0(T)\mathbf{m}[(k-p_0+1)T] + \Theta_1(T)\mathbf{m}[(k-p_1+1)T] \\
&+ \ldots + \Theta_r(T)\mathbf{m}[(k-p_r+1)T]\} \\
&- \ldots - \Phi(-NT)\{\Theta_0(T)\mathbf{m}[(k-p_0+N-1)T] \\
&+ \Theta_1(T)\mathbf{m}[(k-p_1+N-1)T] \\
&+ \ldots + \Theta_r(T)\mathbf{m}[(k-p_r+N-1)T]\}
\end{aligned} \tag{6-287}$$

We notice that the control signals in the last equation all belong to the following categories:

$$\begin{aligned}
&\mathbf{m}[(k-p_r)T],\ \mathbf{m}[(k-p_{r-1})T], \ldots, \mathbf{m}[(k-p_2)T],\ \mathbf{m}[(k-p_1)T], \\
&\qquad \mathbf{m}(kT),\ \mathbf{m}[(k+1)T], \ldots, \mathbf{m}[(k+N-1)T]
\end{aligned} \tag{6-288}$$

where it is assumed that $p_0 = 0$.

Thus, Eqs. (6-286) and (6-287) can be written in the following matrix form:

$$\begin{aligned}
\mathbf{x}(kT) = &- [S_{-p_r} \quad S_{-p_{r-1}} \quad \ldots \quad S_{-p_1}] \begin{bmatrix} \mathbf{m}[(k-p_r)T] \\ \mathbf{m}[(k-p_{r-1})T] \\ \vdots \\ \mathbf{m}[(k-p_1)T] \end{bmatrix} \\
&- [S_0 \quad S_1 \quad \ldots \quad S_{N-1}] \begin{bmatrix} \mathbf{m}(kT) \\ \mathbf{m}[(k+1)T] \\ \vdots \\ \mathbf{m}[(N+k-1)T] \end{bmatrix}
\end{aligned} \tag{6-289}$$

It remains for us to define the matrices $S_{-p_r}, S_{-p_{r-1}}, \ldots, S_{-p_1}$, and $S_0, S_1, \ldots, S_{N-1}$.

Let us simplify the notation by letting $p_j = j$, where $j = 0, 1, 2, \ldots, r$. This is, of course, not necessarily true, since in general, $p_j = i + j$, where i is any integer so that the inequalities in Eq. (6-279) are satisfied.

Substituting $p_j = j$ in Eq. (6-287), we have

$$
\begin{aligned}
\mathbf{x}(kT) = &- \Phi(-T)\{\Theta_0(T)\mathbf{m}(kT) + \Theta_1(T)\mathbf{m}[(k-1)T] \\
&+ \ldots + \Theta_{r-1}(T)\mathbf{m}[(k-r+1)T] + \Theta_r(T)\mathbf{m}[(k-r)T]\} \\
&- \Phi(-2T)\{\Theta_0(T)\mathbf{m}[(k+1)T] + \Theta_1(T)\mathbf{m}(kT) \\
&+ \Theta_2(T)\mathbf{m}[(k-1)T] + \ldots + \Theta_{r-1}(T)\mathbf{m}[(k-r+2)T] \\
&+ \Theta_r(T)\mathbf{m}[(k-r+1)T]\} \\
&- \Phi(-3T)\{\Theta_0(T)\mathbf{m}[(k+2)T] + \Theta_1(T)\mathbf{m}[(k+1)T] \\
&+ \ldots + \Theta_{r-1}(T)\mathbf{m}[(k-r+1)T] + \Theta_r(T)\mathbf{m}[(k-r+2)T]\} \\
&- \ldots - \Phi(-rT)\{\Theta_0(T)\mathbf{m}[(k+r-1)T] \\
&+ \Theta_1(T)\mathbf{m}[(k+r-2)T] + \ldots + \Theta_{r-1}(T)\mathbf{m}[(k-2)T] \\
&+ \Theta_r(T)\mathbf{m}[(k-1)T]\} \\
&- \ldots - \Phi[(-N+1)T]\{\Theta_0(T)\mathbf{m}[(k+N-2)T] \\
&+ \Theta_1(T)\mathbf{m}[(k+N-3)T] + \ldots + \Theta_{r-1}(T)\mathbf{m}[(k+N-r-1)T] \\
&+ \Theta_r(T)\mathbf{m}[(k+N-r-2)T]\} \\
&- \Phi(-NT)\{\Theta_0(T)\mathbf{m}[(k+N-1)T] + \Theta_1(T)\mathbf{m}[(k+N-2)T] \\
&+ \ldots + \Theta_{r-1}(T)\mathbf{m}[(k+N-r)T] + \Theta_r(T)\mathbf{m}[(k+N-r-1)T]\}
\end{aligned}
\tag{6-290}
$$

Rearranging and collecting terms, Eq. (6-290) is written

$$
\begin{aligned}
\mathbf{x}(kT) = &- \Phi(-T)\Theta_r(T)\mathbf{m}[(k-r)T] \\
&- \{\Phi(-T)\Theta_{r-1}(T) + \Phi(-2T)\Theta_r(T)\}\mathbf{m}[(k-r+1)T] \\
&- \{\Phi(-T)\Theta_{r-2}(T) + \Phi(-2T)\Theta_{r-1}(T) \\
&+ \Phi(-3T)\Theta_r(T)\}\mathbf{m}[(k-r+2)T] \\
&- \ldots - \{\Phi(-T)\Theta_2(T) + \Phi(-2T)\Theta_3(T) \\
&+ \ldots + \Phi[(-r+1)T]\Theta_r(T)\}\mathbf{m}[(k-2)T] \\
&- \{\Phi(-T)\Theta_1(T) + \Phi(-2T)\Theta_2(T) \\
&+ \ldots + \Phi(-rT)\Theta_r(T)\}\mathbf{m}[(k-1)T] \\
&- \{\Phi(-T)\Theta_0(T) + \Phi(-2T)\Theta_1(T) + \ldots + \Phi(-rT)\Theta_{r-1}(T) \\
&+ \Phi[(-r-1)T]\Theta_r(T)\}\mathbf{m}(kT) \\
&- \{\Phi(-2T)\Theta_0(T) + \Phi(-3T)\Theta_1(T) \\
&+ \ldots + \Phi(-rT)\Theta_{r-2}(T) + \Phi[(-r-1)T]\Theta_{r-1}(T) \\
&+ \Phi[(-r-2)T]\Theta_r(T)\}\mathbf{m}[(k+1)T] \\
&- \ldots - \{\Phi[(-r-1)T]\Theta_0(T) + \Phi[(-r-2)T]\Theta_1(T) \\
&+ \ldots + \Phi(-2T)\Theta_{r-1}(T) + \Phi(-T)\Theta_r(T)\}\mathbf{m}[(k+r)T] \\
&- \ldots - \{\Phi[(-N+r-1)T]\Theta_0(T) + \Phi[(-N+r-2)T]\Theta_1(T)
\end{aligned}
$$

$$+ \ldots + \Phi[(-N+1)T]\Theta_{r-2}(T)$$
$$+ \Phi(-NT)\Theta_{r-1}(T)\}\mathbf{m}[(k+N-r)T]$$
$$- \ldots - \{\Phi(-N+2)\Theta_0(T) + \Phi[(-N+1)T]\Theta_1(T)$$
$$+ \Phi(-NT)\Theta_2(T)\}\mathbf{m}[(k+N-3)T]$$
$$- \{\Phi[(-N+1)T]\Theta_0(T) + \Phi(-NT)\Theta_1(T)\}\mathbf{m}[(k+N-2)T]$$
$$- \Phi(NT)\Theta_0(T)\mathbf{m}[(k+N-1)T] \quad (6\text{-}291)$$

Replacing $j = 0, 1, 2, \ldots, r$ by p_j and comparing Eq. (6-291) with the elements of the matrix equation of Eq. (6-289), the following expressions are obtained:

$$S_{-u} = \sum_{p_j=u}^{p_r} \Phi[(-p_j + u - 1)T]\Theta_j(T) \quad (6\text{-}292)$$
$$u = p_1, p_2, \ldots, p_r$$
$$S_v = \sum_{p_j=p_0}^{p_r} \Phi[(-p_j - v - 1)T]\Theta_j(T) \quad (6\text{-}293)$$
$$v = 0, 1, 2, \ldots, N - p_r - 1$$

and

$$S_v = \sum_{p_j=p_0}^{N-v-1} \Phi[(-p_j - v - 1)T]\Theta_j(T) \quad (6\text{-}294)$$
$$v = N - p_r, N - p_r + 1, \ldots, N - 1$$

Knowledge of the S_{-u} and S_v matrices in Eqs. (6-292), (6-293), and (6-294), the initial state $\mathbf{x}(kT)$, and the past inputs is sufficient to determine the optimum control sequence for minimum-time control from Eq. (6-289). The optimum control sequence is obtained by selecting from the matrix $[S_0 \quad S_1 \ldots S_{N-1}]$ the first n linearly independent columns and forming with them the $n \times r$ nonsingular matrix Q. In general, the matrix $[S_0 \quad S_1 \ldots S_{N-1}]$ is $n \times (r+1)N$.

From Eq. (6-289) the optimum input vector is obtained as

$$\begin{bmatrix} \mathbf{m}(kT) \\ \mathbf{m}[(k+1)T] \\ \cdot \\ \cdot \\ \cdot \\ \mathbf{m}[(k+N-1)T] \end{bmatrix}_Q = -Q^{-1}\left\{\mathbf{x}(kT) + [S_{-p_r} \quad S_{-p_{r-1}} \ldots S_{-p_1}] \begin{bmatrix} \mathbf{m}[(k-p_r)T] \\ \mathbf{m}[(k-p_{r-1})T] \\ \cdot \\ \cdot \\ \cdot \\ \mathbf{m}[(k-p_1)T] \end{bmatrix}\right\} \quad (6\text{-}295)$$

where $[\ ]_Q$ denotes the $n \times 1$ column matrix which is formed by selecting the control inputs that correspond to the columns of Q.

If the first L columns of $[S_0 \quad S_1 \ldots S_{N-1}]$ contain n linearly independent columns, then a P step control strategy evolves where P is a positive integer such that

$$\frac{L}{r} \leq P < \frac{L}{r} + 1 \tag{6-296}$$

Let $\mathbf{m}_j[(k + P)T]$ be the row of $[m(kT), m[(k + 1)T] \ldots m[(k + N - 1)T]'$ which is multiplied by the Lth column of S. Then the required settling time is $P + p_r - 1$ sampling periods if $p_j < p_r$, and $P + p_r$ sampling periods if $p_j = p_r$.

In the multivariable case we can again adopt the state feedback concept of design if at least the first r columns of $[S_0 \quad S_1 \ldots S_{N-1}]$ are used to form the nonsingular Q. Then, from Eq. (6-295) we write

$$\mathbf{m}(kT) = \begin{bmatrix} m_0(kT) \\ m_1(kT) \\ \cdot \\ \cdot \\ \cdot \\ m_r(kT) \end{bmatrix}$$

$$= -[I \quad 0]Q^{-1}\left\{\mathbf{x}(kT) + [S_{-p_r} \quad S_{-p_{r-1}} \ldots S_{-p_1}]\begin{bmatrix} \mathbf{m}[(k - p_r)T] \\ \mathbf{m}[(k - p_{r-1})T] \\ \cdot \\ \cdot \\ \cdot \\ \mathbf{m}[(k - p_1)T] \end{bmatrix}\right\} \tag{6-297}$$

where I is an $(r + 1) \times (r + 1)$ unity matrix.

EXAMPLE 6-10

Consider that a linear multivariable process with input delays is described by the state equation

$$\begin{bmatrix} \dfrac{dx_1(t)}{dt} \\ \dfrac{dx_2(t)}{dt} \end{bmatrix} = \begin{bmatrix} -1 & 0 \\ 0 & -2 \end{bmatrix}\begin{bmatrix} x_1(t) \\ x_2(t) \end{bmatrix} + \begin{bmatrix} 1 \\ 0 \end{bmatrix} m_0(t) + \begin{bmatrix} 0.5 \\ 1 \end{bmatrix} m_1(t - T) \tag{6-298}$$

The state equation is written in accordance with Eq. (6-280)

$$\begin{bmatrix} \dfrac{dx_1(t)}{dt} \\ \dfrac{dx_2(t)}{dt} \end{bmatrix} = \begin{bmatrix} -1 & 0 \\ 0 & -2 \end{bmatrix}\begin{bmatrix} x_1(t) \\ x_2(t) \end{bmatrix} + \begin{bmatrix} 1 & 0 \\ 0 & 0 \end{bmatrix}\begin{bmatrix} m_0(t) \\ m_1(t) \end{bmatrix} + \begin{bmatrix} 0 & 0.5 \\ 0 & 1 \end{bmatrix}\begin{bmatrix} m_0(t - T) \\ m_1(t - T) \end{bmatrix} \tag{6-299}$$

The initial state is

$$\mathbf{x}(0) = \begin{bmatrix} 1.5 \\ 0.5 \end{bmatrix} \tag{6-300}$$

The past control signal is given as

$$\mathbf{m}(-T) = \begin{bmatrix} m_0(-T) \\ m_1(-T) \end{bmatrix} = \begin{bmatrix} 0 \\ -1 \end{bmatrix} \tag{6-301}$$

and the sampling period is $T = 0.4$ second.

It is desired that the process can be driven from the initial state to the equilibrium state $\mathbf{x}(NT) = \mathbf{0}$ for minimum N. The following transition matrices are calculated:

$$\Phi(T) = \begin{bmatrix} e^{-T} & 0 \\ 0 & e^{-2T} \end{bmatrix} = \begin{bmatrix} 0.67 & 0 \\ 0 & 0.45 \end{bmatrix}$$

$$\Phi(-T) = \begin{bmatrix} 1.49 & 0 \\ 0 & 2.22 \end{bmatrix}$$

$$\Phi(-2T) = \begin{bmatrix} 2.23 & 0 \\ 0 & 4.95 \end{bmatrix}$$

$$\Theta_0(T) = \int_0^T \Phi(T-\tau)B_0\,d\tau = \begin{bmatrix} 0.33 & 0 \\ 0 & 0 \end{bmatrix}$$

$$\Theta_1(T) = \int_0^T \Phi(T-\tau)B_1\,d\tau = \begin{bmatrix} 0 & 0.165 \\ 0 & 0.275 \end{bmatrix}$$

For this problem, there are two control inputs, so $r = 0$ and $r = 1$. Therefore, in Eq. (6-292)

$$S_{-p_1} = \Phi(-T)\Theta_1(T) \tag{6-302}$$

In Eq. (6-293), if $N = 2$,

$$S_0 = \sum_{p_j=0}^{p_1} \Phi[(-p_j - 1)T]\Theta_j(T)$$
$$= \Phi(-T)\Theta_0(T) + \Phi(-2T)\Theta_1(T) \tag{6-303}$$

In Eq. (6-294),

$$S_1 = \Phi(-2T)\Theta_0(T) \tag{6-304}$$

Therefore,

$$S_{-p_1} = \begin{bmatrix} 0 & 0.246 \\ 0 & 0.611 \end{bmatrix}$$

$$S_0 = \begin{bmatrix} 0.49 & 0 \\ 0 & 0 \end{bmatrix} + \begin{bmatrix} 0 & 0.368 \\ 0 & 1.36 \end{bmatrix} = \begin{bmatrix} 0.49 & 0.368 \\ 0 & 1.36 \end{bmatrix}$$

$$S_1 = \begin{bmatrix} 0.735 & 0 \\ 0 & 0 \end{bmatrix}$$

and

$$[S_0 \quad S_1] = \begin{bmatrix} 0.49 & 0.368 & 0.735 & 0 \\ 0 & 1.36 & 0 & 0 \end{bmatrix} \tag{6-305}$$

We notice that the first two columns of the last matrix form a nonsingular $n \times n$ matrix. Therefore, we let

$$Q = \begin{bmatrix} 0.49 & 0.368 \\ 0 & 1.36 \end{bmatrix} \tag{6-306}$$

The optimal control strategy is now obtained by using Eq. (6-295).

$$\begin{bmatrix} \mathbf{m}(kT) \\ m[(k+1)T] \end{bmatrix}_Q = \mathbf{m}(kT) = -\begin{bmatrix} 0.49 & 0.368 \\ 0 & 1.36 \end{bmatrix}^{-1} \left\{ \mathbf{x}(kT) + \begin{bmatrix} 0 & 0.246 \\ 0 & 0.611 \end{bmatrix} \mathbf{m}[(k-1)T] \right\} \tag{6-307}$$

With $k = 0$, Eq. (6-307) gives

$$\begin{bmatrix} m_0(0) \\ m_1(0) \end{bmatrix} = -\begin{bmatrix} 0.49 & 0.368 \\ 0 & 1.36 \end{bmatrix}^{-1} \left\{ \begin{bmatrix} 1.5 \\ 0.5 \end{bmatrix} + \begin{bmatrix} 0 & 0.246 \\ 0 & 0.611 \end{bmatrix} \begin{bmatrix} 0 \\ 1 \end{bmatrix} \right\} = \begin{bmatrix} -2.43 \\ -1.51 \end{bmatrix} \tag{6-308}$$

6.12 DESIGN OF TIME-OPTIMAL SYSTEM BY MAXIMUM PRINCIPLE[16]

The maximum principle of Pontryagin has proved to be valuable in the design of optimal control systems. In this section we shall apply the maximum principle to the design of time-optimal sampled-data control systems. Since the development and proof of the maximum principle already appear in many current texts, we shall consider only the application of the technique to the current design problem.

The model of the system under consideration is shown in Figure 6-26.

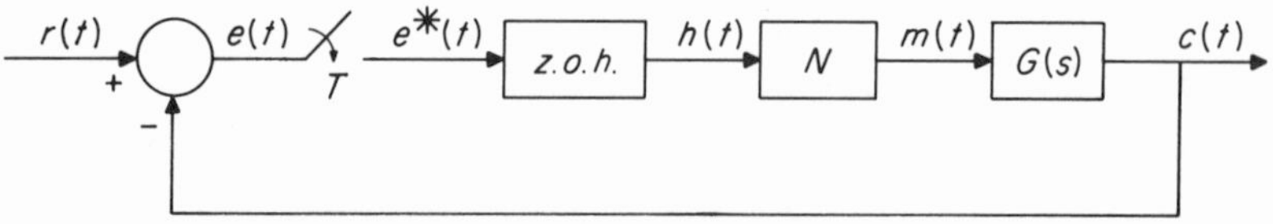

FIG. 6-26. Block diagram of nonlinear sampled-data control system.

The linear process is described by the dynamic equations

$$\frac{d\mathbf{x}(t)}{dt} = A\mathbf{x}(t) + Bm(t) \tag{6-309}$$

$$c(t) = D\mathbf{x}(t) \tag{6-310}$$

where $\mathbf{x}(t)$ is the $n \times 1$ state vector, $m(t)$ is the scalar control signal, and $c(t)$

is the scalar output of the system. The matrices A, B, and D all contain constant elements. The characteristics of the nonlinear element N are described by

$$\begin{aligned} m(t) &= +M \qquad h(t) \geq 1 \\ m(t) &= -M \qquad h(t) \leq -1 \\ m(t) &= h(t) \qquad -1 \leq h(t) \leq 1 \end{aligned} \tag{6-311}$$

The sampling period T is to be specified after the determination of the optimal control $m(t)$. The problem is to determine the optimal input $r(t)$ over the time interval $0 \leq t \leq t_f$ so that the system is brought from an initial state $\mathbf{x}(0)$ to the final state $\mathbf{x}(t_f)$ in the shortest possible time. Once the system has reached $\mathbf{x}(t_f)$, it should remain at that state.

Since the nonlinear element is amplitude-dependent, for analytical purposes, its position in the block diagram of Figure 6-26 can be interchanged with that of the zero-order hold. The block diagram of Figure 6-27 results.

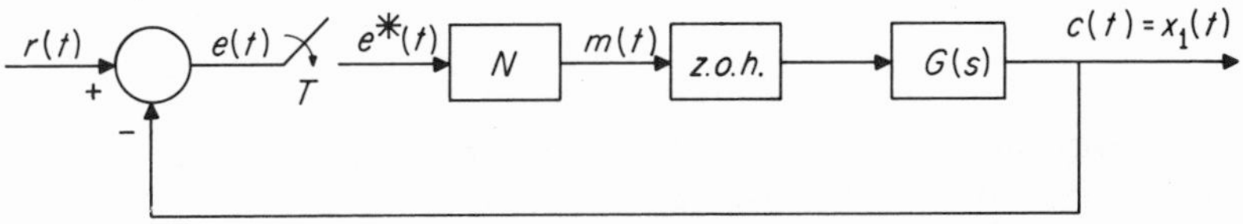

FIG. 6-27. Block diagram of nonlinear sampled-data control system with the nonlinearity and zero-order hold interchanged.

A second-order process and a third-order process are considered in this discussion.

A Second-Order System

Consider that the linear process is described by the transfer function

$$G(s) = \frac{K}{s(s+a)} \tag{6-312}$$

A state diagram of the overall system with this second-order process for the time interval $t_k \leq t \leq t_{k+1}$ is drawn as shown as shown in Figure 6-28. The dynamic equations of the system are written

$$\frac{dx_1(t)}{dt} = x_2(t) \tag{6-313}$$

$$\frac{dx_2(t)}{dt} = -ax_2(t) + Km(t) \tag{6-314}$$

and

$$c(t) = x_1(t) \tag{6-315}$$

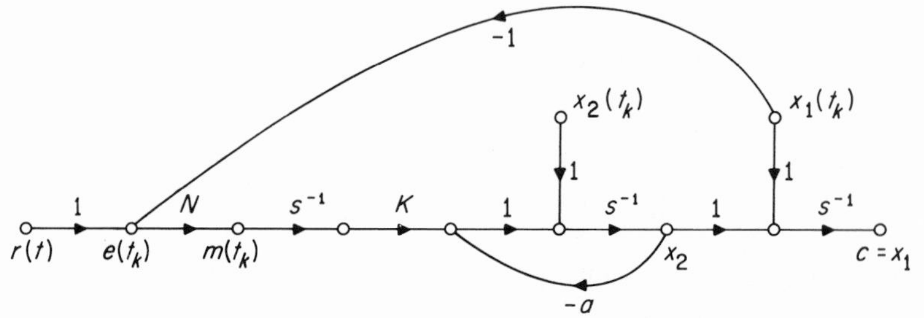

FIG. 6-28. A state diagram of the system shown in Fig. 6-27 with a second-order process.

Using the matrix equation form of Eq. (6-309), we have

$$A = \begin{bmatrix} 0 & 1 \\ 0 & -a \end{bmatrix} \tag{6-316}$$

and

$$B = \begin{bmatrix} 0 \\ K \end{bmatrix} \tag{6-317}$$

For convenience of solution, the state equations of Eqs. (6-313) and (6-314) are transformed by the similarity transformation into

$$\frac{d\mathbf{y}(t)}{dt} = \Lambda\mathbf{y}(t) + \Gamma m(t) \tag{6-318}$$

where Λ is a diagonal matrix which contains the eigenvalues of A as elements on its main diagonal. Let us select the matrix P, such that

$$P = \begin{bmatrix} 1 & -1/a \\ 0 & 1 \end{bmatrix} \tag{6-319}$$

Then

$$\Lambda = P^{-1}AP = \begin{bmatrix} 0 & 0 \\ 0 & -a \end{bmatrix} \tag{6-320}$$

Let us define the vector $\mathbf{z}(t)$ as

$$\mathbf{z}(t) = P^{-1}\mathbf{x}(t) \tag{6-321}$$

Then, Eq. (6-309) becomes

$$\frac{d\mathbf{z}(t)}{dt} = \Lambda\mathbf{z}(t) + P^{-1}Bm(t) \tag{6-322}$$

or in scalar form,

$$\frac{dz_1}{dt} = \frac{K}{a}m(t) \tag{6-323}$$

$$\frac{dz_2}{dt} = -\,az_2(t) + Km(t) \tag{6-324}$$

Now letting

$$y_1(t) = \frac{a^2}{K} z_1(t) \tag{6-325}$$

$$y_2(t) = \frac{a}{K} z_2(t) \tag{6-326}$$

we have

$$\frac{dy_1(t)}{dt} = am(t) \tag{6-327}$$

$$\frac{dy_2(t)}{dt} = -ay_2(t) + am(t) \tag{6-328}$$

Therefore, in Eq. (6-318),

$$\mathbf{\Gamma} = \begin{bmatrix} a \\ a \end{bmatrix} \tag{6-329}$$

Note that $x_1 = x_2 = 0$ implies $y_1 = y_2 = 0$, and the relation between $\mathbf{x}$ and $\mathbf{y}$ is

$$\mathbf{y}(t) = \begin{bmatrix} \frac{a^2}{K} & \frac{a}{K} \\ 0 & \frac{a}{K} \end{bmatrix} \mathbf{x}(t) \tag{6-330}$$

Now we proceed with the formulation of the time-optimal control of the linear open-loop part of the system using Eqs. (6-327) and (6-328).

The Hamiltonian H is formed as

$$H = \sum_{i=1}^{n} p_i(t) f_i[\mathbf{y}(t), m(t)] \tag{6-331}$$

where

$$f_i = \dot{y}_i(t) \qquad i = 1, 2 \tag{6-332}$$

and $p_i(t)$ denotes the costate variables. For the second-order system under consideration, $n = 2$.

Substitution of Eqs. (6-327) and (6-328) into Eq. (6-331) gives

$$\begin{aligned} H &= p_1(t)\frac{dy_1(t)}{dt} + p_2(t)\frac{dy_2(t)}{dt} \\ &= am(t)[p_1(t) + p_2(t)] - ap_2(t)y_2(t) \end{aligned} \tag{6-333}$$

The maximum principle asserts that when the system is optimal, the Hamiltonian H of the variable $m(t) \in \mathscr{M}$, where $\mathscr{M}$ denotes the admissible region of the control, attains its maximum, and there exists a nonezero continuous vector

$$\mathbf{p}(t) = \begin{bmatrix} p_1(t) \\ p_2(t) \end{bmatrix} \tag{6-334}$$

From Eq. (6-333), it is observed that the function H attains its maximum when

$$m(t) = \text{sign}\,[p_1(t) + p_2(t)] \tag{6-335}$$

The costate variables $p_1(t)$ and $p_2(t)$ satisfy the following relations:

$$\frac{dp_1(t)}{dt} = -\frac{\partial H}{\partial y_1} \tag{6-336}$$

$$\frac{dp_2(t)}{dt} = -\frac{\partial H}{\partial y_2} \tag{6-337}$$

Therefore, substituting Eq. (6-333) into Eqs. (6-336) and (6-337), we have

$$\frac{dp_1(t)}{dt} = 0 \tag{6-338}$$

$$\frac{dp_2(t)}{dt} = ap_2(t) \tag{6-339}$$

The solutions of the last two equations are

$$p_1(t) = c_1 \tag{6-340}$$

and

$$p_2(t) = c_2 e^{at} \tag{6-341}$$

with the notation

$$c_1 = p_1(0) \tag{6-342}$$

$$c_2 = p_2(0) \tag{6-343}$$

Then, Eq. (6-335) becomes

$$m(t) = \text{sign}\,[c_1 + c_2 e^{at}] \tag{6-344}$$

The function $c_1 + c_2 e^{at}$ can change its sign at most once; hence the maximum number of switching is one.

We next solve Eqs. (6-327) and (6-328) by setting $m(t) = \sigma = \pm 1$, and the solutions are

$$y_1(t) = y_1(t_0) + \sigma a(t - t_0) \tag{6-345}$$

$$y_2(t) = [y_2(t_0) - \sigma]e^{-a(t-t_0)} + \sigma \tag{6-346}$$

Eliminating $t - t_0$ from the last two equations, we have

$$y_2(t) = [y_2(t_0) - \sigma]e^{-(1/\sigma)[y_1(t)-y_1(t_0)]} + \sigma \tag{6-347}$$

The trajectories for various values of the initial states and $\sigma = +1, -1$, are plotted in the y_1-y_2 plane, as shown in Figure 6-29. Notice that the trajectories are independent of the value of a.

Let us assume that the desired equilibrium state is the origin. We are then particularly interested in the two trajectories, which pass through the origin. One of these trajectories is obtained with $\sigma = +1$ and the other is with

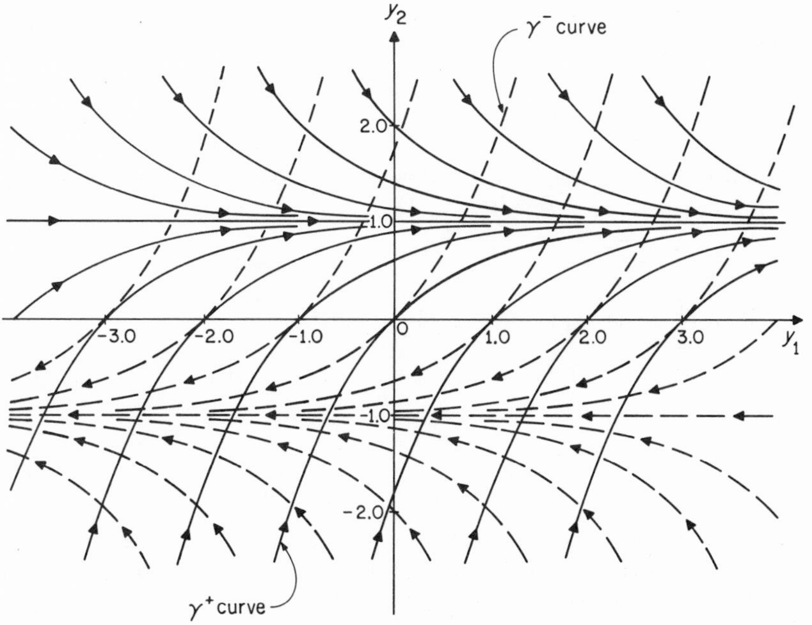

FIG. 6-29. The trajectories generated by the constant control $m = +1$ or -1 in the y_1-y_2 plane. The solid curves are due to $m = +1$ and the dashed ones are due to $m = -1$. The arrow indicates the direction of the motion of the state. For third-order systems, change y_1 to y_2 and y_2 to y_3.

$\sigma = -1$. Let $\gamma+$ and $\gamma-$ denote the set of all initial states which are forced to the origin by the control sequence $m = \{+1\}$ and $m = \{-1\}$, respectively. Then, $\gamma+$ and $\gamma-$ are given by

$$\gamma + = \{(y_1, y_2): \; y_2 = 1 - e^{-y_1}, y_1 < 0\} \tag{6-348}$$

$$\gamma - = \{(y_1, y_2): \; y_2 = -1 + e^{y_1}, y_1 > 0\} \tag{6-349}$$

Let γ denote the union of $\gamma+$ and $\gamma-$. Then, from Figure 6-29, the control law necessary to drive a given initial state to $\mathbf{y} = \mathbf{0}$ is described as follows:

$$m = \{+1, -1\} \quad \text{if } (y_1, y_2) \text{ is above the } \gamma \text{ curve} \tag{6-350}$$

$$m = \{-1, +1\} \quad \text{if } (y_1, y_2) \text{ is below the } \gamma \text{ curve} \tag{6-351}$$

$$m = \{+1\} \quad \text{if } (y_1, y_2) \text{ is on the } \gamma+ \text{ curve} \tag{6-352}$$

$$m = \{-1\} \quad \text{if } (y_1, y_2) \text{ is on the } \gamma- \text{ curve} \tag{6-353}$$

The control sequence $m = \{+1, -1\}$ implies that $m = +1$ is applied first until the trajectory intersects the $\gamma-$ curve and then the control is switched to $m = -1$. Once the optimal control sequence is determined, the optimal trajectory may be constructed in the y_1-y_2 plane.

The total transition time t_f is determined as follows. Let t_1 denote the switching time, and t_f the final time; and let

$$T_1 = t_1 - t_0 \tag{6-354}$$

$$T_2 = t_f - t_1 \tag{6-355}$$

Then, from Eq. (6-345), if there is no switching, the time required to bring $y_1(t_0)$ to $y_1(t_f) = 0$ is ($\sigma = \pm 1$)

$$t_f - t_0 = T_1 = \left| \frac{1}{a} y_1(t_0) \right| \tag{6-356}$$

If there is one switching, Eq. (6-345) gives

$$t_1 - t_0 = T_1 = \left| \frac{1}{a} [y_1(t_1) - y(t_0)] \right| \tag{6-357}$$

and

$$t_f - t_1 = T_2 = \left| \frac{1}{a} y_1(t_1) \right| \tag{6-358}$$

Next, we shall show an alternative method of obtaining the time intervals between two successive switchings and the control sequences. We first consider the case when there is only one switching. Assuming that the control sequence is

$$m = \{\sigma, -\sigma\}$$

the initial state, $[y_1(t_0), y_2(t_0)]$, is first brought to the switching point, $[y_1(t_1), y_2(t_1)]$, by the control $m = \sigma$. Using the notation defined in Eqs. (6-354) and (6-355), the switching point is obtained from Eqs. (6-345) and (6-346) as follows:

$$y_1(t_1) = y_1(t_0) + \sigma a T_1 \tag{6-359}$$

$$y_2(t_1) = [y_2(t_0) - \sigma]e^{-aT_1} + \sigma \tag{6-360}$$

Then by switching m to $-\sigma$ at t_1, the system is brought to the equilibrium point $\mathbf{y}(t) = \mathbf{0}$ at $t = t_f$.

Thus, using Eqs. (6-345) and (6-346), at $t - t_f$, we have

$$y_1(t_f) = 0 = y_1(t_0) + \sigma a(T_1 - T_2) \tag{6-361}$$

$$y_2(t_f) = 0 = [y_2(t_0) - \sigma]e^{-a(T_1+T_2)} + 2\sigma e^{-aT_2} + \sigma \tag{6-362}$$

From Eq. (6-361) we obtain the relation

$$T_2 = T_1 + \frac{1}{\sigma a} y_1(t_0) \tag{6-363}$$

Substituting the last equation into Eq. (6-362) and solving for T_1, we get

$$T_1 = \frac{1}{a\sigma} y_1(t_0) - \frac{1}{a} \ln \frac{\sigma}{\sigma - y_2(t_0)} \left[1 \pm \sqrt{1 - \frac{1}{\sigma} [y_2(t_0) - \sigma] e^{-(1/\sigma) y_1(t_0)}} \right] \tag{6-364}$$

for

$$y_2(t_0) \neq \sigma$$

and

$$T_1 = \frac{1}{a} \ln 2 - \frac{1}{a\sigma} y_1(t_0) \tag{6-365}$$

for

$$y_2(t_0) = \sigma$$

After determining a set of T_1 and T_2 for $\sigma = \pm 1$, we substitute them back into Eqs. (6-361) and (6-362) to determine the true values of T_1 and T_2 and σ.

For the case when there is no switching, the time interval T_1 is obtained from Eqs. (6-359) and (6-360) by setting $y_1(t_1) = y_2(t_1) = 0$. Therefore,

$$T_1 = -\frac{1}{a\sigma} y_1(t_0) \tag{6-366}$$

and

$$T_1 = \frac{1}{a} \ln \left[1 - \frac{1}{\sigma} y_2(t_0) \right] \tag{6-367}$$

These last two equations single out one pair of T_1 and σ for any initial state which can be forced to the origin by the control sequence $m = \{\sigma\}$.

Once the control sequence and the time intervals between the two successive switchings are determined, the optimal trajectory and the optimal transient process are determined from Eqs. (6-345) and (6-347). Once the system reaches the origin, we set $m(t) = 0$ for $t \geq t_f$ and the system will remain there. The entire switching process for the second-order system is summarized in Table 6-1.

We have completed the first part of the problem, i.e., the determination of the optimal control to the linear process. Now we must determine the desired input to the system, $r(t)$.

From the characteristics of the nonlinear element, the following relations for $e^*(t)$ are written:

$$e^*(t) \geq 1 \qquad \text{if } m(t) = 1 \tag{6-368}$$

$$e^*(t) \leq 1 \qquad \text{if } m(t) = -1 \tag{6-369}$$

$$e^*(t) = 0 \qquad \text{if } m(t) = 0 \tag{6-370}$$

and from the system configuration shown in Figure 6-27, we have

$$r(t) = e(t) + x_1(t) \tag{6-371}$$

Because of the sampling operation, we are interested only in the values of $r(t)$ at the sampling instants, and the function $r(t)$ can take any form so long as it maintains the proper values at the sampling instants. However, for simplicity, we realize $r(t)$ by a series of step functions. Then from Eqs. (6-368) and (6-371), we have

Table 6-1

Time Intervals	$t_0 \le t \le t_1 \quad (m = \sigma)$	$t_1 \le t \le t_f \quad (m = -\sigma)$
Transient Process	$y_1(t) = y_1(t_0) + \sigma a(t - t_0)$ $y_2(t) = [y_2(t_0) - \sigma]e^{-a(t-t_0)} + \sigma$ $y_1(t_1) = y_1(t_0) + \sigma a T_1$ $y_2(t_1) = [y_2(t_0) - \sigma]e^{-aT_1} + \sigma$	$y_1(t) = y_1(t_1) - \sigma a(t - t_1)$ $y_2(t) = [y_2(t_1) + \sigma]e^{-a(t-t_1)} - \sigma$ $y_1(t_f) = y_1(t_1) - \sigma a T_2$ $y_2(t_f) = [y_2(t_1) + \sigma]e^{-aT_2} - \sigma$
Trajectory	$y_2(t) = [y_2(t_0) - \sigma]e^{-(1/\sigma)[y_1(t)-y_1(t_0)]} + \sigma$	$y_2(t) = [y_2(t_0 + T_1) + \sigma]e^{(1/\sigma)[y_1(t)-y_1(t_0+T_1)]} - \sigma$

Remarks:
1. $t_f = T_1 + T_2$
2. If there is no switching, $T_2 = 0$, $T_1 = -\dfrac{1}{a\sigma}y_1(t_0) = \dfrac{1}{a}\ln\left[1 - \dfrac{1}{\sigma}y_2(t_0)\right]$
3. If there is one switching,

$$T_1 = \frac{1}{a}\ln 2 - \frac{1}{a\sigma}y_1(t_0) \quad \text{for } y_2(t_0) = \sigma$$

$$T_1 = \frac{1}{a\sigma}y_1(t_0) - \frac{1}{a}\ln\frac{\sigma}{\sigma - y_2(t_0)}\left[1 \pm \sqrt{1 - \frac{1}{\sigma}[y_2(t_0) - \sigma]e^{-(1/\sigma)y_1(t_0)}}\right] \quad \text{for } y_2(t_0) \ne \sigma$$

$$T_2 = T_1 + \frac{1}{a\sigma}y_1(t_0)$$

$$r(t) \ge 1 + \max x_1(t) \qquad \text{if } m(t) = 1 \tag{6-372}$$

$$r(t) \le -1 + \min x_1(x) \qquad \text{if } m(t) = -1 \tag{6-373}$$

$$r(t) = 0 \qquad \text{if } m(t) = 0 \tag{6-374}$$

The value of $x_1(t)$ is obtained from Eq. (6-330); that is

$$x_1(t) = \frac{K}{a^2}[y_1(t) - y_2(t)] \tag{6-375}$$

Thus, we may determine the maximum and minimum values of $x_1(t)$ for each interval.

The sampling period T is obtained as a common factor of T_1 and T_2.

EXAMPLE 6-11

Consider that the linear controlled process for the system shown in Figure 6-26 is described by the transfer function

$$G(s) = \frac{1}{s(s + 1)} \tag{6-376}$$

The initial state is given as

$$\mathbf{x}(0) = \begin{bmatrix} 1.0 \\ 0.5 \end{bmatrix} \tag{6-377}$$

It is desired to find the optimal input $r(t)$ so that the system is driven to the equilibrium state $\mathbf{x}(t_f) = \mathbf{0}$ in the shortest possible time.

The problem is solved in the following fashion:

1. Convert $\mathbf{x}(0)$ into $\mathbf{y}(0)$ using Eq. (6-330).
2. Determine the time intervals between the two successive switchings, the control sequence, and the sampling period T.
3. Determine the maximum or minimum value of $x_1(t)$ corresponding to the value of $m(t)$ for each time interval.
4. Determine the desired system input $r(t)$.
5. Sketch the time-optimal trajectory in the y_1-y_2 plane and the time response of the system.

Using Eq. (6-330), the initial state $\mathbf{y}(0)$ is obtained

$$\mathbf{y}(0) = \begin{bmatrix} 1 & 1 \\ 0 & 1 \end{bmatrix} \quad \mathbf{x}(0) = \begin{bmatrix} 1.5 \\ 0.5 \end{bmatrix} \tag{6-378}$$

From Figure 6-29, since $\mathbf{y}(0)$ is situated below the γ curve, the optimal control sequence is

$$m(t) = \{-1, +1\} \tag{6-379}$$

From Eq. (6-330),

$$\mathbf{y}(0) = \begin{bmatrix} 1.5 \\ 0.5 \end{bmatrix} \tag{6-380}$$

and from Table 6-1, with $\sigma = -1$, T_1 and T_2 are determined as follows:

$$T_1 \cong 2.1, \qquad T_2 \cong 0.6 \tag{6-381}$$

Thus the total transition time is $T_1 + T_2 = 2.7$ seconds, and the sampling period can be made equal to 0.1 second.

Next, the desired reference input to the system, $r(t)$, is determined using Eqs. (6-372) through (6-374). Therefore,

$$r(t) \geq 1 + \max x_1(t) \qquad 2.1 \leq t \leq 2.7 \tag{6-382}$$

$$r(t) \leq -1 + \max x_1(t) \qquad 0 \leq t \leq 2.1 \tag{6-383}$$

$$r(t) = 0 \qquad t \geq 2.7 \tag{6-384}$$

The state plane trajectories for the various switching periods are described by the following equations which are obtained using Eqs. (6-330) and Table 6-1:

$$x_1(t) = 2.5 - t - 1.5e^{-t} \tag{6-385}$$

$$0 \leq t \leq 2.1$$

$$x_2(t) = 1.5e^{-t} - 1 \tag{6-386}$$

$$x_1(t) = -1.6 + (t - 2.1) + 1.82e^{-(t-2.1)} \tag{6-387}$$

$$2.1 \leq t \leq 2.7$$

$$x_2(t) = 1 - 1.82e^{-(t-2.1)} \tag{6-388}$$

Thus the minimum value of $x_1(t)$ for the time interval $0 \le t \le 2.1$ and the maximum value of $x_1(t)$ for the interval $2.1 \le t \le 2.7$ are equal to $x_1(2.1) = 0.216$. Then, $r(t)$ is determined as follows:

$$r(t) = -0.8 \qquad 0 \le t \le 2.1$$
$$r(t) = 1.3 \qquad 2.1 \le t \le 2.7$$
$$r(t) = 0 \qquad t \ge 2.7$$

The optimal trajectory in the y_1-y_2 plane is shown in Figure 6-30. The optimal

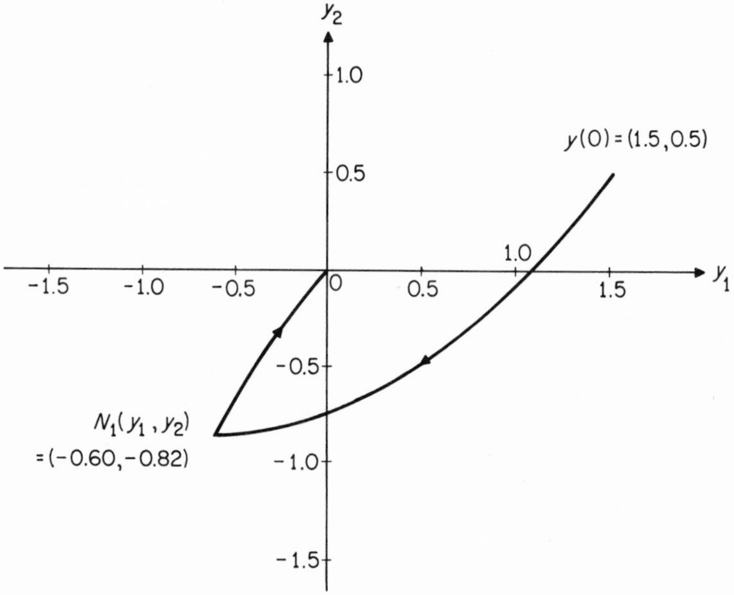

FIG. 6-30. Time-optimal trajectory in the y_1-y_2 plane for Example 6-11.

time responses of $x_1(t)$ and $x_2(t)$, the desired input $r(t)$, and the control signal $m(t)$, are sketched as shown in Figure 6-31.

A Third-Order System

For a third-order system we use the same system configuration as in Figure 6-26, except that $G(s)$ is now described by

$$G(s) = \frac{K}{s^2(s + a)} \tag{6-389}$$

A state diagram for the system for the time interval $t_k \le t \le t_{k+1}$ is drawn as shown in Figure 6-32. The state equations of the third-order process are written as

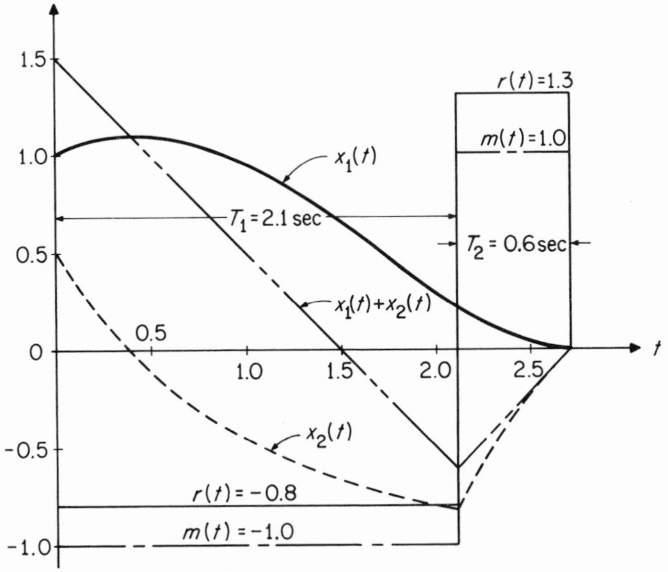

FIG. 6-31. Time-optimal transient process of $\mathbf{x}(t)$, the desired input $m(t)$, and the desired system imput $r(t)$ for Example 6-11.

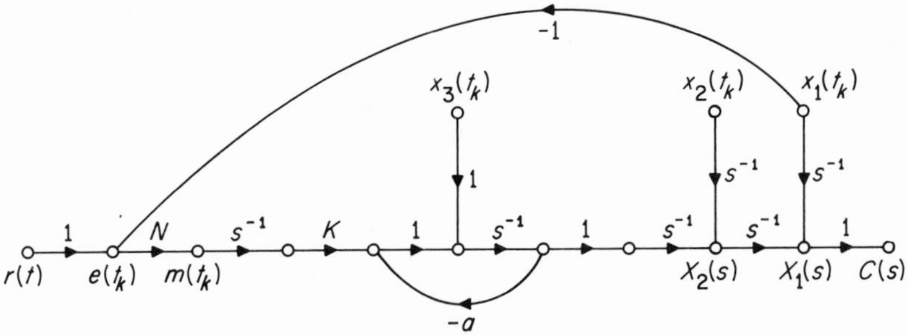

FIG. 6-32. A third-order system.

$$\begin{bmatrix} \dfrac{dx_1}{dt} \\ \dfrac{dx_2}{dt} \\ \dfrac{dx_3}{dt} \end{bmatrix} = \begin{bmatrix} 0 & 1 & 0 \\ 0 & 0 & 1 \\ 0 & 0 & -a \end{bmatrix} \begin{bmatrix} x_1 \\ x_2 \\ x_3 \end{bmatrix} + \begin{bmatrix} 0 \\ 0 \\ K \end{bmatrix} m(t_k) \tag{6-390}$$

or

$$\frac{d\mathbf{x}}{dt} = A\mathbf{x} + Bm \tag{6-391}$$

where

$$m(t_k) = m(t) \qquad t_k \leq t < t_{k+1}$$

Using the similarity transformation with

$$P = \begin{bmatrix} 1 & 0 & \frac{1}{a^2} \\ 0 & 1 & -\frac{1}{a} \\ 0 & 0 & 1 \end{bmatrix} \tag{6-392}$$

we get the Jordan canonical form for A.

$$\Lambda = P^{-1}AP = \begin{bmatrix} 0 & 1 & 0 \\ 0 & 0 & 0 \\ 0 & 0 & -a \end{bmatrix} \tag{6-393}$$

Then Eq. (6-390) is transformed into

$$\frac{d\mathbf{z}}{dt} = \Lambda\mathbf{z} + \Gamma m \tag{6-394}$$

where Λ is given by Eq. (6-393) and Γ is given by

$$\Gamma = P^{-1}B = \begin{bmatrix} -\frac{K}{a^2} \\ \frac{K}{a} \\ K \end{bmatrix} \tag{6-395}$$

For simplicity, we introduce another set of new variables, y_1, y_2, y_3, such that

$$\mathbf{y} = \begin{bmatrix} y_1 \\ y_2 \\ y_3 \end{bmatrix} = \begin{bmatrix} \frac{a^3}{K} z_1 \\ \frac{a^2}{K} z_2 \\ \frac{a}{K} z_3 \end{bmatrix} = \begin{bmatrix} \frac{a^3}{K} & 0 & 0 \\ 0 & \frac{a^2}{K} & 0 \\ 0 & 0 & \frac{a}{K} \end{bmatrix} \mathbf{z} \tag{6-396}$$

Then Eq. (6-394) becomes

$$\frac{d\mathbf{y}}{dt} = \begin{bmatrix} 0 & a & 0 \\ 0 & 0 & 0 \\ 0 & 0 & -a \end{bmatrix} \mathbf{y} + \begin{bmatrix} -a \\ a \\ a \end{bmatrix} m \tag{6-397}$$

The relation between $\mathbf{x}$ and $\mathbf{y}$ is described by

$$\begin{bmatrix} y_1 \\ y_2 \\ y_3 \end{bmatrix} = \begin{bmatrix} \frac{a^3}{K} & 0 & -\frac{a}{K} \\ 0 & \frac{a^2}{K} & \frac{a}{K} \\ 0 & 0 & \frac{a}{K} \end{bmatrix} \begin{bmatrix} x_1 \\ x_2 \\ x_3 \end{bmatrix} \tag{6-398}$$

and thus $\mathbf{x}(t) = \mathbf{0}$ implies $\mathbf{y}(t) = \mathbf{0}$.

Applying the maximum principle to the system described by Eq. (6-397) to determine the proper input signal $m(t)$, the Hamiltonian H is formed as follows:

$$H = a\,p_1(t)y_2(t) - a\,p_3(t)y_3(t) + a\,m(t)[-p_1(t) + p_2(t) + p_3(t)] \tag{6-399}$$

Therefore, the control which minimizes H is

$$m(t) = \text{sign}\,[-p_1(t) + p_2(t) + p_3(t)] \tag{6-400}$$

The costate vector $\mathbf{p}(t)$ satisfies the differential equations

$$\frac{dp_1(t)}{dt} = 0 \tag{6-401}$$

$$\frac{dp_2(t)}{dt} = -a\,p_1(t) \tag{6-402}$$

$$\frac{dp_3(t)}{dt} = a\,p_3(t) \tag{6-403}$$

Solving Eqs. (6-401) through (6-403), we get

$$p_1(t) = c_1 \tag{6-404}$$

$$p_2(t) = c_2 - ac_1 t \tag{6-405}$$

$$p_3(t) = c_3 e^{at} \tag{6-406}$$

with $c_i = p_i(0)$, $i = 1, 2, 3$. Substitution of the last three equations into Eq. (6-400), we have

$$m(t) = \text{sign}\,(-c_1 + c_2 - a\,c_1 t + c_3 e^{at}) \tag{6-407}$$

The function $(-c_1 + c_2 - a\,c_1 t + c_3 e^{at})$ can change its sign at most twice, which means that the maximum number of switchings is two, and the sign of the control $m(t)$ alternates between two successive switchings.

Now Eq. (6-397) is solved by setting

$$m(t) = \sigma = \pm 1 \tag{6-408}$$

The solutions are:

$$y_1(t) = y_1(0) + a[y_2(0) - \sigma]t + \tfrac{1}{2}\sigma\,a^2 t^2 \tag{6-409}$$

$$y_2(t) = y_2(0) + \sigma a t \tag{6-410}$$

$$y_3(t) = [y_3(0) - \sigma]e^{-at} + \sigma \tag{6-411}$$

Eliminating t from the equations above, we have

$$y_1 = y_1(0) - [y_2 - y_2(0)] + \frac{\sigma}{2}[y_2^2 - y_2^2(0)] \tag{6-412}$$

$$y_3 = [y_3(0) - \sigma]e^{-\sigma[y_2 - y_2(0)]} + \sigma \tag{6-413}$$

$$y_1 = y_1(0) + [\sigma - y_2(0)] \ln\left[\frac{y_3 - \sigma}{y_3(0) - \sigma}\right] + \frac{1}{2}\left\{\ln\left[\frac{y_3 - \sigma}{y_3(0) - \sigma}\right]\right\}^2 \tag{6-414}$$

The last three equations represent the projections of the state space trajectories generated by the constant control $m = \sigma$ onto the y_1-y_2, y_3-y_2, and y_1-y_3 planes, respectively. Notice that all these equations are independent of a.

The projections of trajectories onto the y_1-y_2 plane for various values of initial conditions are shown in Figure 6-33. The projections of trajectories onto the y_2-y_3 plane are the same as the curves shown in Figure 6-29, since Eq. (6-413) is identical to Eq. (6-347) except that y_2 is changed to y_1 and y_3 is changed to y_2. The projections of trajectories onto the y_1-y_3 plane cannot be drawn unless the initial values of y_2, $y_2(0)$, are given.

Next, we consider the determination of the time intervals between two

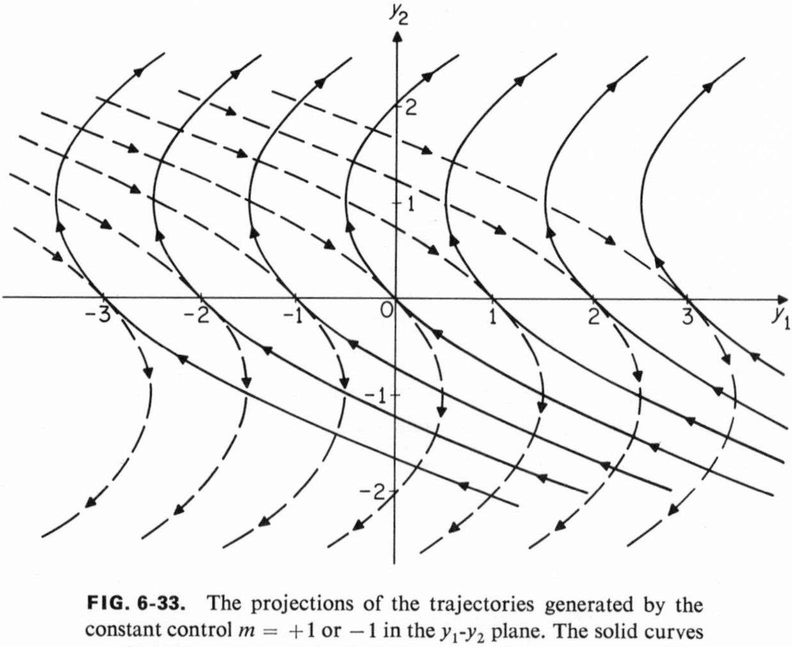

FIG. 6-33. The projections of the trajectories generated by the constant control $m = +1$ or -1 in the y_1-y_2 plane. The solid curves are due to $m = +1$ and the dashed ones are due to $m = -1$. The arrow indicates the direction of motion of the state.

consecutive switchings and the control sequence, following the same procedure as for the second-order case.

Let

$$t_1 = \text{first switching time}$$
$$t_2 = \text{second switching time}$$
$$t_f = \text{final time}$$

and

$$T_1 = t_1$$
$$T_2 = t_2 - t_1$$
$$T_3 = t_f - t_2$$

assuming that the first control $m = \sigma$ is applied at $t = 0$. If there are a total of two switchings, the control sequence is given by

$$m(t) = \{\sigma, -\sigma, \sigma\} \tag{6-415}$$

Thus, from Eqs. (6-409) to (6-411) the first switching point, $N_1(y_1(t_1), y_2(t_1), y_3(t_1))$, is given by the following set of equations:

$$y_1(t_1) = y_1(0) + a[y_2(0) - \sigma]T_1 + \tfrac{1}{2}\sigma a^2 T_1^2 \tag{6-416}$$
$$y_2(t_1) = y_2(0) + \sigma a T_1 \tag{6-417}$$
$$y_3(t_1) = [y_3(0) - \sigma]e^{-aT_1} + \sigma \tag{6-418}$$

and the second switching point, $N_2(y_1(t_2), y_2(t_2), y_3(t_2))$, is given by

$$y_1(t_2) = y_1(t_1) + a[y_2(0) + \sigma]T_2 - \tfrac{1}{2}\sigma a^2 T_2^2 \tag{6-419}$$
$$y_2(t_2) = y_2(t_1) - \sigma a T_2 \tag{6-420}$$
$$y_3(t_2) = [y_3(t_1) + \sigma]e^{-aT_2} - \sigma \tag{6-421}$$

Since the final point must be the origin, we have

$$y_1(t_f) = 0 = y_1(t_2) + a[y_2(t_2) - \sigma]T_3 + \tfrac{1}{2}\sigma a^2 T_3^2 \tag{6-422}$$
$$y_2(t_f) = 0 = y_2(t_2) + \sigma a T_3 \tag{6-423}$$
$$y_3(t_f) = 0 = [y_3(t_2) - \sigma]e^{-aT_3} + \sigma \tag{6-424}$$

Eliminating $\mathbf{y}(t_1)$ and $\mathbf{y}(t_2)$ from Eqs. (6-416) to (6-424), we have

$$y_1(0) = \sigma a(T_1 - T_2 + T_3) = 0 \tag{6-425}$$
$$[y_3(0) - \sigma]e^{-a(T_1+T_2+T_3)} + 2\sigma e^{-a(T_2+T_3)} - 2\sigma e^{-aT_3} + \sigma = 0 \tag{6-426}$$
$$y_1(0) + a[y_2(0) - \sigma](T_1 + T_3) + a[y_2(0) + \sigma]T_2 + \tfrac{1}{2}\sigma a^2(T_1^2 - T_2^2 + T_3^2) + \sigma a^2(T_1T_2 + T_1T_3 - T_2T_3) = 0 \tag{6-427}$$

From Eqs. (6-425) to (6-427), we have the relations

$$T_1 = T_2 - T_3 - \frac{\sigma}{a} y_2(0) \tag{6-428}$$

$$T_3 = \frac{1}{2\sigma a^2 T_2}\left[y_1(0) + y_2(0) - \frac{1}{2}\sigma y_2^2(0) + \sigma a^2 T_2^2\right] \tag{6-429}$$

$$e^{-aT_3} = \frac{[\sigma y_3(0) - 1]e^{-a(2T_2-(\sigma/a)y_2(0))} + 1}{2(1 - e^{-aT_2})} \tag{6-430}$$

The last three equations may be solved graphically or by a trial-and-error method. For the case when there is one switching, the time intervals T_1 and T_2 are obtained from Eqs. (6-428) and (6-429) by setting $T_3 = 0$ as follows:

$$T_1 = T_2 - \frac{\sigma}{a} y_2(0) \tag{6-431}$$

$$T_2 = \sqrt{\frac{1}{a^2}\left[\frac{1}{2} y_2^2(0) - \sigma(y_1(0) + y_2(0))\right]} \tag{6-432}$$

and for the case when there is no switching, T_1 is obtained from Equation (6-428) by setting $T_2 = T_3 = 0$,

$$T_1 = -\frac{\sigma}{a} y_2(0) \tag{6-433}$$

After determining the T's and σ's for each case, we substitute the back into Eqs. (6-425) to (6-427) to find out the true values of T_1, T_2, T_3, and σ. Once T_1, T_2, T_3, and σ are determined, the optimal trajectory and the optimal transient process are determined from Eqs. (6-409) to (6-414) using different initial conditions for each interval. The control sequence is determined from Eq. (6-415). Once the system has reached the origin, we set $m(t) = 0$ so that it remains there. The entire process is summarized in Table 6-2.

Next, we proceed to determine the desired system input $r(t)$. Since the system configuration is the same as that of the second-order case, and $m(t)$ is $+1$, -1, or zero, $r(t)$ can be obtained from Eqs. (6-372) to (6-374). The vector, $\mathbf{x}(t)$, is given in scalar form as follows:

$$x_1(t) = \frac{k}{a^3}(y_1(t) + y_3(t)) \tag{6-434}$$

$$x_2(t) = \frac{k}{a^2}(y_2(t) - y_3(t)) \tag{6-435}$$

$$x_3(t) = \frac{k}{a} y_3(t) \tag{6-436}$$

The sampling period, T, is given by the common factor of T_1, T_2, and T_3. If there is no common factor, the partial uniform sampling scheme may be used as in the second-order case.

EXAMPLE 6-12

Consider that the linear controlled process for the system shown in Figure 6-26 is described by the transfer function

$$G(s) = \frac{K}{s^2(s + a)}, \qquad K = a = 1 \tag{6-437}$$

Table 6-2

Time	*Transient Process*	*Projections of Trajectory*
$t = 0$	$(y_1(0), y_2(0), y_3(0))$	
Interval I $0 \leq t \leq t_1$	$y_1(t) = y_1(0) + a[y_2(0) - \sigma]t + \frac{1}{2}\sigma a^2 t^2$ $y_2(t) = y_2(0) + \sigma at$ $y_3(t) = (y_3(0) - \sigma)e^{-at} + \sigma$	$y_1 = y_1(0) - [y_2 - y_2(0)] + \frac{\sigma}{2}[y_2^2 - y_2(0)]$ $y_3 = [y_3(0) - \sigma]e^{-(y_2 - y_2(0))} + \sigma$ $y_1 = y_1(0) + [\sigma - y_2(0)] \ln\left(\frac{y_3 - \sigma}{y_3(0) - \sigma}\right) - \frac{1}{2}\sigma\left[\ln\left(\frac{y_3 - \sigma}{y_3(0) - \sigma}\right)\right]^2$
$t = t_1$	$(y_1(t_1), y_2(t_1), y_3(t_1))$	
Interval II $t_1 \leq t \leq t_2$	$y_1(t) = y_1(t_1) + a(y_2(t_1) + \sigma)(t - t_1) - \frac{1}{2}\sigma a^2(t - t_1)^2$ $y_2(t) = y_2(t_1) - \sigma a(t - t_1)$ $y_3(t) = (y_3(t_1) + \sigma)e^{-a(t-t_1)} - \sigma$	$y_1 = y_1(t_1) - [y_2 - y_2(t_1)] - \frac{\sigma}{2}[y_2^2 - y_2^2(t_1)]$ $y_3 = [y_3(0) + \sigma]e^{-(y_2 - y_2(t_1))} - \sigma$ $y_1 = y_1(t_1) - [\sigma + y_2(t_1)] \ln\left(\frac{y_3 + \sigma}{y_3(0) + \sigma}\right) + \frac{1}{2}\sigma\left[\ln\left(\frac{y_3 + \sigma}{y_3(0) + \sigma}\right)\right]^2$
$t = t_2$	$(y_1(t_2), y_2(t_2), y_3(t_2))$	
Interval III $t_2 \leq t \leq t_f$	$y_1(t) = y_1(t_2) + a(y_2(t_2) - \sigma)(t - t_2) + \frac{1}{2}\sigma a^2(t - t_2)^2$ $y_2(t) = y_2(t_2) + \sigma a(t - t_2)$ $y_3(t) - (y_3(t_2) - \sigma)e^{-a(t-t_2)} + \sigma$	$y_1 = y_1(t_2) - (y_2 - (y_2(t)) + \frac{1}{2}\sigma(y_2^2 - y_2^2(t_2))$ $y_3 = [y_3(t_2) - \sigma]e^{-(y_2 - y_2(t_2))} + \sigma$ $y_1 = y_1(t_1) + (\sigma - y_2(t_2)) \ln\left(\frac{y_3 - \sigma}{y_3(t_2) - \sigma}\right) - \frac{1}{2}\sigma\left[\ln\left(\frac{y_3 - \sigma}{y_3(t_2) - \sigma}\right)\right]^2$
$t = t_f$	$(0, 0, 0)$	

Remarks:

1. $t_1 = T_1, t_2 = T_1 + T_2, t_f = T_1 + T_2 + T_3$
2. If there is no switching

$$T_2 = T_3 = 0, \quad T_1 = -\frac{\sigma}{a} y_2(0)$$

3. If there is one switching

$$T_3 = 0, \quad T_1 = T_2 - \frac{\sigma}{a} y_2(0)$$

$$T_2 = \sqrt{\frac{1}{a^2}[\tfrac{1}{2}y_2^2(0) - \sigma(y_1(0) - y_2(0))]}$$

4. If there are two switchings

$$T_1 = T_2 - T_3 - \frac{\sigma}{a} y_2(0)$$

$$T_3 = \frac{1}{2\sigma a^2 T_2}\sqrt{y_1(0) + y_2(0) - \tfrac{1}{2}\sigma y_2^2(0) + \sigma a^2 T_2^2}$$

$$e^{-aT_3} = \frac{\sigma(y_3(0) - 1)e^{-a(2T_2 - (\sigma/a)y_2(0))} + 1}{2(1 - e^{-aT_2})}$$

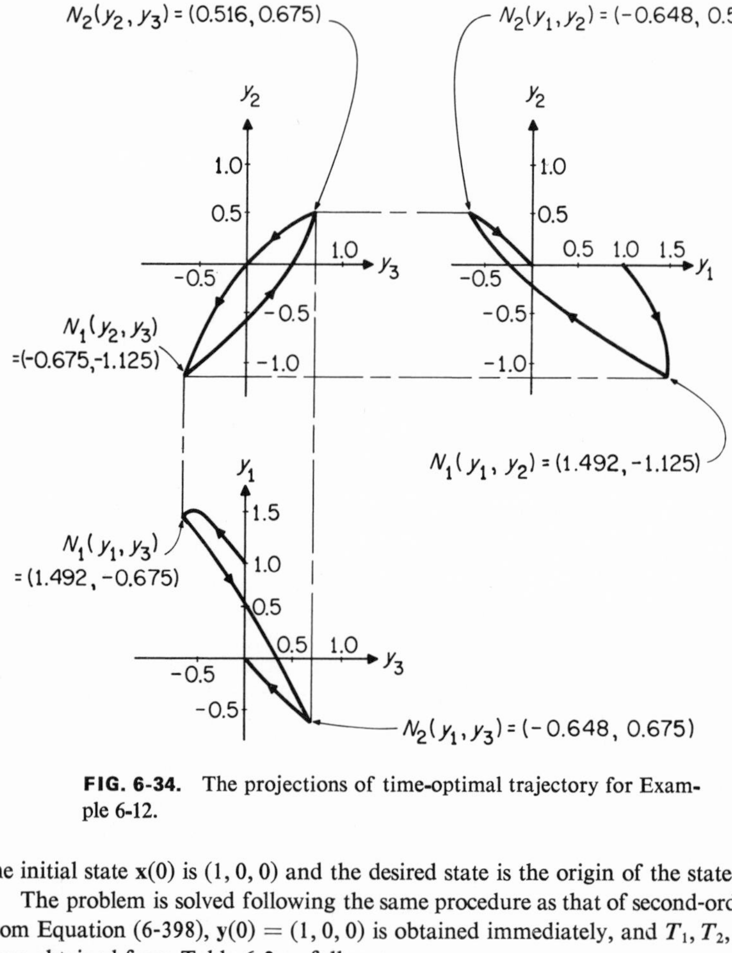

FIG. 6-34. The projections of time-optimal trajectory for Example 6-12.

The initial state $\mathbf{x}(0)$ is $(1, 0, 0)$ and the desired state is the origin of the state space.

The problem is solved following the same procedure as that of second-order case. From Equation (6-398), $\mathbf{y}(0) = (1, 0, 0)$ is obtained immediately, and T_1, T_2, T_3, and σ are obtained from Table 6-2 as follows:

$$\sigma = -1,\ T_1 \cong 1.125,\ T_2 \cong 1.641,\ T_3 \cong 0.516 \tag{6-438}$$

Thus, the total transition time is $T_1 + T_2 + T_3 = 3.282$ seconds, and the sampling period T is 0.003 second. The control sequence is given by

$$m(t) = \{-1, +1, -1\} \tag{6-439}$$

From Eqs. (6-372) to (6-374), $r(t)$ is written as

$$r(t) \leq -1 + \min x_1(t) \qquad 0 \leq t \leq 1.125 \tag{6-440}$$

$$r(t) \geq 1 + \max x_1(t) \qquad 1.125 \leq t \leq 2.766 \tag{6-441}$$

$$r(t) \leq -1 + \min x_1(t) \qquad 2.766 \leq t \leq 3.282 \tag{6-442}$$

$$r(t) = 0 \qquad t \geq 3.282 \tag{6-443}$$

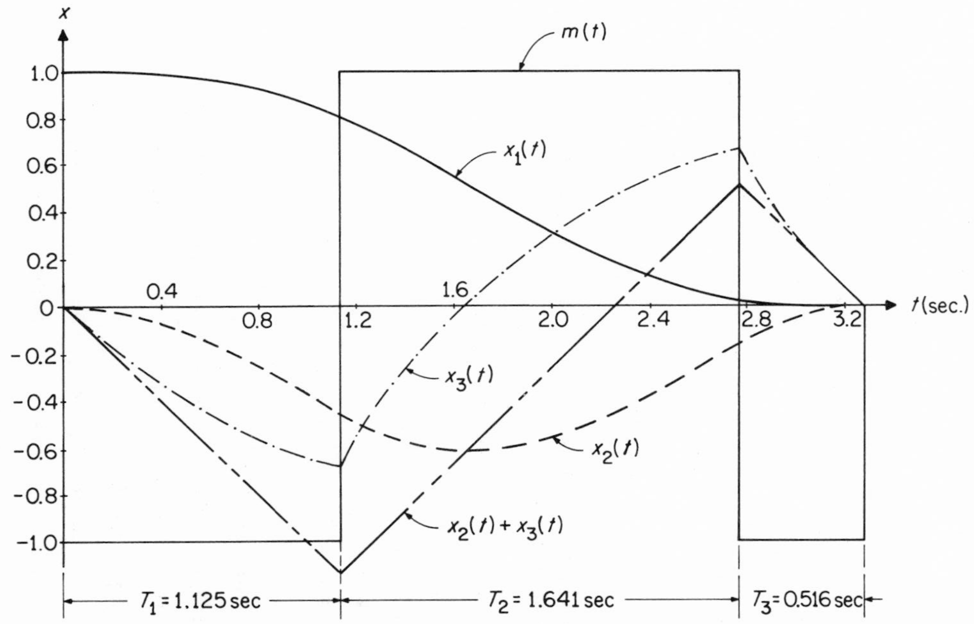

FIG. 6-35. Time-optimal transient process of $\mathbf{x}(t)$ and the desired input $m(t)$ for Example 6-12.

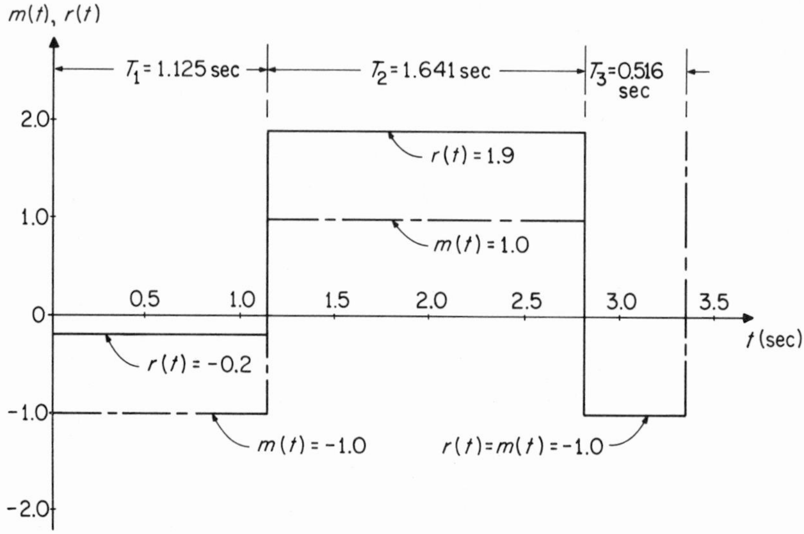

FIG. 6-36. The desired input $m(t)$ and the desired system input $r(t)$ for Example 6-12.

and from Eqs. (6-434) to (6-436), $\mathbf{x}(t)$ is obtained as

$$x_1(t) = t - \tfrac{1}{2}t^2 + e^{-t} \tag{6-444}$$

$$x_2(t) = 1 - t - e^{-t} \qquad 0 \leq t \leq 1.125 \tag{6-445}$$

$$x_3(t) = e^{-t} - 1 \tag{6-446}$$

$$x_1(t) = 2.492 - 1.125\,(t - 1.125) + \tfrac{1}{2}(t - 1.125)^2 + 1.675e^{-(t-0.125)} \tag{6-447}$$

$$x_2(t) = -2.125 + (t - 1.125) + 1.675e^{-(t-1.125)} \qquad 1.125 \leq t \leq 2.766 \tag{6-448}$$

$$x_3(t) = 1 - 1.675e^{-(t-1.125)} \tag{6-449}$$

$$x_1(t) = 1.648 + 1.516(t - 2.766) - \tfrac{1}{2}(t - 2.766)^2 + 1.675e^{-(t-2.766)} \tag{6-450}$$

$$x_2(t) = 1.516 - (t - 2.766) - 1.675e^{-(t-2.766)} \qquad 2.766 \leq t \leq 3.282 \tag{6-451}$$

$$x_3(t) = -1 + 1.675e^{-(t-2.766)} \tag{6-452}$$

The minimum or maximum value of $x_1(t)$ for each interval is

$$\text{min. } x_1(t) = 0.817 \qquad \text{for the first interval} \tag{6-453}$$

$$\text{max. } x_1(t) = -0.817 \qquad \text{for the second interval} \tag{6-454}$$

$$\text{min. } x_1(t) = 0 \qquad \text{for the third interval} \tag{6-455}$$

Thus, we can set $r(t)$ as follows:

$$r(t) = -0.2 \qquad 0 \leq t \leq 1.125 \tag{6-456}$$

$$r(t) = 1.9 \qquad 1.125 \leq t \leq 2.766 \tag{6-457}$$

$$r(t) = -1.0 \qquad 2.766 \leq t \leq 3.282 \tag{6-458}$$

$$r(t) = 0 \qquad t \geq 3.282 \tag{6-459}$$

The projections of the time-optimal trajectory in y plane are shown in Figure 6-34. The time-optimal transient process of $\mathbf{x}(t)$ is shown in Figure 6-35. The desired input $m(t)$ and the desired system input $r(t)$ are shown in Figure 6-36.

REFERENCES

1. Kuo, B. C., *Automatic Control Systems*, 2nd edition. Prentice-Hall, Englewood Cliffs, N. J., 1967.
2. Truxal, J. G., *Automatic Feedback Control System Synthesis*. McGraw-Hill, New York, 1955.
3. Kalman, R. E., "On the General Theory of Control Systems," *Proc. I. F. A. C.*, Vol. 1, Butterworths, London, 1961, pp. 481-492.
4. Kreindler, E., and Sarachik, P. E., "On the Concepts of Controllability and Observability of Linear Systems," *IEEE Trans. on Automatic Control*, Vol. AC-9, April 1964, pp. 129-136.
5. Kalman, R. E., "Mathematical Description of Linear Dynamical Systems," *Journal on Control*, Series A, Vol. 1, No. 2, SIAM, 1963.
6. Chidambara, M. R., and Wells, C. H., "State Variable Determination for Digital Control," *IEEE Trans. on Automatic Control*, Vol. AC-11, April 1966, p. 326.
7. Desoer, C. A., and Wing, J., "A Minimal Time Discrete System," *IRE Trans. on Automatic Control*, Vol. AC-6, May 1961, pp. 111-125.
8. Desoer, C. A., and Wing, J., "The Minimum Time Regulator Problem for Linear Sampled-Data Systems: General Theory," *J. Franklin Inst.*, Vol. 272, September 1961, pp. 208-228.
9. Polak, E., "Optimal Time Control of Some Pulse-Width Modulated Sampled-Data Systems," University of California (Berkeley), ERL Report, Series No. 60, August 1961.
10. Polak, E., "Minimum Time Control of Second Order Pulse-Width Modulated Sampled-Data Systems," *Trans. ASME. J. Basic Eng.*, Vol. 84, March 1962, pp. 101-110.
11. Kurzweil, F., "The Control of Multivariable Processes in the Presence of Pure Transport Delays," *IEEE Trans. on Automatic Control*, Vol. AC-8, January 1963, pp. 27-34.
12. Koepcke, R. W., "On the Control of Linear Systems with Pure Time-Delay," *Journal of Basic Engineering*, March 1965, pp. 74-80.
13. Mueller, T. E., "Optimal Control of Linear Systems With Time Lag," *Technical Report No. R-254*, Coordinated Science Laboratory, University of Illinois, Urbana, Ill., June 1965.
14. Tou, J. T., *Modern Control Theory*. McGraw-Hill, New York, 1964.
15. Nagata, A., Kodama, S., and Kumagai, S., "Time Optimal Discrete Control System with Bounded State Variable," *IEEE Trans. on Automatic Control*, Vol. AC-10, April 1965, pp. 155-164.

16. Rozonoer, L. I., "The Maximum Principle of L. S. Pontryagin in Optimal System Theory," *Automation and Remote Control*, Parts I, II, III, Vol. 20, 1959.

17. Chang, S. S. L., "Digitized Maximum Principle," *Proc. IRE*, Vol. 48, December 1960, pp. 2030–2031.

18. Butkovskii, A. G., "The Necessary and Sufficient Conditions for Optimality of Discrete Control Systems," *Automation and Remote Control*, Vol. 24, August 1963, pp. 1056–1064.

19. Katz, S., "A Discrete Version of Pontryagin's Maximum Principle," *J. Electron. Control*, Vol. 13, No. 2, 1962, pp. 179–184.

20. Ogata, K., *State Space Analysis of Control Systems*. Prentice-Hall, Englewood Cliffs, N. J., 1967.

21. Halkin, H., "Optimal Control for Systems Described by Difference Equations," *Advances in Control Systems*, Chapter 4, C. T. Leondes, ed. Academic Press, New York, 1964.

22. Tou, T., and Vadhanaphuti, B., "Optimum Control of Nonlinear Discrete-Data Systems," *AIEE Trans.*, Part II, September 1961, pp. 166–170.

23. Desoer, A., and Wing, J., "An Optimal Strategy for a Saturating Sampled-Data System," *IRE Trans. on Automatic Control*, Vol. AC-6, February 1961, pp. 5–15.

24. Nelson, W. L., "Optimal Control Methods for On-Off Sampling Systems," *Journal of Basic Engineering, Trans. of ASME*, March 1962, pp. 91–100.

25. Kalman, R. E., "Optimal Nonlinear Control of Saturating Systems by Intermittent Action," *1957 IER WESCON Records*, Pt. 4, pp. 130–135.

26. Moroz, A. I., "Synthesis of an Optimally Rapid Control For a Linear Discrete Third-Order Object," *Automation and Remote Control*, Vol. 26, No. 2, 1965.

27. Pavlov, A. A., "Optimum Transient Processes in Systems with a Restricted Third Derivative," *Automation and Remote Control*, Vol. 20, No. 8, 1959.

28. Athans, M., and Falb, P. L., *Optimal Control*. McGraw-Hill, New York, 1966.

29. Pontryagin, L. S., Boltyanskii, V. G., Gamkrelidze, R. V., and Mishchenko, E. F., *The Mathematical Theory of Optimal Processes*. John Wiley & Sons, New York, 1962.

7

OPTIMAL DESIGN OF DISCRETE-DATA SYSTEMS BY PERFORMANCE INDEX

7.1 INTRODUCTION

A wider class of optimum design problems is defined by use of the performance index.[1] The basic design problem can be described by referring to the block diagram of Figure 7-1. The figure shows a linear controlled process whose out-

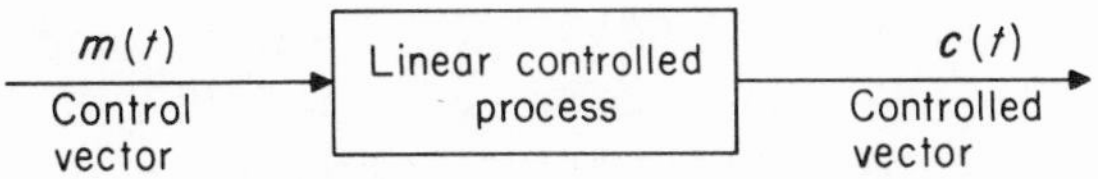

FIG. 7-1. Block diagram of a linear controlled process.

put $\mathbf{c}(t)$ is the controlled vector, and the input $\mathbf{m}(t)$ is the control vector. The design problem involves the determination of the optimal control $\mathbf{m}(t)$ so that the controlled vector $\mathbf{c}(t)$ behaves according to some prescribed fashion over a given time interval. From the mathematical viewpoint, it is convenient to establish a performance index I so that the optimization of the control system is achieved by finding the control $\mathbf{m}(t)$ which maximizes or minimizes the performance index I. For continuous-data systems, the performance index can assume a general form of

$$I = \int_{t_0}^{t_f} f[\mathbf{x}(t), \mathbf{m}(t)]\, dt \tag{7-1}$$

where f is a functional relationship, and $\mathbf{x}(t)$ is the state vector of the linear controlled process. For discrete-data systems, the components of $\mathbf{m}(t)$ are discontinuous with respect to time, for they are either digital or outputs of zero-order hold devices. Therefore, the performance index may be written as

$$I = \sum_{k=0}^{N} f[\mathbf{x}(kT), \mathbf{m}(kT)] \tag{7-2}$$

The advantages of using the performance index as a measuring criterion in optimal control design are that the index permits the use of certain established mathematical principles which allow the design to be carried out in a unique analytical manner. The system designed is the optimal in the sense that the chosen performance index is extremized (maximized or minimized); and deterministic as well as random signals may be considered.

7.2 MINIMUM SUMMED-ERROR-SQUARE DESIGN[2]

One of the most important performance indices used in the analytical design of control systems is the integral-square error,

$$I_e = \int_0^\infty e^2(t)\, dt \tag{7-3}$$

where $e(t)$ is defined as the difference between the desired output and the actual output variables; that is,

$$e(t) = c(t) - c_d(t) \tag{7-4}$$

The significance of using I_e is that any positive or negative error contributes a positive amount to I_e which increases with the amplitude and time duration of $e(t)$. An important advantage of the integral-square error criterion is that the time integral of Eq. (7-3) can be expressed in terms of a line integral of the Laplace transform of $e(t)$ by use of the Parseval's theorem. The design is then carried out analytically in the transform domain.

Similar to the continuous-data system design, the analytical design of discrete-data systems is formulated by referring to the block diagram of Figure 7-2.

The transfer function of the linear controlled process is $G_p(s)$. The control signal at the input of the zero-order hold is denoted by $m^*(t)$. The error signal is defined as the difference between the desired output $c_d(t)$ and the actual output $c(t)$. For discrete-data systems, the performance index I is considered to include the following two summations:

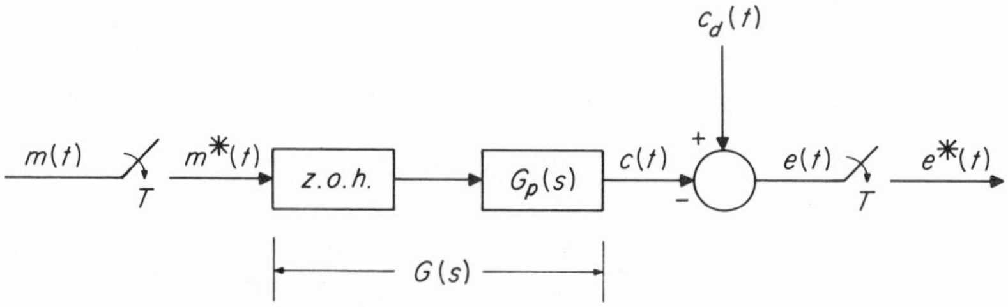

FIG. 7-2. Block diagram of analytic design of discrete-data systems.

$$I_e = \sum_{i=0}^{\infty} e^2(iT) \tag{7-5}$$

$$I_m = \sum_{i=0}^{\infty} m^2(iT) \tag{7-6}$$

The second summation puts a constraint on the control signal, since all physical signals are subject to saturation.

In general, the overall performance index I can be formulated according to the following ways:

1. $I_e =$ Minimum $\quad I_m \leq K$ (constant)

In this case, the design problem is to determine the optimal control signal $m^*(t)$ so that

$$I = I_e + k^2 I_m = \text{Minimum} \tag{7-7}$$

2. $I_e \leq K$ (constant), $I_m =$ Minimum

In this case, the design problem is to find the optimal control signal $m^*(t)$ so that

$$I = I_m + k^2 I_e = \text{Minimum} \tag{7-8}$$

In these last two expressions, k^2 is the well-known *Lagrangian multiplier*.[2,3]

The design procedure is developed by obtaining an equivalent frequency domain expression for Eq. (7-7) or Eq. (7-8) in terms of the transfer function and variables of the system.

With reference to Figure 7-2, the z-transform of the error signal is written

$$\begin{aligned} E(z) &= C_d(z) - C(z) \\ &= C_d(z) - G(z)M(z) \end{aligned} \tag{7-9}$$

where

$$\begin{aligned} G(z) &= \mathscr{Z}\left[\frac{1 - e^{-Ts}}{s} G_p(s)\right] \\ &= (1 - z^{-1})\mathscr{Z}\left[\frac{G_p(s)}{s}\right] \end{aligned} \tag{7-10}$$

Consider the performance index of Eq. (7-7), and using Eq. (3-106), we have

$$I = \sum_{i=0}^{\infty} [e^2(iT) + k^2 m^2(iT)] = \mathscr{Z}[e^2(iT) + k^2 m^2(iT)]|_{z=1} \tag{7-11}$$

From Eq. (3-105), the performance index is written

$$I = \frac{1}{2\pi j}\oint [E(z)E(z^{-1}) + k^2 M(z)M(z^{-1})]\frac{dz}{z} \tag{7-12}$$

Substituting Eq. (7-9) into Eq. (7-12), we get

$$I = \frac{1}{2\pi j}\oint [C_d(z) - G(z)M(z)][C_d(z^{-1}) - G(z^{-1})M(z^{-1})]z^{-1}\,dz$$
$$+ \frac{k^2}{2\pi j}\oint M(z)M(z^{-1})z^{-1}\,dz \tag{7-13}$$

The design requires that we find the optimum $M(z)$ so that the performance index of Eq. (7-8) is minimized.

Let $M_{0k}(z)$ be the optimum for a given k^2. Then we can write

$$M(z) = M_{0k}(z) + \lambda M_1(z) \tag{7-14}$$

where $M_1(z)$ is any arbitrary but physically realizable stable transfer function, and λ is a constant. Substituting Eq. (7-14) into Eq. (7-13), we get

$$I = I_1 + \lambda(I_2 + I_3) + \lambda^2 I_4 = I_{0k} + \Delta I \tag{7-15}$$

where

$$I_1 = \frac{1}{2\pi j}\oint [(C_d - GM_{0k})(\bar{C}_d - \bar{G}\bar{M}_{0k}) + k^2 M_{0k}\bar{M}_{0k}]z^{-1}\,dz \tag{7-16}$$

$$I_2 = \frac{1}{2\pi j}\oint M_1[G\bar{G}\bar{M}_{0k} - G\bar{C}_d + k^2\bar{M}_{0k}]z^{-1}\,dz \tag{7-17}$$

$$I_3 = \frac{1}{2\pi j}\oint \bar{M}_1[G\bar{G}M_{0k} - \bar{G}C_d + k^2 M_{0k}]z^{-1}\,dz \tag{7-18}$$

$$I_4 = \frac{1}{2\pi j}\oint [G\bar{G}M_1\bar{M}_1 + k^2 M_1\bar{M}_1]z^{-1}\,dz \tag{7-19}$$

Then

$$\Delta I = \lambda(I_2 + I_3) + \lambda^2 I_4 \tag{7-20}$$

and

$$I_{0k} = I_1 \tag{7-21}$$

For simplicity, we have used M_{0k} for $M_{0k}(z)$, G for $G(z)$, M_1 for $M_1(z)$, $\bar{M}_{0k}$ for $M_{0k}(z^{-1})$, etc.

Since $I_2(z) = I_3(z^{-1})$, and the poles of the integrand of I_2 are symmetrical to those of the integrand of I_3, with respect to the unit circle, $I_2 = I_3$. Also, I_4 is always positive.

The necessary and sufficient conditions for I in Eq. (7-15) to be a minimum are

$$\left.\frac{d\Delta I}{d\lambda}\right|_{\lambda=0} = 0 \tag{7-22}$$

and

$$\left.\frac{d^2\Delta I}{d\lambda^2}\right|_{\lambda=0} > 0 \tag{7-23}$$

Applying Eq. (7-22) to Eq. (7-20), we have

$$\left.\frac{d\Delta I}{d\lambda}\right|_{\lambda=0} = I_2 + I_3 = 2I_3 = 0 \tag{7-24}$$

Applying Eq. (7-23) to Eq. (7-20), we have

$$\left.\frac{d^2\Delta I}{d\lambda^2}\right|_{\lambda=0} = 2I_4 > 0 \tag{7-25}$$

Since I_4 is always positive, the necessary and sufficient condition for M_{0k} to give minimum I is

$$I_3 = \frac{1}{2\pi j}\oint \bar{M}_1[G\bar{G}M_{0k} - \bar{G}C_d + k^2 M_{0k}]z^{-1}\,dz = 0 \tag{7-26}$$

Since M_1 has poles only inside the unit circle of the z-plane, the poles of $\bar{M}_1$ are all outside the unit circle. Therefore, Eq. (7-26) is satisfied if the integrand of the integral in Eq. (7-26) has only poles which are outside the unit circle and the integral is performed along the circle in the counterclockwise direction. Therefore, this condition leads to the necessary and sufficient requirement for I = minimum to be that the following expression has poles only outside the unit circle.

$$X(z) = (\bar{G}GM_{0k} - \bar{G}C_d + k^2 M_{0k})z^{-1} \tag{7-27}$$

The poles and zeros of the function $(k^2 + G\bar{G})$ are symmetrical with respect to the unit circle. Thus, we let

$$Y\bar{Y} = k^2 + G\bar{G} \tag{7-28}$$

where

$$Y = Y(z) = \{k^2 + G(z)G(z^{-1})\}^{+} \tag{7-29}$$

and

$$\bar{Y} = Y(z^{-1}) = \{k^2 + G(z)G(z^{-1})\}^{-} \tag{7-30}$$

Y denotes the part of $Y\bar{Y}$ which has poles and zeros that are inside the unit circle, and $\bar{Y}$ is that part which has poles and zeros outside the unit circle.

Equation (7-27) is now written as

$$X(z) = [Y\bar{Y}M_{0k} - \bar{G}C_d]z^{-1} \tag{7-31}$$

Now we try to write Eq. (7-31) as the sum of two parts, one with poles only inside the unit circle and one with poles outside the unit circle. Dividing both sides of Eq. (7-31) by $\bar{Y}$, we have

$$\frac{X(z)}{\bar{Y}} = YM_{0k}z^{-1} - \frac{\bar{G}C_d}{\bar{Y}}z^{-1} \tag{7-32}$$

The last term on the right side of the last equation is expanded by partial fraction expansion into

$$\frac{\bar{G}C_d}{\bar{Y}}z^{-1} = \left[\frac{\bar{G}C_d}{\bar{Y}}z^{-1}\right]_+ + \left[\frac{\bar{G}C_d}{\bar{Y}}z^{-1}\right]_- \tag{7-33}$$

where

$$\left[\frac{\bar{G}C_d}{\bar{Y}}z^{-1}\right]_+ = \text{Terms with poles inside the unit circle} \tag{7-34}$$

$$\left[\frac{\bar{G}C_d}{\bar{Y}}z^{-1}\right]_- = \text{Terms with poles outside the unit circle} \tag{7-35}$$

Now Eq. (7-32) is written

$$\frac{X}{\bar{Y}} + \left[\frac{\bar{G}C_d}{\bar{Y}}z^{-1}\right]_- = YM_{0k}z^{-1} - \left[\frac{\bar{G}C_d}{\bar{Y}}z^{-1}\right]_+ = 0 \tag{7-36}$$

where the left side terms have poles outside the unit circle, and the right side terms have poles only inside the unit circle. Therefore,

$$YM_{0k}z^{-1} = \left[\frac{\bar{G}C_d}{\bar{Y}}z^{-1}\right]_+ \tag{7-37}$$

Solving for the optimal control signal from the last equation, we get

$$M_{0k} = \frac{z}{Y}\left[\frac{\bar{G}C_d}{\bar{Y}}z^{-1}\right]_+ \tag{7-38}$$

The expression of M_{0k} given by the last equation is still a function of k^2, and it is the optimal control signal which minimizes the performance index $I = I_e + k^2 I_m$ a for given value of k^2. To find the absolute optimal $M_0(z)$, we substitute M_{0k} into the constraint criterion

$$I_m = \sum_{i=0}^{\infty} m^2(iT) = \frac{1}{2\pi j}\oint M_{0k}(z)M_{0k}(z^{-1})z^{-1}\,dz \leq K \tag{7-39}$$

The worst case is considered by taking the equal sign in Eq. (7-39), and determining the value of k^2 so that $I_m = K$. Then, $M_0(z)$ is obtained by substituting this k^2 into Eq. (7-38).

EXAMPLE 7-1

For the discrete-data system shown in Figure 7-2, given

$$G_p(s) = \frac{1}{s}$$

find the optimal control signal $M_0(z)$ so that the output of the system will follow a unit-step input with

1. $I_e = \sum_{i=0}^{\infty} e^2(iT) = \sum_{i=0}^{\infty} [c_d(iT) - c(iT)]^2 = \text{Minimum}$ (7-40)

and

2. $I_m = \sum_{i=0}^{\infty} m^2(iT) \leq 2$ (7-41)

The sampling period is one second.

Since the desired output is a unit-step function,

$$C_d(z) = \frac{z}{z-1} \tag{7-42}$$

Using Eq. (7-10),

$$G(z) = (1 - z^{-1})\mathscr{Z}\left[\frac{G_p(s)}{s}\right] = (1 - z^{-1})\mathscr{Z}\left(\frac{1}{s^2}\right) = \frac{1}{z-1} \tag{7-43}$$

Substitution of Eq. (7-43) into Eq. (7-28) yields

$$\begin{aligned} Y\bar{Y} = Y(z)Y(z^{-1}) &= k^2 + \frac{1}{(z-1)(z^{-1}-1)} \\ &= \frac{k^2(z-1)(z^{-1}-1)+1}{(z-1)(z^{-1}-1)} \end{aligned} \tag{7-44}$$

The numerator of the last equation is written as

$$\begin{aligned} (1 + 2k^2) - k^2 z - k^2 z^{-1} &= (a_1 + a_2 z)(a_1 + a_2 z^{-1}) \\ &= a_1^2 + a_1 a_2 z + a_1 a_2 z^{-1} + a_2^2 \end{aligned} \tag{7-45}$$

where a_1 and a_2 are constants, and $a_1 < a_2$.

Equating like coefficients, Eq. (7-45) leads to

$$a_1^2 + a_2^2 = 1 + 2k^2 \tag{7-46}$$

$$a_1 a_2 = -k^2 \tag{7-47}$$

Solving for a_1 and a_2 from the last two equations, we get

$$a_1 = \frac{1}{2}\left(1 \mp \sqrt{1 - 4k^2}\right) \tag{7-48}$$

$$a_2 = \frac{1}{2}\left(1 \pm \sqrt{1 - 4k^2}\right) \tag{7-49}$$

Since $a_1 < a_2$, the signs in Eqs. (7-48) and (7-49) are uniquely determined. Therefore,

$$\begin{aligned} Y(z) = \left\{\frac{k^2(z-1)(z^{-1}-1)+1}{(z-1)(z^{-1}-1)}\right\}^+ &= \frac{a_1 + a_2 z}{z-1} \\ &= \frac{(1 - \sqrt{1+4k^2}) + (1 + \sqrt{1+4k^2})z}{2(z-1)} \end{aligned} \tag{7-50}$$

In factorizing $Y\bar{Y}$ it is assumed that the zero of $(z-1)$ is slightly inside the unit circle, and that of $(z^{-1}-1)$ is just outside the unit circle. Similarly,

$$\bar{Y} = Y(z^{-1}) = \frac{(1-\sqrt{1+4k^2}) + (1+\sqrt{1+4k^2})z^{-1}}{2(z^{-1}-1)} \tag{7-51}$$

Now substitution of $Y, \bar{Y}, \bar{G}$, and C_d into Eq. (7-38) yields

$$M_{0k}(z) = \frac{2z(z-1)}{(1-\sqrt{1+4k^2}) + (1+\sqrt{1+4k^2})z} \times \left[\frac{\dfrac{1}{z^{-1}-1}\cdot\dfrac{z}{z-1}}{z\dfrac{(1-\sqrt{1+4k^2}) + (1+\sqrt{1+4k^2})z^{-1}}{z^{-1}-1}}\right]_+$$

$$= \frac{2z}{(1-\sqrt{1+4k^2}) + (1+\sqrt{1+4k^2})z} \tag{7-52}$$

Thus, Eq. (7-52) gives the z-transform of the optimal control signal as a function of k^2. To determine the optimal value of k^2 so that Eqs. (7-40) and (7-41) are satisfied simultaneously, we substitute Eq. (7-52) into Eq. (7-41), and we have

$$I_m = \frac{1}{2\pi j}\oint M_{0k}\bar{M}_{0k}z^{-1}\,dz$$

$$= \frac{1}{2\pi j}\oint \frac{2z}{(1+\sqrt{1+4k^2})z + (1-\sqrt{1+4k^2})} \times \frac{2z^{-1}}{(1+\sqrt{1+4k^2})z^{-1} + (1-\sqrt{1+4k^2})}z^{-1}\,dz = 2 \tag{7-53}$$

Evaluating the integral in the last equation by the residue theorem, we get

$$I_m = \frac{1}{\sqrt{1+4k^2}} = 2 \tag{7-54}$$

or

$$k^2 = \frac{-3}{16} \tag{7-55}$$

Substituting this value of k^2 into Eq. (7-52), the optimal control signal is obtained as

$$M_0(z) = \frac{4z}{3z+1} \tag{7-56}$$

The z-transform of the output of the optimal system is determined from

$$C(z) = G(z)M_0(z) = \frac{4z}{(z-1)(3z+1)} \tag{7-57}$$

The sampled output $c^*(t)$ is obtained by expanding $C(z)$ into a power series in z^{-1}; then

$$c^*(t) = \frac{4}{3}\delta(t-1) + \frac{8}{9}\delta(t-2) + \frac{28}{27}\delta(t-3) + \dots \tag{7-58}$$

The output approaches the final desired value of unity quite rapidly.

The summed-error-square criterion for the optimum system is given by

$$I_e = \frac{1}{2\pi j}\oint [C_d(z) - C(z)][C_d(z^{-1}) - C(z^{-1})]z^{-1}\,dz = \frac{9}{8} \tag{7-59}$$

In completing the design problem, it is usually desirable to find the transfer function of the digital controller so that the optimum control signal described by Eq. (7-56) is realized. Let us select the configuration of the closed-loop system as shown in Figure 7-3. The transfer function of the digital controller is written as

$$D(z) = \frac{M_0(z)}{E(z)} = \frac{M_0(z)}{C_d(z) - C(z)} \tag{7-60}$$

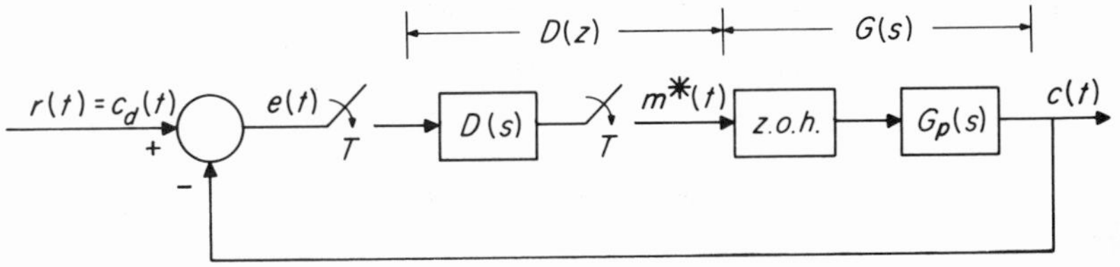

FIG. 7-3. Closed-loop discrete-data system for Example 7-1.

Therefore,

$$D(z) = \frac{\dfrac{4z}{3z+1}}{\dfrac{z}{z-1} - \dfrac{4z}{(z-1)(3z-1)}} = \frac{4}{3} \tag{7-61}$$

Thus, the optimum controller in this case is simply an amplifier with a gain of $\frac{4}{3}$.

The design technique formulated above is for single input systems only. For multivariable systems, the equations would have to be modified, and one is faced with the problem of spectral factorizing a matrix, since $Y\bar{Y}$ is now a matrix.

7.3 THE QUADRATIC PERFORMANCE INDEX

One of the most important performance indices used in the design of multivariable optimal control systems is the quadratic performance index. Just as the summed-error-square criterion forms the basis of the analytical design of discrete-data systems, the quadratic performance index forms a more general measure of performance, and makes the design of discrete-data systems more convenient by dynamic programming.[4]

The controlled process is described by the following state equation:

$$\frac{d\mathbf{x}(t)}{dt} = \mathbf{A}\mathbf{x}(t) + \mathbf{B}\mathbf{m}(t) \tag{7-62}$$

where

$\mathbf{x}(t) = n \times 1$ state vector

$\mathbf{m}(t) = p \times 1$ input vector

$A = n \times n$ matrix whose elements may be time-varying

$B = n \times p$ matrix whose elements may be time-varying

The elements of $\mathbf{m}(t)$ are constant over one sampling period; that is,

$$m_i(t) = m_i(k) = \text{constant} \qquad k \leq t < (k+1)$$
$$i = 1, 2, \ldots, p$$

The state transition equation of the process is written

$$\mathbf{x}(k+1) = \mathbf{\Phi}(1)\mathbf{x}(k) + \mathbf{\Theta}(1)\mathbf{m}(k) \tag{7-63}$$

where

$$\mathbf{\Phi}(1) = e^{A} \quad (n \times n) \tag{7-64}$$

and

$$\mathbf{\Theta}(1) = \int_{k}^{k+1} \mathbf{\Phi}[(k+1) - \tau]B\,d\tau \quad (n \times p) \tag{7-65}$$

In general, the design objective can be stated as to determine the optimal $\mathbf{m}(k)$ for $k = 0, 1, 2, \ldots, N-1$, so that the quadratic performance index I_N is minimized, subject to an initial state $\mathbf{x}(0)$. In one form, the quadratic performance index is written as

$$I_N = \sum_{k=1}^{N} [\mathbf{x}'(k)Q(k)\mathbf{x}(k) + \lambda\mathbf{m}'(k-1)H(k-1)\mathbf{m}(k-1)] \tag{7-66}$$

where

$Q(k) = n \times n$ symmetric and positive-semidefinite matrix

$H(k-1) = p \times p$ symmetric and positive-semidefinite matrix

$\lambda =$ scalar constant

$N =$ positive integer

$\mathbf{x}'(k) =$ transpose of $\mathbf{x}(k)$, $1 \times n$ row vector

$\mathbf{m}'(k) =$ transpose of $\mathbf{m}(k)$, $1 \times p$ row vector

In the literature it is common to define the performance index as

$$I_N = \sum_{k=1}^{N} [\mathbf{e}'(k)Q(k)\mathbf{e}(k) + \lambda\mathbf{m}'(k-1)H(k-1)\mathbf{m}(k-1)] \tag{7-67}$$

where

$\mathbf{e}(k) = \mathbf{x}_d(k) - \mathbf{x}(k) =$ error vector

$\mathbf{x}_d(k) =$ desired state vector

7.4 MINIMUM QUADRATIC PERFORMANCE INDEX DESIGN[4-11]

The optimum design of discrete-data control systems with quadratic performance index can be carried out using the dynamic programming. The method of dynamic programming for solving multi-stage allocation problems was first introduced by R. Bellman.[4] The application of dynamic programming to the optimum design of discrete-data control systems was due to Kalman, Koepcke, Tou, Joseph, and others.

The basic design problem is stated as:

Given a linear process described by

$$\mathbf{x}(k+1) = \mathbf{\Phi}(1)\mathbf{x}(k) + \mathbf{\Theta}(1)\mathbf{m}(k) \tag{7-68}$$

find the optimal control law $\mathbf{m}(k)$ for $k = 0, 1, 2, \ldots, N-1$, so that the quadratic performance index

$$I_N = \sum_{k=1}^{N} [\mathbf{x}'(k)Q(k)\mathbf{x}(k) + \lambda\mathbf{m}'(k-1)H(k-1)\mathbf{m}(k-1)] \tag{7-69}$$

is minimized, subject to an initial state

$$\mathbf{x}(0) = \mathbf{x}_0 \tag{7-70}$$

Kalman and Koepcke[7] first suggested the use of dynamic programming for the determination of the optimal control law. It is well known that dynamic programming is a method of solving multi-stage decision problems. It is based on the *principle of invariant imbedding*, and it leads to the *principle of optimality* which has found broad use in optimal control theory.

In order to lay the necessary foundation for the minimum quadratic performance index design, a brief introduction of dynamic programming and the principle of optimality is given below.

A multi-stage allocation process is best described by the following example on investment.

Assume that we have an initial amount of capital C to be invested in the television rental business, with

C = Initial amount of capital to be invested

x = Amount invested in black and white televisions

$C - x$ = Amount invested in color televisions

$g(x)$ = Return or yield from investment in black and white televisions in one year

$h(C - x)$ = Return or yield from color televisions for one year

Therefore, the total return for one year is

$$R_1(C, x) = g(x) + h(C - x) \tag{7-71}$$

Assume that at the end of each year the used televisions are traded in for new ones, and

ax = Trade-in value of the black and white televisions at the end of one year $(0 < a < 1)$

$b(C - x)$ = Trade-in value of color televisions at the end of one year $(0 < b < 1)$

If the return from rentals is not reinvested, the total capital available for reinvestment at the end of one year is

$$C_1 = ax + b(C - x) \tag{7-72}$$

The maximum yield for one year is written as

$$f_1(C) = \max_{0 \le x \le C} R_1(C, x) = \max_{0 \le x \le C} [g(x) + h(C - x)] \tag{7-73}$$

For a two-year process, the following quantities are defined:

$C_1 = ax + b(C - x)$ = Capital available at the beginning of the second year

x_1 = Amount invested in black and white television at the beginning of the second year

$C_1 - x_1$ = Amount invested in color television at the beginning of the second year.

$$R_2(C, x, x_1) = g(x) + h(C - x) + g(x_1) + h(C_1 - x_1)$$
$$= \text{Total return at the end of two years} \tag{7-74}$$

ax_1 = Trade-in value of black and white televisions at the end of the second year $(0 < a < 1)$

$b(C_1 - x_1)$ = Trade-in value of color televisions at the end of the second year $(0 < b < 1)$

The maximum yield at the end of the second year is

$$f_2(C) = \max_{\substack{0 \le x \le C \\ 0 \le x_1 \le C_1}} R_2(C, x, x_1) = \max_{\substack{0 \le x \le C \\ 0 \le x_1 \le C_1}} [g(x) + h(C - x) + g(x_1) + h(C_1 - x_1)] \tag{7-75}$$

In general, for an N-year investment process as described above, the total return is

$$R_N(C, x, x_1, x_2, \ldots, x_{N-1}) = g(x) + h(C - x) + g(x_1) + h(C_1 - x_1)$$
$$+ g(x_{N-1}) + h(C_{N-1} - x_{N-1})$$
$$= g(x) + h(C - x) + \sum_{i=1}^{N-1} [g(x_i) + h(C_i - x_i)] \tag{7-76}$$

where

C_i = Capital available at the end of the ith year

x_i = Amount spent on black and white television at the end of the ith year

$C_i - x_i$ = Amount spent on color television at the end of the ith year.

The maximum yield at the end of the Nth year is

$$f_N(C) = \max_{\substack{0 \le x \le C \\ 0 \le x_1 \le C_1 \\ \vdots \\ 0 \le x_{N-1} \le C_{N-1}}} R_N(C, x, x_1, x_2, \ldots, x_{N-1})$$

$$= \text{Maximum total return over } N \text{ years} \tag{7-77}$$

One obvious method of solving the N-dimensional maximization problem is to equate to zero each of the partial derivatives of R_N with respect to the N variables $x_1, x_2, \ldots, x_{N-1}$. This will yield N simultaneous equations, nonlinear in general, in the N variables. The solution of these N equations, however, is difficult to obtain. Furthermore, the solution is useful only if the absolute maximum occurs in an interior point and not on the boundary of the set of N variables. The amount of computation involved for any reasonable sized N is prohibitive.

The optimization problem can be carried out using the dynamic programming method based on the principle of optimality. For a single-stage problem, the maximum yield for one year is given by Eq. (7-73); that is,

$$f_1(C) = \max_{0 \le x \le C} R_1(C, x) = \max_{0 \le x \le C} [g(x) + h(C - x)] \tag{7-78}$$

We can carry out the maximization problem simply by taking the derivative of $R_1(C, x)$ with respect to x and setting it equal to zero. The optimum value of x is then substituted into $R_1(C, x)$ to yield $f_1(C)$.

For a two-stage problem, the return after two years is

$$R_2(C, x, x_1) = g(x) + h(C - x) + g(x_1) + h(C_1 - x_1) \tag{7-79}$$

The maximum return is

$$f_2(C) = \max_{\substack{0 \le x \le C \\ 0 \le x_1 \le C_1}} R_2(C, x, x_1) = \max_{\substack{0 \le x \le C \\ 0 \le x_1 \le C_1}} [g(x) + h(C - x) + g(x_1) + h(C_1 - x_1)] \tag{7-80}$$

Notice that whatever the value of x chosen initially, the remaining amount of return, $C_1 = ax + b(C - x)$ should be allocated in the optimal manner in order to maximize the two-stage process. Therefore, if x_1 is chosen optimally, we shall obtain a return of

$$f_1(C_1) = f_1[ax + b(C - x)] = \max_{0 \le x_1 \le C_1} R_1(C_1, x_1) \tag{7-81}$$

The total return for the two-stage process is given by Eq. (7-79), and for a given optimal x_1,

$$R_2(C, x) = g(x) + h(C - x) + f_1(C_1)$$
$$= g(x) + h(C - x) + f_1[ax + b(C - x)] \tag{7-82}$$

Therefore, the optimal return for a two-stage process is

$$f_2(C) = \max_{\substack{0 \leq x \leq C \\ 0 \leq x_1 \leq C_1}} R_2(C, x, x_1) = \max_{0 \leq x \leq C} R_2(C, x) \tag{7-83}$$

Hence, we have reduced the two dimensional maximization problem to a recursive relationship which requires only two single dimensional maximizations. The procedure can be extended to an N-stage process. Therefore, for an N-stage optimization problem, the optimal return can be written as

$$f_N(C) = \max_{0 \leq x \leq C} \{g(x) + h(C - x) + f_{N-1}[ax + b(C - x)]\} \tag{7-84}$$

The multi-stage decision problem described above leads to the *principle of optimality*, which states that *an optimal policy has the property that whatever the initial state and initial decision are, the remaining decisions must constitute an optimal policy with regard to the state resulting from the first decision.*

Single-Input Systems

Now we return to the design problem described by Eq. (7-69) and Eq. (7-70). We shall first consider a system with a single input. Then, Eqs. (7-69) and (7-70) become

$$\mathbf{x}(k + 1) = \mathbf{\Phi}(1)\mathbf{x}(k) + \mathbf{\Theta}(1)m(k) \tag{7-85}$$

where $m(k)$ is a scalar input, and

$$I_N = \sum_{k=1}^{N} [\mathbf{x}'(k)Q(k)\mathbf{x}(k) + \lambda m^2(k - 1)] \tag{7-86}$$

The problem is to find the optimal control law $m(k)$ for $k = 0, 1, 2, \ldots, N - 1$, so that I_N of Eq. (7-86) is minimized, subject to an initial state $\mathbf{x}(0)$.

Let the minimum value of I_N be written as

$$f_N[\mathbf{x}(0)] = \min_{\substack{m(0) \\ m(1) \\ \vdots \\ m(N-1)}} I_N = \min_{\substack{m(0) \\ m(1) \\ \vdots \\ m(N-1)}} \sum_{k=1}^{N} [\mathbf{x}'(k)Q(k)\mathbf{x}(k) + \lambda m^2(k - 1)] \tag{7-87}$$

Equation (7-87) gives the minimum return for the entire N stages. In general, the minimum return for the last $N - j$ stages of the N-stage process can be written as

$$f_{N-j}[\mathbf{x}(j)] = \min_{\substack{m(j) \\ m(j+1) \\ \vdots \\ m(N-1)}} I_{N-j} = \min_{\substack{m(j) \\ m(j+1) \\ \vdots \\ m(N-1)}} \sum_{k=j+1}^{N} [\mathbf{x}'(k)Q(k)\mathbf{x}(k) + \lambda m^2(k - 1)] \tag{7-88}$$

$j = 0, 1, 2, \ldots, N - 1.$

When $j = 0$, it is apparent that we have $f_N[\mathbf{x}(0)]$ as in Eq. (7-87). The principle of optimality states that regardless of what the first stage does, the remaining $N - 1$ stages must constitute an optimal policy with regard to the state resulting from the first decision. The output from the first stage is $\mathbf{x}'(1)Q(1)\mathbf{x}(1) + \lambda m^2(0)$, and the optimal output from the remaining $N - 1$ stages is $f_{N-1}[\mathbf{x}(1)]$. Therefore, the total return of the N-stage process under the situation is

$$I_N = \mathbf{x}'(1)Q(1)\mathbf{x}(1) + \lambda m^2(0) + f_{N-1}[\mathbf{x}(1)] \tag{7-89}$$

Then, according to the principle of optimality, the minimum return for the N-stage process is

$$f_N[\mathbf{x}(0)] = \min_{m(0)} I_N = \min_{m(0)} \{\mathbf{x}'(1)Q(1)\mathbf{x}(1) + \lambda m^2(0) + f_{N-1}[\mathbf{x}(1)]\} \tag{7-90}$$

Therefore, in the last equation we have only one parameter, $m(0)$, to determine.

In general, starting from the jth stage, the output from the last $N - j$ stages is equal to the output from the $(j + 1)$th stage plus the optimum output from the remaining $N - (j + 1)$ stages. Therefore, under the condition the return from the last $N - j$ stages is written as

$$I_{N-j} = \mathbf{x}'(j + 1)Q\mathbf{x}(j + 1) + \lambda m^2(j) + f_{N-(j+1)}[\mathbf{x}(j + 1)] \tag{7-91}$$

The minimum return for the last $N - j$ stages is

$$f_{N-j}[\mathbf{x}(j)] = \min_{m(j)} \{\mathbf{x}'(j + 1)Q(j + 1)\mathbf{x}(j + 1) + \lambda m^2(j) + f_{N-(j+1)}[\mathbf{x}(j + 1)]\} \tag{7-92}$$

The minimum returns for $j = 0, 1, 2, \ldots, N - 1$, are tabulated below:

$$j = 0, \qquad f_N[\mathbf{x}(0)] = \min_{m(0)} \{\mathbf{x}'(1)Q\mathbf{x}(1) + \lambda m^2(0) + f_{N-1}[\mathbf{x}(1)]\} \tag{7-93}$$

$$j = 1, \qquad f_{N-1}[\mathbf{x}(1)] = \min_{m(1)} \{\mathbf{x}'(2)Q\mathbf{x}(2) + \lambda m^2(1) + f_{N-2}[\mathbf{x}(2)]\} \tag{7-94}$$

$$\vdots \qquad \vdots \qquad \vdots$$

$$j = N - 2, \quad f_2[\mathbf{x}(N - 2)] = \min_{m(N-2)} \{\mathbf{x}'(N - 1)Q\mathbf{x}(N - 1) + \lambda m^2(N - 2) + f_1[\mathbf{x}(N - 1)]\} \tag{7-95}$$

$$\begin{aligned} j = N - 1, \quad f_1[\mathbf{x}(N - 1)] &= \min_{m(N-1)} \{\mathbf{x}'(N)Q\mathbf{x}(N) + \lambda m^2(N - 1) + f_0[\mathbf{x}(N)]\} \\ &= \min_{m(N-1)} \{\mathbf{x}'(N)Q\mathbf{x}(N) + \lambda m^2(N - 1)\} \end{aligned} \tag{7-96}$$

since

$$f_0[\mathbf{x}(N)] = 0 \tag{7-97}$$

In these equations, for simplicity, the weighting matrix Q is assumed to be constant with respect to time.

From Eqs. (7-93) through (7-96), it is clear that the N-stage decision process is now reduced to a sequence of N single-stage optimization problems.

Starting with Eq. (7-96), the optimum $m(N-1)$ is found to yield the minimum return $f_1[\mathbf{x}(N-1)]$ which is then substituted into Eq. (7-95). The process is repeated until the last optimal control $m(0)$ is determined from Eq. (7-93). Therefore, a characteristic of the principle of optimality of dynamic programming is that the solution is carried from the last stage back to the first stage in a recursive manner.

Let us consider a one-stage process, that is, $N = 1$. From Eq. (7-89), we have

$$I_1 = \mathbf{x}'(1)Q\mathbf{x}(1) + \lambda m^2(0) \tag{7-98}$$

since $f_0[\mathbf{x}(1)] = 0$, and it is assumed that Q is a constant matrix. Using Eq. (7-69), we get

$$\mathbf{x}(1) = \Phi(1)\mathbf{x}(0) + \Theta(1)m(0) \tag{7-99}$$

Therefore, Eq. (7-98) becomes

$$\begin{aligned} I_1 &= [\Phi(1)\mathbf{x}(0) + \Theta(1)m(0)]'Q[\Phi(1)\mathbf{x}(0) + \Theta(1)m(0)] + \lambda m^2(0) \\ &= [\mathbf{x}'(0)\Phi'(1) + m(0)\Theta'(1)]Q[\Phi(1)\mathbf{x}(0) + \Theta(1)m(0)] + \lambda m^2(0) \end{aligned} \tag{7-100}$$

To find the optimum $m(0)$ for minimum I_1, we take the derivative on both sides of Eq. (7-100) with respect to $m(0)$, and set the result equal to zero. We have

$$\begin{aligned} \frac{dI_1}{dm(0)} &= \mathbf{x}'(0)\Phi'(1)Q\Theta(1) + \Theta'(1)Q\Phi(1)\mathbf{x}(0) \\ &\quad + 2\Theta'(1)Q\Theta(1)m(0) + 2\lambda m(0) = 0 \end{aligned} \tag{7-101}$$

Solving for $m(0)$ from the last equation, we get

$$m(0) = -\frac{\mathbf{x}'(0)\Phi'(0)Q\Theta(1) + \Theta'(1)Q\Phi(1)\mathbf{x}(0)}{2[\Theta'(1)Q\Theta(1) + \lambda]} \tag{7-102}$$

Since Q is a symmetric matrix,

$$\mathbf{x}'(0)\Phi'(1)Q\Theta(1) = \Theta'(1)Q\Phi(1)\mathbf{x}(0) \tag{7-103}$$

Equation (7-102) is simplified to

$$m(0) = -\frac{\Theta'(1)Q\Phi(1)}{\Theta'(1)Q\Theta(1) + \lambda}\mathbf{x}(0) \tag{7-104}$$

Let us define

$$H(1) = -\frac{\Theta'(1)Q\Phi(1)}{\Theta'(1)Q\Theta(1) + \lambda} \tag{7-105}$$

Then, Eq. (7-104) becomes

$$m(0) = H(1)\mathbf{x}(0) \tag{7-106}$$

which shows that the optimal control is a function of the state variable, and the control can be achieved theoretically through state variable feedback. Now substituting Eq. (7-106) into Eq. (7-100) and simplifying, we get

$$f_1[\mathbf{x}(0)] = \min_{m(0)} I_1 = \mathbf{x}'(0)\{[\Phi(1) + \Theta(1)H(1)]'Q[\Phi(1) + \Theta(1)H(1)] + \lambda H'(1)H(1)\}\mathbf{x}(0) \tag{7-107}$$

or simply

$$f_1[\mathbf{x}(0)] = \mathbf{x}'(0)P(1)\mathbf{x}(0) \tag{7-108}$$

where

$$P(1) = [\Phi(1) + \Theta(1)H(1)]'Q[\Phi(1) + \Theta(1)H(1)] + \lambda H'(1)H(1) \tag{7-109}$$

It is apparent that $P(1)$ is a symmetric matrix. Also of significance is that $f_1[\mathbf{x}(0)]$ is of quadratic form.

For a two-stage process, $N = 2$, Eqs. (7-93) and (7-94) become

$$f_2[\mathbf{x}(0)] = \min_{m(0)} \{\mathbf{x}'(1)Q\mathbf{x}(1) + \lambda m^2(0) + f_1[\mathbf{x}(1)]\} \tag{7-110}$$

and

$$f_1[\mathbf{x}(1)] = \min_{m(1)} \{\mathbf{x}'(2)Q\mathbf{x}(2) + \lambda m^2(1)\} \tag{7-111}$$

respectively. From Eq. (7-111), the optimal control $m(1)$ is determined as

$$m(1) = H(1)\mathbf{x}(1) \tag{7-112}$$

where $H(1)$ is given by Eq. (7-105). Then, the optimum return in Eq. (7-111) is

$$f_1[\mathbf{x}(1)] = \mathbf{x}'(1)P(1)\mathbf{x}(1) \tag{7-113}$$

where $P(1)$ is as given in Eq. (7-109).

Now substituting Eq. (7-113) into Eq. (7-110), we have

$$f_2[\mathbf{x}(0)] = \min_{m(0)} I_2[\mathbf{x}(0)] \tag{7-114}$$

where

$$I_2[\mathbf{x}(0)] = [\mathbf{x}'(0)\Phi'(1) + m(0)\Theta'(1)][Q + P(1)][\Phi(1)\mathbf{x}(0) + \Theta(1)m(0)] + \lambda m^2(0) \tag{7-115}$$

Taking the derivative of $I_2[\mathbf{x}(0)]$ with respect to $m(0)$ and setting the result equal to zero, we get the optimal control

$$m(0) = -\frac{\Theta'(1)[Q + P(1)]\Phi(1)}{\Theta'(1)[Q + P(1)]\Theta(1) + \lambda}\mathbf{x}(0) \tag{7-116}$$

The last equation can be written as

$$m(0) = H(2)\mathbf{x}(0) \tag{7-117}$$

where

$$H(2) = -\frac{\Theta'(1)[Q + P(1)]\Phi(1)}{\Theta'(1)[Q + P(1)]\Theta(1) + \lambda} \tag{7-118}$$

The optimal control $m(0)$ of Eq. (7-117) when substituted into Eq. (7-114) yields

$$f_2[\mathbf{x}(0)] = \min_{m(0)} I_2[\mathbf{x}(0)] = \mathbf{x}'(0)\{[\Phi(1) + \Theta(1)H(2)]'[Q + P(1)][\Phi(1) + \Theta(1)H(2)] + \lambda H'(2)H(2)\}\mathbf{x}(0) \qquad (7\text{-}119)$$

Letting

$$P(2) = [\Phi(1) + \Theta(1)H(2)]'[Q + P(1)][\Phi(1) + \Theta(1)H(2)] + \lambda H'(2)H(2) \qquad (7\text{-}120)$$

Equation (7-119) is written

$$f_2[\mathbf{x}(0)] = \mathbf{x}'(0)P(2)\mathbf{x}(0) \qquad (7\text{-}121)$$

If the same iterative procedure as described above is carried out for $N = 3, 4, \ldots$, it can be shown that the optimal control at the jth sampling instant for an N-stage process is given by

$$m(j) = H(N - j)\mathbf{x}(j) \qquad (7\text{-}122)$$

for $j = 0, 1, 2, \ldots, N - 1$, and

$$H(N - j) = -\frac{\Theta'(1)[Q + P(N - j - 1)]\Phi(1)}{\Theta'(1)[Q + P(N - j - 1)]\Theta(1) + \lambda} \qquad (7\text{-}123)$$

The minimum return $f_{N-j}[\mathbf{x}(j)]$ is

$$f_{N-j}[\mathbf{x}(j)] = \min_{m(j)} I_{N-j}[\mathbf{x}(j)] = \mathbf{x}'(j)P(N - j)\mathbf{x}(j) \qquad (7\text{-}124)$$

where

$$P(N - j) = [\Phi(1) + \Theta(1)H(N - j)]'[Q + P(N - j - 1)][\Phi(1) + \Theta(1)H(N - j)] + \lambda H'(N - j)H(N - j) \qquad (7\text{-}125)$$

Equations (7-123) and (7-125) form the recursive relations for the determination of the H and the P matrices. Starting with $j = N - 1$, $P(N - j - 1) = P(0) = 0$; Eq. (7-123) gives $H(1)$. Then, Eq. (7-125) is used to obtain $P(1)$, and then $H(2)$, and so on.

EXAMPLE 7-2

Consider that a linear controlled process is described by the first-order difference equation

$$x_1(k + 1) = \Phi(1)x_1(k) + \Theta(1)m(k) \qquad (7\text{-}126)$$

where $\Phi(1) = 0.368$ and $\Theta(1) = 0.632$.

It is desired to determine the optimal control $m(0)$, $m(1)$, $m(2)$, and $m(3)$, so that the performance index

$$I_4 = \sum_{k=1}^{4} [x_1^2(k) + m^2(k - 1)] \qquad (7\text{-}127)$$

is minimized.

Comparing Eq. (7-127) with Eq. (7-86), we notice that $N = 4$, $Q = 1$, and $\lambda = 1$. Using Eq. (7-122), the optimal control sequence is obtained through state variable feedback. Therefore,

$$m(0) = H(4)\mathbf{x}(0) \tag{7-128}$$

$$m(1) = H(3)\mathbf{x}(1) \tag{7-129}$$

$$m(2) = H(2)\mathbf{x}(2) \tag{7-130}$$

$$m(3) = H(1)\mathbf{x}(3) \tag{7-131}$$

The feedback matrices $H(N - j)$ are determined from Eq. (7-123). Thus,

$$H(1) = -\frac{\Theta'(1)Q\Phi(1)}{\Theta'(1)Q\Theta(1) + \lambda} = -\frac{(0.632)(1)(0.368)}{(0.632)(1)(0.632) + 1}$$

$$= -0.166 \tag{7-132}$$

In order to determine $H(2)$ we must first calculate $P(1)$ using Eq. (7-125). We have

$$P(1) = \Phi'(1)[Q + P(0)][\Phi(1) + \Theta(1)H(1)]$$

$$= 0.368[0.368 + 0.632(-0.166)]$$

$$= 0.097 \tag{7-133}$$

Then, using Eq. (7-123) with $N = 4$ and $j = 2$, we get

$$H(2) = -\frac{\Theta'(1)[Q + P(1)]\Phi(1)}{\Theta'(1)[Q + P(1)]\Theta(1) + \lambda} = \frac{-0.632(1.097)0.368}{0.632(1.097)0.632 + 1}$$

$$= -0.178 \tag{7-134}$$

Continuing the same process, $P(2)$, $H(3)$, $P(3)$, and $H(4)$ are determined,

$$P(2) = 0.1$$

$$H(3) = -0.178$$

$$P(3) = 0.11$$

$$H(4) = -0.1782$$

The block diagram of the optimum system is shown in Figure 7-4. The optimal control is implemented by the feedback of the state variable through a time-varying gain element.

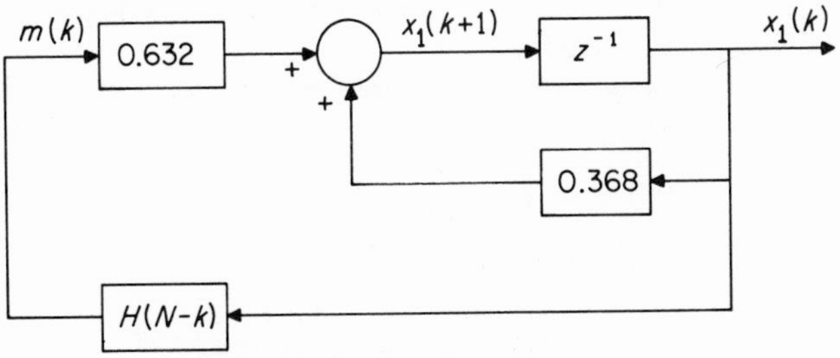

FIG. 7-4. Block diagram of optimum control system in Example 7-2.

The optimal control sequence at the sampling instants is listed below:

$$\begin{aligned} m(0) &= -0.1782x_1(0) \\ m(1) &= -0.178x_1(1) \\ m(2) &= -0.178x_1(2) \\ m(3) &= -0.166x_1(3) \end{aligned} \tag{7-135}$$

In order to evaluate the actual performance of the optimized system, let us assign an arbitrary initial condition, $x_1(0) = 1$. The response and control of the system are computed from Eqs. (7-126) and (7-135). We have

$$\begin{aligned} m(0) &= -0.1782x_1(0) = -0.1782 \\ x_1(1) &= 0.368x_1(0) + 0.632m(0) = 0.255 \\ m(1) &= -0.178x_1(1) = -0.0454 \\ x_1(2) &= 0.368x_1(1) + 0.632m(1) = 0.00653 \\ m(2) &= -0.178x_1(2) = -0.0116 \\ x_1(3) &= 0.0167 \\ m(3) &= -0.0277 \\ x_1(4) &= 0.0044 \end{aligned}$$

The response of the system is shown in Figure 7-5. Notice that the response

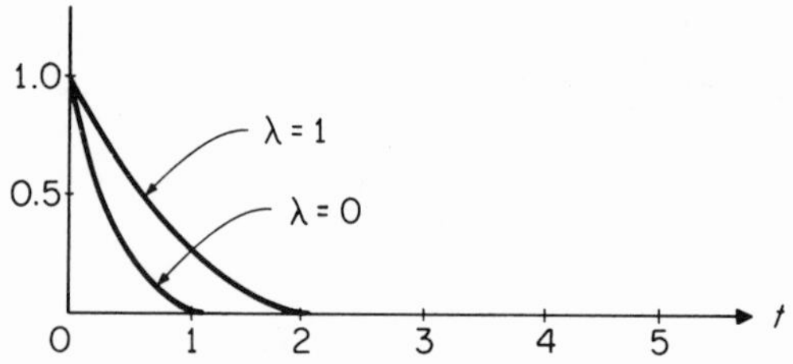

FIG. 7-5. Responses of optimum control system in Example 7-2.

decays toward zero as time increases, since the design objective calls for the minimization of the summed square value of $x_1(k)$ for the first four periods, with constraint placed on $m^2(k-1)$. This type of performance index assures that the amplitude of the control signal will not be excessive.

Let us now consider the same design problem but with $\lambda = 0$; that is, there is no constraint on m. Then,

$$H(1) = -\frac{\Theta'(1)Q\Phi(1)}{\Theta'(1)Q\Theta(1)} = -0.581 \tag{7-136}$$

$$P(1) = 0.0692$$

$$H(2) = -\frac{\Theta'(1)[Q + P(1)]\Phi(1)}{\Theta'(1)[Q + P(1)]\Theta(1)} = -0.581 \tag{7-137}$$

$$m(0) = -0.581x_1(0) = -0.581 \tag{7-138}$$

$$x_1(1) = 0.368 + 0.632(-0.581) = 0.0008 \tag{7-139}$$

Therefore, $m(1)$ becomes very small and $x_1(2)$ is practically zero. Figure 7-5 shows that when $\lambda = 0$, the response decays to zero very rapidly. However, the control $m(0)$, which is -0.581, is far greater than that for $\lambda = 1$ in magnitude.

Multivariable Systems

For a multivariable system, $\mathbf{m}(k)$ denotes the $p \times 1$ input vector. The state equation is written

$$\mathbf{x}(k+1) = \Phi(1)\mathbf{x}(k) + \Theta(1)\mathbf{m}(k) \tag{7-140}$$

The performance index for an N-stage process is now written as

$$I_N = \sum_{k=1}^{N} \mathbf{x}'(k)Q(k)\mathbf{x}(k) + \lambda \mathbf{m}'(k-1)R(k-1)\mathbf{m}(k-1) \tag{7-141}$$

where $Q(k)$ and $R(k-1)$ are symmetric and positive definite matrices.

Let the minimum for I_N be written as

$$f_N[\mathbf{x}(0)] = \min_{\substack{\mathbf{m}(0)\\ \vdots \\ \mathbf{m}(N-1)}} I_N = \min_{\substack{\mathbf{m}(0)\\ \vdots \\ \mathbf{m}(N-1)}} \sum_{k=1}^{N} [\mathbf{x}'(k)Q(k)\mathbf{x}(k) + \lambda \mathbf{m}'(k-1)R(k-1)\mathbf{m}(k-1)] \tag{7-142}$$

and the minimum return for the last $N - j$ stages of the N-stage process be

$$f_{N-j}[\mathbf{x}(j)] = \min_{\substack{\mathbf{m}(j)\\ \vdots \\ \mathbf{m}(N-1)}} I_{N-j} = \min_{\substack{\mathbf{m}(j)\\ \vdots \\ \mathbf{m}(N-1)}} \sum_{k=j+1}^{N} [\mathbf{x}'(k)Q(k)\mathbf{x}(k) + \lambda \mathbf{m}'(k-1)R(k-1)\mathbf{m}(k-1)] \tag{7-143}$$

$j = 0, 1, 2, \ldots, N-1$.

Now applying the principle of optimality, we can write the last equation as

$$f_{N-j}[\mathbf{x}(j)] = \min_{\substack{\mathbf{m}(j)\\ \vdots \\ \mathbf{m}(N-1)}} I_{N-j} = \min_{\mathbf{m}(j)} \{\mathbf{x}'(j+1)Q(j+1)\mathbf{x}(j+1) + \lambda \mathbf{m}'(j)R(j)\mathbf{m}(j) + f_{N-(j+1)}[\mathbf{x}(j+1)]\} \tag{7-144}$$

As in the single-input case, we can again show by induction that $f_{N-j}[\mathbf{x}(j)]$ can be expressed as

$$f_{N-j}[\mathbf{x}(j)] = \mathbf{x}'(j)P(N-j)\mathbf{x}(j) \tag{7-145}$$

where $P(N-j)$ is a symmetric matrix. Then,

$$f_{N-(j+1)}[\mathbf{x}(j+1)] = \mathbf{x}'(j+1)P(N-j-1)\mathbf{x}(j+1) \tag{7-146}$$

Substituting Eq. (7-146) into Eq. (7-144), we get

$$f_{N-j}[\mathbf{x}(j)] = \min_{\mathbf{m}(j)} \{\mathbf{x}'(j+1)V(N-j-1)\mathbf{x}(j+1) + \lambda \mathbf{m}'(j)R(j)\mathbf{m}(j)\} \tag{7-147}$$

where

$$V(N-j-1) = Q(j+1) + P(N-j-1) \tag{7-148}$$

Now substituting Eq. (7-140) into Eq. (7-147) yields

$$\begin{aligned} f_{N-j}[\mathbf{x}(j)] &= \min_{\mathbf{m}(j)} \{[\Phi(1)\mathbf{x}(j) + \Theta(1)\mathbf{m}(j)]'V(N-j-1)[\Phi(1)\mathbf{x}(j) \\ &\quad + \Theta(1)\mathbf{m}(j)] + \lambda\mathbf{m}'(j)R(j)\mathbf{m}(j)\} \\ &= \min_{\mathbf{m}(j)} \{\mathbf{x}'(j)\Phi'(1)V(N-j-1)\Phi(1)\mathbf{x}(j) \\ &\quad + \mathbf{m}'(j)\Theta'(1)V(N-j-1)\Phi(1)\mathbf{x}(j) \\ &\quad + \mathbf{x}'(j)\Phi'(1)V(N-j-1)\Theta(1)\mathbf{m}(j) \\ &\quad + \mathbf{m}'(j)\Theta'(1)V(N-j-1)\Theta(1)\mathbf{m}(j) \\ &\quad + \lambda\mathbf{m}'(j)R(j)\mathbf{m}(j)\} \end{aligned} \tag{7-149}$$

The last equation is simplified if we let

$$V_{\Phi\Phi}(N-j-1) = \Phi'(1)V(N-j-1)\Phi(1) \tag{7-150}$$

$$V_{\Theta\Phi}(N-j-1) = \Theta'(1)V(N-j-1)\Phi(1) \tag{7-151}$$

$$V_{\Phi\Theta}(N-j-1) = \Phi'(1)V(N-j-1)\Theta(1) \tag{7-152}$$

$$V_{\Theta\Theta}(N-j-1) = \Theta'(1)V(N-j-1)\Theta(1) \tag{7-153}$$

Then,

$$\begin{aligned} f_{N-j}[\mathbf{x}(j)] = \min_{\mathbf{m}(j)} I_{N-j} &= \min_{\mathbf{m}(j)} \{\mathbf{x}'(j)V_{\Phi\Phi}(N-j-1)\mathbf{x}(j) \\ &\quad + \mathbf{x}'(j)V_{\Phi\Theta}(N-j-1)\mathbf{m}(j) + \mathbf{m}'(j)V_{\Theta\Phi}(N-j-1)\mathbf{x}(j) \\ &\quad + \mathbf{m}'(j)[V_{\Theta\Theta}(N-j-1) + \lambda R(j)]\mathbf{m}(j)\} \end{aligned} \tag{7-154}$$

To find the optimal $\mathbf{m}(j)$ for minimum I_{N-j}, we take the gradient of the last expression with respect to $\mathbf{m}(j)$ and set the result equal to zero. The gradient of $\mathbf{m}'(j)V_{\Theta\Phi}(N-j-1)\mathbf{x}(j)$ with respect to $\mathbf{m}(j)$ is written

$$\begin{aligned} &\frac{\partial}{\partial \mathbf{m}(j)} \mathbf{m}'(j)V_{\Theta\Phi}(N-j-1)\mathbf{x}(j) \\ &= \begin{bmatrix} \dfrac{\partial}{\partial m_1(j)} \\ \vdots \\ \dfrac{\partial}{\partial m_p(j)} \end{bmatrix} \{[m_1(j) \quad m_2(j) \ldots m_p(j)]V_{\Theta\Phi}(N-j-1)\mathbf{x}(j)\} \\ &= V_{\Theta\Phi}(N-j-1)\mathbf{x}(j) \end{aligned} \tag{7-155}$$

Similarly,

$$\frac{\partial}{\partial \mathbf{m}(j)} [\mathbf{x}'(j)V_{\Phi\Theta}(N-j-1)\mathbf{m}(j)] = \frac{\partial}{\partial \mathbf{m}(j)} [\mathbf{m}'(j)V'_{\Phi\Theta}(N-j-1)\mathbf{x}(j)]' \tag{7-156}$$

However, since $\mathbf{m}'(j)V'_{\Phi\Theta}(N-j-1)\mathbf{x}(j)$ is a scalar, the last equation is written

$$\frac{\partial}{\partial \mathbf{m}(j)}[\mathbf{x}'(j)V_{\Phi\Theta}(N-j-1)\mathbf{m}(j)] = \frac{\partial}{\partial \mathbf{m}(j)}[\mathbf{m}'(j)V'_{\Phi\Theta}(N-j-1)\mathbf{x}(j)]$$

$$= \begin{bmatrix} \dfrac{\partial}{\partial m_1(j)} \\ \vdots \\ \dfrac{\partial}{\partial m_p(j)} \end{bmatrix} \{[m_1(j) \quad m_2(j) \ldots m_p(j)]V'_{\Phi\Theta}(N-j-1)\mathbf{x}(j)\}$$

$$= V'_{\Phi\Theta}(N-j-1)\mathbf{x}(j)$$

$$= V_{\Theta\Phi}(N-j-1)\mathbf{x}(j) \tag{7-157}$$

Therefore,

$$\frac{\partial}{\partial \mathbf{m}(j)} I_{N-j} = 2V_{\Theta\Phi}(N-j-1)\mathbf{x}(j) + 2[V_{\Theta\Theta}(N-j-1) + \lambda R(j)]\mathbf{m}(j) = 0 \tag{7-158}$$

Now solving for $\mathbf{m}(j)$, we have the optimal control vector

$$\mathbf{m}^0(j) = -[V_{\Theta\Theta}(N-j-1) + \lambda R(j)]^{-1}V_{\Theta\Phi}(N-j-1)\mathbf{x}(j) \tag{7-159}$$

or

$$\mathbf{m}^0(j) = H(N-j)\mathbf{x}(j) \tag{7-160}$$

where

$$H(N-j) = -[V_{\Theta\Theta}(N-j-1) + \lambda R(j)]^{-1}V_{\Theta\Phi}(N-j-1) \tag{7-161}$$

Substituting $\mathbf{m}^0(j)$ from Eq. (7-159) into Eq. (7-154), the optimal return from the last $N-j$ stages is

$$f_{N-j}[\mathbf{x}(j)] = \mathbf{x}'(j)\{V_{\Phi\Phi}(N-j-1) + V_{\Phi\Theta}(N-j-1)H(N-j) + H'(N-j)V_{\Theta\Phi}(N-j-1) + H'(N-j)[V_{\Theta\Theta}(N-j-1) + \lambda R(j)]H(N-j)\}\mathbf{x}(j) \tag{7-162}$$

Using the identity in Eq. (7-161), the last expression is simplified to

$$f_{N-j}[\mathbf{x}(j)] = \mathbf{x}'(j)[V_{\Phi\Phi}(N-j-1) + V_{\Phi\Theta}(N-j-1)H(N-j)]\mathbf{x}(j) \tag{7-163}$$

or simply

$$f_{N-j}[\mathbf{x}(j)] = \mathbf{x}'(j)P(N-j)\mathbf{x}(j) \tag{7-164}$$

where

$$P(N-j) = V_{\Phi\Phi}(N-j-1) + V_{\Phi\Theta}(N-j-1)H(N-j) \tag{7-165}$$

or

$$P(N-j) = \Phi'(1)[Q(j+1) + P(N-j-1)][\Phi(1) + \Theta(1)H(N-j)] \tag{7-166}$$

Equation (7-166) and Eq. (7-161) which is expanded below, form a pair of recursive relationships for the determination of $P(N-j)$ and $H(N-j)$.

$$H(N-j) = -\{\Theta'(1)[Q(j+1) + P(N-j-1)]\Theta(1) + \lambda R(j)\}^{-1}\Theta'(1) \times [Q(j+1) + P(N-j-1)]\Phi(1) \tag{7-167}$$

At this point let us investigate a special condition when $\lambda = 0$.

When $\lambda = 0$, that is, there is no constraint on the control $\mathbf{m}(k)$ in the performance index, let us also assume that $\Theta(1)$ is a nonsingular square matrix. Equation (7-167) becomes

$$\begin{aligned} H(N-j) &= -\{\Theta'(1)[Q(j+1) + P(N-j-1)]\Theta(1)\}^{-1}\Theta'(1)[Q(j+1) \\ &\quad + P(N-j-1)]\Phi(1) \\ &= -\Theta^{-1}(1)[Q(j+1) + P(N-j-1)]^{-1}[\Theta'(1)]^{-1}\Theta'(1)[Q(j+1) \\ &\quad + P(N-j-1)]\Phi(1) = -\Theta^{-1}(1)\Phi(1) \end{aligned} \tag{7-168}$$

Substitution of the last equation into Eq. (7-166) gives

$$\begin{aligned} P(N-j) &= \Phi'(1)[Q(j+1) + P(N-j-1)][\Phi(1) - \Theta(1)\Theta^{-1}(1)\Phi(1)] \\ &= \mathbf{0} \end{aligned} \tag{7-169}$$

The significance of these results is that when $\lambda = 0$, and $\Theta(1)$ is a nonsingular square matrix, the optimum return $f_{N-j}[\mathbf{x}(j)]$ is zero for all N and j. Consequently, the feedback matrix for the optimal control is given by Eq. (7-168) and is a time-invariant matrix. Therefore, we can write

$$\mathbf{m}^0(j) = H(N-j)\mathbf{x}(j) = H\mathbf{x}(j) = -\Theta^{-1}(1)\Phi(1)\mathbf{x}(j) \tag{7-170}$$

Let us consider the following numerical example for the optimal design of a multivariable control system.

EXAMPLE 7-3

Assume that a linear process is described by the following discrete state equation:

$$\mathbf{x}(k+1) = \Phi(1)\mathbf{x}(k) + \Theta(1)\mathbf{m}(k) \tag{7-171}$$

where

$$\Phi(1) = \begin{bmatrix} 0 & 0.632 \\ -0.632 & 0.368 \end{bmatrix}$$

$$\Theta(1) = \begin{bmatrix} 0.368 & 0 \\ 0.632 & 1 \end{bmatrix}$$

$$\mathbf{x}(k) = \begin{bmatrix} x_1(k) \\ x_2(k) \end{bmatrix}$$

and

$$\mathbf{m}(k) = \begin{bmatrix} m_1(k) \\ m_2(k) \end{bmatrix}$$

It is desired to determine the optimal control law $\mathbf{m}(k)$ so that the performance index

$$I_5 = \sum_{k=1}^{5} [\mathbf{x}'(k)Q\mathbf{x}(k)] + \mathbf{m}'(k-1)R\mathbf{m}(k-1)] \tag{7-172}$$

is minimized. The weighting matrices are given as

$$Q = \begin{bmatrix} 0.5 & 0 \\ 0 & 1 \end{bmatrix}$$

and

$$R = \begin{bmatrix} 1 & 0 \\ 0 & 1 \end{bmatrix}$$

From Eq. (7-160) the optimal control vector is

$$\mathbf{m}^0(j) = H(5-j)\mathbf{x}(j) \tag{7-173}$$

where $j = 0, 1, 2, 3, 4$.

The optimal control matrix $H(5-j)$ is given by Eq. (7-167), and together with $P(5-j)$ of Eq. (7-166) they form the recursion relationships for the computation of the H and P matrices.

When $j = 4$,

$$\begin{aligned} H(1) &= -\{\Theta'(1)[Q + P(0)]\Theta(1) + R\}^{-1}\Theta'(1)[Q + P(0)]\Phi(1) \\ &= -\{\Theta'(1)Q\Theta(1) + R\}^{-1}\Theta'(1)Q\Phi(1) \\ &= \begin{bmatrix} 0.158 & -0.184 \\ 0.265 & -0.126 \end{bmatrix} \end{aligned} \tag{7-174}$$

where it has been established that $P(0) = 0$. Then,

$$\begin{aligned} P(1) &= \Phi'(1)[Q + P(0)][\Phi(1) + \Theta(1)H(1)] \\ &= \begin{bmatrix} 0.169 & -0.0796 \\ -0.08 & 0.224 \end{bmatrix} \end{aligned} \tag{7-175}$$

Continuing with $j = 3, 2, 1$, and 0, we obtain the results of $P(2)$, $P(3)$, $P(4)$, $H(2)$, $H(3)$, $H(4)$, and $H(5)$, as follows:

$$H(2) = \begin{bmatrix} 0.137 & -0.172 \\ 0.245 & -0.094 \end{bmatrix} \qquad P(2) = \begin{bmatrix} 0.235 & -0.1 \\ -0.1037 & 0.254 \end{bmatrix}$$

$$H(3) = \begin{bmatrix} 0.136 & -0.174 \\ 0.248 & -0.0875 \end{bmatrix} \qquad P(3) = \begin{bmatrix} 0.24 & -0.098 \\ -0.098 & 0.311 \end{bmatrix}$$

$$H(4) = \begin{bmatrix} 0.136 & -0.177 \\ 0.25 & -0.0892 \end{bmatrix} \qquad P(4) = \begin{bmatrix} 0.248 & -0.103 \\ -0.102 & 0.315 \end{bmatrix}$$

$$H(5) = \begin{bmatrix} 0.136 & -0.177 \\ 0.25 & -0.0892 \end{bmatrix}$$

Therefore, the optimal control vector can be expressed in terms of state feedback,

$$\mathbf{m}(0) = H(5)\mathbf{x}(0) \tag{7-176}$$

$$\mathbf{m}(1) = H(4)\mathbf{x}(1) \tag{7-177}$$

$$\mathbf{m}(2) = H(3)\mathbf{x}(2) \tag{7-178}$$

$$\mathbf{m}(3) = H(2)\mathbf{x}(3) \tag{7-179}$$

$$\mathbf{m}(4) = H(1)\mathbf{x}(4) \tag{7-180}$$

Notice that in this case the feedback matrix $H(N - j)$ converges rapidly to a constant matrix in four stages.

Assuming an arbitrary initial state

$$\mathbf{x}(0) = \begin{bmatrix} 1 \\ 0 \end{bmatrix}$$

the optimal control vectors and the state vectors at the first five instants are determined from Eqs. (7-171) and (7-176) through (7-180).

$$\mathbf{m}(0) = \begin{bmatrix} 0.136 \\ 0.25 \end{bmatrix}$$

$$\mathbf{m}(1) = \begin{bmatrix} 0.0592 \\ 0.0101 \end{bmatrix} \qquad \mathbf{x}(1) = \begin{bmatrix} 0.05 \\ -0.296 \end{bmatrix}$$

$$\mathbf{m}(2) = \begin{bmatrix} -0.0062 \\ -0.03 \end{bmatrix} \qquad \mathbf{x}(2) = \begin{bmatrix} -0.165 \\ -0.093 \end{bmatrix}$$

$$\mathbf{m}(3) = \begin{bmatrix} -0.014 \\ -0.018 \end{bmatrix} \qquad \mathbf{x}(3) = \begin{bmatrix} -0.0611 \\ 0.033 \end{bmatrix}$$

$$\mathbf{m}(4) = \begin{bmatrix} -0.00196 \\ 0.00111 \end{bmatrix} \qquad \mathbf{x}(4) = \begin{bmatrix} 0.0156 \\ 0.024 \end{bmatrix}$$

$$\mathbf{x}(5) = \begin{bmatrix} -0.0158 \\ -0.00115 \end{bmatrix}$$

The optimal response of the system is shown in Figure 7-6.

In this example if we set $\lambda = 0$, the performance index becomes

$$I_5 = \sum_{k=1}^{5} \mathbf{x}'(k)Q\mathbf{x}(k) \tag{7-181}$$

In this case, the feedback matrix $H(N - j)$ is stationary and stays constant from instant to instant, and is given by Eq. (7-168).

$$H = -\Theta^{-1}(1)\Phi(1) = -\begin{bmatrix} 0.368 & 0 \\ 0.632 & 1 \end{bmatrix}^{-1} \begin{bmatrix} 0 & 0.632 \\ -0.632 & 0.368 \end{bmatrix}$$

$$= \begin{bmatrix} 0 & -1.72 \\ 0.632 & 0.72 \end{bmatrix} \tag{7-182}$$

With the same initial state, the control vector for $j = 0$ is

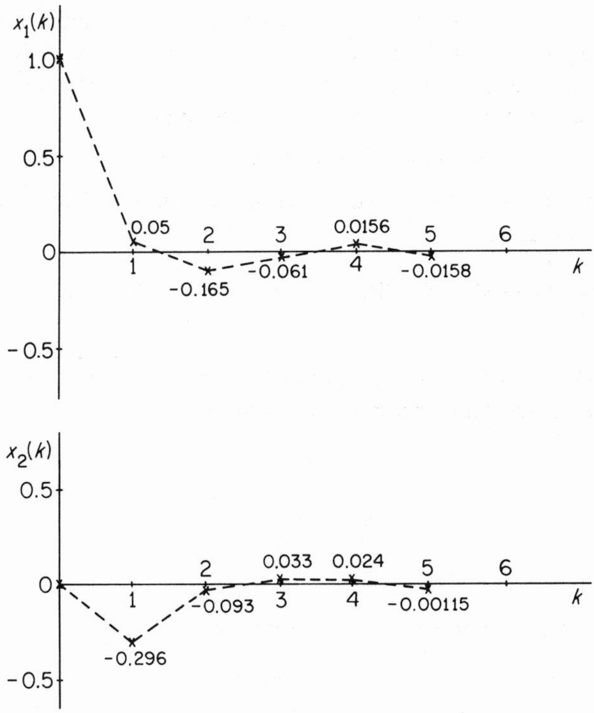

FIG. 7-6. Responses of optimal system in Example 7-3.

$$\mathbf{m}(0) = H\mathbf{x}(0) = \begin{bmatrix} 0 & -1.72 \\ 0.632 & 0.72 \end{bmatrix} \begin{bmatrix} 1 \\ 0 \end{bmatrix} = \begin{bmatrix} 0 \\ 0.632 \end{bmatrix} \tag{7-183}$$

Substituting $\mathbf{m}(0)$ into Eq. (7-171), with $k = 0$, we have

$$\mathbf{x}(1) = \begin{bmatrix} 0 & 0.632 \\ -0.632 & 0.368 \end{bmatrix} \begin{bmatrix} 1 \\ 0 \end{bmatrix} + \begin{bmatrix} 0.368 & 0 \\ 0.632 & 1 \end{bmatrix} \begin{bmatrix} 0 \\ 0.632 \end{bmatrix}$$

$$= \begin{bmatrix} 0 \\ 0 \end{bmatrix} \tag{7-184}$$

It is interesting to note that with no constraint placed on the control vector, the optimal system reaches the equilibrium state in one sampling period, and it is easy to show that all subsequent controls are zero.

It should be pointed out that the recursive relations of Eq. (7-166) and Eq. (7-167) for the determination of the optimal control vector are applicable only to the performance index of Eq. (7-141). They are not valid for the more general performance index given in Eq. (7-67). Although we can follow through a similar derivation as that given above to find the recursive relations for the determination of the optimal control law for this more general performance index, we can easily modify the state vector and the performance index slightly so that the previous recursive relations are still applicable.

Consider that the state equation is expressed as

$$\mathbf{x}(k+1) = \mathbf{\Phi}(1)\mathbf{x}(k) + \mathbf{\Theta}(k)\mathbf{m}(k) \tag{7-185}$$

where $\mathbf{x}(k)$ is the $n \times 1$ state vector, and $\mathbf{m}(k)$ is the $p \times 1$ control vector. The performance index is

$$I_N = \sum_{k=1}^{N} \{[\mathbf{x}_d(k) - \mathbf{x}(k)]'Q(k)[\mathbf{x}_d(k) - \mathbf{x}(k)] + \lambda \mathbf{m}'(k-1)R(k-1)\mathbf{m}(k-1)\} \tag{7-186}$$

where $Q(k)$ and $R(k-1)$ are symmetric positive definite matrices; $\mathbf{x}_d(k)$ denotes the desired state vector and is given as

$$\mathbf{x}_d(k) = \begin{bmatrix} \mathbf{x}_{d1}(k) \\ \mathbf{x}_{d2}(k) \\ \vdots \\ \mathbf{x}_{dn}(k) \end{bmatrix} \tag{7-187}$$

Let us form a new state vector in the form of

$$\mathbf{y}(k) = \begin{bmatrix} \mathbf{x}_{d1}(k) \\ \mathbf{x}_{d2}(k) \\ \vdots \\ \mathbf{x}_{dn}(k) \\ \mathbf{x}_n(k) \\ \mathbf{x}_{n-1}(k) \\ \vdots \\ \mathbf{x}_1(k) \end{bmatrix} \tag{7-188}$$

Let the weighting matrix $Q(k)$ be designated as

$$Q(k) = \begin{bmatrix} q_{11}(k) & q_{12}(k) & \cdots & q_{1n}(k) \\ q_{21}(k) & q_{22}(k) & \cdots & q_{2n}(k) \\ \vdots & \vdots & & \vdots \\ q_{n1}(k) & q_{n2}(k) & \cdots & q_{nn}(k) \end{bmatrix} \tag{7-189}$$

Then, the performance index of Eq. (7-186) can be written

$$I_N = \sum_{k=1}^{N} [\mathbf{y}'(k)S(k)\mathbf{y}(k) + \lambda \mathbf{m}'(k-1)R(k-1)\mathbf{m}(k-1)] \tag{7-190}$$

where

$$S(k) =$$

$$\left|\begin{array}{cccc|cccc}
q_{11}(k) & q_{12}(k) & \cdots & q_{1n}(k) & -q_{1n}(k) & \cdots & -q_{12}(k) & -q_{11}(k) \\
q_{21}(k) & q_{22}(k) & \cdots & q_{2n}(k) & -q_{2n}(k) & \cdots & -q_{22}(k) & -q_{21}(k) \\
\vdots & \vdots & & \vdots & \vdots & & \vdots & \vdots \\
q_{n1}(k) & q_{n2}(k) & \cdots & q_{nn}(k) & -q_{nn}(k) & \cdots & -q_{n2}(k) & -q_{n1}(k) \\
\hline
-q_{n1}(k) & -q_{n2}(k) & \cdots & -q_{nn}(k) & q_{nn}(k) & \cdots & q_{n2}(k) & q_{n1}(k) \\
\vdots & \vdots & & \vdots & \vdots & & \vdots & \vdots \\
-q_{21}(k) & -q_{22}(k) & \cdots & -q_{2n}(k) & q_{2n}(k) & \cdots & q_{22}(k) & q_{21}(k) \\
-q_{11}(k) & -q_{12}(k) & \cdots & -q_{1n}(k) & q_{1n}(k) & \cdots & q_{12}(k) & q_{11}(k)
\end{array}\right| \quad (7\text{-}191)$$

However, in this case $S(k)$ is only semipositive definite, although it is symmetric, as it should be.

The state equation in Eq. (7-185) is transformed into

$$\mathbf{y}(k+1) = \mathbf{\Gamma}(1)\mathbf{y}(k) + \mathbf{\Omega}(1)\mathbf{m}(k) \quad (7\text{-}192)$$

where

$$\mathbf{\Gamma}(1) = \left[\begin{array}{cccc|cccc}
 & & & & \cdot & \cdot & \cdot & 0 \\
 & \mathbf{\Psi} & & & \cdot & \cdot & \cdot & \vdots \\
 & & & & \cdot & \cdot & \cdot & 0 \\
0 & 0 & \cdots & 0 & \mathbf{\Phi}_{nn} & \mathbf{\Phi}_{n(n-1)} & \cdots & \mathbf{\Phi}_{n1} \\
0 & 0 & \cdots & 0 & \mathbf{\Phi}_{(n-1)n} & \cdot & \cdots & \mathbf{\Phi}_{(n-1)} \\
\vdots & \vdots & \vdots & \vdots & \vdots & \vdots & \vdots & \vdots \\
0 & 0 & \cdots & 0 & \mathbf{\Phi}_{1n} & \mathbf{\Phi}_{1(n-1)} & \cdots & \mathbf{\Phi}_{11}
\end{array}\right] \begin{array}{l} \updownarrow n \\ \updownarrow n \end{array} \quad (7\text{-}193)$$

$$\longleftarrow n \longrightarrow | \longleftarrow n \longrightarrow$$

and

$$\mathbf{\Omega}(1) = \left[\begin{array}{cccccc}
0 & 0 & 0 & \cdot & \cdot & 0 \\
\vdots & \vdots & \vdots & \cdot & \cdot & \cdot \\
0 & 0 & 0 & \cdot & \cdot & 0 \\
\mathbf{\Theta}_{n1} & \mathbf{\Theta}_{n2} & \cdot & \cdot & \cdot & \mathbf{\Theta}_{np} \\
\vdots & \vdots & \vdots & \cdot & \cdot & \cdot \\
\mathbf{\Theta}_{21} & \mathbf{\Theta}_{22} & \cdot & \cdot & \cdot & \mathbf{\Theta}_{2p} \\
\mathbf{\Theta}_{11} & \mathbf{\Theta}_{12} & \cdot & \cdot & \cdot & \mathbf{\Theta}_{1p}
\end{array}\right] \begin{array}{l} \updownarrow n \\ \updownarrow n \end{array} \quad (7\text{-}194)$$

$$\longleftarrow p \longrightarrow$$

where $\mathbf{\Psi}$ is a matrix which describes the relationship between $\mathbf{x}_d(k+1)$ and $\mathbf{x}_d(k)$.

Once the transformations into Eqs. (7-190) and (7-192) have been established, the optimal design of the system can again be carried out by use of Eqs. (7-166) and (7-167).

As an illustrative example, consider that given a linear process which is described by Eq. (7-185); it is desired to minimize the performance index

$$I_N = \sum_{k=1}^{N} [\mathbf{x}_d(k) - \mathbf{x}(k)]' Q [\mathbf{x}_d(k) - \mathbf{x}(k)] \tag{7-195}$$

where

$$\mathbf{x}_d(k) = \begin{bmatrix} 1 \\ 0 \end{bmatrix} \quad \text{for all } k$$

$$\mathbf{x}(k) = \begin{bmatrix} x_1(k) \\ x_2(k) \end{bmatrix}$$

$$Q = \begin{bmatrix} 2 & 0 \\ 0 & 1 \end{bmatrix}$$

$$\mathbf{m}(k) = \begin{bmatrix} \mathbf{m}_1(k) \\ \mathbf{m}_2(k) \end{bmatrix}$$

Then, the performance index of Eq. (7-196) can be transformed to

$$I_N = \sum_{k=1}^{N} \mathbf{y}'(k) S \mathbf{y}(k) \tag{7-196}$$

where

$$\mathbf{y}(k) = \begin{bmatrix} 1 \\ 0 \\ x_2(k) \\ x_1(k) \end{bmatrix} \tag{7-197}$$

and

$$S = \left[\begin{array}{cc|cc} 2 & 0 & 0 & 2 \\ 0 & 1 & -1 & 0 \\ \hline 0 & -1 & 1 & 0 \\ -2 & 0 & 0 & 2 \end{array}\right] \tag{7-198}$$

The original state equation is transformed into Eq. (7-192), where

$$\Gamma(1) = \left[\begin{array}{cc|cc} 1 & 0 & 0 & 0 \\ 0 & 0 & 0 & 0 \\ \hline 0 & 0 & \boldsymbol{\Phi}_{22}(1) & \boldsymbol{\Phi}_{21}(1) \\ 0 & 0 & \boldsymbol{\Phi}_{12}(1) & \boldsymbol{\Phi}_{11}(1) \end{array}\right] \tag{7-199}$$

and

$$\boldsymbol{\Omega}(1) = \begin{bmatrix} 0 & 0 \\ 0 & 0 \\ \boldsymbol{\Theta}_{21} & \boldsymbol{\Theta}_{22} \\ \boldsymbol{\Theta}_{11} & \boldsymbol{\Theta}_{12} \end{bmatrix} \tag{7-200}$$

Notice that in this case since $x_{d2}(k) = 0$, the problem may be simplified by reducing the dimensions of all the vectors and matrices. Therefore, $\mathbf{y}(k)$ can be written

$$\mathbf{y}(k) = \begin{bmatrix} 1 \\ x_2(k) \\ x_1(k) \end{bmatrix} \tag{7-201}$$

and accordingly, the second row and the second column of S are removed to give

$$S = \begin{bmatrix} 2 & 0 & -2 \\ 0 & 1 & 0 \\ -2 & 0 & 2 \end{bmatrix} \tag{7-202}$$

Similarly,

$$\Gamma(1) = \begin{bmatrix} 1 & 0 & 0 \\ 0 & \boldsymbol{\Phi}_{22}(1) & \boldsymbol{\Phi}_{21}(1) \\ 0 & \boldsymbol{\Phi}_{12}(1) & \boldsymbol{\Phi}_{11}(1) \end{bmatrix} \tag{7-203}$$

$$\boldsymbol{\Omega}(1) = \begin{bmatrix} 0 & 0 \\ \boldsymbol{\Theta}_{21} & \boldsymbol{\Theta}_{22} \\ \boldsymbol{\Theta}_{11} & \boldsymbol{\Theta}_{12} \end{bmatrix} \tag{7-204}$$

7.5 CONCLUSIONS

In this chapter we have discussed two methods of optimal design of discrete-data systems using performance index. The first type of problem deals with the minimization of the summed-error-square criterion, which is a logical adaptation of the integral-square error criterion for continuous-data systems. The design is carried out by use of the calculus of variation and contour integration in the z-plane. Only the design for single-variable systems is discussed. Although the method can be extended to the multivariable case, the complexity involved with spectral factoring a matrix makes the problem much more tedious even for the simple systems. One may regard the design with the summed-error-square criterion as a transition between the conventional and the modern design principles. The design is carried out in a completely analytical fashion, and is independent of the initial conditions of the system. The minimum quadratic performance index design has the advantages that it is readily applicable to multivariable systems and the procedure is convenient for digital computer computation. However, the design does depend on the initial state of the system.

The two design methods discussed in this chapter can also be extended to systems with random inputs. However, for the summed-error-square criterion, it is generally more convenient to use the mean-square sampled-error [Eq. (8-109)], since then the random signals and the system transfer characteristics can readily be described by standard statistical means. For the quadratic performance index, the extension to the statistical case is simple.[5] The performance index is written

$$I_N = E \sum_{k=1}^{N} [\mathbf{x}'(k)Q(k)\mathbf{x}(k) + \lambda \mathbf{m}'(k-1)R(k-1)\mathbf{m}(k-1)]$$

where E denotes the expectation or the mean value (see Chapter 8), and the elements of $\mathbf{m}(k-1)$ are now members of stationary random processes. Carrying out the optimization procedure similar to those presented in the last section, except now with operations on statistical expectations, the same results as in Eqs. (7-122) to (7-125) are obtained for the optimal system.

It was pointed out in the literature that the quadratic performance index design can also be carried out by means of ordinary calculus.[10,11]

REFERENCES

1. Newton, G. C., Gould, L. A., and Kaiser, J. F., *Analytical Design of Linear Feedback Controls*. John Wiley & Sons, New York, 1957.

2. Chang, S. S. L., *Synthesis of Optimum Control Systems*. McGraw-Hill, New York, 1961.

3. Eveleigh, V. W., *Adaptive Control and Optimization Techniques*. McGraw-Hill, New York, 1967.

4. Bellman, R., *Dynamic Programming*. Princeton University Press, Princeton, N. J., 1957.

5. Tou, J. T., *Optimum Design of Digital Control Systems*. Academic Press, New York, 1963.

6. Tou, J. T., *Modern Control Theory*. McGraw-Hill, New York, 1964.

7. Kalman, R. E., and Koepcke, R. W., "Optimal Synthesis of Linear Sampling Control Systems Using Generalized Performance Indices," *Trans. ASME*, Vol. 80, November 1958, pp. 1812–1820.

8. Gunckel, T. L., and Franklin, G. F., "A General Solution for Linear Sampled-Data Control," *Trans. ASME*, Series D, Vol. 85, 1963, pp. 197–203.

9. Joseph, P. D., and Tou, J. T., "On Linear Control Theory," *AIEE Trans.*, Vol. 80, September 1961, pp. 193–196.

10. Tou, J. T., Liu, T. T., and Meksawan, T., "A Study of Digital Control Systems, "*ONR Technical Report No. 103*, Computer Science Lab., Northwestern University, 1963.

11. Reynolds, P. A., Cadzow, J. A., and Tou, J. T., "Solution of an Optimization Problem for Linear Discrete System Through Ordinary Calculus," *IEEE Trans. on Automatic Control*, Vol. AC-10, April 1965, pp. 209–211.

12. Kleindorfer, G. B., and Kleindorfer, P. R., "Quadratic Performance Criteria with Linear Terms in Discrete-Time Control," *IEEE Trans. on Automatic Control*, Vol. AC-12, June 1967, pp. 320–321.

13. Soliman, J. I., and Al-Shaikh, A., "Weighted Performance Criteria for the Synthesis of Discrete Systems," *IEEE Trans. on Automatic Control*, Vol. AC-11, April 1966, pp. 227–281.

14. Larson, R. E., "Optimum Quantization in Dynamic Systems," *IEEE Trans. on Automatic Control*, Vol. AC-12, April 1967, pp. 162–168.

15. Kirk, D. E., "Optimization of Systems with Pulse-width Modulated Control," *IEEE Trans. on Automatic Control*, Vol. AC-12, June 1967, pp. 307–309.

16. Hsieh, H. C., and Leondes, C. T., "System Synthesis with Random Inputs," Chapter 2, *Modern Control Systems Theory*, C. T. Leondes, ed. McGraw-Hill, New York, 1965.

17. Youla, D. C., "On the Factorization of Rational Matrices," *IRE Trans. on Information Theory*, Vol. IT-7, July 1961, pp. 172–189.

8

STATISTICAL DESIGN: WIENER FILTER

8.1 INTRODUCTION

In many applications of control systems, the actual signals and disturbances received by the control system are random in nature, unlike the deterministic signals such as the step, ramp, or sinusoidal functions considered in the ordinary analysis. These signals and disturbances which are random functions of time can often be described only by statistical means.

The problems in communication and control systems with statistical considerations can generally be classified as: (1) prediction of random signals, and (2) separation of random signals from random noise. These problems are also known as the *prediction and filtering* problems, respectively, in random processes.

The prediction and filtering problems for continuous-data systems were first solved by Wiener.[1] The criterion used by Wiener was the mean-square error. In the design of control systems with random signals, the main objective is the creation of an optimum system which will render a minimum mean-square error between the actual output and the desired output. Of course, the mean-square error criterion may not be ideal for all control systems. However, the criterion leads to the so-called Wiener-Hopf integral equation, which can be solved by spectral factorization. Furthermore, if the minimization of the mean-square error is used as the design criterion, the random signals and disturbances are described by their correlation functions and power spectral density functions.

Since its inception, Wiener's statistical theory has experienced a rapid development in a comparatively short time, and many extensions and generalizations have been made. These developments are now available in many standard texts.[2,3,4]

Although Wiener's theory has provided valuable insight into the analysis and design of systems subjected to random signals and disturbances, the practical applications of the theory are subject to several limitations which often cause difficulties. The most apparent of these difficulties are that the mathematics of the derivations of the theory is complex, and important generalizations such as to nonstationary and nonlinear systems require new derivations.

Kalman[5] in 1960 presented a new look to the prediction and filtering problem by introducing the Kalman filter theory. Essentially, Kalman has solved the same problem but showed that Wiener's basic results could be obtained in a simpler and more unified fashion. In contrast to the Fourier transforms, integral equations, and power spectral density functions which are used in Wiener filter theory, Kalman's theory employs the state variable concept of linear dynamic systems, and conditional expectations. Therefore, we have in Wiener and Kalman filters, two contrasting approaches, representing the conventional transform method and the modern state transition method.

The applications of Wiener filter and Kalman filter techniques to the design of discrete-data systems will be discussed in the ensuing sections. A comparison of the two methods is given. A brief review of some elementary definitions and theorems on random signals and processes is given in the next section.

8.2 BASIC CONCEPTS OF RANDOM PROCESSES

In this section we shall briefly review some of the important properties and definitions of probability and random processes. Some previous knowledge of basic probability theory and statistical theory on the part of the reader is assumed. Therefore, the material presented here consists of merely the simple descriptions of some important definitions and concepts often used in the study of random processes.

Sample space and sample points. The aggregation of all possible outcomes of a random experiment is called a *sample space*. With each basic outcome of the experiment, a *sample point* may be assigned. A sample space may consist of a finite number of sample points, and the sample space is said to be discrete. For instance, as a die is thrown, the sample space consists of six discrete sample points; as a coin is thrown, the sample space consists of two discrete points, namely heads or tails. On the other hand, a discrete sample space may consist of an infinite number of points and is called a continuous sample space.

A well-known example of a continuous sample space is a thermal noise voltage caused by thermally excited electrons in a conductor.

Random variable. The sample points in a sample space are designated a variable that takes on the values of the points. A random variable may be discrete or continuous in correspondence with the discrete sample space or the continuous sample space. If the random variable is denoted by x, then, in the coin throwing experiment x is discrete, that is, $x =$ heads or $x =$ tails. In a thermal noise voltage, x is a continuous random variable which may take on all possible values of the noise voltage.

In general, the values of a random variable may be real or complex, and multiple variables are represented by vectors.

Probability density function. The probability density function, denoted by $p(\xi)$, describes the probabilistic characteristics of a random variable. By definition, the probability that a random variable x lies in an infinitesimal range between ξ and $\xi + d\xi$ is

$$\Pr(\xi < x \leq \xi + d\xi) = p(\xi)\, d\xi \tag{8-1}$$

Therefore, the probability that the random variable x lies in the range ξ_1 to ξ_2 is

$$\Pr(\xi_1 < x \leq \xi_2) = \int_{\xi_1}^{\xi_2} p(x)\, dx \tag{8-2}$$

When the random variable represents points of a random signal or a random process that is a function of time, probability density functions of various orders may be conveniently defined. The first-order probability density function $p(\xi)$ can also be written as $p(\xi, t_1)$, which denotes the probability that the random variable x lies between ξ and $\xi + d\xi$ at the time $t = t_1$. The second-order probability density function, $p(\xi_1, t_1; \xi_2, t_2)$, denotes the probability that the variable x lies between ξ_1 and $\xi_1 + d\xi_1$ at t_1, and *also* lies between ξ_2 and $\xi_2 + d\xi_2$ at t_2. Higher-order probability density functions may be defined in similar fashion.

The higher-order probability density function is also referred to as the *joint probability density function.* For instance, the joint probability density function, $p(\xi_1, \xi_2, \ldots, \xi_n)$, simply implies the simultaneous occurrence of

$$\xi_1 < x(t_1) \leq \xi_1 + d\xi_1,\ \xi_2 < x(t_2) \leq \xi_2 + d\xi_2, \ldots,$$
$$\xi_n < x(t_n) \leq \xi_n + d\xi_n.$$

Or

$$\begin{aligned}\Pr[\xi_1 < x(t_1) \leq \xi_1 + d\xi_1,\ \xi_2 < x(t_2) \leq \xi_2 + d\xi_2, \ldots,\\ \xi_n < x(t_n) \leq \xi_n + d\xi_n] = p(\xi_1, \xi_2, \ldots, \xi_n)\, d\xi_1\, d\xi_2 \ldots d\xi_n\end{aligned} \tag{8-3}$$

Probability distribution function. The probability distribution function, $P(\xi, t)$, is defined as the probability that the random variable x is less than some value ξ at time t. Therefore,

$$P(\xi, t) = \Pr[x(t) \leq \xi] = \int_{-\infty}^{\xi} p(x, t)\, dx \tag{8-4}$$

If $P(\xi, t)$ possesses a first-order derivative with respect to ξ, the probability distribution function is related to the probability density function by

$$p(\xi, t) = \frac{dP(\xi, t)}{d\xi} \tag{8-5}$$

It is simple to see that the probability distribution function has the following properties:

$$P(-\infty, t) = \Pr[x(t) \leq -\infty] = 0 \tag{8-6}$$

and

$$P(\infty, t) = \Pr[x(t) \leq \infty] = 1 \tag{8-7}$$

The *joint probability distribution function* of the random variables $x(t_1), x(t_2), \ldots, x(t_n)$ is defined as the probability of the simultaneous occurrence of

$$x(t_1) \leq \xi_1, x(t_2) \leq \xi_2, \ldots, x(t_n) \leq \xi_n$$

and is denoted by $P(\xi_1, \xi_2, \ldots, \xi_n)$.

The joint probability density function and the joint probability distribution function are related by

$$p(\xi_1, \xi_2, \ldots, \xi_n) = \frac{\partial^n P(\xi_1, \xi_2, \ldots, \xi_n)}{\partial\xi_1 \partial\xi_2 \ldots \partial\xi_n} \tag{8-8}$$

The conditional probability density and distribution functions. Often it is of interest to consider the probability distribution of a random variable x of a random process, given the actual value η with which the random variable y has occurred. The conditional probability density function of x given y is denoted by $p(\xi|\eta)$, and is defined by the following relationship:

$$\Pr(\xi < x \leq \xi + d\xi | y = \eta) = p(\xi|\eta)\, d\xi \tag{8-9}$$

In general, the conditional probability density of x given $y_1 = \eta_1, y_2 = \eta_2, \ldots, y_n = \eta_n$ is denoted by $p(\xi|\eta_1, \eta_2, \ldots, \eta_n)$.

The conditional probability distribution function of x given $y_1 = \eta_1, y_2 = \eta_2, \ldots, y_n = \eta_n$ is written

$$\begin{aligned} P(\xi|\eta_1, \eta_2, \ldots, \eta_n) &= \Pr(x \leq \xi | y_1 = \eta_1, y_2 = \eta_2, \ldots, y_n = \eta_n) \\ &= \int_{-\infty}^{\xi} p(x|\eta_1, \eta_2, \ldots, \eta_n)\, dx \end{aligned} \tag{8-10}$$

Let

$$\begin{aligned} \Pr(x \leq \xi, y_1 \leq \eta_1, \ldots, y_n \leq \eta_n) &= P(\xi, \eta_1, \eta_2, \ldots, \eta_n) \\ &= \text{Marginal probability distribution function of random variables} \\ &\quad\; x, y_1, y_2, \ldots, y_n \end{aligned} \tag{8-11}$$

Then, the conditional probability distribution function is written

$$P(\xi|\eta_1, \eta_2, \ldots, \eta_n) = \frac{P(\xi, \eta_1, \eta_2, \ldots, \eta_n)}{P(\eta_1, \eta_2, \ldots, \eta_n)} \tag{8-12}$$

where $P(\eta_1, \eta_2, \ldots, \eta_n)$ is the joint probability distribution function defined in Eq. (8-8).

Time approach and ensemble approach. In general, the statistical properties of a random signal may be measured in two ways: the time approach and the ensemble approach. In the time approach, measurements of the random signal are made beginning at some time t_1. An ensemble approach involves the measurements of a large number of random signals, all with similar characteristics, simultaneously at time t_1. Therefore, the time approach implies the measurement of the output of a single random signal generator over a time interval, whereas the ensemble approach makes use of the measurements of the outputs of a large number of random signal generators, all with similar statistical characteristics, at one given instant.

Stationary and ergodic processes. If in a random process, the statistical properties of the process are constant and are independent of the time origin at which they are measured, the process is said to be stationary with respect to time. In the time approach, if the measurements of a random signal made beginning at time t_1 produce identical results to the measurements made beginning at t_2, or at any other time, the process is stationary. In the ensemble approach, a random process is defined as stationary in time if the statistical properties measured on a large number of random signals from identical sources are the same, regardless of the time at which the measurements are made.

A random process is said to be *ergodic* if its statistical properties are identical under the time and the ensemble measurements. The definitions imply that *an ergodic process is always stationary, since the stationary process is a presupposition of ergodicity. However, not all stationary processes are ergodic.* A random process may be shown to be stationary under the time measurements, but the two results may not be identical.

Moments and averages. Moments, average, mean, mathematical expectation, or expected value, are synonymous in the literature on random processes.

The nth moment of a random variable $x(t)$ is defined to be the statistical average of the nth power of $x(t)$:

$$m_n = \text{av}\,[x^n(t)] = E[x^n(t)] = \overline{x^n(t)} = \langle x^n(t) \rangle$$

$$= \int_{-\infty}^{\infty} x^n(t) p(x, t)\, dx \tag{8-13}$$

where $p(x, t)$ is the probability density function of x.

The first moment of $x(t)$ is just the average value of $x(t)$:

$$E[x(t)] = \text{av}\,[x(t)] = \overline{x(t)} = \langle x(t) \rangle = \int_{-\infty}^{\infty} x(t)p(x, t)\, dx \tag{8-14}$$

Equation (8-14) gives the average or expected value of $x(t)$ from the ensemble approach. We can, however, perform a time averaging process. The time average of $x(t)$ is

$$\overline{x(t)} = \lim_{T\to\infty} \frac{1}{2T} \int_{-T}^{T} x(t)\, dt \tag{8-15}$$

Naturally, if $x(t)$ belongs to a random process possessing the ergodic property, the time average and the ensemble average are identical.

The second moments of $x(t)$ in the time average and the ensemble average are given by

$$E[x^2(t)] = \overline{x^2(t)} = \langle x^2(t) \rangle = \lim_{T\to\infty} \frac{1}{2T} \int_{-T}^{T} x^2(t)\, dt \tag{8-16}$$

and

$$E[x^2(t)] = \overline{x^2(t)} = \langle x^2(t) \rangle = \int_{-\infty}^{\infty} x^2(t)p(x, t)\, dx \tag{8-17}$$

respectively. From Eq. (8-16) we see that the second moments of $x(t)$ are simply the mean-square value of $x(t)$. Therefore, when $x(t)$ represents the error signal of a control system, Eq. (8-16) or Eq. (8-17) gives the mean-square error for Wiener filter design.

The concept of moments may be extended to two random variables. The $(n + m)$th order *joint* moment of two scalar random variables x and y is

$$\begin{aligned} E[x^n(t_1)y^m(t_2)] &= \langle x^n(t_1)y^m(t_2) \rangle \\ &= \int_{-\infty}^{\infty}\int_{-\infty}^{\infty} x^n(t)y^m(t)p(x, t_1; y, t_2)\, dx\, dy \end{aligned} \tag{8-18}$$

where $p(x, t_1; y, t_2)$ denotes the *joint probability density function* of x and y; that is, the probability that $x(t)$ has a value in the range between x and $x + dx$ at time t_1, and $y(t)$ has a value in the range between y and $y + dy$ at t_2.

In a similar manner, the expectation or average of any nonrandom function $f(x)$ of a random variable x can be written as

$$E[f(x)] = \langle f(x) \rangle = \overline{f(x)} = \int_{-\infty}^{\infty} f(x)p(x, t)\, dx \tag{8-19}$$

Correlation functions. In statistical theory, correlation functions (or covariance) are measurements of the statistical dependence of one random signal upon another, or upon itself. If $x(t)$ and $y(t)$ are two random signals or variables, the *crosscorrelation* between x and y is represented by the crosscorrelation function

$$\phi_{xy}(t_1, t_2) = \int_{-\infty}^{\infty}\int_{-\infty}^{\infty} xyp(x, t_1; y, t_2)\, dx\, dy \tag{8-20}$$

This expression clearly indicates that the crosscorrelation function of x and y is equal to the first-order joint moment (or mean) of x and y. If x and y are stationary processes, the probability density function does not depend upon the origin of time of the measurements. Then the crosscorrelation function given by Eq. (8-20) is independent of time t_1 and can be written as

$$\begin{aligned}\phi_{xy}(\tau) &= E[x(t)y(t+\tau)] = \langle x(t)y(t+\tau)\rangle \\ &= \int_{-\infty}^{\infty}\int_{-\infty}^{\infty} xyp(x, y, \tau)\,dx\,dy \end{aligned} \tag{8-21}$$

For an ergodic process, the crosscorrelation function can be determined from a time average. Thus,

$$\begin{aligned}\phi_{xy}(\tau) &= E[x(t)y(t+\tau)] = \langle x(t)y(t+\tau)\rangle \\ &= \lim_{T\to\infty}\frac{1}{2T}\int_{-T}^{T} x(t)y(t+\tau)\,dt \end{aligned} \tag{8-22}$$

If the process is ergodic, the results of Eqs. (8-21) and (8-22) are identical.

When $x(t) = y(t)$, or when only one random signal is involved, the correlation is given by the autocorrelation function, and is defined to be

$$\begin{aligned}\phi_{xx}(\tau) = \langle x(t)x(t+\tau)\rangle &= E[x(t)x(t+\tau)] \\ &= \lim_{T\to\infty}\frac{1}{2T}\int_{-T}^{T} x(t)x(t+\tau)\,dt \end{aligned} \tag{8-23}$$

or

$$\phi_{xx}(\tau) = \int_{-\infty}^{\infty} x^2 p(x, \tau)\,dx \tag{8-24}$$

Some of the important properties of correlation functions are given below without proofs:

1. $\phi_{xx}(\tau)$ is an even function, i.e., $\phi_{xx}(\tau) = \phi_{xx}(-\tau)$.
2. $|\phi_{xx}(\tau)| \leq \phi_{xx}(0)$
3. $\phi_{xy}(\tau) = \phi_{yx}(-\tau)$
4. $\phi_{xx}(0) = E[x^2(t)] = \langle x^2(t)\rangle =$ mean-square value of x.

A random process $x(t)$ is said to be independent or uncorrelated if

$$E[x(t_1)x(t_2)] = E[x(t_1)]E[x(t_2)] \quad (t_1 \neq t_2) \tag{8-25}$$

Furthermore, if

$$E[x(t_1)x(t_2)] = 0 \quad (t_1 \neq t_2) \tag{8-26}$$

the random process is zero mean and independent.

Covariance matrix. Since we will be dealing with multivariable systems, the random variables encountered may be represented by vector quantities. The expectation of a random vector $\mathbf{x}(t)$, where

$$\mathbf{x}(t) = [x_1(t)\ x_2(t) \ldots x_n(t)]' \tag{8-27}$$

is

$$E[\mathbf{x}(t)] = \begin{bmatrix} E[x_1(t)] \\ E[x_2(t)] \\ \vdots \\ E[x_n(t)] \end{bmatrix} = \begin{bmatrix} \int x_1 p(x_1)\, dx_1 \\ \int x_2 p(x_2)\, dx_2 \\ \vdots \\ \int x_n p(x_n)\, dx_n \end{bmatrix} \tag{8-28}$$

For the scalar case, the covariance of the random variable $x(t)$ is defined as

$$\text{cov}\,[x(t)] = E\left[\{x(t_1) - E[x(t_1)]\}\{x(t_2) - E[x(t_2)]\}\right] \tag{8-29}$$

If the random variable has a zero mean, that is,

$$E[x(t_1)] = 0 \tag{8-30}$$

$$E[x(t_2)] = 0 \tag{8-31}$$

Eq. (8-29) becomes

$$\text{cov}\,[x(t)] = E[x(t_1)x(t_2)] \tag{8-32}$$

Furthermore, if the process is stationary,

$$\begin{aligned} \text{cov}\,[x(t)] &= E[x(t)x(t+\tau)] = \phi_{xx}(\tau) \\ &= \text{Autocorrelation of } x(t) \end{aligned} \tag{8-33}$$

For the random vector $\mathbf{x}(t)$ of Eq. (8-21), the *covariance matrix* of $\mathbf{x}(t)$ is written

$$\text{cov}\,[\mathbf{x}] = E[\{\mathbf{x} - E[\mathbf{x}]\}\{\mathbf{x}' - E[\mathbf{x}']\}] \tag{8-34}$$

If $E[\mathbf{x}] = \mathbf{0}$, Eq. (8-28) becomes

$$\text{cov}\,[\mathbf{x}] = E[\mathbf{x}\,\mathbf{x}'] = \begin{bmatrix} E[x_1x_1] & E[x_1x_2] & \cdots & E[x_1x_n] \\ E[x_2x_1] & E[x_2x_2] & \cdots & E[x_2x_n] \\ \cdots & \cdots & & \cdots \\ E[x_nx_1] & E[x_nx_2] & \cdots & E[x_nx_n] \end{bmatrix} \tag{8-35}$$

Conditional expectation. The conditional expectation of a function of random variable x, $f(x)$, given the values of $y_1, y_2, \ldots, y_n$, is denoted by

$$E_x[f(x)|y_1, y_2, \ldots, y_n] = \int_{-\infty}^{\infty} f(x)p(x|y_1, y_2, \ldots, y_n)\, dx \tag{8-36}$$

where E_x denotes the expectation taken with respect to x. Applying the expectation on both sides of Eq. (8-36) with respect to $y_1, y_2, \ldots, y_n$, we have

$$E_y\{E_x[f(x)|y_1, y_2, \ldots, y_n]\} = E_y\left\{\int_{-\infty}^{\infty} f(x)p(x|y_1, y_2, \ldots y_n)\, dx\right\} \quad (8\text{-}37)$$

Or,

$$E_y\{E_x[f(x)|y_1, y_2, \ldots, y_n]\} = \int_{-\infty}^{\infty} \cdots \int_{-\infty}^{\infty} \int_{-\infty}^{\infty} f(x)p(x|y_1, y_2, \ldots, y_n)\, dx \times p(y_1, y_2, \ldots, y_n)\, dy_1 \ldots dy_n \quad (8\text{-}38)$$

Using the expression of the probability density functions which corresponds to Eq. (8-12), the last equation is written

$$E_y\{E_x[f(x)|y_1, y_2, \ldots, y_n]\} = \int_{-\infty}^{\infty} \cdots \int_{-\infty}^{\infty} f(x)p(x, y_1, y_2, \ldots, y_n)\, dx\, dy_1, \ldots, dy_n \quad (8\text{-}39)$$

If $x, y_1, y_2, \ldots, y_n$ are all independent variables, we have

$$E_y\{E_x[f(x)|y_1, y_2, \ldots, y_n]\} = \int_{-\infty}^{\infty} \cdots \int_{-\infty}^{\infty} f(x)p(x)p(y_1) \ldots p(y_n)\, dx\, dy_1, \ldots, dy_n \quad (8\text{-}40)$$

Since the values of $y_1, y_2, \ldots, y_n$ are all given,

$$\int_{-\infty}^{\infty} p(y_1)\, dy_1 = \int_{-\infty}^{\infty} p(y_2)\, dy_2 = \ldots = \int_{-\infty}^{\infty} p(y_n)\, dy_n = 0 \quad (8\text{-}41)$$

Therefore,

$$E_y\{E_x[f(x)|y_1, y_2, \ldots, y_n]\} = \int_{-\infty}^{\infty} f(x)p(x)\, dx = E[f(x)] \quad (8\text{-}42)$$

Power spectral density function. The correlation functions defined in the last paragraph essentially give descriptions of random signals in terms of time domain characteristics. On the other hand, it is often desirable to characterize random signals in terms of frequency-domain characteristics. If the correlation functions are looked upon as time functions (of τ), it is natural to take the Laplace transform of $\phi_{xx}(\tau)$ or $\phi_{xy}(\tau)$ in order to obtain the frequency-domain characteristics. Since the correlation functions theoretically extend from $-\infty$ to $+\infty$ in τ, a two-sided Laplace transform is carried out. Therefore, the auto-power spectral density function is defined to be

$$\Phi_{xx}(s) = \int_{-\infty}^{\infty} \phi_{xx}(\tau)\, e^{-s\tau}\, d\tau \quad (8\text{-}43)$$

and the cross-power spectral density function is

$$\Phi_{xy}(s) = \int_{-\infty}^{\infty} \phi_{xy}(\tau)\, e^{-s\tau}\, d\tau \quad (8\text{-}44)$$

The correlation functions can be obtained by taking the inverse Laplace transform of the power spectral density functions in the usual manner. Thus,

$$\phi_{xx}(\tau) = \frac{1}{2\pi j}\int_{-j\infty}^{j\infty} \Phi_{xx}(s)\, e^{s\tau}\, ds \quad (8\text{-}45)$$

Some important properties of the power spectral density functions are given below:

1. $\Phi_{xx}(j\omega)$ is an even function of frequency. Thus,

$$\Phi_{xx}(j\omega) = \Phi_{xx}(-j\omega)$$

2. $\Phi_{xy}(j\omega) = \Phi_{yx}(-j\omega)$

3. If $x(t)$ and $y(t)$ are the input and output signals, respectively, of a linear system with transfer function $G(s)$, the power spectral densities of $x(t)$ and $y(t)$ are related by

$$\Phi_{yy}(s) = G(s)G(-s)\Phi_{xx}(s) \tag{8-46}$$

8.3 STATISTICAL CHARACTERISTICS OF DISCRETE RANDOM SIGNALS

Correlation functions. In the preceding section we have defined mathematical ways of characterizing continuous-data random signals. These concepts can be extended to the statistical descriptions of sampled or discrete random signals. As has been pointed out, there are two ways of determining the average of a random process: the time average and the ensemble average. However, in sampled-data systems, in addition to the two ways of averaging, there are at least two possible ways of considering a sampled random signal. First of all, we can consider the values of a signal at the discrete sampling instants only. In this case, the signal is treated as a sampled sequence or a sequence of numbers. Another way is to treat the signal as a sampled signal which consists of a train of amplitude modulated pulses. Referring to Figure 8-1(a), $r(t)$ is assumed to be a member of a stationary random process which is sampled periodically every T second. The time during which the sampler is closed is p which is assumed to be very short compared with the sampling period T. The waveforms of the continuous and the sampled signals are shown in Figure 8-1(b). If $r(t)$ is a stationary ergodic process, the autocorrelation function of $r(t)$ may be obtained as either a time average or an ensemble average. That is,

time average:
$$\phi_{rr}(\tau) = \lim_{T_0 \to \infty} \frac{1}{2T_0} \int_{-T_0}^{T_0} r(t)r(t+\tau)\,dt \tag{8-47}$$

ensemble average:
$$\phi_{rr}(\tau) = \int_{-\infty}^{\infty} \int_{-\infty}^{\infty} x_1 x_2 p(x_1, x_2; \tau)\,dx_1\,dx_2 \tag{8-48}$$

where $p(x_1, x_2; \tau)$ is the joint probability density function or the probability that $r(t)$ lies between x_1 and $x_1 + dx_1$ at any time and between x_2 and $x_2 + dx_2$ τ seconds later.

First, let us consider that the signal $r(t)$ is characterized by the values at the sampling instants only. Then, the sampled sequence is denoted by $r(nT)$,

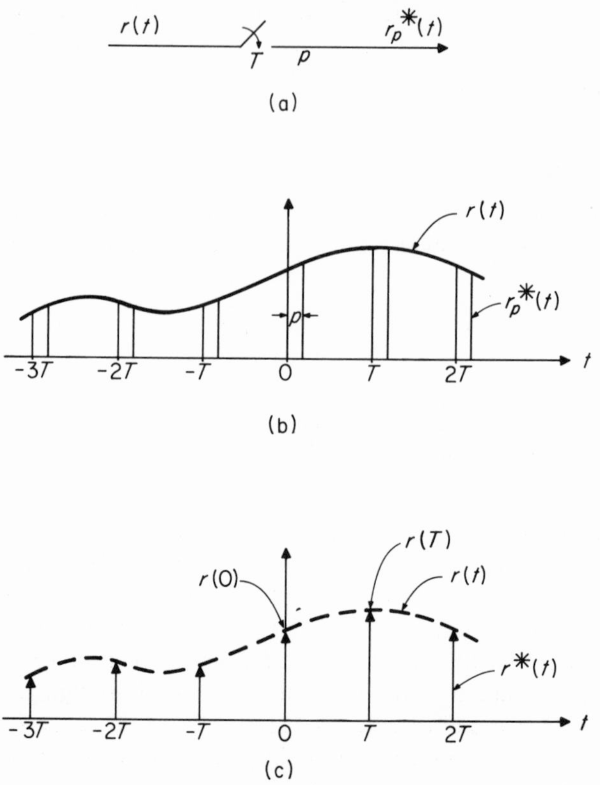

FIG. 8-1. (a) A sampler; (b) continuous and sampled random signals; (c) sampled random sequence and impulse train.

where $n = 0, \pm 1, \pm 2, \ldots$. We shall now show that if $r(t)$ is assumed to be a member of a stationary ergodic process, $r(nT)$ is also a stationary ergodic random sequence, and therefore the time averaging and the ensemble averaging are equivalent for this discrete case. Furthermore, the sampled sequence averages are equal to the time average of the continuous time function $r(t)$.

Since $r(t)$ and $r(nT)$ are all ergodic, the second probability density function of $r(t)$ is equal to that of $r(nT)$ when $\tau = kT$. Therefore,

$$p_r(x_1, x_2; \tau)|_{\tau=kT} = p_{r_n}(x_1, x_2; kT) \tag{8-49}$$

where $p_r(x_1, x_2; \tau)$ denotes the probability density of $r(t)$ and $p_{r_n}(x_1, x_2; kT)$ denotes the same of $r(nT)$, and k is an integer.

The autocorrelation function of $r(nT)$ may be expressed as an ensemble average.

$$\phi_{r_n r_n}(kT) = \int_{-\infty}^{\infty}\int_{-\infty}^{\infty} x_1 x_2 p_{r_n}(x_1, x_2; kT)\, dx_1\, dx_2 \tag{8-50}$$

However, referring to Eq. (8-49), the last equation is written

$$\phi_{r_n r_n}(kT) = \int_{-\infty}^{\infty}\int_{-\infty}^{\infty} x_1 x_2 p_r(x_1, x_2; kT)\, dx_1\, dx_2 \tag{8-51}$$

Now comparing Eq. (8-51) with Eq. (8-48) we see that

$$\phi_{r_n r_n}(kT) = \phi_{rr}(\tau)|_{\tau = kT} = \phi_{rr}(kT) \tag{8-52}$$

proving that when $r(t)$ is an ergodic prosess, the sampled sequence average and the continuous time average are equivalent. Equation (8-52) also provides an interesting result in that the autocorrelation function of a stationary random sequence $r(nT)$ may be obtained by sampling the autocorrelation function $\phi_{rr}(\tau)$ of the continuous ergodic process $r(t)$. For instance, assuming that the autocorrelation function of a certain ergodic process $r(t)$ is as shown in Figure 8-2(a), the autocorrelation function of $r(nT)$, $\phi_{r_n r_n}(kT)$ can be thought of as the output of an ideal sampler at $\tau = kT$ when $\phi_{rr}(\tau)$ is the input to the sampler.

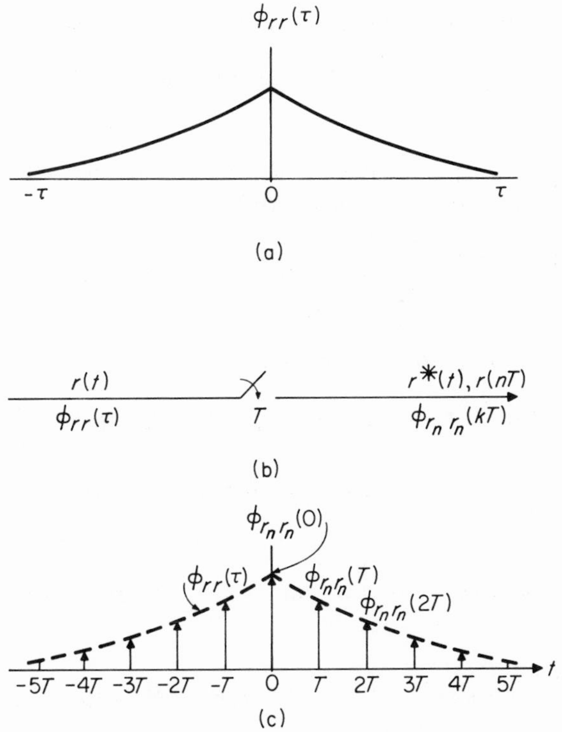

FIG. 8-2. (a) Autocorrelation function of $r(t)$; (b) sampling of autocorrelation function $\phi_{rr}(\tau)$; (c) autocorrelation function of $r(nT)$, $\phi_{r_n r_n}(kT)$.

The autocorrelation function of $r(nT)$ can also be written as a time average:

$$\phi_{r_n r_n}(kT) = \lim_{N\to\infty} \frac{1}{2N+1} \sum_{n=-N}^{N} r(nT)r(nT + kT) \tag{8-53}$$

Since $r(nT)$ is ergodic, Eqs. (8-51), (8-52), and (8-53) are all equivalent.

When the pulse train of Figure 8-1(b) is replaced by an impulse train, we write

$$r^*(t) = \sum_{m=-\infty}^{\infty} r(mT)\delta(t - mT) \tag{8-54}$$

The shifted impulse train is expressed as

$$r^*(t + \tau) = \sum_{m=-\infty}^{\infty} r(mT)\delta(t - mT + \tau) \tag{8-55}$$

The autocorrelation function of $r^*(t)$ is

$$\phi_{r^*r^*}(\tau) = \lim_{T_0\to\infty} \frac{1}{2T_0} \int_{-T_0}^{T_0} r^*(t)r^*(t + \tau)\,dt \tag{8-56}$$

Substitution of Eqs. (8-54) and (8-55) into Eq. (8-56) yields

$$\phi_{r^*r^*}(\tau) = \lim_{T_0\to\infty} \frac{1}{2T_0} \int_{-T_0}^{T_0} \sum_{m=-\infty}^{\infty} \sum_{n=-\infty}^{\infty} r(mT)r(nT)\delta(t - mT)\delta(t - nT + \tau)\,dt \tag{8-57}$$

The integral of Eq. (8-57) may be broken up into a sum of integrals with each integral performed over one sampling period.

$$\phi_{r^*r^*}(\tau) = \lim_{N\to\infty} \frac{1}{2N+1} \sum_{m=-N}^{N} \int_{(m-1/2)T}^{(m+1/2)T} \sum_{m=-\infty}^{\infty} \sum_{n=-\infty}^{\infty} r(mT)r(nT) \times \delta(t - mT)\delta(t - nT - \tau)\,dt \tag{8-58}$$

Letting $\sigma = t - mT$, the last expression becomes

$$\phi_{r^*r^*}(\tau) = \lim_{N\to\infty} \frac{1}{2N+1} \sum_{m=-N}^{N} \sum_{m=-\infty}^{\infty} \sum_{n=-\infty}^{\infty} r(mT)r(nT) \times \frac{1}{T} \int_{-T/2}^{T/2} \delta(\sigma)\delta(\sigma + mT - nT + \tau)\,dt \tag{8-59}$$

Since

$$\phi_{r^*r^*}[(n - m)T] = \lim_{N\to\infty} \frac{1}{2N+1} \sum_{m=-N}^{N} r(mT)r(nT) \tag{8-60}$$

and

$$\int_{-T/2}^{T/2} \delta(\sigma)\delta(\sigma + mT - nT + \tau)\,dt = \delta[\tau - (n - m)T] \tag{8-61}$$

Eq. (8-59) is written as

$$\phi_{r^*r^*}(\tau) = \frac{1}{T} \sum_{m=-\infty}^{\infty} \sum_{n=-\infty}^{\infty} \phi_{r_n r_n}[(n - m)T]\delta[\tau - (n - m)T] \tag{8-62}$$

Letting $k = n - m$, Eq. (8-62) is finally written as

$$\phi_{r^*r^*}(\tau) = \frac{1}{T}\sum_{k=-\infty}^{\infty} \phi_{r_n r_n}(kT)\delta(\tau - kT) \tag{8-63}$$

The last equation shows that the autocorrelation function of a stationary sampled impulse train $r^*(t)$ is an impulse train in τ modulated by $1/T$ times the autocorrelation function of the stationary continuous signal $r(t)$.

However, it must be pointed out that the sampled random process $r_p^*(t)$ and the impulse modulated signal $r^*(t)$ are not stationary processes, even though $r(t)$ is stationary. In this case the correlation functions derived from the ensemble approach are not equal to those of the time averages. For instance, the autocorrelation function of $r^*(t)$ based upon the ensemble average is

$$\phi_{r^*r^*}(\tau, t) = \lim_{N\to\infty}\frac{1}{2N+1}\sum_{m=-N}^{N}\sum_{m=-\infty}^{\infty}\sum_{n=-\infty}^{\infty} r(mT)r(nT)\delta(t - mT)\delta(t - nT + \tau) \tag{8-64}$$

Simplifying, the last expression becomes

$$\phi_{r^*r^*}(\tau, t) = \sum_{m=-\infty}^{\infty}\sum_{n=-\infty}^{\infty} \phi_{r_n r_n}[(n - m)T]\delta(t - mT)\delta(t - nT + \tau) \tag{8-65}$$

In addition to being a function of the shift variable τ, the autocorrelation function determined from the ensemble approach is also a periodic function of time. This fact checks with the physical interpretation of ensemble averaging. Since any number of the ensemble of $r^*(t)$ has values only at the sampling instants $t = kT$, the average of the ensemble has to be evaluated at the sampling instants and has zero values at all other times. The difference between Eqs. (8-63) and (8-65) is illustrated by the typical correlation of $r^*(t)$ shown in Figure 8-3.

The time-average autocorrelation functions of $r^*(t)$ appear as impulse trains in a two-dimensional t versus τ plane. In the following sections only the time-invariant correlation functions will be considered.

In a similar manner we can show that the crosscorrelation function of two stationary sampled random sequences $r(nT)$ and $c(nT)$ is given by

$$\phi_{r_n c_n}(kT) = \phi_{rc}(kT) = \lim_{N\to\infty}\frac{1}{2N+1}\sum_{n=-N}^{N} r(nT)c(nT + kT) \tag{8-66}$$

and the crosscorrelation function of two sampled impulse trains $r^*(t)$ and $c^*(t)$ is

$$\begin{aligned}\phi_{r^*c^*}(\tau) &= \lim_{T_0\to\infty}\frac{1}{2T_0}\int_{-T_0}^{T_0} r^*(t)c^*(t + \tau)\,dt \\ &= \frac{1}{T}\sum_{k=-\infty}^{\infty}\phi_{r_n c_n}(kT)\delta(\tau - kT) \quad \text{(time average)}\end{aligned} \tag{8-67}$$

where $\phi_{r_n c_n}(kT)$ is given by Eq. (8-66).

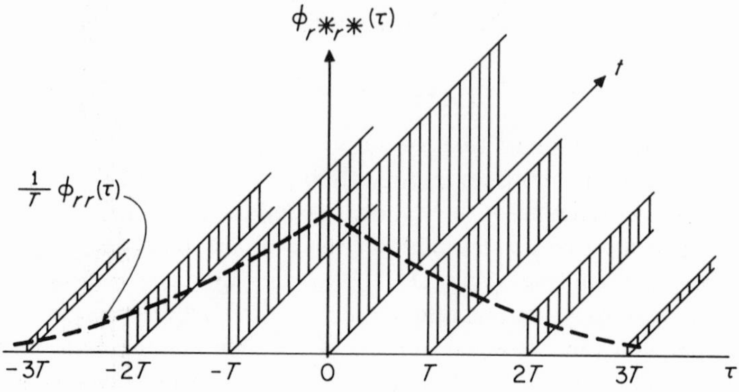

(a)

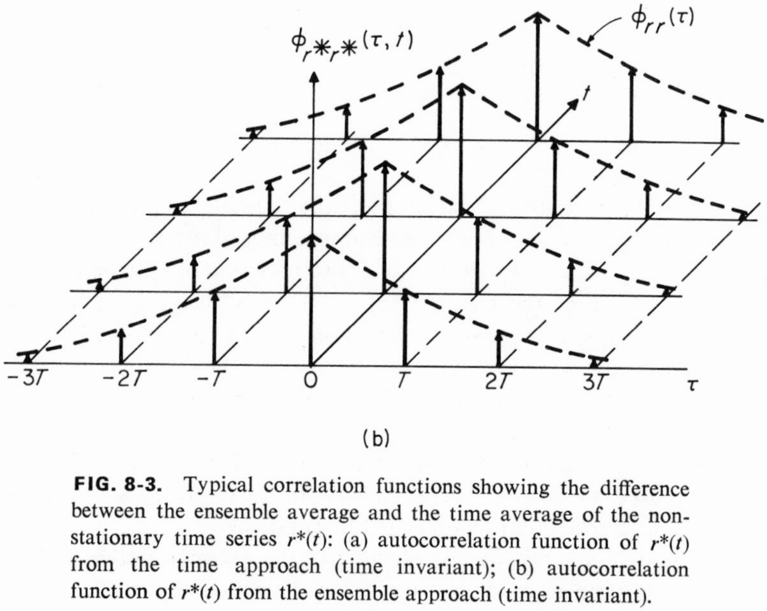

(b)

FIG. 8-3. Typical correlation functions showing the difference between the ensemble average and the time average of the nonstationary time series $r^*(t)$: (a) autocorrelation function of $r^*(t)$ from the time approach (time invariant); (b) autocorrelation function of $r^*(t)$ from the ensemble approach (time invariant).

Sampled power spectral density function. As in the continuous case, the *sampled power spectral density function* of a sampled signal $r^*(t)$ is defined as the Laplace transform or the z-transform of the correlation function of $r^*(t)$. Therefore,

$$\Phi_{r^*r^*}(s) = \mathscr{L}[\phi_{r^*r^*}(\tau)] = \frac{1}{T}\int_{-\infty}^{\infty} \sum_{k=-\infty}^{\infty} \phi_{r_n r_n}(kT)\delta(\tau - kT)\, e^{-s\tau}\, d\tau \tag{8-68}$$

or

$$\Phi_{r^*r^*}(s) = \frac{1}{T} \sum_{k=-\infty}^{\infty} \phi_{r_n r_n}(kT)\, e^{-kTs} \tag{8-69}$$

Taking the z-transform on both sides of Eq. (8-69) yields

$$\Phi_{r^*r^*}(z) = \mathscr{Z}[\Phi_{r^*r^*}(s)] = \frac{1}{T} \sum_{k=-\infty}^{\infty} \phi_{r_n r_n}(kT) z^{-k} \tag{8-70}$$

Since $\phi_{rr}(kT) = \phi_{r_n r_n}(kT)$, Eq. (8-70) becomes

$$\Phi_{r^*r^*}(z) = \frac{1}{T} \sum_{k=-\infty}^{\infty} \phi_{rr}(kT) z^{-k} \tag{8-71}$$

Notice that except for the multiplying factor $1/T$ and the lower limit on the summation, Eq. (8-71) is analogous to the relation of the z-transform of the time series $r^*(t)$; that is,

$$R(z) = \sum_{k=0}^{\infty} r(kT) z^{-k} \tag{8-72}$$

where it is assumed that $r(t)$ is nonzero only for $t \geq 0$. Equation (8-71) also implies that whenever the correlation function of a stationary random process $r(t)$ is given, the sampled power spectral density function of $r(t)$ or of $r^*(t)$ can be obtained with the aid of the z-transform table.

To show how $\Phi_{r^*r^*}(z)$ can be determined with the use of the z-transform table, let us rewrite Eq. (8-71) as

$$\Phi_{r^*r^*}(z) = \frac{1}{T}\left[\sum_{k=-\infty}^{0} \phi_{rr}(kT) z^{-k} + \sum_{k=0}^{\infty} \phi_{rr}(kT) z^{-k} - \phi_{rr}(0)\right] \tag{8-73}$$

or

$$\Phi_{r^*r^*}(z) = \frac{1}{T}\left[\sum_{k=0}^{\infty} \phi_{rr}(-kT) z^{k} + \sum_{k=0}^{\infty} \phi_{rr}(kT) z^{-k} - \phi_{rr}(0)\right] \tag{8-74}$$

Since $\phi_{rr}(\tau)$ is an even function, $\phi_{rr}(-kT) = \phi_{rr}(kT)$, and Eq. (8-74) is written

$$\Phi_{r^*r^*}(z) + \frac{1}{T}[A(z^{-1}) + A(z) - \phi_{rr}(0)] \tag{8-75}$$

where

$$A(z) = \sum_{k=0}^{\infty} \phi_{rr}(kT) z^{-k} = \mathscr{Z}[\phi_{rr}(s)]^{+} \tag{8-76}$$

and

$$A(z^{-1}) = \sum_{k=0}^{\infty} \phi_{rr}(kT) z^{k} = \mathscr{Z}[\Phi_{rr}(s)]^{-}$$

$$= \mathscr{Z}[\Phi_{rr}(s)]^{+}_{z=z^{-1}} \tag{8-77}$$

It is important to note that $[\Phi_{rr}(s)]^{+}$ has all the poles of $\Phi_{rr}(s)$ that lie inside the left-half s-plane, and $[\Phi_{rr}(s)]^{-}$ has all the poles of $\Phi_{rr}(s)$ that lie inside the right-half s-plane, and

$$\Phi_{rr}(s) = [\Phi_{rr}(s)]^+ + [\Phi_{rr}(s)]^- \tag{8-78}$$

Furthermore, $\phi_{rr}(0)$ is given by

$$\phi_{rr}(0) = \phi_{rr}(\tau)|_{\tau=0} = \mathscr{L}^{-1}\{[\Phi_{rr}(s)]^+\}_{\tau=0} = \mathscr{L}^{-1}\{[\Phi_{rr}(s)]^-\}_{\tau=0} \tag{8-79}$$

The following example illustrates how the sampled spectral density of a stationary random process may be obtained by use of Eq. (8-75) and the z-transform table.

EXAMPLE 8-1

Assuming that the auto-power spectral density of a stationary random process $r(t)$ is given by

$$\Phi_{rr}(s) = \frac{36}{-s^2 + 36} \tag{8-80}$$

we wish to find the sampled power spectral density function $\Phi_{r^*r^*}(z)$. Using Eq. (8-78), we have

$$\Phi_{rr}(s) = [\Phi_{rr}(s)]^+ + [\Phi_{rr}(s)]^- = \frac{3}{6+s} + \frac{3}{6-s} \tag{8-81}$$

From the z-transform table,

$$A(z) = \mathscr{Z}[\Phi_{rr}(s)]^+ = \frac{3z}{z - e^{-6T}} \tag{8-82}$$

$$A(z^{-1}) = \mathscr{Z}[\Phi_{rr}(s)]^- = \frac{3z^{-1}}{z^{-1} - e^{-6T}} \tag{8-83}$$

Also

$$\phi_{rr}(0) = \mathscr{L}^{-1}\{[\Phi_{rr}(s)]^+\}_{\tau=0} = \frac{1}{2\pi j}\int_{c-j\infty}^{c+j\infty} [\Phi_{rr}(s)]^+ \, ds$$

$$= \frac{1}{2\pi j}\int_{c-j\infty}^{c+j\infty} \frac{3}{6+s}\, ds = 3 \tag{8-84}$$

Substitution of Eqs. (8-82), (8-83), and (8-84) into Eq. (8-75) yields

$$\Phi_{r^*r^*}(z) = \frac{3}{T}\left[\frac{z}{z - e^{-6T}} + \frac{z^{-1}}{z^{-1} - e^{-6T}} - 1\right] = \frac{3}{T}\,\frac{1 - e^{-12T}}{(1 - e^{-6T}z^{-1})(1 - e^{-6T}z)} \tag{8-85}$$

8.4 RANDOM SIGNALS IN DISCRETE-DATA SYSTEMS

In this section we shall conduct the analysis of discrete-data systems with random signals. Let us consider the open-loop discrete-data system of Figure 8-4(a). The input $r(t)$ is assumed to be a member of a stationary random

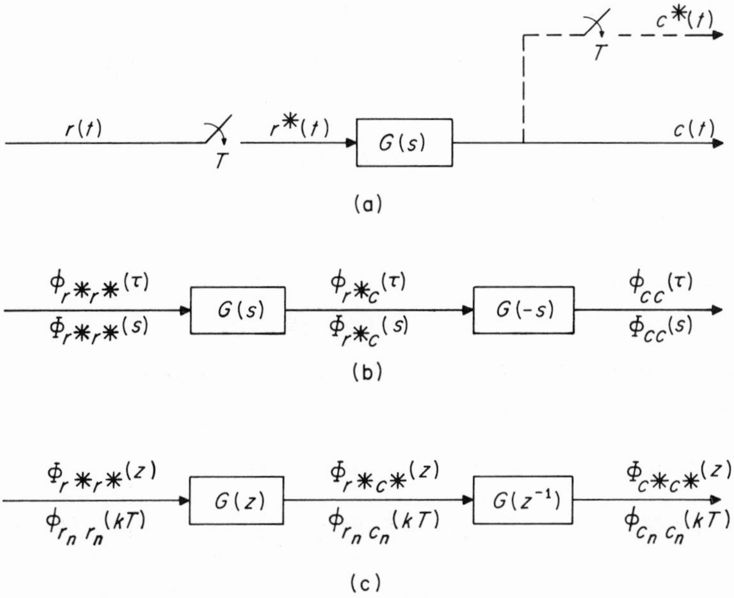

FIG. 8-4. (a) Open-loop discrete-data system with random input. (b) Block diagram showing the relationship between power spectral functions in a discrete-data system. (c) Block diagram showing the z-transform relationship between sampled power spectral functions of a discrete-data system.

process. We shall see that the power spectral function and the sampled power spectral function of the system output are given by

$$\Phi_{cc}(s) = G(s)G(-s)\Phi_{r^*r^*}(s) \tag{8-86}$$

and

$$\Phi_{c^*c^*}(z) = G(z)G(z^{-1})\Phi_{r^*r^*}(z) \tag{8-87}$$

respectively.

With reference to Figure 8-4(b), Eq. (8-86) implies that if $\phi_{r^*r^*}(\tau)$ is the input of the system with $G(s)G(-s)$ as the transfer function, the output of the system is $\phi_{cc}(\tau)$. Furthermore, when $\phi_{r^*r^*}(\tau)$ is the input signal, the output of $G(s)$ is $\phi_{r^*c}(\tau)$, the crosscorrelation function of the sampled signal $r^*(t)$ and continuous output $c(t)$.

By use of the convolution integral, the output of $G(s)$ is written

$$\int_{-\infty}^{\infty} g(\sigma)\phi_{r^*r^*}(\tau - \sigma)\, d\sigma = \int_{-\infty}^{\infty} g(\sigma)\, d\sigma \lim_{T_0 \to \infty} \frac{1}{2T_0} \int_{-T_0}^{T_0} r^*(t) r^*[t + (\tau - \sigma)]\, dt \tag{8-88}$$

Interchanging the order of integration, Eq. (8-88) becomes

$$\lim_{T_0\to\infty}\frac{1}{2T_0}\int_{-T_0}^{T_0} r^*(t)\,dt\int_{-\infty}^{\infty} g(\sigma)r^*[t+(\tau-\sigma)]\,d\sigma$$

$$=\lim_{T_0\to\infty}\frac{1}{2T_0}\int_{-T_0}^{T_0} r^*(t)c(t+\tau)\,dt \tag{8-89}$$

which is the definition of $\phi_{r^*c}(\tau)$. Thus,

$$\Phi_{r^*c}(s)=G(s)\Phi_{r^*r^*}(s) \tag{8-90}$$

In order to prove that

$$\Phi_{cc}(s)=G(-s)\Phi_{r^*c}(s) \tag{8-91}$$

we use the properties of $\Phi_{cc}(s)$ and $\Phi_{r^*c}(s)$ as a basis and write

$$\Phi_{cc}(s)=\Phi_{cc}(-s)=G(s)\Phi_{r^*c}(-s)=G(s)\Phi_{cr^*}(s) \tag{8-92}$$

Thus, when $\phi_{cr^*}(\tau)$ is the input signal, the output of $G(s)$ is

$$\int_{-\infty}^{\infty} g(\sigma)\phi_{cr^*}(\tau-\sigma)\,d\sigma=\int_{-\infty}^{\infty} g(\sigma)\,d\sigma\lim_{T_0\to\infty}\frac{1}{2T_0}\int_{-T_0}^{T_0} c(t)r^*[t+(\tau-\sigma)]\,dt \tag{8-93}$$

Interchanging the order of integration in the last equation, we have

$$\lim_{T_0\to\infty}\frac{1}{2T_0}\int_{-T_0}^{T_0} c(t)\,dt\int_{-\infty}^{\infty} g(\sigma)r^*[t+(\tau-\sigma)]\,d\sigma$$

$$=\lim_{T_0\to\infty}\frac{1}{2T_0}\int_{-T_0}^{T_0} c(t)c(t+\tau)\,dt \tag{8-94}$$

which is $\phi_{cc}(\tau)$. Combining Eqs. (8-90) and (8-91), we have Eq. (8-86).

Equation (8-87) is the counterpart of Eq. (8-86) in the z-domain. The proof of Eq. (8-87) is also divided into two steps. First, we shall prove that

$$\Phi_{r^*c^*}(z)=G(z)\Phi_{r^*r^*}(z) \tag{8-95}$$

When the input to $G(s)$ is $\phi_{r_nr_n}(kT)$, the output is written from the superposition principle.

$$\sum_{k=-\infty}^{\infty}\phi_{r_nr_n}(kT)g(mT-kT)$$

$$=\sum_{k=-\infty}^{\infty}\lim_{N\to\infty}\frac{1}{2N+1}\sum_{q=-N}^{N} r(qT)r(qT+kT)g(mT-kT) \tag{8-96}$$

Interchanging the order of summation in the last equation, we get

$$\sum_{k=-\infty}^{\infty}\phi_{r_nr_n}(kT)g(mT-kT)$$

$$=\lim_{N\to\infty}\frac{1}{2N+1}\sum_{q=-N}^{N} r(qT)\sum_{k=-\infty}^{\infty} r(qT+kT)g(mT-kT) \tag{8-97}$$

Letting $q+k=n$, Eq. (8-97) reads

$$\lim_{N\to\infty} \frac{1}{2N+1} \sum_{q=-N}^{N} r(qT) \sum_{n=-\infty}^{\infty} r(nT)g(mT + qT - nT)$$

$$= \lim_{N\to\infty} \frac{1}{2N+1} \sum_{q=-N}^{N} r(qT)c(mT + qT) \quad (8\text{-}98)$$

which is $\phi_{r_n c_n}(mT)$. Therefore,

$$\phi_{r_n c_n}(mT) = \sum_{k=-\infty}^{\infty} \phi_{r_n r_n}(kT)g(mT - kT) \quad (8\text{-}99)$$

Multiplying both sides of Eq. (8-99) by z^{-m}/T and taking the summation from $m = -\infty$ to $+\infty$, we have

$$\frac{1}{T}\sum_{m=-\infty}^{\infty} \phi_{r_n c_n}(mT)z^{-m} = \frac{1}{T}\sum_{m=-\infty}^{\infty}\sum_{k=-\infty}^{\infty} \phi_{r_n r_n}(kT)g(mT - kT)z^{-m}$$

$$= \frac{1}{T}\sum_{k=-\infty}^{\infty} \phi_{r_n r_n}(kT)z^{-k} \sum_{m=-\infty}^{\infty} g(mT - kT)z^{-(m-k)}$$

$$(8\text{-}100)$$

Therefore, we have proved that

$$\Phi_{r^*c^*}(z) = G(z)\Phi_{r^*r^*}(z) \quad (8\text{-}101)$$

In a similar manner, we can show that

$$\Phi_{c^*c^*}(z) = G(z^{-1})\Phi_{r^*c^*}(z) \quad (8\text{-}102)$$

Combining Eqs. (8-101) and (8-102) yields

$$\Phi_{c^*c^*}(z) = G(z)G(z^{-1})\Phi_{r^*r^*}(z) \quad (8\text{-}103)$$

which is the expression in Eq. (8-87).

Before proceeding with further discussion on signal relations in discrete-data systems, a comparison of the statistical descriptions of continuous data and discrete data may prove helpful. Table 8-1 gives a comparison of these relationships.

Thus far we have limited our illustrations to the derivations of the correlation relations between signals in a single-path system. In the design of optimum systems using the mean-square-error criterion, the crosscorrelation between signals in two independent systems is desired. Shown in Figure 8-5 are two independent open-loop discrete-data systems. The cross-spectral densities $\Phi_{c_1c_2}(z)$ and $\Phi_{c_1c_2}(s)$ are to be expressed as functions of the system transfer functions and the power spectral densities of the other signals. In a manner similar to the method used in proving Eqs. (8-86) and (8-87), we can show that

$$\Phi_{c_1c_2}(z) = G_1(z^{-1})\Phi_{r_1r_2}(z) = G_1(z^{-1})G_2(z)\Phi_{r_1r_2}(z) \quad (8\text{-}104)$$

and

$$\Phi_{c_2c_1}(z) = \Phi_{c_1c_2}(z^{-1}) = G_2(z^{-1})\Phi_{r_2r_1}(z)$$

$$= G_1(z)G_2(z^{-1})\Phi_{r_2r_1}(z) \quad (8\text{-}105)$$

Table 8-1

CORRESPONDENCE BETWEEN STATISTICAL PROPERTIES OF DISCRETE-DATA AND CONTINUOUS-DATA SYSTEMS

	Continuous-Data Systems	*Discrete-Data Systems*
Ergodic random process	$r(t)$ or $R(s)$	$r(nT)$, $r^*(t)$, $R^*(s)$, or $R(z)$
Correlation functions	$\Phi_{rr}(\tau) = \lim_{T_0\to\infty} \frac{1}{2T_0}\int_{-T_0}^{T_0} r(t)r(t+\tau)\,d\tau$	$\Phi_{rr}(kT) = \lim_{N\to\infty} \frac{1}{2N+1}\sum_{n=-N}^{N} r(nT)r(nT+kT)$ $\Phi_{r^*r^*}(\tau) = \lim_{T_0\to\infty} \frac{1}{2T_0}\int_{T_0}^{T_0} r^*(t)r^*(t+\tau)\,dt$ $\Phi_{r^*r^*}(\tau) = \frac{1}{T}\sum_{k=-\infty}^{\infty} \Phi_{rr}(kT)\delta(\tau - kT)$
Spectral density	$\Phi_{rr}(s) = \int_{-\infty}^{\infty} \phi_{rr}(\tau)\,e^{-s\tau}\,d\tau$	$\Phi_{r^*r^*}(s) = \frac{1}{T}\sum_{k=-\infty}^{\infty} \phi_{rr}(kT)\,e^{-kTs}$ $\Phi_{r^*r^*}(z) = \frac{1}{T}\sum_{k=-\infty}^{\infty} \phi_{rr}(kT)z^{-k}$
Input-output relations	$C(s) = G(s)R(s)$ $C(j\omega) = G(j\omega)R(j\omega)$ 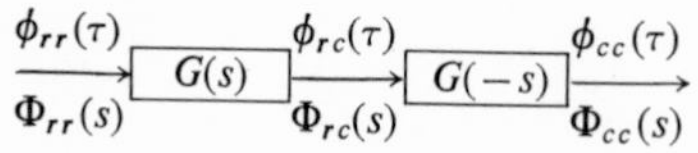 $\Phi_{cc}(s) = G(s)G(-s)\Phi_{rr}(s)$	$C(z) = G(z)R(z)$ $C^*(s) = G^*(s)R^*(s)$ 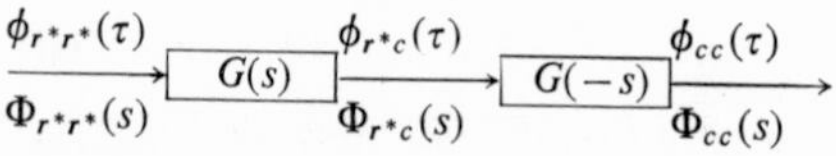 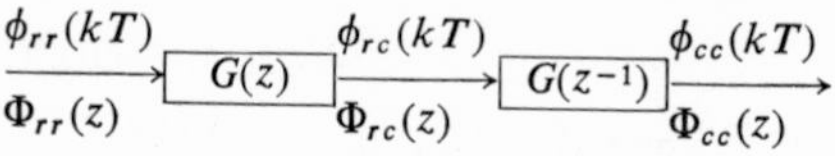 $\Phi_{cc}(z) = G(z)G(z^{-1})\Phi_{rr}(z)$ $\Phi_{c^*c^*}(j\omega) = \lvert G^*(j\omega)\rvert^2\Phi_{r^*r^*}(j\omega)$

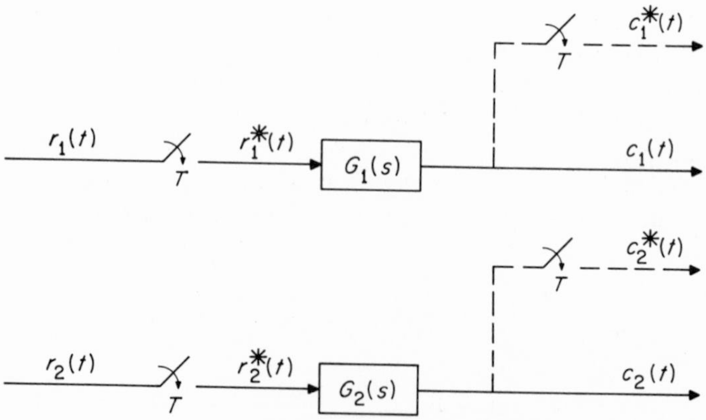

FIG. 8-5. Block diagrams illustrating statistical relationship between inputs and outputs of multichannel discrete-data systems.

Similarly, the cross-spectral density between the two continuous outputs $c_1(t)$ and $c_2(t)$ may be written as

$$\Phi_{c_1c_2}(s) = G_1(-s)G_2(s)\Phi_{r_1^*r_2^*}(s) \tag{8-106}$$

Since these transfer relationships for discrete-data systems are analogous to those for continuous-data systems, if one is familiar with the relations between signals for continuous-data systems, Eqs. (8-104), (8-105), and (8-106) may be written out directly. A comparison of the spectral density functions for continuous-data and discrete-data systems is made in Figure 8-6.

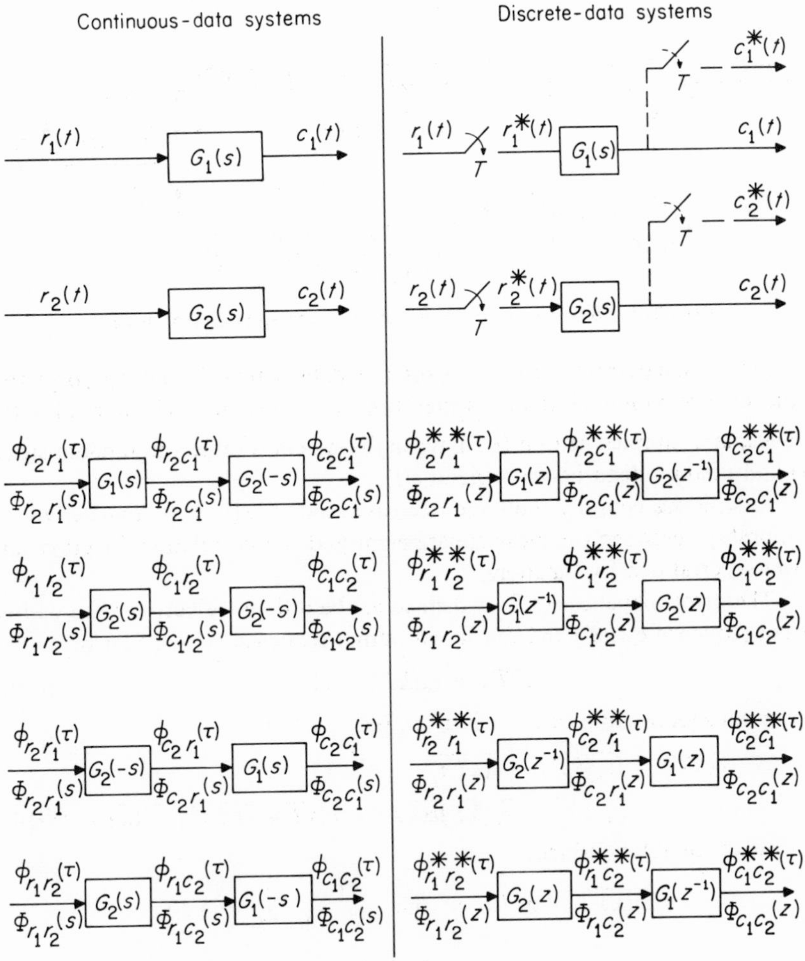

FIG. 8-6. Block diagrams illustrating the relationships between input and output signals of two systems.

8.5 THE MEAN-SQUARE ERROR

The main objective of Wiener's design of control systems with random inputs is the design of an optimum system that will render a minimum mean-square error. It should be noted that the mean-square-error criterion is not ideal for all control systems. However, the popular use of the criterion in statistical design stems from the fact that the use of the mean-square error as the design criterion allows the description of the random signals by their correlation function and power spectral density functions.

Shown in Figure 8-7 is a block diagram which illustrates the formulation

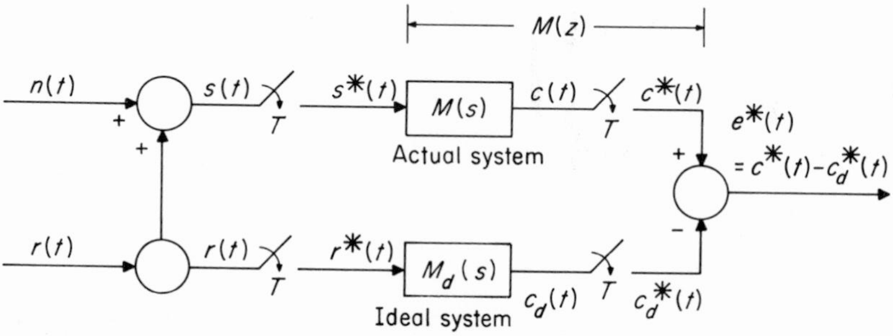

FIG. 8-7. Block diagram for minimum mean-square error design.

of the mean-square-error design of a discrete-data system. The design objective is the determination of the z-transfer function $M(z)$ that will minimize the mean-square sampled error $\overline{e^2(kT)}$, when the system is subject to random signal $r(t)$ and random disturbance or noise $n(t)$.

First, let us consider the noise-free case, that is, $n(t) = 0$, and derive the relationships between the mean-square error and the correlation function and power spectral density functions.

The error sequence $e(kT)$ is defined as the difference between the desired output sequence $c_d(kT)$ and the actual output sequence $c(kT)$; that is,

$$e(kT) = c_d(kT) - c(kT) \tag{8-107}$$

The mean-square sampled error is given by

$$\begin{aligned}\overline{e(kT)} &= \overline{[c_d(kT) - c(kT)]^2} \\ &= \overline{c_d^2(kT)} - \overline{c_d(kT)c(kT)} - \overline{c(kT)c_d(kT)} + \overline{c^2(kT)}\end{aligned} \tag{8-108}$$

In view of the relationships

$$\begin{aligned}\overline{e^2(kT)} = \phi_{ee}(0) &= \lim_{N\to\infty} \frac{1}{2N+1} \sum_{k=-N}^{N} e^2(kT) \\ &= \frac{1}{2\pi j} \oint_\Gamma \Phi_{ee}(z) z^{-1}\, dz\end{aligned} \tag{8-109}$$

and

$$\overline{c_d(kT)c(kT)} = \phi_{c_dc}(0) = \lim_{N\to\infty} \frac{1}{2N+1} \sum_{k=-N}^{N} c_d(kT)c(kT) \tag{8-110}$$

the sampled spectral density of $e(kT)$ may be written as

$$\Phi_{ee}(z) = \Phi_{c_dc_d}(z) - \Phi_{c_dc}(z) - \Phi_{cc_d}(z) + \Phi_{cc}(z) \tag{8-111}$$

From Figure 8-6 and Table 8-1, the following equations are written:

$$\Phi_{c_dc_d}(z) = M_d(z)M_d(z^{-1})\Phi_{rr}(z) \tag{8-112}$$

$$\Phi_{cc}(z) = M(z)M(z^{-1})\Phi_{rr}(z) \tag{8-113}$$

$$\Phi_{c_dc}(z) = M(z)M_d(z^{-1})\Phi_{rr}(z) \tag{8-114}$$

$$\Phi_{cc_d}(z) = M(z^{-1})M_d(z)\Phi_{rr}(z) \tag{8-115}$$

Substitution of Eqs. (8-112) through (8-115) into Eq. (8-111) yields

$$\Phi_{ee}(z) = [M_d(z)M_d(z^{-1}) - M(z)M_d(z^{-1}) - M(z^{-1})M_d(z) + M(z)M(z^{-1})]\Phi_{rr}(z) \tag{8-116}$$

Therefore, Eq. (8-116) shows that the sampled spectral density function $\Phi_{ee}(z)$ is represented by the system transfer functions $M(z)$, $M_d(z)$, and the sampled input spectral density function $\Phi_{rr}(z)$. The mean-square sampled error is obtained by substituting $\Phi_{ee}(z)$ into Eq. (8-109). The design objective is the determination of the transfer function $M(z)$ so that the mean-square error $\overline{e^2(nT)}$ is a minimum.

In general, most discrete-data systems have discrete-data as well as continuous-data signals. Shown in Figure 8-8 is a discrete-data system which is

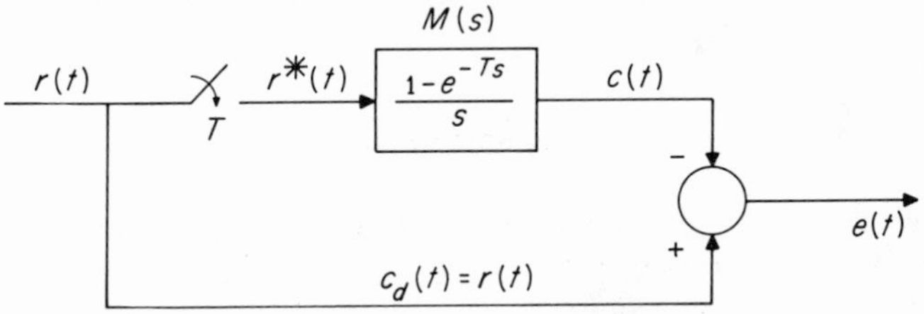

FIG. 8-8. Zero-order hold system for Example 8-2.

compared with a continuous-data system model. The mean-square error of the system is defined as

$$\begin{aligned} \overline{e^2(t)} &= \overline{[c_d(t) - c(t)]^2} \\ &= \overline{c_d^2(t)} - \overline{c_d(t)c(t)} - \overline{c(t)c_d(t)} + \overline{c^2(t)} \end{aligned} \tag{8-117}$$

Also

$$\begin{aligned} \overline{e^2(t)} &= \lim_{T_0\to\infty} \frac{1}{2T_0} \int_{-T_0}^{T_0} e^2(t)\,dt = \phi_{ee}(0) \\ &= \frac{1}{2\pi j} \int_{c-j\infty}^{c+j\infty} \Phi_{ee}(s)\,ds \end{aligned} \tag{8-118}$$

Similar expressions can also be written for $\overline{c_d^2(t)}$, $\overline{c_d(t)c(t)}$, and the other terms in Eq. (8-117). Therefore, the spectral density of $e(t)$ is written as

$$\Phi_{ee}(s) = \Phi_{c_dc_d}(s) - \Phi_{c_dc}(s) - \Phi_{cc_d}(s) + \Phi_{cc}(s) \tag{8-119}$$

From Eq. (8-8) the following relationships between the spectral density functions may be written:

$$\Phi_{c_dc_d}(s) = M_d(s)M_d(-s)\Phi_{rr}(s) \tag{8-120}$$

$$\Phi_{cc}(s) = M(s)M(-s)\Phi_{r^*r^*}(s) \tag{8-121}$$

$$\Phi_{c_dc}(s) = M(s)M_d(-s)\Phi_{rr^*}(s) \tag{8-122}$$

$$\Phi_{cc_d}(s) = M(-s)M_d(s)\Phi_{r^*r}(s) \tag{8-123}$$

where $\Phi_{rr^*}(s)$ and $\Phi_{r^*r}(s)$ involve the crosscorrelation between a continuous-data signal and a discrete-data signal. It can be shown that

$$\phi_{r^*r}(\tau) = \phi_{rr^*}(\tau) = \frac{1}{T}\phi_{rr}(\tau) \tag{8-124}$$

Now substitution of Eqs. (8-119) through (8-124) into Eq. (8-118) yields

$$\overline{e^2(t)} = \frac{1}{2\pi j}\int_{c-j\infty}^{c+j\infty}\left\{\left[M_d(s)M_d(-s) - \frac{1}{T}M(s)M_d(-s) - \frac{1}{T}M(-s)M_d(s)\right]\Phi_{rr}(s) + M(s)M(-s)\Phi_{r^*r^*}(s)\right\}ds \tag{8-125}$$

which clearly indicates that the mean-square error may be expressed as a function of the power spectral densities of $r(t)$ and $r^*(t)$ and the transfer functions $M(s)$ and $M_d(s)$.

EXAMPLE 8-2

In this example we shall illustrate the use of the expressions just derived on the evaluation of the mean-square error.

Figure 8-8 shows a discrete-data system whose mean-square error is to be evaluated. The system M is a zero-order hold, and the ideal system has a transfer function of $M_d(s) = 1$.

The power spectral density function of the input signal is given by

$$\Phi_{rr}(s) = \frac{2\omega_0}{\omega_0^2 - s^2} \tag{8-126}$$

The mean-square error of this system will give a measure of the ability of data-reconstruction of the zero-order hold. From Eqs. (8-118) and (8-119), the mean-square error is written as

$$\begin{aligned}\overline{e^2(t)} &= \frac{1}{2\pi j}\int_{c-j\infty}^{c+j\infty}[\Phi_{c_dc_d}(s) - \Phi_{c_dc}(s) - \Phi_{cc_d}(s) + \Phi_{cc}(s)]\,ds \\ &= \Phi_{c_dc_d}(0) - \Phi_{c_dc}(0) - \Phi_{cc_d}(0) + \Phi_{cc}(0)\end{aligned} \tag{8-127}$$

Substituting Eqs. (8-120) through (8-123) into Eq. (8-127), we have

$$\overline{e^2(t)} = \frac{1}{2\pi j}\int_{c-j\infty}^{c+j\infty}\left\{\left[1 - \frac{1}{T}M(s) - \frac{1}{T}M(-s)\right]\Phi_{rr}(s) + M(s)M(-s)\Phi_{r^*r^*}(s)\right\} ds \tag{8-128}$$

where $\Phi_{r^*r^*}(s)$ is the sampled spectrum of $r(t)$ and can be determined by the method described in Sec. 8.3. Substituting $M(s)$, $M(-s)$, $\Phi_{rr}(s)$, and $\Phi_{r^*r^*}(s)$ into the right-hand side terms of Eq. (8-127), and using contour integration, we get

$$\begin{aligned}\phi_{c_dc_d}(0) &= \phi_{c_dc}(0) - \phi_{cc_d}(0)\\ &= \frac{1}{2\pi j}\int_{c-j\infty}^{c+j\infty}\left(1 - \frac{1-e^{-Ts}}{Ts} - \frac{1-e^{Ts}}{-Ts}\right)\frac{2\omega_0}{\omega_0^2 - s^2}\,ds = 1 - \frac{2(1-e^{-\omega_0 T})}{\omega_0 T}\end{aligned} \tag{8-129}$$

and

$$\begin{aligned}\phi_{cc}(0) &= \frac{1}{2\pi j}\int_{c-j\infty}^{c+j\infty}\frac{(1-e^{-Ts})(1-e^{Ts})}{-s^2}\,\frac{1-e^{-2\omega_0 T}}{T(1-e^{-\omega_0 T}e^{-Ts})(1-e^{\omega_0 T}e^{Ts})}\,ds\\ &= 1\end{aligned} \tag{8-130}$$

Therefore, Eq. (8-127) gives

$$\overline{e^2(t)} = 2\left(1 - \frac{1-e^{-\omega_0 T}}{\omega_0 T}\right) \tag{8-131}$$

The mean-square error of the system is plotted in Figure 8-9 as a function of

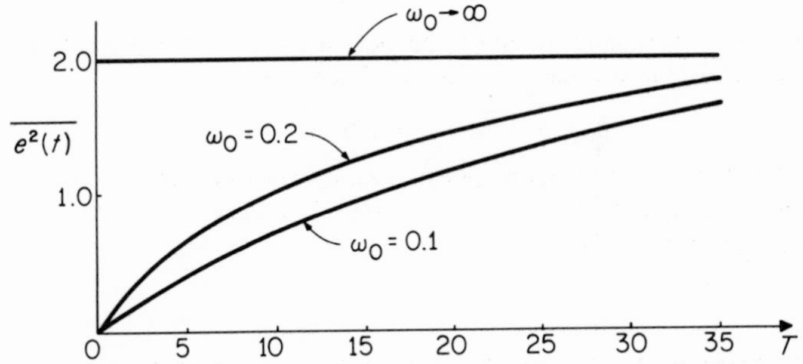

FIG. 8-9. Mean-square error versus T of the zero-order hold system in Example 8-2.

ω_0 and T. When T is equal to zero, $\overline{e^2(t)}$ is also zero; when T approaches infinity, $\overline{e^2(t)}$ approaches the value 2. When T is held constant, increasing ω_0 will increase the mean-square error as expected.

Since many discrete-data systems utilize zero-order hold for data-reconstruction, further investigation must be conducted on the statistical behavior of the hold device. Referring to Figure 8-8, we see that the actual system consists of an ideal sampler and a zero-order hold. As Figure 8-3(a) and Eq. (8-63) indicate,

$$\phi_{r^*r^*}(kT) = \frac{1}{T}\phi_{r_nr_n}(kT) \tag{8-132}$$

When $k = 0$,

$$\phi_{r^*r^*}(0) = \frac{1}{T}\phi_{r_n r_n}(0) = \frac{1}{T}\phi_{rr}(0) \tag{8-133}$$

An important result shown in Eq. (8-133) is that the mean-square value of a sampled signal is equal to $1/T$ times the mean-square value of the unsampled signal. In other words,

$$\overline{[r^*(t)]^2} = \frac{1}{T}\overline{r^2(t)} \tag{8-134}$$

When the impulse $\Phi_{r^*r^*}(0)$ is applied to the zero-order hold, we can show by taking the inverse transform of $M(s)M(-s)$ that the two-sided impulse response of the zero-order hold in the τ domain is as shown in Figure 8-10. From this, we con-

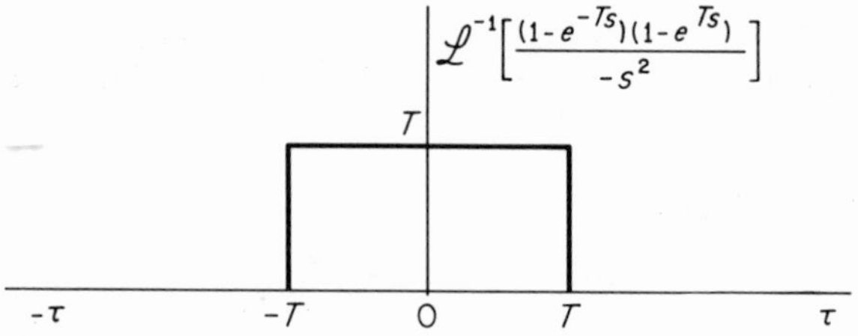

FIG. 8-10. Two-sided impulse response of the zero-order hold in the τ-domain.

clude that the mean-square value of the output of the zero-order hold must equal T times the mean-square value of the input. Therefore,

$$\phi_{cc}(0) = \overline{c^2(t)} = T\overline{[r^*(t)]^2} = \overline{r^2(t)} \tag{8-135}$$

or the mean-square value of the output of a sampler-zero-order hold combination is equal to the mean-square value of the input signal. This is also verified by the result of the integral of Eq. (8-130).

8.6 OPTIMUM DESIGN USING MEAN-SQUARE-ERROR CRITERION

The principal objectives of the preceding section are the introduction of the mean-square error criterion and the establishment of the relationship between the statistical properties of the input and output signals of discrete-data systems. In this section, we turn our attention from analysis to design. Furthermore, in all the discussions of the previous section, we have neglected noise and interference in the system model considered. However, noise and disturbance are present in most practical control systems, and they tend to obscure the signal used for the optimal control and are the basic restriction in the design of optimal control systems.

The statistical design of discrete-data systems in the sense of minimum mean-square error can be readily visualized by referring to the block diagram of Figure 8-11(a). In this diagram, $M(z)$ denotes the closed-loop transfer func-

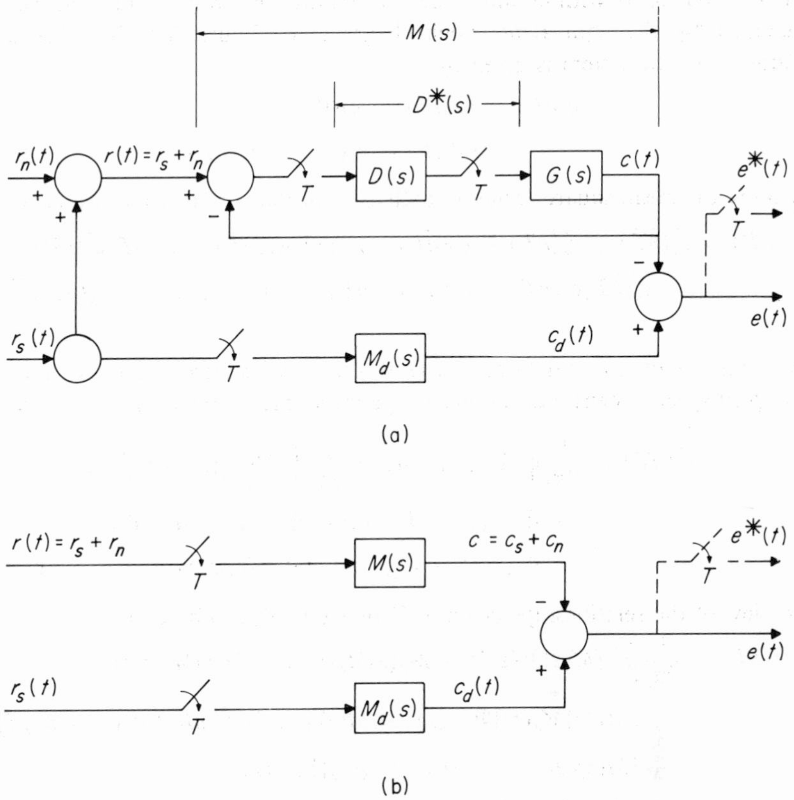

FIG. 8-11. (a) Discrete-data system to be designed in the sense of minimum mean-square error; (b) an equivalent system block diagram of the system in (a).

tion of the actual system which is to be optimized in the sense of minimum mean-square error, and $M_d(z)$ is the desired overall system transfer function. The transfer function $G(s)$ represents the continuous-data portion of the system, and is assumed to be known. $D^*(s)$ or $D(z)$ denotes the transfer function of the discrete compensator to be designed. The input signal $r(t)$ is assumed to consist of a noise component $r_n(t)$ and a signal component $r_s(t)$. The input signal and the noise are assumed to be functions of stationary processes. The system output also consists of two components: one due to $r_n(t)$ and one due to $r_s(t)$. The system error is defined as

$$e(t) = c_d(t) - c(t) = c_d(t) - c_s(t) - c_n(t) \tag{8-136}$$

If the system is well-behaved between sampling instants, design may be carried out based on the mean-square-sampled error $\overline{e^2(nT)}$. Since the signal $r(t) - c(t)$ is sampled in the actual system, we can insert a fictitious sampler at the input to M without any effect. Therefore, the entire system can be represented by the equivalent block diagram of Figure 8-11(b). The error sequence of the system is given by

$$\begin{aligned} e(nT) &= c_d(nT) - c(nT) \\ &= c_d(nT) - c_s(nT) - c_n(nT) \end{aligned} \tag{8-137}$$

Taking the mean-square value on both sides of the last equation, we have

$$\begin{aligned} \overline{e^2(nT)} = {} & \overline{c_d^2(nT)} + \overline{c_s^2(nT)} + \overline{c_n^2(nT)} + \overline{c_s(nT)c_n(nT)} + \overline{c_n(nT)c_s(nT)} \\ & - \overline{c_d(nT)c_s(nT)} - \overline{c_s(nT)c_d(nT)} - \overline{c_d(nT)c_n(nT)} - \overline{c_n(nT)c_d(nT)} \end{aligned} \tag{8-138}$$

Making use of the relations between the mean-square sequence and the sampled spectral densities, the mean-square-sampled error may be written as

$$\begin{aligned} \overline{e^2(nT)} &= \frac{1}{2\pi j}\oint_\Gamma \Phi_{ee}(z)z^{-1}\,dz = \frac{1}{2\pi j}\oint_\Gamma [\Phi_{c_dc_d}(z) + \Phi_{c_sc_s}(z) \\ & \quad + \Phi_{c_nc_n}(z) + \Phi_{c_sc_n}(z) + \Phi_{c_nc_s}(z) - \Phi_{c_dc_s}(z) \\ & \quad - \Phi_{c_sc_d}(z) - \Phi_{c_dc_n}(z) - \Phi_{c_nc_d}(z)]z^{-1}\,dz \end{aligned} \tag{8-139}$$

In view of the relationships given in Figure 8-6, $\overline{e^2(nT)}$ becomes

$$\begin{aligned} \overline{e^2(nT)} &= \frac{1}{2\pi j}\oint_\Gamma \{[M(z)M(z^{-1}) + M_d(z)M_d(z^{-1}) - M(z)M_d(z^{-1}) \\ & \quad - M(z^{-1})M_d(z)]\Phi_{r_sr_s}(z) + [M(z)M(z^{-1}) - M(z)M_d(z^{-1})]\Phi_{r_sr_n}(z) \\ & \quad + [M(z)M(z^{-1}) - M(z^{-1})M_d(z)]\Phi_{r_nr_s}(z) \\ & \quad + M(z)M(z^{-1})\Phi_{r_nr_n}(z)\}z^{-1}\,dz \end{aligned} \tag{8-140}$$

where

$$M(z) = \frac{D(z)G(z)}{1 + D(z)G(z)} \tag{8-141}$$

The synthesis objective is to determine the transfer function $M(z)$ so that the mean-square-sampled error of Eq. (8-140) is a minimum. Once the optimum $M(z)$ is determined, the transfer function of the digital compensator is given by

$$D(z) = \frac{M(z)}{G(z)[1 - M(z)]} \tag{8-142}$$

The synthesis procedure developed below is essentially an extension of the theory of optimum design of linear continuous-data filters developed by

N. Wiener and A. Kolmogoroff and is also a parallel of the work of J. T. Tou. The application of the Wiener-Kolmogoroff criterion to the design of linear continuous-data control systems is well known. The design principle of Wiener and Kolmogoroff is based on the application of the calculus of variations to the integral of Eq. (8-140).

If $M(z)$ is an optimum transfer function, $\overline{e^2(nT)}$ must be at a minimum. Replacing the optimum $M(z)$ by a function $M(z) = \epsilon\eta(z)$, where ϵ is a parameter and $\eta(z)$ is an arbitrary but physically realizable function of z, would increase the mean-square-sampled error to $\overline{e^2(nT)} + \overline{\Delta e^2(nT)}$, where $\overline{\Delta e^2(nT)}$ is the change in $\overline{e^2(nT)}$. Then, a necessary condition that $\overline{e^2(nT)}$ is a minimum is

$$\frac{\partial[\overline{\Delta e^2(nT)}]}{\partial\epsilon}\bigg|_{\epsilon=0} = 0 \tag{8-143}$$

To insure that $\overline{e^2(nT)}$ is a minimum, the second derivative of $\overline{\Delta e^2(nT)}$ must be positive with respect to ϵ when $\epsilon = 0$. Substituting $M(z) + \epsilon\eta(z)$ and $M(z^{-1}) + \epsilon\eta(z^{-1})$ for $M(z)$ and $M(z^{-1})$, respectively, in Eq. (8-140) and simplifying, we have

$$\begin{aligned}\overline{e^2(nT)} + \overline{\Delta e^2(nT)} &= \overline{e^2(nT)} + \frac{1}{2\pi j}\oint_\Gamma \epsilon\eta(z)\{[M(z^{-1}) - M_d(z^{-1})]\\ &\quad\times \Phi_{r_sr_s}(z) + [M(z^{-1}) - M_d(z^{-1})]\Phi_{r_sr_n}(z) + M(z^{-1})\Phi_{r_nr_s}(z) + M(z^{-1})\\ &\quad\times \Phi_{r_nr_n}(z)\}z^{-1}\,dz + \frac{1}{2\pi j}\oint_\Gamma \epsilon\eta(z^{-1})\{[M(z) - M_d(z)]\Phi_{r_sr_s}(z)\\ &\quad+ M(z)\Phi_{r_sr_n}(z) + M(z)\Phi_{r_nr_s}(z) + M(z)\Phi_{r_nr_n}(z) - M_d(z)\Phi_{r_nr_s}(z)\}z^{-1}\,dz\\ &\quad+ \frac{1}{2\pi j}\oint_\Gamma \epsilon\eta(z)\epsilon\eta(z^{-1})\{\Phi_{r_sr_s}(z) + \Phi_{r_sr_n}(z) + \Phi_{r_nr_s}(z) + \Phi_{r_nr_n}(z)\}z^{-1}\,dz\end{aligned} \tag{8-144}$$

Thus,

$$\begin{aligned}\overline{\Delta e^2(nT)} &= \frac{1}{2\pi j}\oint_\Gamma \epsilon\eta(z)\{M(z^{-1})[\Phi_{r_sr_s}(z) + \Phi_{r_sr_n}(z) + \Phi_{r_nr_s}(z) + \Phi_{r_nr_n}(z)]\\ &\quad- M_d(z^{-1})[\Phi_{r_sr_s}(z) + \Phi_{r_sr_n}(z)]\}z^{-1}\,dz\\ &\quad+ \frac{1}{2\pi j}\oint_\Gamma \epsilon\eta(z^{-1})\{M(z)[\Phi_{r_sr_s}(z) + \Phi_{r_sr_n}(z) + \Phi_{r_nr_s}(z) + \Phi_{r_nr_n}(z)]\\ &\quad- M_d(z)[\Phi_{r_sr_s}(z) + \Phi_{r_nr_s}(z)]\}z^{-1}\,dz + \frac{1}{2\pi j}\oint_\Gamma \epsilon\eta(z)\epsilon\eta(z^{-1})\\ &\quad\times [\Phi_{r_sr_s}(z) + \Phi_{r_sr_n}(z) + \Phi_{r_nr_s}(z) + \Phi_{r_nr_n}(z)]z^{-1}\,dz\end{aligned} \tag{8-145}$$

Differentiating with respect to ϵ and setting the resultant equal to zero yields

$$\left.\frac{\partial[\overline{\Delta e^2(nT)}]}{\partial\epsilon}\right|_{\epsilon=0} = \frac{1}{2\pi j}\oint_\Gamma \eta(z)\{M(z^{-1})\Phi_{rr}(z) - M_d(z^{-1})[\Phi_{r_sr_s}(z) + \Phi_{r_sr_n}(z)]\}z^{-1}dz + \frac{1}{2\pi j}\oint_\Gamma \eta(z^{-1})\{M(z)\Phi_{rr}(z) - M_d(z)[\Phi_{r_sr_s}(z) + \Phi_{r_nr_s}(z)]\}z^{-1}dz = 0 \qquad (8\text{-}146)$$

where

$$\Phi_{rr}(z) = \Phi_{r_sr_s}(z) + \Phi_{r_sr_n}(z) + \Phi_{r_nr_s}(z) + \Phi_{r_nr_n}(z) \qquad (8\text{-}147)$$

When Eq. (8-146) is satisfied, $M(z)$ is optimum. However, before the optimum $M(z)$ can be obtained from Eq. (8-146), the equation must be first modified. It is evident that $\Phi_{rr}(z)$ is an even function of z. Therefore, we assume that $\Phi_{rr}(z)$ can be factored into two parts, one part having poles and zeros inside the unit circle of the z-plane and the other part having poles and zeros outside the unit circle. Thus $\Phi_{rr}(z)$ can be written

$$\Phi_{rr}(z) = \Phi_{rr}^+(z)\Phi_{rr}^-(z) \qquad (8\text{-}148)$$

Equation (8-146) is now written

$$\frac{1}{2\pi j}\oint_\Gamma \eta(z)\Phi_{rr}^+(z)\left\{M(z^{-1})\Phi_{rr}^-(z) - \frac{M_d(z^{-1})[\Phi_{r_sr_s}(z) + \Phi_{r_sr_n}(z)]}{\Phi_{rr}^+(z)}\right\}z^{-1}\,dz + \frac{1}{2\pi j}\oint_\Gamma \eta(z^{-1})\Phi_{rr}^-(z)\left\{M(z)\Phi_{rr}^+(z) - \frac{M_d(z)[\Phi_{r_sr_s}(z) + \Phi_{r_nr_s}(z)]}{\Phi_{rr}^-(z)}\right\}z^{-1}\,dz = 0 \qquad (8\text{-}149)$$

We can separate each of the two functions

$$\frac{M_d(z^{-1})[\Phi_{r_sr_s}(z) + \Phi_{r_sr_n}(z)]}{\Phi_{rr}^+(z)}z^{-1}$$

and

$$\frac{M_d(z)[\Phi_{r_sr_s}(z) + \Phi_{r_nr_s}(z)]}{\Phi_{rr}^-(z)}z^{-1}$$

into two parts; that is,

$$\frac{M_d(z^{-1})[\Phi_{r_sr_s}(z) + \Phi_{r_sr_n}(z)]}{\Phi_{rr}^+(z)} = F_1(z) = [F_1(z)]_+ + [F_2(z)]_- \qquad (8\text{-}150)$$

and

$$\frac{M_d(z)[\Phi_{r_sr_s}(z) + \Phi_{r_nr_s}(z)]}{\Phi_{rr}^-(z)} = F_2(z) = [F_2(z)]_+ + [F_2(z)]_- \qquad (8\text{-}151)$$

where $[F_1(z)]_+$ and $[F_2(z)]_+$ denote the parts with singularities only inside the unit circle, and $[F_1(z)]_-$ and $[F_2(z)]_-$ denote the parts with singularities only outside the unit circle. Note that these notations are not to be confused with those of Eq. (8-148).

Substitution of Eqs. (8-150) and (8-151) into Eq. (8-149) yields

$$\frac{1}{2\pi j}\oint_{\Gamma} \eta(z)\Phi_{rr}^{+}(z)\{M(z^{-1})\Phi_{rr}^{-}(z)z^{-1} - [F_1(z)]_{+} - [F_1(z)]_{-}\}\, dz$$
$$+ \frac{1}{2\pi j}\oint_{\Gamma} \eta(z^{-1})\Phi_{rr}^{-}(z)\{M(z)\Phi_{rr}^{+}(z)z^{-1} - [F_2(z)]_{+} - [F_2(z)]_{-}\}\, dz = 0 \tag{8-152}$$

For reasons of stability, $M(z)$ and $\eta(z)$ must have poles only inside the unit circle in the z-plane. Thus, $M(z^{-1})$ and $\eta(z^{-1})$ are functions with poles only outside the unit circle. Therefore, in the integrals of Eq. (8-152),

$$\frac{1}{2\pi j}\oint_{\Gamma} \eta(z)\Phi_{rr}^{+}(z)[F_1(z)]_{+}\, dz = 0 \tag{8-153}$$

and

$$\frac{1}{2\pi j}\oint_{\Gamma} \eta(z^{-1})\Phi_{rr}^{-}(z)[F_2(z)]_{-}\, dz = 0 \tag{8-154}$$

since the integrand of the integral in Eq. (8-153) has poles only inside the unit circle and the contour integral can be taken around the unit circle Γ in the clockwise direction, and for similar reasons the integral of Eq. (8-154) vanishes. Equation (8-152) now reads

$$\frac{1}{2\pi j}\oint_{\Gamma} \eta(z)\Phi_{rr}^{+}(z)\{M(z^{-1})\Phi_{rr}^{-}(z)z^{-1} - [F_1(z)]_{-}\}\, dz$$
$$+ \frac{1}{2\pi j}\oint_{\Gamma} \eta(z^{-1})\Phi_{rr}^{-}(z)\{M(z)\Phi_{rr}^{+}(z)z^{-1} - [F_2(z)]_{+}\}\, dz = 0 \tag{8-155}$$

Now if $M(z)$ is indeed the transfer function of the optimum system, Eq. (8-155) must be satisfied for arbitrary $\eta(z)$. Thus, the quantities inside the brackets of Eq. (8-155) must vanish. The optimum transfer function is given by

$$M(z) = \frac{z[F_2(z)]_{+}}{\Phi_{rr}^{+}(z)} = \left\{\frac{M_d(z)[\Phi_{r_sr_s}(z) + \Phi_{r_nr_s}(z)]}{z\Phi_{rr}^{-}(z)}\right\}_{+} \frac{z}{\Phi_{rr}^{+}(z)} \tag{8-156}$$

Once $M(z)$ is determined, the transfer function of the digital controller, $D(z)$, is obtained directly from Eq. (8-142).

Clearly, the properties of the optimum system are determined entirely by the specified system $M_d(z)$ and the spectra $\Phi_{r_sr_s}(z)$, $\Phi_{r_nr_s}(z)$, $\Phi_{r_sr_n}(z)$, and $\Phi_{r_nr_n}(z)$ of the signal and noise. In the event that the noise is absent, $\Phi_{r_nr_s}(z) = \Phi_{r_sr_n}(z) = \Phi_{r_nr_n}(z) = 0$, and

$$\Phi_{r_sr_s}(z) = \Phi_{rr}(z) = \Phi_{rr}^{+}(z)\Phi_{rr}^{-}(z) \tag{8-157}$$

then Eq. (8-156) gives $M(z) = M_d(z)$ as expected, and the error $e(t)$ is always zero.

To illustrate the synthesis of discrete-data systems by means of the mean-square-error design described in the preceding section, a numerical example is presented below.

EXAMPLE 8-3

Consider the discrete-data control system shown in Figure 8-11. Assume that

$$G(s) = \frac{1}{s(s+1)} \qquad \Phi_{r_s r_s}(s) = \frac{2}{1-s^2}$$

$$\Phi_{r_n r_n}(s) = 1 \text{ (white noise)} \qquad T = 1 \text{ second}$$

and the input and noise are uncorrelated; that is, $\Phi_{r_n r_s}(s) = \Phi_{r_s r_n}(s) = 0$. Design a digital controller to minimize the mean-square-sampled error when $M_d(s) = e^{-Ts}$.

For the sampling period given, the z-transform of $G_{h0}(s)G(s)$ is

$$G_{h0}G(z) = \frac{0.005z}{(z-1)(z-0.905)} \tag{8-158}$$

The sampled spectrum of the input signal is determined using Eq. (8-75).

$$\Phi_{r_s r_s}(z) = \frac{-2z}{(z-0.905)(z-1.1)} \tag{8-159}$$

and the sampled spectrum of the noise is simply

$$\Phi_{r_n r_n}(z) = 1 \tag{8-160}$$

Therefore,

$$\Phi_{rr}(z) = \Phi_{r_s r_s}(z) + \Phi_{r_n r_n}(z) = \frac{(z-3.732)(z-0.268)}{(z-0.905)(z-1.1)} \tag{8-161}$$

Dividing $\Phi_{rr}(z)$ into two parts according to Eq. (8-148), we have

$$\Phi_{rr}^{+}(z) = \frac{z-0.268}{z-0.905} \tag{8-162}$$

$$\Phi_{rr}^{-}(z) = \frac{z-3.732}{z-1.1} \tag{8-163}$$

Therefore,

$$F_2(z) = \frac{M_d(z)[\Phi_{r_s r_s}(z) + \Phi_{r_n r_s}(z)]}{z\Phi_{rr}^{-}(z)} = \frac{-2}{z(z-0.905)(z-3.732)} \tag{8-164}$$

Now dividing Eq. (8-164) into two parts according to Eq. (8-151), we have

$$F_2(z) = [F_2(z)]_+ + [F_2(z)]_- = \frac{0.78}{z-0.905} - \frac{0.19}{z-3.732} - \frac{0.58}{z} \tag{8-165}$$

Making use of Eq. (8-156), the optimum transfer function is

$$M(z) = \frac{z[F_2(z)]_+}{\Phi_{rr}^{+}(z)} = \frac{0.2(z+2.62)}{z-0.268} \tag{8-166}$$

The transfer function of the digital controller is determined from Eq. (8-142).

$$D(z) = \frac{M(z)}{G_{h0}G(z)[1-M(z)]} = \frac{50(z-1)(z-0.905)(z+2.62)}{z(z-0.992)} \tag{8-167}$$

However, since the order of the numerator of $D(z)$ in z exceeds that of the denominator of $D(z)$, the transfer function is not physically realizable. This is a common difficulty with the Wiener filter design when the system configuration is fixed. To overcome this difficulty, we can obtain a suboptimal solution by either adding a pole to $M(z)$ at the origin in the z-plane which corresponds to $s \to -\infty$ in the s-plane, or by removing the zero which is far away from the dominating poles and zeros. In this case, we can remove the zero at $z = -2.62$ from $M(z)$. Thus, $M(z)$ is written

$$M(z) = \frac{0.724}{z - 0.268} \tag{8-168}$$

Notice that in modifying $M(z)$ its behavior at $z = 1$ should be maintained.

The transfer function of the suboptimal digital contour system is now

$$D(z) = \frac{145(z-1)(z-0.905)}{z(z-0.992)} \tag{8-169}$$

By substituting $M(z)$ and other known functions of the suboptimal system into Eq. (8-140), the mean-square-sampled error of the optimal system is found to be

$$\overline{e^2(nT)} = 1.16 \tag{8-170}$$

The difficulty with the physical realizability can be traced to the z terms in Eq. (8-156). Therefore, a suboptimal solution for $M(z)$ can be obtained by modifying Eq. (8-156) to read

$$M(z) = \left\{\frac{M_d(z)[\Phi_{r_s r_s}(z) + \Phi_{r_n r_s}(z)]}{\Phi_{rr}^-(z)}\right\}_+ \frac{1}{\Phi_{rr}^+(z)} \tag{8-171}$$

REFERENCES

1. Wiener, N., *The Extrapolation, Interpolation, and Smoothing of Stationary Time Series.* John Wiley & Sons, New York, 1948.

2. Laning, J. H., Jr., and Battin, R. H., *Random Processes in Automatic Control.* McGraw-Hill, New York, 1956.

3. Newton, G. C., Gould, L. A., and Kaiser, J. F., *Analytical Design of Linear Feedback Controls.* John Wiley & Sons, New York, 1957.

4. Gibson, J. E., *Nonlinear Control System.* McGraw-Hill, New York, 1963.

5. Kalman, R.E., "A New Approach to Linear Filtering and Prediction Problems," *J. Basic Eng.*, Vol. 82, March 1960, pp. 35–45.

6. Johnson, G. N., "Statistical Analysis of Sampled-Data Systems," *IRE WESCON Record,* Part 4, 1957, pp. 187–195.

7. Bergen, A. R., "On the Statistical Design of Linear Random Sampling Systems," *Proc. of the International Federation of Automatic Contral*, Moscow 1960, pp. 126–131.

8. Zadeh, L. A., and Ragazzini, J. R., "An Extension of Wiener's Theory of Prediction," *Journal of Applied Physics*, **21**, July 1950, pp. 645–655.

9. Blum, M., "An Extension of the Minimum Mean Square Prediction Theory For Sampled-Input Signals," *Trans. IRE on Information Theory*, Vol. IT-2, September 1956, pp. S176–S184.

10. Brogen, J., "Filters for Sampled-Signals," *Proc. of the Information Theory Symposium*, Brooklyn Polytechnic Institute, 1954.
11. Lees, A. B., "Interpolation and Extrapolation of Sampled-Data," *IRE Trans. on Information Theory*, Vol. IT-2, Mach 1956, pp. 12–17.
12. Lloyd, S. P., and McMillan, B., "Linear Least Square Filtering and Prediction of Sampled Signals," *Proc. of Symposium on Modern Network Synthesis*, Brooklyn Polytechnic Institute, April 1955.
13. Mori, M., "Statistical Treatment of Sampled-Data Control Systems for Actual Random Inputs," *Trans. ASME*, Vol. 80, No. 2, February 1958, pp. 444–456.
14. Kalman, R. E., "Analysis and Synthesis of Linear Systems Operating on Randomly Sampled-Data," doctoral dissertation, Columbia University, New York, 1957.
15. Franklin, G.F., "Linear Filtering of Sampled-Data," *Technical Report T-51B*, Columbia University Electronics Research Lab., New York, December 1954.
16. Franklin, G.F., "The Optimum Synthesis of Sampled-Data Systems," *Technical Report T-61B*, Columbia University Electronics Research Lab., New York, 1955.
17. Chang, S. S. L., "Statistical Design Theory For Strictly Digital Sampled-Data Systems," *Trans. AIEE*, 77, Part 1, February 1958, pp. 702-709.
18. DeRusso, P. M., "Ultimate Performance Limitations of Linear Sampled-Data Systems," *M. I. T. Electric System Lab.*, Report 7843-R-3, June 1959.
19. Hung, J. C., "Double Measurement With Both Sampled and Continuous Inputs," *IRE Convention Records*, Part 4, 1962.
20. Sittler, R. W., "System Analysis of Discrete Markov Processes," *IRE Trans. on Circuit Theory*, Vol. CT-3, No. 4, December 1956, pp. 257–265.
21. Sklansky, J., "On Closed-form Expansions For Mean Squares in Discrete-continuous Systems," *IRE Trans. on Automatic Control*, Vol. AC-4, 1958.
22. Sobral, M., "Statistical Design of Linear Multivariable Sampled-Data Feedback Control System," *Coordinated Science Lab. Report R-132*, University of Illinois, Urbana, Illinois, February 1962.
23. Stewart, R. M., "Statistical Design and Evaluation of Filters for the Restoration of Sampled-Data," *Proc. IRE*, **44**, No. 2, February 1956, pp. 253–257.
24. Tou, J. T., "Statistical Design of Digital Control Systems," *IRE Trans. on Automatic Control*, September 1960, pp. 290–297.
25. Tou, J. T., "Statistical Design of Linear Discrete-Data Control Systems Via the Modified z-transform Method," *Journal of Franklin Institute*, **271**, No. 4, April 1961, pp. 246–262.
26. Tou, J. T., and Kumar, K. S. P., "Statistical Design of Discrete-Data Control Systems Subject to Power Limitation," *Journal of Franklin Institute*, **272**, No. 3, September 1961, pp. 171–184.

27. Liff, A. I., "Mean Square Reconstruction Error," *IEEE Trans. on Automatic Control*, Vol. AC-10, July 1965, pp. 370–371.

28. Brown, W. A., "Multirate Error Criterion for Digital Control Systems," *IEEE Trans. on Automatic Control*, Vol. AC-11, October 1966, p. 761.

29. Leneman, O. A. Z., "On Some Results in Random Pulse Trains," *IEEE Trans. on Automatic Control*, Vol. AC-11, April 1966, p. 331.

30. Kukhtenko, V. I., "On Designing Correcting Circuits for Automatic Control Systems in Accordance with the Mean Square Error Criterion," *Automation and Remote Control*, Vol. 20, September 1959, pp. 1151–1159.

31. Leneman, O. A. Z., "A Note on the Mean-Square Behavior of a First-Order Random Sampling System," *IEEE Trans. on Automatic Control*, Vol. AC-13, August 1968, pp. 450–451.

32. Leneman, O. A. Z., and Lewis, J. B., "Random Sampling of Random Processes: Mean-Square Comparison of Various Interpolators," *IEEE Trans. on Automatic Control*, Vol. AC-11, July 1966, pp. 396–403.

33. Leneman, O. A. Z., "Random Sampling of Random Process: Optimum Linear Interpolation," *J. Franklin Institute*, Vol. 281, April 1966, pp. 302–314.

34. Leneman, O. A. Z., and Lewis, J. B., "On Mean-Square Reconstruction Error," *IEEE Trans. on Automatic Control*, Vol. AC-11, April 1966, pp. 324–325.

35. Prueher, R. F., Jr., "AnEquivalent Gain and Stochastic Analysis For Nonlinear Sampled-Data Control Systems," Ph.D dissertation, Department of Electrical Engineering, University of Illinois, Urbana, Illinois, 1966.

9

STATISTICAL DESIGN: KALMAN FILTER

9.1 INTRODUCTION

In the last chapter it is shown that the problem of prediction, smoothing, and filtering of a random signal can all be solved by means of the Wiener filtering technique. While Wiener filter theory has provided valuable insight to processes that were not possible to handle by deterministic methods, the practical applications of the theory are often confronted with several difficulties. These difficulties are:

1. The design of the Wiener filter is formulated strictly from a mathematical sense, and is not suitable, in general, for machine computation.
2. The classical Wiener filtering technique provides the solution only to single-variable systems. Although generalization to multivariable cases has been made in isolated cases, it requires new derivations, and the procedure is usually quite complex.
3. The Wiener filter is valid only for stationary random processes. Although extensions to the nonstationary cases have been made by Booton, the generalization again requires new derivations, and its application presents difficulty to the average users.

The work of R. E. Kalman is regarded by many as the most significant contribution to filtering and prediction theory, since the original work of Wiener. The Kalman filtering theory introduces a new look at the classical problems of prediction, smoothing, and filtering. We shall show in the following discussion that the Kalman filtering technique will eliminate most of the difficulties encountered by the Wiener filter. More specifically, Kalman's method has the following features:

1. The linear dynamic system is described by the state variables and state equations. This not only represents a modern approach to the systems problem, but also makes machine computation simpler.
2. The Kalman filtering theory treats stationary and nonstationary random processes, single-variable and multivariable systems, all in a unified manner.
3. The problem is approached from the point of view of conditional probability and expectations. This way, the generalized equations for the Kalman filter are quickly obtained. All statistical data required are presented in a much more simplified form than that of the Wiener filter.
4. A more general class of loss functions or error criteria can be used for the Kalman filter problem. The clasical Wiener filter admits only the mean-square-error criterion.

9.2 STATEMENT OF THE KALMAN FILTERING PROBLEM

In the preceding chapter, the Wiener filtering problem is stated as: given a meassage in the form of a signal corrupted by additive noise, find the optimum filter which when operated on the message will produce an output signal in the sense of minimum mean-square error. The error is defined to be the difference between the actual output of the filter and the signal component of the input message. The block diagram showing the design philosophy of Wiener filtering is shown in Figure 9-1(a). It is assumed that the statistical properties of the signal and noise are available in the form of power spectral density functions.

In the Kalman filtering problem, we assume that a message $c(t) = x(t) + w(t)$ is given, where $x(t)$ is the actual signal and $w(t)$ is the noise. The problem in this case is to determine the value of the signal $x(t)$ at some time $t = t_j$, given the measured or observed $c(\tau)$, $t_0 \leq \tau \leq t_k$. The time t_j can be less than, equal to, or greater than t_k, and the three cases can be defined as

$t_j < t_k$	smoothing (interpolation)
$t_j = t_k$	filtering
$t_j > t_k$	predicting

Since we are interested in discrete-data processes, the Kalman filtering problem can be stated, in general, as: given $\mathbf{c}(t) = \mathbf{x}(t) + \mathbf{w}(t)$, where the measurements on the random vector $\mathbf{c}(t)$ are made as $\mathbf{c}(t_0), \mathbf{c}(t_1), \ldots, \mathbf{c}(t_k)$, find the Kalman filter which will produce the best estimate to the signal $x(t_j)$, $t_j \geq t_k$ or $t_j < t_k$. Notice that the problem is stated here in terms of the multivariable notation using vector quantities for the multidimensional random variables. It will become apparent that the formulation of the Kalman filter problem is more conveniently carried out for discrete-data systems, especially in view of the fact that a digital computer is often used as a tool. The case of the continuous-data process is easily extended by taking the time increment to be infinitesimally small.

Since the equations of the Kalman filter are general enough to include all the problems of smoothing, filtering, and predicting, it is customary to refer to the Kalman filtering problem as *estimation*, and the optimum filter as the optimum *estimator*.

$M_d(s) = e^{-\tau s}$ $(\tau > 0)$ Smoothing (interpolation)
$M_d(s) = 1$ Filtering
$M_d(s) = e^{\tau s}$ $(\tau > 0)$ Predicting

(a)

$M_d(s) = e^{-\tau s}$ $(\tau > 0)$ Smoothing (interpolation)
$M_d(s) = 1$ Filtering
$M_d(s) = e^{\tau s}$ $(\tau > 0)$ Predicting

(b)

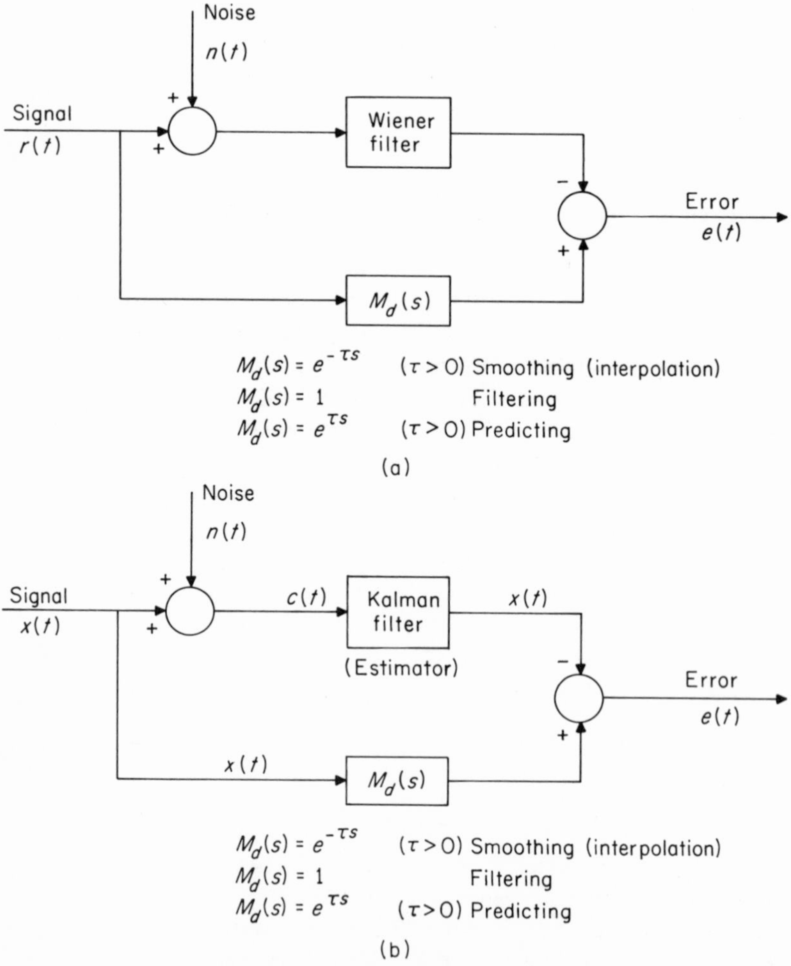

FIG. 9-1. (a) Wiener filter; (b) Kalman filter.

Notice that if $\mathbf{c}(t)$ represents the observable output vector and $\mathbf{x}(t)$ the state vector of a linear process, the Kalman filtering problem gives a solution to the optimum estimate of the inaccessible state variables which are required to form the optimal controller in the design techniques discussed in Chapter 6.

The block diagram illustrating the general philosophy of the Kalman estimator is now shown in Fig. 9-1(b). The similarity between the Wiener problem and the Kalman problem is clearly demonstrated by the two block diagrams in Figure 9-1.

9.3 OPTIMUM ESTIMATION AND CONDITIONAL EXPECTATION

We shall now show that the optimum estimation problem is equivalent to a problem in conditional expectation.

With the knowledge of any given set of measured random variables, $\mathbf{c}(t_0)$, $\mathbf{c}(t_1), \ldots, \mathbf{c}(t_k)$, of the random variable $\mathbf{c}(t)$, we can determine, in principle, the probability of $x_i(t_j)$ assuming a value of ξ_i or less, $i = 1, 2, \ldots, n$. This is the definition of the conditional probability distribution function given in Sec. 8.2.

Let us use the expression

$$\hat{\mathbf{x}}(t_j|t_k) = Pr[x_1(t_j) \leq \xi_1, \ldots, x_n(t_j) \leq \xi_n|\mathbf{c}(t_0), \ldots, \mathbf{c}(t_k)] \tag{9-1}$$

to describe the probability that the random variable $x_i(t)$ will be less than or equal to some value ξ_i simultaneously for $i = 1, 2, \ldots, n$, at the time t_j, given the measurements $\mathbf{c}(t_0)$, $\mathbf{c}(t_1), \ldots, \mathbf{c}(t_k)$, of $\mathbf{c}(t)$. The expression of the conditional expectation distribution function given in Eq. (9-1) is clearly also a statistical estimate of the random variable $\mathbf{x}(t_j)$, given information on $\mathbf{c}(t)$ in the form of $\mathbf{c}(t_0)$, $\mathbf{c}(t_1), \ldots, \mathbf{c}(t_k)$. We have represented this statistical estimate by $\hat{\mathbf{x}}(t_j|t_k)$ which is assumed to be a fixed vector whose elements are known whenever $\mathbf{c}(t_0)$, $\mathbf{c}(t_1), \ldots, \mathbf{c}(t_k)$ are known.

In general, the estimate $\hat{\mathbf{x}}(t_j|t_k)$ will be different from the actual random variable $\mathbf{x}(t_j)$ which is unknown. For the purpose of comparison, it is desirable to define an error function vector $\boldsymbol{\epsilon}$ to measure the error or loss for inaccurate estimates. The estimation error is defined as

$$\boldsymbol{\epsilon}(t_j) = \mathbf{x}(t_j) - \hat{\mathbf{x}}(t_j|t_k) \tag{9-2}$$

It is also natural to define an error criterion or loss function $L(\boldsymbol{\epsilon})$ such as the popular $L(\boldsymbol{\epsilon}) = \boldsymbol{\epsilon}'\boldsymbol{\epsilon}$ used in Wiener filtering. It is generally recognized that the loss function in the Kalman filtering problem can belong to a wider class than the scalar product of $\boldsymbol{\epsilon}$. However,this requires that the random processes $\mathbf{x}(t)$ and $\mathbf{w}(t)$ be gaussian. The gaussian distribution requirement is lifted if the optimal estimate is restricted to be a linear function of the observed random variables, and $L(\boldsymbol{\epsilon}) = \boldsymbol{\epsilon}'\boldsymbol{\epsilon}$.

The key to the Kalman filtering technique is the discovery that minimization of the average or expected value of $L(\boldsymbol{\epsilon})$ is equivalent to minimizing conditional expectation. This is shown by the following derivations.

Given a random variable vector $\mathbf{c}(t) = \mathbf{x}(t) + \mathbf{w}(t)$, where $\mathbf{x}(t)$ is the actual signal vector and $\mathbf{w}(t)$ is the noise vector, and the measured random vectors, $\mathbf{c}(t_0), \mathbf{c}(t_1), \ldots, \mathbf{c}(t_k)$. It is desired to find the best estimate of $\mathbf{x}(t)$ at $t = t_j (t_j \leq t_k$, or $t_j < t_k)$ from knowledge of the measured random variables.

Let the estimated $\mathbf{x}(t)$ at $t = t_j$ be designated by $\hat{\mathbf{x}}(t_j|t_k)$, and let us use the loss function $\boldsymbol{\epsilon}'\boldsymbol{\epsilon}$, where $\boldsymbol{\epsilon}$ is the error between the actual signal $\mathbf{x}(t)$ and the estimated, that is,

$$\boldsymbol{\epsilon}(t_j) = \mathbf{x}(t_j) - \hat{\mathbf{x}}(t_j|t_k) \tag{9-3}$$

The problem is to find the optimum estimate $\hat{\mathbf{x}}^o(t_j|t_k)$, given the measurement on $\mathbf{c}(t_0), \mathbf{c}(t_1), \ldots, \mathbf{c}(t_k)$, so that the expected value of the loss function $\boldsymbol{\epsilon}'(t_j)\boldsymbol{\epsilon}(t_j)$ is minimized, that is,

$$\begin{aligned} &E[\boldsymbol{\epsilon}'(t_j)\boldsymbol{\epsilon}(t_j)|\mathbf{c}(t_0), \mathbf{c}(t_1), \ldots, \mathbf{c}(t_k)] \\ &= E\{[\mathbf{x}(t_j) - \hat{\mathbf{x}}(t_j|t_k)]'[\mathbf{x}(t_j) - \hat{\mathbf{x}}(t_j|t_k)]|\mathbf{c}(t_0), \mathbf{c}(t_1), \ldots, \mathbf{c}(t_k)\} \\ &= \text{Minimum} \end{aligned} \tag{9-4}$$

Expanding the right-hand side of Eq. (9-4), we have

$$\begin{aligned} &E[\boldsymbol{\epsilon}'(t_j)\boldsymbol{\epsilon}(t_j)|\mathbf{c}(t_0), \mathbf{c}(t_1), \ldots, \mathbf{c}(t_k)] \\ &\quad = E[\mathbf{x}'(t_j)\hat{\mathbf{x}}(t_j)|\mathbf{c}(t_0), \mathbf{c}(t_1), \ldots, \mathbf{c}(t_k)] \\ &\qquad - E[\hat{\mathbf{x}}'(t_j|t_k)\mathbf{x}(t_j)|\mathbf{c}(t_0), \mathbf{c}(t_1), \ldots, \mathbf{c}(t_k)] \\ &\qquad - E[\mathbf{x}'(t_j)\hat{\mathbf{x}}(t_j|t_k)|\mathbf{c}(t_0), \mathbf{c}(t_1), \ldots, \mathbf{c}(t_k)] \\ &\qquad + E[\mathbf{x}'(t_j|t_k)\mathbf{x}(t_j|t_k)|\mathbf{c}(t_0), \mathbf{c}(t_1), \ldots, \mathbf{c}(t_k)] \end{aligned} \tag{9-5}$$

Since $\hat{\mathbf{x}}(t_j|t_k)$ is a known constant vector when $\mathbf{c}(t_0), \mathbf{c}(t_1), \ldots, \mathbf{c}(t_k)$ are given, we have

$$\begin{aligned} &E[\boldsymbol{\epsilon}'(t_j)\boldsymbol{\epsilon}(t_j)|\mathbf{c}(t_0), \mathbf{c}(t_1), \ldots, \mathbf{c}(t_k)] \\ &\quad = E[\mathbf{x}'(t_j)\mathbf{x}(t_j)|\mathbf{c}(t_0), \mathbf{c}(t_1), \ldots, \mathbf{c}(t_k)] \\ &\qquad - \hat{\mathbf{x}}'(t_j|t_k)E[\mathbf{x}(t_j)|\mathbf{c}(t_0), \mathbf{c}(t_1), \ldots, \mathbf{c}(t_k)] \\ &\qquad - E[\mathbf{x}'(t_j)|\mathbf{c}(t_0), \mathbf{c}(t_1), \ldots, \mathbf{c}(t_k)]\hat{\mathbf{x}}(t_j|t_k) + \hat{\mathbf{x}}'(t_j|t_k)\hat{\mathbf{x}}(t_j|t_k) \end{aligned} \tag{9-6}$$

Taking the partial derivative of the last equation with respect to $\mathbf{x}(t_j|t_k)$ and set it equal to $\mathbf{0}$, and since

$$\begin{bmatrix} \dfrac{\partial}{\partial \hat{x}_1(t_j|t_k)} \\ \dfrac{\partial}{\partial \hat{x}_2(t_j|t_k)} \\ \vdots \\ \dfrac{\partial}{\partial \hat{x}_n(t_j|t_k)} \end{bmatrix} \hat{\mathbf{x}}'(t_j|t_k)E[\mathbf{x}(t_j)|\mathbf{c}(t_0), \mathbf{c}(t_1), \ldots, \mathbf{c}(t_k)]$$

$$= E[\mathbf{x}(t_j)|\mathbf{c}(t_0), \mathbf{c}(t_1), \ldots, \mathbf{c}(t_k)] \tag{9-7}$$

$$\frac{\partial}{\partial \hat{\mathbf{x}}(t_j|t_k)} E[\mathbf{x}'(t_j)|\mathbf{c}(t_0), \mathbf{c}(t_1), \ldots, \mathbf{c}(t_k)]\hat{\mathbf{x}}(t_j|t_k)$$

$$= \frac{\partial}{\partial \hat{\mathbf{x}}(t_j|t_k)} \left[E[x_1(t_j)|\mathbf{c}(t_0), \ldots, \mathbf{c}(t_k)] \; E[x_2(t_j)|\mathbf{c}(t_0), \ldots, \mathbf{c}(t_k)] \ldots \right] \times \begin{bmatrix} \hat{x}_1(t_j|t_k) \\ \hat{x}_2(t_j|t_k) \\ \vdots \\ \hat{x}_n(t_j|t_k) \end{bmatrix}$$

$$= \frac{\partial}{\partial \hat{\mathbf{x}}(t_j|t_k)} \Big[E[x_1(t_j)|\mathbf{c}(t_0), \ldots, \mathbf{c}(t_k)]\hat{x}_1(t_j|t_k) + \cdots + E[x_n(t_j)|\mathbf{c}(t_0), \ldots, \mathbf{c}(t_k)]\hat{x}_n(t_j|t_k) \Big]$$

$$= E[\mathbf{x}(t_j)|\mathbf{c}(t_0), \mathbf{c}(t_1), \ldots, \mathbf{c}(t_k)] \tag{9-8}$$

and

$$\begin{bmatrix} \dfrac{\partial}{\partial \hat{x}_1(t_j|t_k)} \\ \vdots \\ \dfrac{\partial}{\partial \hat{x}_n(t_j|t_k)} \end{bmatrix} [\hat{\mathbf{x}}'(t_j|t_k)\hat{\mathbf{x}}(t_j|t_k)] = \begin{bmatrix} \dfrac{\partial}{\partial \hat{x}_1(t_j|t_k)} \\ \vdots \\ \dfrac{\partial}{\partial \hat{x}_n(t_j|t_k)} \end{bmatrix} \sum_{i=1}^{n} \hat{x}_i^2(t_j|t_k)$$

$$= 2 \begin{bmatrix} \hat{x}_1(t_j|t_k) \\ \hat{x}_2(t_j|t_k) \\ \vdots \\ \hat{x}_n(t_j|t_k) \end{bmatrix} = 2\hat{\mathbf{x}}(t_j|t_k) \tag{9-9}$$

we have

$$-2E[\mathbf{x}(t_j)|\mathbf{c}(t_0), \mathbf{c}(t_1), \ldots, \mathbf{c}(t_k)] + 2\hat{\mathbf{x}}^o(t_j|t_k) = 0 \tag{9-10}$$

Or,

$$\hat{\mathbf{x}}^o(t_j|t_k) = E[\mathbf{x}(t_j)|\mathbf{c}(t_0), \mathbf{c}(t_1), \ldots, \mathbf{c}(t_k)] \tag{9-11}$$

Therefore, we have shown that the optimum estimate of $\mathbf{x}(t)$ in the sense of the minimum mean-square error is the conditional expectation of $\mathbf{x}(t)$.

9.4 OPTIMUM ESTIMATION AND ORTHOGONAL PROTECTION

The optimal estimation problem can be interpreted geometrically by the vector space concepts.

Consider the set of random vectors, $\mathbf{x}(t_0), \mathbf{x}(t_1), \ldots, \mathbf{x}(t_j)$, where $\mathbf{x}(t)$ is an n-vector. The set of all linear combinations of these random vectors of the form

$$\sum_{i=0}^{j} A_i \mathbf{x}(t_i)$$

where A_i is an $n \times n$ matrix with constant coefficients, forms a vector space which is denoted by $\boldsymbol{\chi}(t_j)$. Similarly, we call the vector space generated by the set of vectors

$$\sum_{i=0}^{k} B_i \mathbf{c}(t_i)$$

$\mathbf{C}(t_k)$. The dimension of the matrix B_i with constant elements is $n \times p$. If $k \leq j$, the vector space $\mathbf{C}(t_k)$ is a subspace of the vector space $\boldsymbol{\chi}(t_j)$.

Now let us consider that the random vector $\mathbf{x}(t_j)$ is composed of two components, that is,

$$\mathbf{x}(t_j) = \bar{\mathbf{x}}(t_j|t_k) + \tilde{\mathbf{x}}(t_j|t_k) \tag{9-12}$$

where $\bar{\mathbf{x}}(t_j|t_k)$ is the orthogonal projection of $\mathbf{x}(t_j)$ on the subspace $\mathbf{C}(t_k)$, and thus is a member of $\mathbf{C}(t_k)$. The vector $\tilde{\mathbf{x}}(t_j|t_k)$ is the component which is orthogonal to $\mathbf{C}(t_k)$, that is, orthogonal to every vector in $\mathbf{C}(t_k)$. We shall now show that the orthogonal projection $\bar{\mathbf{x}}(t_j|t_k)$ is that vector in $\mathbf{C}(t_k)$ which minimizes that loss function defined in Eq. (9-4).

Let $\mathbf{y}$ be any $n \times 1$ vector in the vector space $\mathbf{C}(t_k)$. We form the conditional expectation

$$E\{[\mathbf{x}(t_j) - \mathbf{y}]'[\mathbf{x}(t_j) - \mathbf{y}]|\mathbf{c}(t_0), \mathbf{c}(t_1), \ldots, \mathbf{c}(t_k)\} \tag{9-13}$$

Substituting Eq. (9-12) into Eq. (9-13) yields

$$\begin{aligned}
&E\{[\mathbf{x}(t_j) - \mathbf{y}]'[\mathbf{x}(t_j) - \mathbf{y}]|\mathbf{c}(t_0), \mathbf{c}(t_1), \ldots, \mathbf{c}(t_k)\} \\
&\quad = E\{[\tilde{\mathbf{x}}(t_j|t_k) + \bar{\mathbf{x}}(t_j|t_k) - \mathbf{y}]'[\tilde{\mathbf{x}}(t_j|t_k) + \bar{\mathbf{x}}(t_j|t_k) - \mathbf{y}]|\mathbf{c}(t_0), \ldots, \mathbf{c}(t_k)\} \\
&\quad = E\{\tilde{\mathbf{x}}'(t_j|t_k)\tilde{\mathbf{x}}(t_j|t_k) + \tilde{\mathbf{x}}'(t_j|t_k)[\bar{\mathbf{x}}(t_j|t_k) - \mathbf{y}] + [\bar{\mathbf{x}}(t_j|t_k) - \mathbf{y}]'\tilde{\mathbf{x}}(t_j) \\
&\qquad + [\bar{\mathbf{x}}(t_j|t_k) - \mathbf{y}]'[\bar{\mathbf{x}}(t_j|t_k) - \mathbf{y}]|\mathbf{c}(t_0), \ldots, \mathbf{c}(t_k)\}
\end{aligned} \tag{9-14}$$

Since $\bar{\mathbf{x}}(t_j|t_k) - \mathbf{y}$ is a vector which is in $\mathbf{C}(t_k)$, and $\tilde{\mathbf{x}}(t_j|t_k)$ is orthogonal to $\mathbf{C}(t_k)$ and, therefore, to $\hat{\mathbf{x}}(t_j|t_k) - \mathbf{y}$, Eq. (9-14) is written

$$\begin{aligned}
&E\{[\mathbf{x}(t_j) - \mathbf{y}]'[\mathbf{x}(t_j) - \mathbf{y}]|\mathbf{c}(t_0), \ldots, \mathbf{c}(t_k)\} \\
&\quad = E\{\tilde{\mathbf{x}}'(t_j|t_k)\tilde{\mathbf{x}}(t_j|t_k) + [\bar{\mathbf{x}}(t_j|t_k) - \mathbf{y}]'[\bar{\mathbf{x}}(t_j|t_k) - \mathbf{y}]|\mathbf{c}(t_0), \ldots, \mathbf{c}(t_k)\} \\
&\quad = E\{\tilde{\mathbf{x}}'(t_j|t_k)\tilde{\mathbf{x}}(t_j|t_k)|\mathbf{c}(t_0), \ldots, \mathbf{c}(t_k)\} \\
&\qquad + E\{[\bar{\mathbf{x}}(t_j|t_k) - \mathbf{y}]'[\bar{\mathbf{x}}(t_j|t_k) - \mathbf{y}]|\mathbf{c}(t_0), \ldots, \mathbf{c}(t_k)\}
\end{aligned} \tag{9-15}$$

It is apparent that to minimize the quadratic loss of Eq. (9-15), we must have $\mathbf{y}$ equal to $\bar{\mathbf{x}}(t_j|t_k) = \hat{\mathbf{x}}^o(t_j|t_k)$ so that the minimum loss function is

$$E\{[\mathbf{x}(t_j) - \hat{\mathbf{x}}^o(t_j|t_k)]'[\mathbf{x}(t_j) - \hat{\mathbf{x}}^o(t_j|t_k)]|\mathbf{c}(t_0), \ldots, \mathbf{c}(t_k)\}$$
$$= E\{\tilde{\mathbf{x}}^{o\prime}(t_j|t_k)\tilde{\mathbf{x}}^o(t_j|t_k)|\mathbf{c}(t_0), \ldots, \mathbf{c}(t_k)\} \qquad (9\text{-}16)$$

This shows that the optimal estimate in the sense of minimum quadratic loss is the orthogonal projection of $\mathbf{x}(t_j)$ on the subspace spanned by the vectors $\mathbf{c}(t_0)$, $\mathbf{c}(t_1), \ldots, \mathbf{c}(t_k)$. The component $\tilde{\mathbf{x}}(t_j|t_k)$ also represents the "estimation error."

Based on the discussions carried out in these last two sections, we can now write

$$\begin{aligned}\hat{\mathbf{x}}^o(t_j|t_k) &= \text{Optimal estimate of } \mathbf{x}(t_j) \text{ given } \mathbf{c}(t_0), \mathbf{c}(t_1), \ldots, \mathbf{c}(t_k)\\ &= E[\mathbf{x}(t_j)|\mathbf{c}(t_0), \mathbf{c}(t_1), \ldots, \mathbf{c}(t_k)]\\ &= \text{Conditional expectation of } \mathbf{x}(t_j) \text{ given } \mathbf{c}(t_0), \ldots, \mathbf{c}(t_k)\\ &= \text{Orthogonal projection } \bar{\mathbf{x}}(t_j|t_k), \text{ of } \mathbf{x}(t_j) \text{ on } \mathbf{C}(t_k) \qquad (9\text{-}17)\end{aligned}$$

9.5 THE DISCRETE-TIME KALMAN FILTER

In this section we shall derive the Kalman filter equations for a discrete-data system.

Consider the linear discrete-data system which is described by the following set of dynamic equations:

$$\mathbf{x}(k+1) = \mathbf{\Phi}(k+1, k)\mathbf{x}(k) + \mathbf{\Theta}(k+1, k)\mathbf{m}(k) \qquad (9\text{-}18)$$
$$\mathbf{c}(k) = D(k)\mathbf{x}(k) + \mathbf{w}(k) \qquad (9\text{-}19)$$

where

$$\begin{aligned}\mathbf{x}(k) &= n \times 1 \text{ state vector}\\ \mathbf{c}(k) &= p \times 1 \text{ output vector}\\ \mathbf{m}(k) &= r \times 1 \text{ input vector}\\ \mathbf{\Phi}(k+1, k) &= n \times n \text{ coefficient matrix}\\ \mathbf{\Theta}(k+1, k) &= n \times r \text{ driving matrix}\\ D(k) &= p \times r \text{ measurement matrix}\\ \mathbf{w}(k) &= p \times 1 \text{ noise vector}\end{aligned}$$

Notice that the dynamic equations are written for general time-varying systems. For the time-invariant case, $\mathbf{\Phi}(k+1, k)$ and $\mathbf{\Theta}(k+1, k)$ become $\mathbf{\Phi}(1)$ and $\mathbf{\Theta}(1)$, respectively, and $D(k)$ also becomes a constant matrix. It is assumed that all the rows of $D(k)$ are independent of each other. For simplicity, the sampling time T is dropped from all expressions.

It is assumed that $\mathbf{m}(k)$ and $\mathbf{w}(k)$ are independent random vectors with zero mean, and are uncorrelated with each other, that is,

$$\begin{aligned} E[\mathbf{m}(k)] &= 0 \\ E[\mathbf{w}(k)] &= 0 \\ E[\mathbf{m}'(k)\mathbf{w}(k)] &= 0 \end{aligned} \tag{9-20}$$

Furthermore, the components of $\mathbf{m}(k)$ and $\mathbf{w}(k)$ are constant during the sampling periods, that is,

$$m_i(k) = m_{ik} = \text{constant} \quad kT \leq t < (k+1)T \tag{9-21}$$

$i = 1, 2, \ldots, r$, and

$$w_j(k) = w_{jk} = \text{constant} \quad kT \leq t < (k+1)T \tag{9-22}$$

$j = 1, 2, \ldots, p$.

The covariance matrices of $\mathbf{m}(k)$ and $\mathbf{w}(k)$ are defined as

$$E[\mathbf{m}(k)\mathbf{m}'(k)] = \text{cov}\,[\mathbf{m}(k)] = Q(k) \tag{9-23}$$

and

$$E[\mathbf{w}(k)\mathbf{w}'(k)] = \text{cov}\,[\mathbf{w}(k)] = R(k) \tag{9-24}$$

respectively, where $Q(k)$ and $R(k)$ are symmetric semi-positive definite matrices.

The general Kalman estimation problem is stated as:

Given the linear system described by Eqs. (9-18), (9-19), and the measurements on $\mathbf{c}(0), \mathbf{c}(1), \ldots, \mathbf{c}(k)$, *find the optimal estimate of the state vector* $\mathbf{x}(j)$, $\hat{\mathbf{x}}^o(j|k)$, *where*

$$\begin{aligned} \hat{\mathbf{x}}^o(j|k) &= E[\mathbf{x}(j)|\mathbf{c}(0), \mathbf{c}(1), \ldots, \mathbf{c}(k)] \\ &= \textit{orthogonal projection } \bar{\mathbf{x}}(j|k) \end{aligned} \tag{9-25}$$

which minimizes the quadratic loss function of Eq. (9-4). The error vector is defined as

$$\boldsymbol{\epsilon}(j) = \mathbf{x}(j) - \hat{\mathbf{x}}(j|k) \tag{9-26}$$

$(j \leq k \text{ or } j > k)$.

The solution to the Kalman estimation problem can be effected using the relationships between optimal estimate, conditional expectation, and orthogonal projection.

The optimal estimate of $\mathbf{x}(t)$ at $t = jT$, given $\mathbf{c}(t)$ for $t = 0, T, \ldots, kT$, is written as

$$\hat{\mathbf{x}}^o(j|k) = \bar{\mathbf{x}}(j|k) = E[\mathbf{x}(j)|\mathbf{c}(0), \mathbf{c}(1), \ldots, \mathbf{c}(k)] \tag{9-27}$$

Let us assume that the vector space spanned by the vectors $\mathbf{c}(0), \mathbf{c}(1), \ldots, \mathbf{c}(k-1)$ is designated by $\mathbf{C}(k-1)$. Then, Eq. (9-27) can be written as

$$\mathbf{x}^o(j|k) = E[\mathbf{x}(j)|\mathbf{C}(k)]$$
$$= E[\mathbf{x}(j)|\mathbf{C}(k-1), \mathbf{c}(k)] \tag{9-28}$$

Let the vector $\mathbf{c}(k)$ be composed of two components, i.e.,

$$\mathbf{c}(k) = \bar{\mathbf{c}}(k|k-1) + \tilde{\mathbf{c}}(k|k-1) \tag{9-29}$$

where $\bar{\mathbf{c}}(k|k-1)$ is the orthogonal projection of $\mathbf{c}(k)$ on the vector space $\mathbf{C}(k-1)$, and $\tilde{\mathbf{c}}(k-1)$ is the component which is orthogonal to $\mathbf{C}(k-1)$. Equation (9-28) is written

$$\hat{\mathbf{x}}^o(j|k) = E[\mathbf{x}(j)|\mathbf{C}(k-1), \bar{\mathbf{c}}(k|k-1) + \tilde{\mathbf{c}}(k|k-1)] \tag{9-30}$$

However, since $\bar{\mathbf{c}}(k|k-1)$ is in $\mathbf{C}(k-1)$ and, by definition, it can be characterized by a linear combination of the vectors $\mathbf{c}(0), \mathbf{c}(1), \ldots, \mathbf{c}(k-1)$, Eq. (9-30) becomes

$$\hat{\mathbf{x}}^o(j|k) = E[\mathbf{x}(j)|\mathbf{C}(k-1), \tilde{\mathbf{c}}(k|k-1)] \tag{9-31}$$

Since $\mathbf{c}(k|k-1)$ is orthogonal to every vector in $\mathbf{C}(k-1)$, we have from Eq. (9-31),

$$\hat{\mathbf{x}}^o(j|k) = E[\mathbf{x}(j)|\mathbf{C}(k-1)] + E[\mathbf{x}(j)|\tilde{\mathbf{c}}(k|k-1)] \tag{9-32}$$

or

$$\hat{\mathbf{x}}^o(j|k) = \hat{\mathbf{x}}^o(j|k-1) + E[\mathbf{x}(j)|\tilde{\mathbf{c}}(k|k-1)] \tag{9-33}$$

Let us consider the case when $j = k + 1$. This will represent the problem of estimating $\mathbf{x}(k+1)$ from knowledge of the past input $\mathbf{c}(0), \mathbf{c}(1), \ldots, \mathbf{c}(k)$. Equation (9-33) becomes

$$\hat{\mathbf{x}}^o(k+1|k) = \hat{\mathbf{x}}^o(k+1|k-1) + E[\mathbf{x}(k+1)|\tilde{\mathbf{c}}(k|k-1)] \tag{9-34}$$

From Eq. (9-18), we write

$$\hat{\mathbf{x}}^o(k+1|k-1) = E[\mathbf{x}(k+1)|k-1]$$
$$= E\{[\mathbf{\Phi}(k+1, k)\mathbf{x}(k) + \mathbf{\Theta}(k+1, k)\mathbf{m}(k)]|\mathbf{C}(k-1)\}$$
$$= \mathbf{\Phi}(k+1, k)\hat{\mathbf{x}}^o(k|k-1) + \mathbf{\Theta}(k+1, k)\hat{\mathbf{m}}^o(k|k-1) \tag{9-35}$$

Since $\mathbf{m}(k)$ is uncorrelated for successive time instants, $\hat{\mathbf{m}}^o(k|k-1) = \mathbf{0}$,† and thus Eq. (9-35) becomes

$$\mathbf{x}^o(k+1|k-1) = \mathbf{\Phi}(k+1, k)\hat{\mathbf{x}}^o(k|k-1) \tag{9-36}$$

Similarly, from Eq. (9-19), we write

$$\bar{\mathbf{c}}(k|k-1) = D(k)\bar{\mathbf{x}}(k|k-1) + \bar{\mathbf{w}}(k|k-1) \tag{9-37}$$

† $\mathbf{c}(k) = M(k)\mathbf{x}(k)$, and $\mathbf{x}(k)$ is a function of $\mathbf{m}(0), \mathbf{m}(1), \ldots, \mathbf{m}(k-1)$. Therefore, $\mathbf{c}(k)$ is a function of $\mathbf{m}(0), \mathbf{m}(1), \ldots, \mathbf{m}(k-1)$. Or $\mathbf{c}(0), \mathbf{c}(1), \ldots, \mathbf{c}(k-1)$ are functions only of $\mathbf{m}(k-2), \mathbf{m}(k-3), \ldots$, for $k \geq 0$. Now since $\mathbf{m}(k)$ is independent of $\mathbf{m}(k-2), \mathbf{m}(k-3), \ldots$, we say that $\mathbf{m}(k)$ is independent of $\mathbf{c}(0), \mathbf{c}(1), \ldots, \mathbf{c}(k-1)$.

Since $\mathbf{w}(k)$ is uncorrelated for successive time instants, $\bar{\mathbf{w}}(k|k-1) = \mathbf{0}$. Thus Eq. (9-37) becomes

$$\bar{\mathbf{c}}(k|k-1) = D(k)\bar{\mathbf{x}}(k|k-1) \tag{9-38}$$

We have from Eq. (9-29),

$$\tilde{\mathbf{c}}(k|k-1) = \mathbf{c}(k) - \bar{\mathbf{c}}(k|k-1) \tag{9-39}$$

Substitution of Eq. (9-38) into Eq. (9-39) yields

$$\begin{aligned}\tilde{\mathbf{c}}(k|k-1) &= \mathbf{c}(k) - D(k)\bar{\mathbf{x}}(k|k-1)\\ &= \mathbf{c}(k) - D(k)\hat{\mathbf{x}}^o(k|k-1)\end{aligned} \tag{9-40}$$

Now let us return to the optimal estimate of Eq. (9-34). The last term in Eq. (9-34) can be written as

$$E[\mathbf{x}(k+1)|\tilde{\mathbf{c}}(k|k-1)] = \Delta(k)\tilde{\mathbf{c}}(k|k-1) \tag{9-41}$$

where $\Delta(k)$ is an $n \times p$ matrix which represents the linear dependence of $E[\mathbf{x}(k+1)|\tilde{\mathbf{c}}(k|k-1)]$ on $\tilde{\mathbf{c}}(k|k-1)$. For a gaussian random process it can be shown that

$$\Delta(k) = \{\text{cov}\,[\mathbf{x}(k+1), \tilde{\mathbf{c}}(k|k-1)]\}\{\text{cov}\,[\tilde{\mathbf{c}}(k|k-1)]\}^{-1} \tag{9-42}$$

Thus, Eq. (9-34) is written as

$$\hat{\mathbf{x}}^o(k+1|k) = \hat{\mathbf{x}}^o(k+1|k-1) + \Delta(k)\tilde{\mathbf{c}}(k|k-1) \tag{9-43}$$

Substitution of Eqs. (9-35) and (9-40) into Eq. (9-43) gives

$$\begin{aligned}\hat{\mathbf{x}}^o(k+1|k) &= \Phi(k+1, k)\hat{\mathbf{x}}^o(k|k-1) + \Delta(k)[\mathbf{c}(k) - D(k)\hat{\mathbf{x}}^o(k|k-1)]\\ &= [\Phi(k+1, k) - \Delta(k)D(k)]\hat{\mathbf{x}}^o(k|k-1) + \Delta(k)\mathbf{c}(k)\end{aligned} \tag{9-44}$$

Now letting

$$\Lambda(k+1, k) = \Phi(k+1, k) - \Lambda(k)D(k) \tag{9-45}$$

Eq. (9-44) becomes

$$\hat{\mathbf{x}}^o(k+1|k) = \Lambda(k+1, k)\hat{\mathbf{x}}^o(k|k-1) + \Delta(k)\mathbf{c}(k) \tag{9-46}$$

Once the matrix $\Delta(k)$ is determined, Eq. (9-46) represents the expression for the computation of the optimal estimation of $\mathbf{x}(k+1)$ given $\mathbf{c}(0), \mathbf{c}(1), \ldots, \mathbf{c}(k)$. Of particular significance is that Eq. (9-46) has the same form as that of the system's state equations of Eq. (9-18). Therefore, if the initial optimal estimate $\mathbf{x}^o(0|-1) = \mathbf{x}^o(0)$ and the initial output $\mathbf{c}(0)$ are known, Eq. (9-46) provides a recursive relationship for the computation of the optimal estimate of $\mathbf{x}(k)$ for all $k \geq 1$.

Now we must find an explicit expression for $\Delta(k)$. The vector $\mathbf{x}(k+1)$ is represented as

$$\mathbf{x}(k+1) = \bar{\mathbf{x}}[k+1|\tilde{\mathbf{c}}(k|k-1)] + \mathbf{x}[k+1|\tilde{\mathbf{c}}(k|k-1)] \tag{9-47}$$

where $\bar{\mathbf{x}}[k+1|\mathbf{c}(k|k-1)]$ is the orthogonal projection of $\mathbf{x}(k+1)$ on $\tilde{\mathbf{c}}(k|k-1)$, and $\tilde{\mathbf{x}}[k+1|\tilde{\mathbf{c}}(k|k-1)]$ is its normal component.

In view of Eq. (9-41), we can rewrite Eq. (9-47) as

$$\begin{aligned}\tilde{\mathbf{x}}[k+1|\tilde{\mathbf{c}}(k|k-1)] &= \mathbf{x}(k+1) - \bar{\mathbf{x}}[k+1|\tilde{\mathbf{c}}(k|k-1)] \\ &= \mathbf{x}(k+1) - E[\mathbf{x}(k+1)|\tilde{\mathbf{c}}(k|k-1)] \\ &= \mathbf{x}(k+1) - \Delta(k)\tilde{\mathbf{c}}(k|k-1) \qquad (9\text{-}48)\end{aligned}$$

Since $\tilde{\mathbf{x}}[k+1|\tilde{\mathbf{c}}(k|k-1)]$ is orthogonal to $\tilde{\mathbf{c}}(k|k-1)$, the following equation results by first post-multiplying both sides of Eq. (9-48) by $\tilde{\mathbf{c}}'(k|k-1)$ and then taking the expected value:

$$\begin{aligned}&E\{[\tilde{\mathbf{x}}(k+1|\tilde{\mathbf{c}}(k|k-1)]\tilde{\mathbf{c}}'(k|k-1)\} = 0 \\ &= E[\mathbf{x}(k+1)\tilde{\mathbf{c}}'(k|k-1)] - \Delta(k)E[\tilde{\mathbf{c}}(k|k-1)\tilde{\mathbf{c}}'(k|k-1)] \qquad (9\text{-}49)\end{aligned}$$

Next, the vector $\mathbf{x}(k+1)$ is written as

$$\mathbf{x}(k+1) = \bar{\mathbf{x}}(k+1|k-1) + \tilde{\mathbf{x}}(k+1|k-1) \qquad (9\text{-}50)$$

where $\bar{\mathbf{x}}(k+1|k-1)$ denotes the orthogonal projection of $\mathbf{x}(k+1)$ on the subspace $\mathbf{C}(k-1)$, and $\tilde{\mathbf{x}}(k+1|k-1)$ is orthogonal to $\mathbf{C}(k-1)$.

Since $\bar{\mathbf{x}}(k+1|k-1)$ is in $\mathbf{C}(k-1)$, and $\tilde{\mathbf{c}}(k|k-1)$ is orthogonal to $\mathbf{C}(k-1)$, $\bar{\mathbf{x}}(k+1|k-1)$ is orthogonal to $\tilde{\mathbf{c}}(k|k-1)$. Therefore,

$$E[\bar{\mathbf{x}}(k+1|k-1)\tilde{\mathbf{c}}'(k|k-1)] = 0 \qquad (9\text{-}51)$$

Substituting Eqs. (9-50) and (9-51) into Eq. (9-49), we have

$$E[\tilde{\mathbf{x}}(k+1|k-1)\tilde{\mathbf{c}}'(k|k-1)] - \Delta(k)E[\tilde{\mathbf{c}}(k|k-1)\tilde{\mathbf{c}}'(k|k-1)] = 0 \qquad (9\text{-}52)$$

From Eqs. (9-18) and (9-19), and using Eq. (9-36), we get

$$\begin{aligned}\tilde{\mathbf{x}}(k+1|k-1) &= \Phi(k+1,k)\tilde{\mathbf{x}}(k|k-1) + \Theta(k+1,k)\tilde{\mathbf{m}}(k|k-1) \\ &= \Phi(k+1,k)\tilde{\mathbf{x}}(k|k-1) + \Theta(k+1,k)\mathbf{m}(k) \qquad (9\text{-}53)\end{aligned}$$

since $\bar{\mathbf{m}}(k|k-1) = \hat{\mathbf{m}}^o(k|k-1) = \mathbf{0}$, and $\tilde{\mathbf{m}}(k|k-1) = \mathbf{m}(k)$.

Also, using Eq. (9-40),

$$\begin{aligned}\tilde{\mathbf{c}}(k|k-1) &= -D(k)\bar{\mathbf{x}}(k|k-1) + \mathbf{c}(k) \\ &= -D(k)\bar{\mathbf{x}}(k|k-1) + D(k)\mathbf{x}(k) + \mathbf{w}(k) \qquad (9\text{-}54)\end{aligned}$$

Therefore,

$$\tilde{\mathbf{c}}(k|k-1) = D(k)\tilde{\mathbf{x}}(k|k-1) + \mathbf{w}(k) \qquad (9\text{-}55)$$

Substituting Eqs. (9-53) and (9-55) into Eq. (9-52), we have

$$\begin{aligned}&E\{[\Phi(k+1,k)\tilde{\mathbf{x}}(k|k-1) + \Theta(k+1,k)\mathbf{m}(k)][D(k)\tilde{\mathbf{x}}(k|k-1) \\ &\quad + \mathbf{w}(k)]'\} - \Delta(k)E\{[D(k)\tilde{\mathbf{x}}(k|k-1) \\ &\quad + \mathbf{w}(k)][D(k)\tilde{\mathbf{x}}(k|k-1) + \mathbf{w}(k)]'\} = 0 \qquad (9\text{-}56)\end{aligned}$$

After simplification, and using the fact that $\mathbf{x}(k)$, $\mathbf{m}(k)$, and $\mathbf{w}(k)$ are all independent of each other, Eq. (9-56) becomes

$$\begin{aligned}\Phi(k+1, k)E[\tilde{\mathbf{x}}(k|k-1)\tilde{\mathbf{x}}'(k|k-1)]D'(k) \\ - \Delta(k)\{D(k)E[\tilde{\mathbf{x}}(k|k-1)\tilde{\mathbf{x}}'(k|k-1)]D'(k) \\ + E[\mathbf{w}(k)\mathbf{w}'(k)]\} = 0\end{aligned} \tag{9-57}$$

In order to further simplify the notation, let

$$P(k) = E[\tilde{\mathbf{x}}(k|k-1)\tilde{\mathbf{x}}'(k|k-1)] \tag{9-58}$$

and using Eq. (9-24), Eq. (9-57) is written

$$\Phi(k+1, k)P(k)D'(k) - \Delta(k)[D(k)P(k)D'(k) + R(k)] = 0 \tag{9-59}$$

where $R(k)$ is the covariance of $\mathbf{w}(k)$.

Now from Eq. (9-59) $\Delta(k)$ can be obtained as

$$\Delta(k) = \Phi(k+1, k)P(k)D'(k)[D(k)R(k)D'(k) + R(k)]^{-1} \tag{9-60}$$

where the matrix inverse of $D(k)P(k)D'(k)$ exists whenever $P(k)$ is positive definite (since it has been assumed that all the rows of $D(k)$ are linearly independent).

The expression of $\Delta(k)$ in Eq. (9-60) contains all known functions in $\Phi(k+1, k)$, $D(k)$, and $R(k)$. The matrix $P(k)$ can be determined from the following derivation:

From Eq. (9-58), we have

$$P(k+1) = E[\tilde{\mathbf{x}}(k+1|k)\tilde{\mathbf{x}}'(k+1|k)] \tag{9-61}$$

where $\tilde{\mathbf{x}}(k+1|k)$ can be written as

$$\tilde{\mathbf{x}}(k+1|k) = \mathbf{x}(k+1) - \bar{\mathbf{x}}(k+1|k) \tag{9-62}$$

Since $\bar{\mathbf{x}}(k+1|k)$ is also the best estimate, $\hat{\mathbf{x}}^o(k+1|k)$, from Eq. (9-46), we can write Eq. (9-62) as

$$\tilde{\mathbf{x}}(k+1|k) = \mathbf{x}(k+1) - \Lambda(k+1, k)\hat{\mathbf{x}}^o(k|k-1) - \Delta(k)\mathbf{c}(k) \tag{9-63}$$

Substituting Eqs. (9-18) and (9-19) into Eq. (9-63), and simplifying, we get

$$\begin{aligned}\tilde{\mathbf{x}}(k+1|k) = \Phi(k+1, k)\mathbf{x}(k) + \Theta(k+1, k)\mathbf{m}(k) \\ - \Lambda(k+1, k)\bar{\mathbf{x}}(k|k-1) \\ - \Delta(k)D(k)\mathbf{x}(k) - \Delta(k)\mathbf{w}(k)\end{aligned} \tag{9-64}$$

Using Eq. (9-45), the last equation becomes

$$\begin{aligned}\tilde{\mathbf{x}}(k+1|k) &= \Lambda(k+1, k)[\mathbf{x}(k) - \bar{\mathbf{x}}(k|k-1)] \\ &\quad + \Theta(k+1, k)\mathbf{m}(k) - \Delta(k)\mathbf{w}(k) \\ &= \Lambda(k+1, k)\tilde{\mathbf{x}}(k|k-1) + \Theta(k+1, k)\mathbf{m}(k) - \Delta(k)\mathbf{w}(k)\end{aligned} \tag{9-65}$$

Now substitution of Eq. (9-65) into Eq. (9-61) yields

$$P(k+1) = E[\tilde{\mathbf{x}}(k+1|k)\tilde{\mathbf{x}}'(k+1|k)] = \Lambda(k+1, k)P(k)\Lambda'(k+1, k) + \Theta(k+1, k)Q(k)\Theta'(k+1, k) + \Delta(k)R(k)\Delta'(k) \tag{9-66}$$

where $Q(k)$ and $R(k)$ are the covariances of $\mathbf{m}(k)$ and $\mathbf{w}(k)$, respectively.

The functions $\Lambda(k+1, k)$ and $\Delta(k)$ can be eliminated from Eq. (9-66) by using Eqs. (9-45) and (9-60). Then, Eq. (9-66) becomes

$$\begin{aligned} P(k+1) &= [\Phi(k+1, k) - \Delta(k)D(k)]P(k)[\Phi(k+1, k) - \Delta(k)D(k)]' \\ &\quad + \Theta(k+1, k)Q(k)\Theta'(k+1, k) + \Delta(k)R(k)\Delta'(k) \\ &= \Phi(k+1, k)P(k)\Phi'(k+1, k) - \Delta(k)D(k)P(k)\Phi'(k+1, k) \\ &\quad - \Phi(k+1, k)P(k)D'(k)\Delta'(k) + \Delta(k)D(k)P(k)D'(k)\Delta'(k) \\ &\quad + \Theta(k+1, k)Q(k)\Theta'(k+1, k) + \Delta(k)R(k)\Delta'(k) \\ &= \Phi(k+1, k)P(k)\Phi'(k+1, k) - \Delta(k)D(k)P(k)\Phi'(k+1, k) \\ &\quad + \Theta(k+1, k)Q(k)\Theta'(k+1, k) \end{aligned} \tag{9-67}$$

Using Eq. (9-59), the last equation is simplified to

$$\begin{aligned} P(k+1) &= \Phi(k+1, k)\{P(k) - P(k)D'(k)[D(k)P(k)D'(k) + R(k)]^{-1} \\ &\quad \times D(k)P(k)\}\Phi'(k+1, k) + \Theta(k+1, k)Q(k)\Theta'(k+1, k) \end{aligned} \tag{9-68}$$

This equation is a nonlinear difference equation (the Riccati's equation) whose general closed-form solution is difficult to obtain. The equation would be linear only if $D(k)$ is invertible and $R(k) = \mathbf{0}$, but then, the problem is trivial since all components of $\mathbf{x}(k)$ are observable, and then,

$$P(k+1) = \Theta(k+1, k)Q(k)\Theta'(k+1, k) \tag{9-69}$$

In arriving at Eq. (9-68), we have used Eqs. (9-23) and (9-58), and the fact that $\mathbf{x}(k)$, $\mathbf{m}(k)$, and $\mathbf{w}(k)$ are independent of each other. It is apparent that Eq. (9-68) represents a recursive relation from which $P(k)$ can be computed once the initial value $P(0)$ is known. With $k = 0$, Eq. (9-58) gives

$$P(0) = E[\tilde{\mathbf{x}}(0|-1)\tilde{\mathbf{x}}'(0|-1)] \tag{9-70}$$

Let the initial state vector $\tilde{\mathbf{x}}(0)$ be written as

$$\mathbf{x}(0) = \bar{\mathbf{x}}(0|-1) + \tilde{\mathbf{x}}(0|-1) \tag{9-71}$$

Since the observation of the process is assumed to begin at $t = 0$, we have no way of estimating $\mathbf{x}(0)$ from $\mathbf{c}(-1)$, and therefore, $\bar{\mathbf{x}}(0|-1) = 0$. Then,

$$\tilde{\mathbf{x}}(0|-1) = \mathbf{x}(0) \tag{9-72}$$

and Eq. (9-70) gives

$$P(0) = E[\mathbf{x}(0)\mathbf{x}'(0)] = \text{cov}\,[\mathbf{x}(0)] \tag{9-73}$$

We have now solved the Kalman filtering problem posted in these sections. The key equations and the procedure are summarized as follows:

The optimal estimate of $\mathbf{x}(k+1)$, $\hat{\mathbf{x}}^o(k+1|k)$, given $\mathbf{c}(0), \mathbf{c}(1), \ldots, \mathbf{c}(k)$, is given by the equation [Eq. (9-44)]

$$\hat{\mathbf{x}}^o(k+1|k) = [\Phi(k+1,k) - \Delta(k)D(k)]\hat{\mathbf{x}}^o(k|k-1) + \Delta(k)\mathbf{c}(k) \tag{9-74}$$

Furthermore, from Eq. (9-60),

$$\Delta(k) = \Phi(k+1,k)P(k)D'(k)[D(k)P(k)D'(k) + R(k)]^{-1} \tag{9-75}$$

and, [Eq. (9-68]

$$P(k+1) = \Lambda(k+1,k)P(k)\Lambda'(k+1,k) + \Theta(k+1,k)Q(k)\Theta'(k+1,k) + \Delta(k)R(k)\Delta'(k) \tag{9-76}$$

The following matrices should be specified initially:

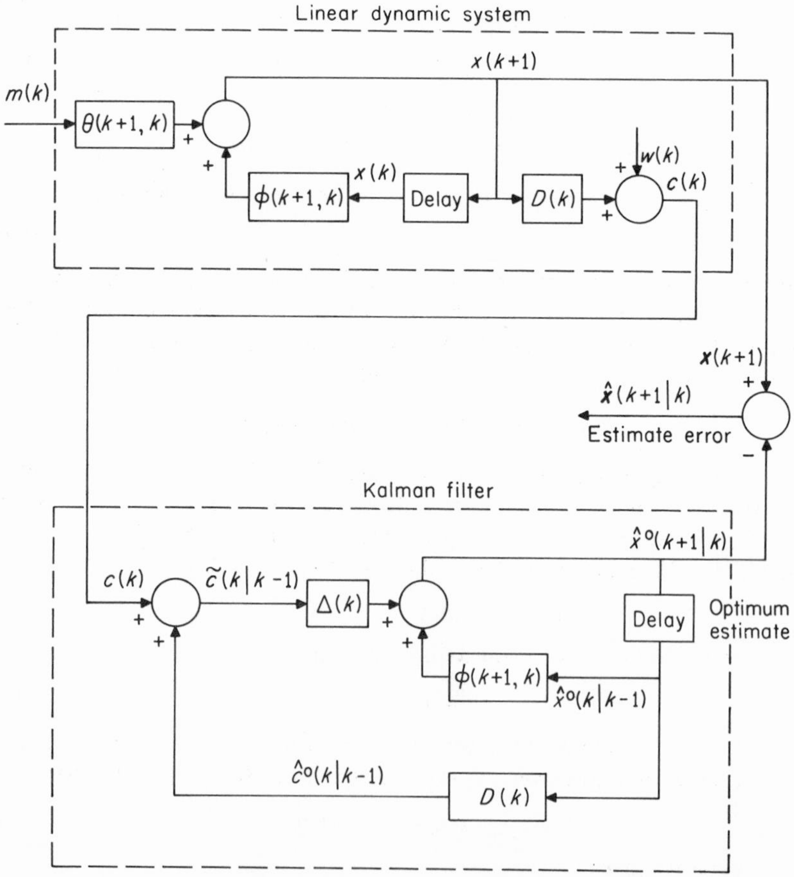

FIG. 9-2. Block diagram of Kalman filter.

$$R(k) = \text{cov}\,[\mathbf{w}(k)]$$

$$Q(k) = \text{cov}\,[\mathbf{m}(k)]$$

$$P(0) = \text{cov}\,[\mathbf{x}(0)]$$

The matrix $P(k)$ must be positive definite for $k = 0, 1, 2, \ldots$. The structure of the solution of the Kalman filtering problem is shown by the block diagram of Fig. 9-2.

Although the Kalman filter design is carried out with $j = k + 1$ in Eq. (9-25), that is, $\mathbf{x}(k + 1)$ is estimated given $\mathbf{c}(0), \mathbf{c}(1), \ldots, \mathbf{c}(k)$, the result can be easily extended to cases with $j = k + N$, where $N = 0, 1, 2, \ldots$. Starting from Eq. (9-28),

$$\hat{\mathbf{x}}^o(j|k) = E[\mathbf{x}(j)|\mathbf{C}(k)] \tag{9-77}$$

it can be shown that for $j = k + N$, the optimal estimate of $\mathbf{x}(k + N)$, given $\mathbf{c}(0), \mathbf{c}(1), \ldots, \mathbf{c}(k)$, is

$$\hat{\mathbf{x}}^o(k + N|k) = \phi(k + N, k + 1)\hat{\mathbf{x}}^o(k + 1|k) \tag{9-78}$$

where $N = 0, 1, 2, \ldots$. The proof of Eq. (9-78) is left as an exercise for the reader (see Prob. 9-1).

EXAMPLE 9-1

As an illustrative example on the design of the Kalman filter, let us consider the system model shown in Figure 9-3. The block diagram shows a first-order discrete-

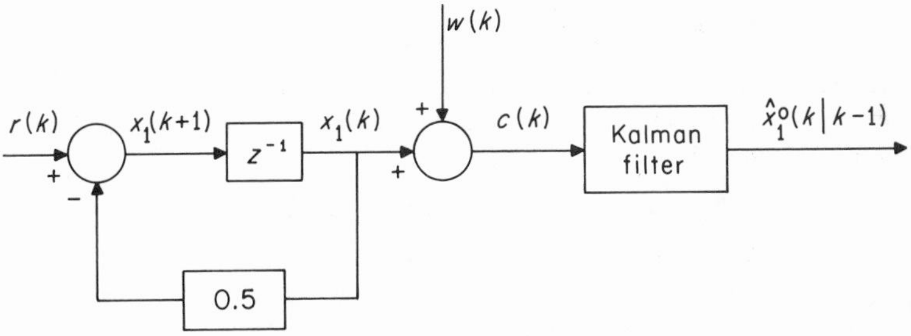

FIG. 9-3. System model for Example 9-1.

data system with an input sequence $m(k)$. The problem is to design a Kalman filter for the optimum estimation of the state variable $x_1(k)$, in the sense of minimum quadratic loss, given the measurements on $c(k)$. The measured output $c(k)$ is shown to be contaminated by additive noise $w(k)$.

Let the input and the noise be gaussian white noise, so that the covariance matrices for $\mathbf{r}(k)$ and $\mathbf{w}(k)$ are constant matrices, and in the present scalar case they are given as

$$Q(k) = \text{cov}\,[r(k)] = 0.25 \tag{9-79}$$

$$R(k) = \text{cov}\,[w(k)] = 0.5 \tag{9-80}$$

The dynamic equations of the system are written

$$x_1(k+1) = \Phi x_1(k) + \Theta r(k) \tag{9-81}$$

$$c(k) = Dx_1(k) + w(k) \tag{9-82}$$

where $\Phi = 0.5$, $\Theta = 1.0$, and $D = 1.0$. The initial state at $t = 0$ is $x_1(0) = 0$.

Substitution of the values of Φ, Θ, and D into Eq. (9-67) gives

$$\begin{aligned} P(k+1) &= \Phi P(k)\Phi' - \Delta(k)DP(k)\Phi' + \Theta Q(k)\Theta' \\ &= 0.25[P(k) - \Delta(k)P(k)] + Q(k) \end{aligned} \tag{9-83}$$

where from Eq. (9-75),

$$\begin{aligned} \Delta(k) &= \Phi P(k)D'[DP(k)D' + R(k)]^{-1} \\ &= 0.25P(k)[P(k) + R(k)]^{-1} \\ &= \frac{0.25P(k)}{P(k) + 0.5} \end{aligned} \tag{9-84}$$

Substituting Eq. (9-84) into Eq. (9-83), we get

$$\begin{aligned} P(k+1) &= 0.25\left[P(k) - \frac{P^2(k)}{P(k) + 0.5}\right] + 0.25 \\ &= 0.25\,\frac{0.5P(k)}{P(k) + 0.5} + 0.25 \end{aligned} \tag{9-85}$$

Now starting with $P(0) = \text{cov}\,[x_1(0)] = 0$, the values of $P(k)$, $k = 1, 2, \ldots$, are determined from Eq. (9-85).

$$P(1) = 0.25, \quad P(2) = 0.2917, \quad P(3) = 0.2960,$$
$$P(4) = 0.2965, \quad P(5) = 0.2970, \ldots$$

As k is increased, $P(k)$ approaches the final value of 0.312.

Substitution of the values of $P(k)$ into Eq. (9-84) gives the corresponding values of $\Delta(k)$ for $k = 0, 1, 2, \ldots$

$$\Delta(0) = 0, \quad \Delta(1) = 0.167, \quad \Delta(2) = 0.184,$$
$$\Delta(3) = 0.186, \quad \Delta(4) = 0.187, \ldots$$

The final value of $\Delta(k)$ as k approaches infinity is 0.188.

The block diagram of the Kalman filter is shown in Figure 9-4. Notice that this is a time-varying filter since $\Delta(k)$ changes its value at the discrete sampling instants.

It is of interest to evaluate the transfer function relationship of the Kalman filter in Figure 9-4. Let us represent the z-transform of the optimal estimate $\hat{x}_1^o(k+1|k)$ by $\hat{X}_1^o(k+1|k)(z)$. Then, from Figure 9-4,

$$\frac{\hat{X}_1^o(k+1|k)(z)}{C(z)} = \frac{\Delta(k)z}{z + \Delta(k) - 0.5} \tag{9-86}$$

Similarly,

$$\frac{X_1^o(k|k-1)(z)}{C(z)} = \frac{\Delta(k)}{z + \Delta(k) - 0.5} \tag{9-87}$$

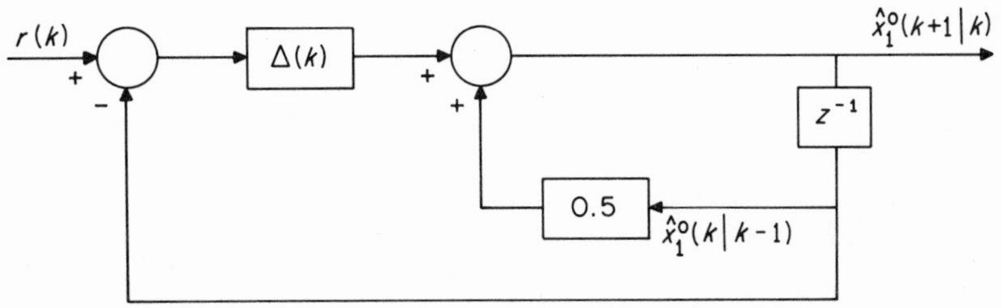

FIG. 9-4. Block diagram of Kalman filter in Example 9-1.

where it is known that the value of $\Delta(k)$ changes at different sampling instants. In the steady state, $\Delta(k)$ approaches 0.188 as k approaches infinity, and Eq. (9-87) becomes

$$\frac{X_1^o(k|k-1)(z)}{C(z)} = \frac{0.188}{z - 0.312} \tag{9-88}$$

Now let us determine the Wiener filter for the same problem, and compare the solution with that of the Kalman filter. Since the Kalman filter problem is stated as the estimation of $x_1(k)$ given the values of $c(0), c(1), \ldots, c(k-1)$, the transfer function of the model system for the Wiener filter can be regarded as

$$M_d(z) = 1 \tag{9-89}$$

The block diagram illustrating the design principle of the Wiener filter is shown in Figure 9-5.

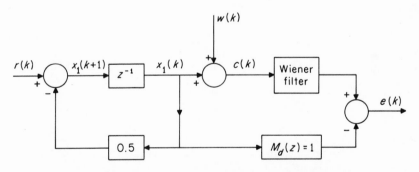

FIG. 9-5. Block diagram for Wiener filter design in Example 9-1.

Since the input and noise are assumed to be white, their power spectral density functions are

$$\Phi_{rr}(z) = 0.25 \tag{9-90}$$

and

$$\Phi_{ww}(z) = 0.5 \tag{9-91}$$

The power spectral density function of $x_1(k)$ is written by use of Eq. (8-87),

$$\Phi_{x_1x_1}(z) = \frac{z^{-1}}{1 - 0.5z^{-1}} \frac{z}{1 - 0.5z} \Phi_{rr}(z)$$

$$= \frac{-0.5z}{(z - 0.5)(z - 2)} \tag{9-92}$$

Therefore,

$$\Phi_{cc}(z) = \Phi_{x_1x_1}(z) + \Phi_{ww}(z)$$

$$= \frac{-0.5z}{(z - 0.5)(z - 2)} + 0.5$$

$$= \frac{0.5(z - 3.188)(z - 0.312)}{(z - 0.5)(z - 0.2)} \tag{9-93}$$

The function $\Phi_{cc}(z)$ is now factorized into the following form:

$$\Phi_{cc}(z) = \Phi_{cc}^{+}(z)\Phi_{cc}^{-}(z) \tag{9-94}$$

where

$$\Phi_{cc}^{+}(z) = \frac{\sqrt{0.5}\,(z - 0.312)}{z - 0.5} \tag{9-95}$$

$$\Phi_{cc}^{-}(z) = \frac{\sqrt{0.5}\,(z - 3.188)}{z - 2} \tag{9-96}$$

From Eq. (8-151) we have

$$F_2(z) = \frac{M_d(z)\Phi_{x\,x}(z)}{\Phi_{cc}^{-}(z)} = \frac{-\sqrt{0.5}z}{(z - 0.5)(z - 3.188)} \tag{9-97}$$

Now factor $F_2(z)$ as

$$F_2(z) = [F_2(z)]_{+} + [F_2(z)]_{-} \tag{9-98}$$

then

$$[F_2(z)]_{+} = \frac{0.5\sqrt{0.5}}{2.688(z - 0.5)} \tag{9-99}$$

Therefore, the transfer function of the optimal Wiener filter is given by [Eq. (8-156)]

$$M(z) = \frac{[F_2(z)]_{+}}{\Phi_{cc}^{+}(z)}$$

$$= \frac{0.188}{z - 0.312} \tag{9-100}$$

It is interesting to note that this solution is also the steady-state transfer function of the Kalman filter obtained in Eq. (9-88).

9.6 KALMAN FILTER EQUATIONS FOR STATE VARIABLE FEEDBACK OPTIMAL CONTROL

The Kalman filter can be used to estimate the inaccessible state variables which are needed for optimal control with state variable feedback. The optimal control $\mathbf{m}^o(k)$ is related to the state variable by

$$\mathbf{m}^o(k) = B(N-k)\mathbf{x}(k) \tag{9-101}$$

where $B(N-k)$ is the time-varying feedback matrix. Figure 9-6 shows the optimal control with state feedback when all the state variables are accessible. However, when the state variables are inaccessible, as in most practical situations, a Kalman filter may be used as shown in Figure 9-7 to obtain the best estimate of the state variables in the sense of minimum quadratic loss.

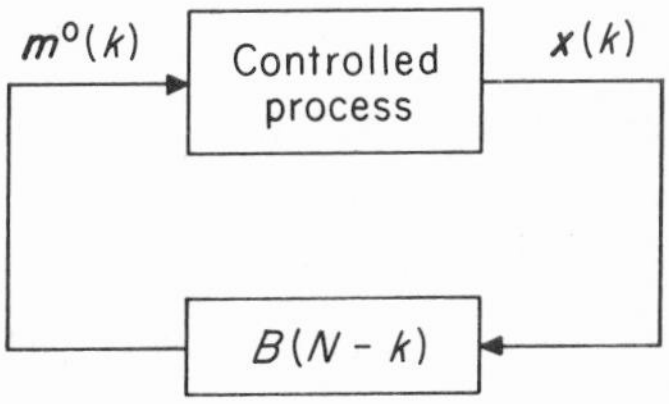

FIG. 9-6. Block diagram showing the optimal control scheme with accessible state variable feedback.

Let the best estimate of $\mathbf{x}(k)$, given the values of $\mathbf{c}(0), \mathbf{c}(1), \ldots, \mathbf{c}(k-1)$, be denoted by $\mathbf{x}^o(k|k-1)$. Then, the control vector obtained according to the estimated state variable is

$$\hat{\mathbf{m}}^o(k|k-1) = B(N-k)\hat{\mathbf{x}}^o(k|k-1) \tag{9-102}$$

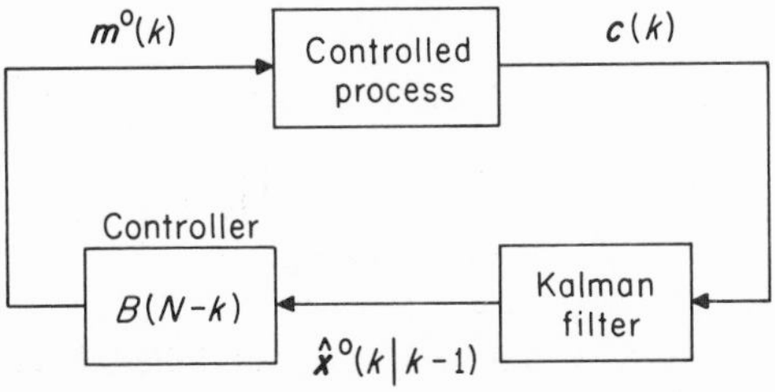

FIG. 9.7. Kalman filter for the estimation of inaccessible state variables for optimal control.

The natural question is, of course, whether this optimal control based on the estimated state variable would still minimize the quadratic performance index. The problem was solved by Tou and Joseph[5,8] and the result is known as the separation theorem. The significance of the separation theorem is that the optimization problem and the estimation problem can be treated separately.†

Substitution of Eq. (9-102) into Eq. (9-35) gives

$$\begin{aligned}\hat{\mathbf{x}}^o(k+1|k-1) &= \Phi(k+1,k)\hat{\mathbf{x}}^o(k|k-1) \\ &\quad + \Theta(k+1,k)B(N-k)\hat{\mathbf{x}}^o(k|k-1) \\ &= [\Phi(k+1,k) + \Theta(k+1,k)B(N-k)\mathbf{x}^o(k|k-1)\end{aligned} \tag{9-103}$$

†The separation theorem proved by Tou and Joseph applies only to the optimization problem presented in this chapter and Chapter 7. Larson[10] has extended the solution to systems with quantization, and Chang[9] treated the case with random sampling.

Or,

$$\hat{\mathbf{x}}^o(k+1|k-1) = \Psi(k+1, k)\hat{\mathbf{x}}^o(k|k-1) \tag{9-104}$$

where

$$\Psi(k+1, k) = \Phi(k+1, k) + \Theta(k+1, k)B(N-k) \tag{9-105}$$

Notice that in this case it is not necessary to make the assumption that $E[\mathbf{m}(k)] = 0$, as in the derivation of Eq. (9-36).

The expression which corresponds to Eq. (9-46) is

$$\hat{\mathbf{x}}^o(k+1|k) = \Gamma(k+1, k)\hat{\mathbf{x}}^o(k|k-1) + \Delta(k)\mathbf{c}(k) \tag{9-106}$$

where

$$\Gamma(k+1, k) = \Psi(k+1, k) - \Delta(k)D(k) \tag{9-107}$$

Following through similar derivations as in the last section, we have the following Kalman filter equations:

$$\Delta(k) = \Psi(k+1, k)P(k)D'(k)[D(k)P(k)D'(k) + R(k)]^{-1} \tag{9-108}$$

and

$$P(k+1) = \Gamma(k+1, k)P(k)\Gamma'(k+1, k) + \Delta(k)R(k)\Delta'(k) \tag{9-109}$$

REFERENCES

1. Kalman, R. E., "A New Approach to Linear Filtering and Prediction Problems," *Trans. ASME*, Series D, *Journal of Basic Engineering*, Vol. 82, 1960, pp. 35–45.
2. Kalman, R. E., and Bucy, R. S., "New Results in Linear Filtering and Prediction Theory," *Trans. ASME*, Series D, *Journal of Basic Engineering*, March, 1961, pp, 95–108.
3. Cox, H., "On the Estimation of State Variables and Parameters for Noisy Dynamic Systems," *IEEE Trans. on Automatic Control*, Vol. AC-9, 1964, pp. 5–12.
4. Tou, J. T., *Modern Control Theory*. McGraw-Hill, New York, 1964.
5. Tou, J. T., *Optimum Design of Digital Control Systems*. Academic Press, New York, 1963.
6. Deutsch, R., *Estimation Theory*. Prentice-Hall, Englewood Cliffs, New Jersey, 1965.
7. Aoki, M., *Optimization of Stochastic Systems*. Academic Press, New York, 1967.
8. Joseph, P. D., and Tou, J.T., "On Linear Control Theory," *AIEE Trans.*, Vol. 80, September 1961, pp. 193–196.

9. Chang, S. S. L., "Optimum Filtering and Control of Randomly Sampled Systems," *IEEE Trans. on Automatic Control*, Vol. AC-12, October 1967, pp. 537–546.

10. Larson, R. E., "Optimum Quantization in Dynamic Systems," *IEEE Trans. on Automatic Control*, Vol. AC-12, April 1967, pp. 162–168.

10

DIGITAL SIMULATION

10.1 INTRODUCTION

It is well known that digital computers are playing an increasingly important role in the analysis and design of feedback control systems. Digital computers are used not only in the computation and simulation of control systems performance, but also often in direct on-line control of systems. In addition, it is quite common to have airborne and shipboard computers for real-time and on-line controls.

In this chapter the subjects of computer realization of discrete transfer functions, various methods of digital simulation, and the effects of amplitude quantization, are given.

10.2 DIGITAL REALIZATION OF DISCRETE-DATA TRANSFER FUNCTION

The subject of decomposition of digital systems is covered in Section 4.9. It is shown that given a transfer function $D(z)$ of a digital system, we can decompose it by means of one of the three decomposition schemes. These are: (1) direct decomposition, (2) cascade decomposition, and (3) parallel decomposition. In this section we will expand these ideas and carry out a more thorough discussion on digital simulation and implementation with the aid of the state diagram.

Let us assume that an nth order linear time-invariant digital system with single input and output is described by the transfer function

$$D(z) = \frac{C(z)}{R(z)} = \frac{a_0 + a_1 z^{-1} + a_2 z^{-2} + \cdots + a_n z^{-n}}{b_0 + b_1 z^{-1} + b_2 z^{-2} + \cdots + b_n z^{-n}} \tag{10-1}$$

where for physical realizability, $b_0 \neq 0$; $C(z)$ and $R(z)$ are the z-transforms of the output and the input of the system, respectively.

We now consider several possible ways of implementation of the transfer function of Eq. (10-1) on a digital computer.

Programming by Direct Decomposition

Using the direct decomposition scheme described in Section 4.9, a state diagram for the system of Eq. (10-1) is drawn as shown in Figure 10-1. The

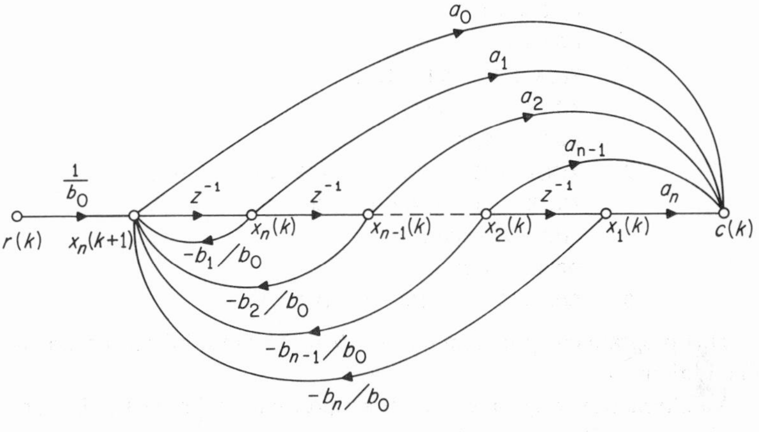

FIG. 10-1. State diagram representation of the transfer function of Eq. (10-1) by direct decomposition.

z^{-1} branch corresponds to a time delay unit with a delay of T second. The outputs of the delay units are designated as the state variables $x_1(k)$, $x_2(k)$, . . . , $x_n(k)$. For simplicity, the sampling period T is dropped in $x(kT)$.

The state equations of the system are written directly from the state diagram of Figure 10-1 by use of the Mason's gain formula. In this process, the z^{-1} branches are deleted (or simply disregarded) since the state equations are not functions of z. The inputs of the delay units correspond to the variables $x_1(k+1)$, $x_2(k+1)$, . . . , $x_n(k+1)$, and, therefore, are treated as output variables in applying the gain formula. The nodes with variables $x_1(k)$, $x_2(k)$, . . . , $x_n(k)$, and $r(k)$ are the input nodes. Therefore, the state equations are

$$x_1(k+1) = x_2(k)$$
$$x_2(k+1) = x_3(k)$$
$$\vdots$$
$$x_n(k+1) = -\frac{b_n}{b_0}x_1(k) - \frac{b_{n-1}}{b_0}x_2(k) - \cdots - \frac{b_2}{b_0}x_{n-1}(k) - \frac{b_1}{b_0}x_n(k) + \frac{1}{b_0}r(k) \tag{10-2}$$

The output equation is

$$c(k) = a_n x_1(k) + a_{n-1}x_2(k) + \cdots + a_1 x_n(k) + \frac{a_0}{b_0}r(k) \tag{10-3}$$

In digital implementation, starting with Eq. (10-3), given $r(k)$, $k = 0, 1, 2, \ldots, N-1$, and $x_i(0)$, $i = 1, 2, \ldots, n$, we can compute $c(k)$ and $x_i(k)$ for $k = 1, 2, \ldots, N$. A FORTRAN program for the above scheme is tabulated below:

```
      DØ 3 K = 0, N
      DØ 1 I = 1, N − 1
1     X(I, K + 1) = X(I + 1, K)
      X(N, K + 1) = R(K)
      DØ 2 I = 1, N
2     X(N, K + 1) = X(N, K + 1) − B(N + 1 − I)*X(I, K)
      X(N, K + 1) = X(N, K + 1)/B(0)
      C(K) = A(0)/B(0)*R(K)
      DØ 3 I = 1, N
3     C(K) = C(K) + A(N + 1 − I)*X(I, K)
```

This program contains a total of $(2nN + 3N)$ multiplications, and $(4nN + 2N)$ additions.

An alternative way of direct decomposition of Eq. (10-1) is carried out through cross-multiplying, and rearranging we get

$$C(z) = \frac{a_0}{b_0}R(z) + \frac{1}{b_0}\sum_{i=1}^{n}[a_i R(z) - b_i C(z)]z^{-i} \tag{10-4}$$

The corresponding time-domain expression for the last equation is

$$c(k) = \frac{a_0}{b_0}r(k) + \frac{1}{b_0}\sum_{i=1}^{n}[a_i r(k-i) - b_i c(k-i)] \tag{10-5}$$

The state diagram portraying these equations is drawn as shown in Figure 10-2. Now applying Mason's gain formula to this state diagram yields

$$x_1(k+1) = \frac{-b_1}{b_0}x_1(k) + x_2(k) + \left(\frac{a_1}{b_0} - \frac{b_1 a_0}{b_0^2}\right)r(k) \tag{10-6}$$

$$x_2(k+1) = -\frac{b_2}{b_0}x_1(k) + x_3(k) + \left(\frac{a_2}{b_0} - \frac{b_2 a_1}{b_0^2}\right)r(k) \tag{10-7}$$

$$\vdots$$

$$x_n(k+1) = -\frac{b_n}{b_0}x_1(k) + \frac{a_n}{b_0}r(k) \tag{10-8}$$

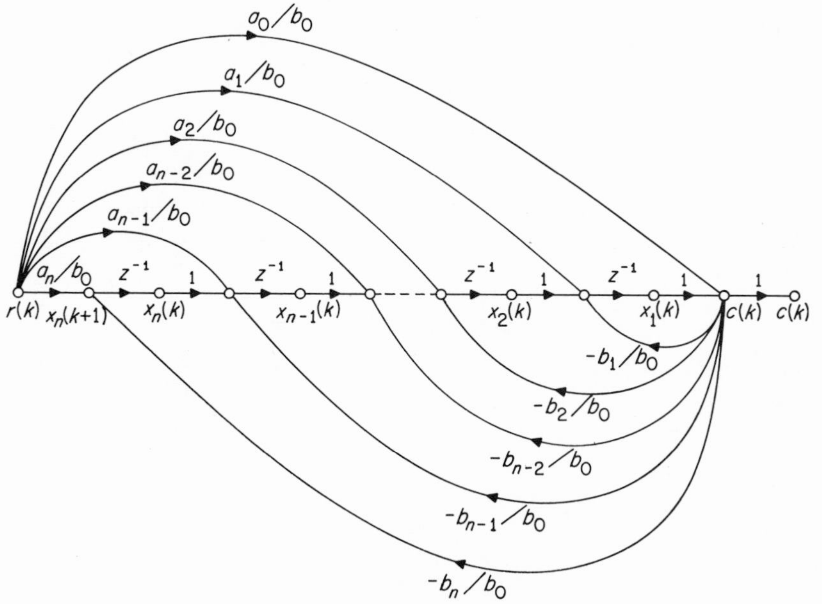

FIG. 10-2. State diagram representation of the transfer function of Eq. (10-1) by direct decomposition (alternate method).

and

$$c(k) = x_1(k) + \frac{a_0}{b_0} r(k) \tag{10-9}$$

These equations can again be used for digital implementation of the transfer function of Eq. (10-1).

A FORTRAN program for the digital implementation of Eqs. (10-6) through (10-9), given the values of $r(k)$, $k = 0, 1, 2, \ldots, N - 1$, and $x_i(0)$, $i = 1, 2, \ldots, n$, are tabulated below:

```
      A0 = A(0)
      B0 = B(0)
      DØ 2 K = 0, NT
      DØ 1 I = 1, N - 1
1     X(I, K + 1) = X(I + 1, K)
                 + (- B(I)*X(1, K) + (A(I) - B(I)*A(I - 1)/B0)*R(K))/B0
      X(N, K + 1) = (- B(N)*X(1, K) + A(N)*R)K))/B0
2     C(K) = X(1, K) + A0/B0*R(K)
```

The above program contains a total of $(5nN + 5N)$ multiplications and $(4nN + 4N)$ additions.

Therefore, comparing the two direct decomposition schemes, the one shown in Figure 10-2 generally requires more multiplication and addition operations and would require more computing time.

Programming by Cascade Decomposition

If the discrete-data transfer function in Eq. (10.1) has simple real poles and zeros, it can be written

$$D(z) = \frac{C(z)}{R(z)} = \frac{K(z + e_1)(z + e_2) \cdots (z + e_m)}{(z + d_1)(z + d_2) \cdots (z + d_n)} \tag{10-10}$$

where for physical realizability, $n \geq m$. Assuming that $n > m$, $D(z)$ can be decomposed as a cascade of n first-order systems as shown in Figure 10-3. The

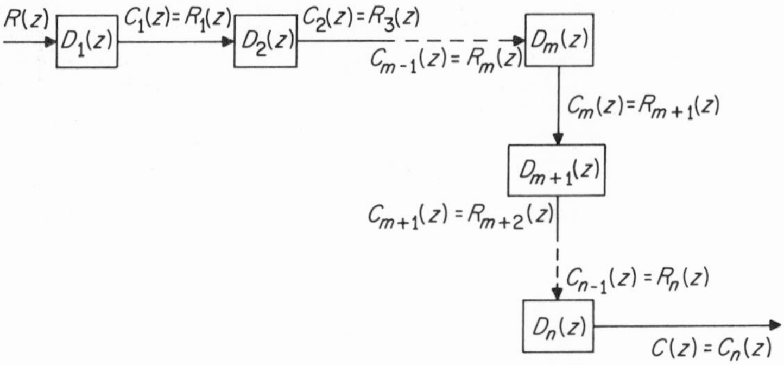

FIG. 10-3. A cascade decomposition of digital systems.

transfer functions of these decomposed systems are:

$$D_1(z) = \frac{C_1(z)}{R(z)} = \frac{z + e_1}{z + d_1} = \frac{1 + e_1 z^{-1}}{1 + d_1 z^{-1}} \tag{10-11}$$

$$D_2(z) = \frac{C_2(z)}{R_2(z)} = \frac{1 + e_2 z^{-1}}{1 + d_2 z^{-1}} \tag{10-12}$$

$$\vdots$$

$$D_m(z) = \frac{C_m(z)}{R_m(z)} = \frac{1 + e_m z^{-1}}{1 + d_m z^{-1}} \tag{10-13}$$

$$D_{m+1} = \frac{C_{m+1}(z)}{R_{m+1}(z)} = \frac{K_{m+1}}{z + d_{m+1}} = \frac{K_{m+1} z^{-1}}{1 + d_{m+1} z^{-1}} \tag{10-14}$$

$$\vdots \qquad \vdots$$

$$D_n(z) = \frac{C(z)}{R_n(z)} = \frac{K_n z^{-1}}{1 + d_n z^{-1}} \tag{10-15}$$

where $K_{m+1} K_{m+2} \cdots \mathrm{K}_n = K$.

The state diagram for the transfer functions in Eqs. (10-11) through (10-13) is shown in Figure 10-4(a), and that for Eqs. (10-14) and (10-15) is shown in Figure 10-4(b). Therefore, the dynamic equations for $D_i(z)$, $i = 1, 2, \ldots, m$, are, from Figure 10-4(a),

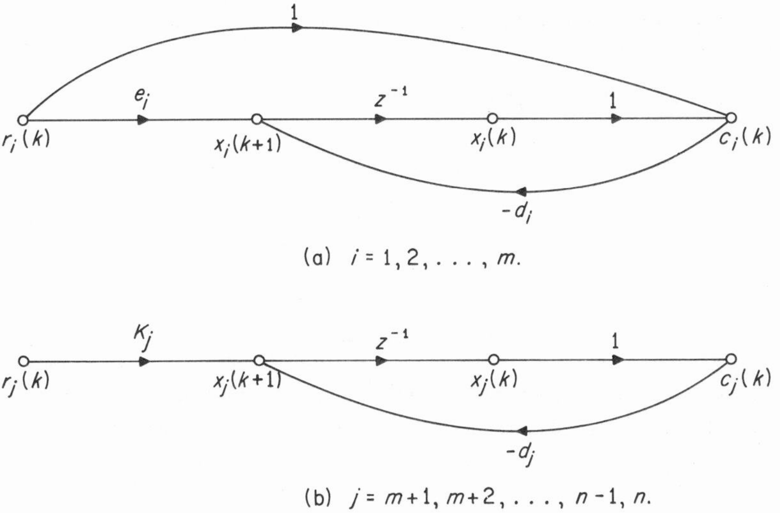

FIG 10-4. State diagrams for components of cascade decomposition.

state equation: $x_i(k+1) = -d_i x_i(k) + (e_i - d_i) r_i(k)$ (10-16)

output equation: $c_i(k) = x_i(k) + r_i(k)$ (10-17)

$i = 1, 2, \ldots, m$, and from Figure 10-4(b),

state equation: $x_j(k+1) = -d_j c_j(k) + K_j r_j(k)$ (10-18)

output equation: $c_j(k) = x_j(k)$ (10-19)

$j = m+1, \mathrm{m}+2, \ldots, n-1, n$, and $c_n(k) = c(k)$, $r_1(k) = r(k)$.

A FORTRAN program for the simulation of transfer function by cascade decomposition is given below for the case $n = m$:

```
     DØ 1 K = 0, NT
     DØ 1 I = 1, N
     X(I, K + 1) = − D(I)*X(I, K) + (E(I) − D(I))*R(I, K)
1    R(I + 1, K) = X(I, K) + R(I, K)
```

Note that $C(K) = R(N+1, K)$. The program contains a total of $2(nN)$ multiplications and $3(nN)$ additions. This makes the digital implementation of cascade decomposition faster than that of the direct decomposition schemes.

Programming by Parallel Decomposition

Still another method of decomposing a transfer function for digital implementation is the parallel decomposition. If the transfer function $D(z)$ has only simple real poles, it can be expanded by partial fraction expansion into the following form:

$$D(z) = \sum_{i=1}^{n} D_i(z) = \sum_{i=1}^{n} \frac{K_i}{z + d_i} \tag{10-20}$$

where

$$D_i(z) = \frac{C_i(z)}{R(z)} = \frac{K_i}{z + d_i} \tag{10-21}$$

and

$$K_i = D(z)(z + d_i)\Big|_{z=-d_i} \tag{10-22}$$

The block diagram portraying the parallel decomposition of $D(z)$ is shown in Figure 10-5. The state diagram of $D_i(z)$ is obtained by performing a direct

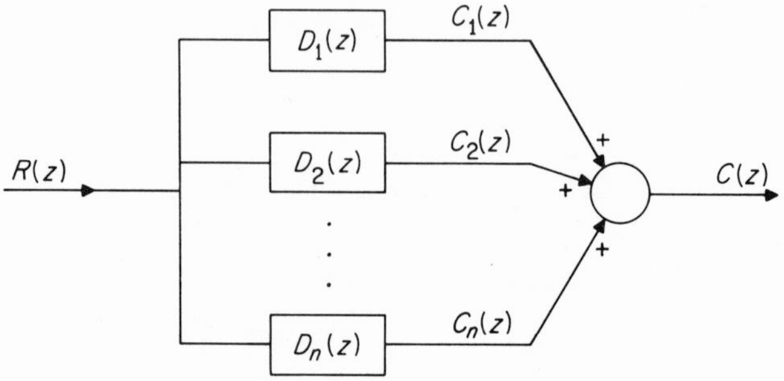

FIG. 10-5. Block diagram showing parallel decomposition.

decomposition to Eq. (10-21), yielding

$$\frac{C_i(z)}{R(z)} = \frac{K_i z^{-1}}{1 + d_i z^{-1}} \tag{10-23}$$

Thus,

$$C_i(z) = K_i z^{-1} R(z) - d_i z^{-1} C_i(z) \tag{10-24}$$

The state diagram representation of the last equation is shown in Figure 10-6. Therefore, the state equations of the system are written

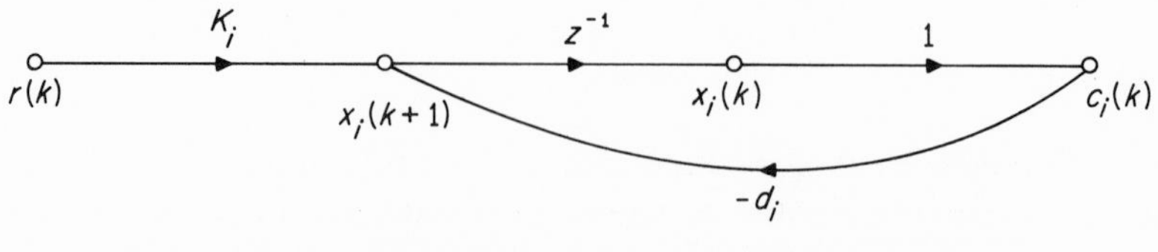

FIG. 10-6. State diagram of componentsof parallel decomposition.

$$x_i(k+1) = -d_i x_i(k) + K_i r(k) \tag{10-25}$$

$i = 1, 2, \ldots, n$, and the output equations are simply

$$c_i(k) = x_i(k)$$

and

$$c(k) = \sum_{i=1}^{n} c_i(k) \tag{10-26}$$

A FORTRAN program is written below to simulate the computation of Eqs. (10-25) through (10-26).

```
      DØ 1 K = 0, N
      C(K) = 0.
      DØ 1 I = 1, N
      X(I, K + 1) = − D(I)*X(I, K) + K(I)*R(I, K)
1     C(K) = C(K) + X(I, K)
```

This program requires a total of $2(nN)$ multiplications and $2(nN)$ additions and, therefore, represents the least time-consuming program among the three decomposition schemes.

The advantage of the direct decomposition is that the numerator and denominator polynomials of $D(z)$ need not be factored into first-order terms. In general, the transfer function may have multiple-order and complex poles and zeros. The computer can be programmed to handle complex numbers if the poles and zeros of $D(z)$ are complex. However, real numbers are always easier to manipulate.

10.3 DIGITAL SIMULATION—DIGITAL MODEL WITH SAMPLE-AND-HOLD

An analog system can be simulated on the digital computer once its system dynamics are approximated by a transfer function in the z-domain. The analysis usually consists of the following two steps:

1. Representation of the continuous-data system by a digital model.
2. Simulation of the digital model on a digital computer.

The second step has been discussed in the last section.

There are many possible ways of representing a continuous-data system by a digital model. In general, the following three methods are used.

1. Insert sample-and-hold devices in the continuous-data system.
2. Numerical integration.
3. The z-form approximation.

The first method is discussed in this section. The simplest way of approximating a continuous-data system by a digital model is to insert fictitious sample-and-hold devices at strategic locations of the system. Then, the system can be described by z-transform transfer functions or difference state equations. For instance, the continuous-data control system shown in Figure 10-7(a) is ap-

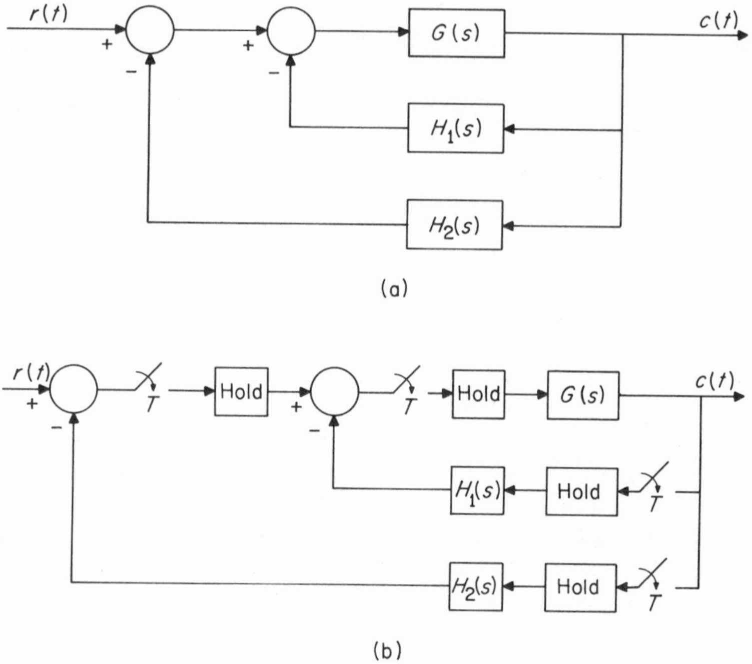

FIG. 10-7. (a) A continuous-data control system; (b) a digital model of the continuous-data system with fictitious sample-and-hold devices.

proximated by the digital model shown in Figure 10-7(b). The hold device can be of any type at the discretion of the analyst. It can even be a polygonal hold which is not physically realizable but nevertheless quite suitable for digital simulation. In Figure 10-7(b), let $G_h(s)$ denote the transfer function of the hold device. Then the discrete transfer function of the digital model is

$$\frac{C(z)}{R(z)} = \frac{G_h(z)G_hG(z)}{1 + G_hG(z)G_hH_1(z) + G_h(z)G_hG(z)G_hH_2(z)} \tag{10-27}$$

where

$$G_h(z) = \mathscr{Z}\,[G_h(s)] \tag{10-28}$$

$$G_hG(z) = \mathscr{Z}\,[G_h(s)G(s)] \tag{10-29}$$

$$G_hH_1(z) = \mathscr{Z}\,[G_h(s)H_1(s)] \tag{10-30}$$

$$G_hH_2(z) = \mathscr{Z}\,[G_h(s)H_2(s)] \tag{10-31}$$

Although in principle the digital modeling method just described is straightforward, it has two important considerations under actual applications. The first concerns the selection of the appropriate sampling period of the fictitious samplers. The sampling period is directly related to the accuracy as well as the time required of the digital simulation. The second consideration is that of stability. It was pointed out earlier that if the continuous-data system is stable, the digital model is not necessarily stable. In fact, sample-and-hold devices generally have an adverse effect on the system stability. Therefore, when inserting sample-and-hold devices in a stable continuous-data system, it is important that the digital model be a stable one.

As an illustrative example, let us refer to the continuous-data system shown in Figure 10-8(a). A digital approximation of the system may be obtained by

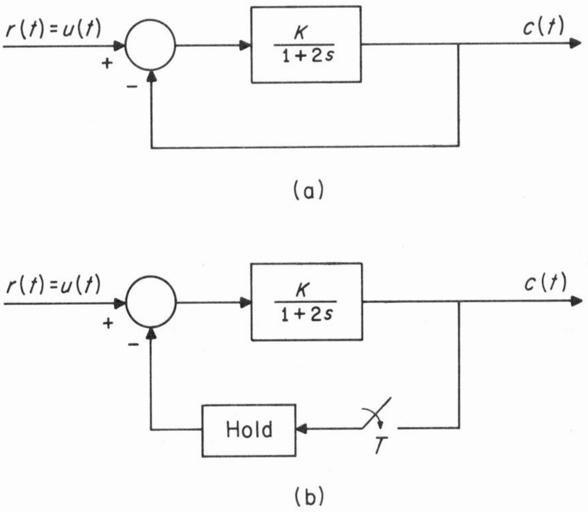

FIG. 10-8. (a) A continuous-data control system; (b) a digital model of the continuous-data system with sample-and-hold inserted in the feedback path.

inserting a sample-and-hold in the feedback path as shown in Figure 10-8(b). The z-transform of the output of the digital model is written

$$C(z) = \frac{RG(z)}{1 + G_{h0}G(z)} \tag{10-32}$$

where

$$RG(z) = \mathscr{Z}\,[R(s)G(s)] \tag{10-33}$$

and

$$G_{h0}G(z) = \mathscr{Z}\,[G_{h0}(s)\,G(s)] \tag{10-34}$$

For a unit-step input,

$$RG(z) = \mathscr{Z}\left[\frac{K}{s(1+2s)}\right] = \frac{K(1-e^{-0.5T})}{(z-1)(z-e^{-0.5T})} \tag{10-35}$$

Let us assume that the hold device is a zero-order hold. Then,

$$G_{h0}G(z) = \mathscr{Z}\left[\frac{1-e^{-Ts}}{s}\frac{K}{1+2s}\right] = \frac{K(1-e^{-0.5T})}{(z-e^{-0.5T})} \tag{10-36}$$

Substitution of Eqs. (10-35) and (10-36) into Eq. (10-32) yields

$$C(z) = \frac{K(1-e^{-0.5T})z}{(z-1)[z-e^{-0.5T}+K(1-e^{-0.5T})]} \tag{10-37}$$

Selecting $T = 0.25$ second and letting $K = 1$, the last equation becomes

$$C(z) = \frac{0.118z}{(z-1)(z-0.764)} \tag{10-38}$$

Expanding $C(z)$ by continued division, we have

$$\begin{aligned} C(z) = {} & 0.118z^{-1} + 0.207z^{-2} + 0.276z^{-3} + 0.329z^{-4} \\ & + 0.369z^{-5} + 0.4z^{-6} + 0.423z^{-7} + 0.441z^{-8} \\ & + 0.455z^{-9} + 0.466z^{-10} + 0.473z^{-11} + 0.48z^{-12} \\ & + 0.485z^{-13} + 0.488z^{-14} + \cdots \end{aligned} \tag{10-39}$$

The final value of $c(kT)$ is 0.5. It is apparent that the digital model is a stable one.

Let us consider that the hold device is a polygonal hold, so that

$$G_h(s) = \frac{e^{Ts}+e^{-Ts}-2}{Ts^2} \tag{10-40}$$

Then, for $T = 0.25$ second and $K = 1$, the z-transform of the output of the digital system is

$$C(z) = \frac{0.1108z}{(z-1)(z-0.778)} \tag{10-41}$$

which is expanded to

$$\begin{aligned} C(z) = {} & 0.1108z^{-1} + 0.197z^{-2} + 0.264z^{-3} + 0.316z^{-4} \\ & + 0.357z^{-5} + 0.388z^{-6} + 0.413z^{-7} + 0.433z^{-8} \\ & + 0.447z^{-9} - 0.459z^{-10} + 0.468z^{-11} + 0.475z^{-12} \\ & + 0.481z^{-13} + 0.485z^{-14} + \cdots \end{aligned} \tag{10-42}$$

As a comparison, the responses of the continuous-data systems and the digital model with zero-order hold and polygonal hold, to a unit-step input, are computed and shown in Figure 10-9. Figure 10-10 shows the responses when the sampling period in the digital simulation model is changed to one second. With this larger sampling period, the deviation between the actual re-

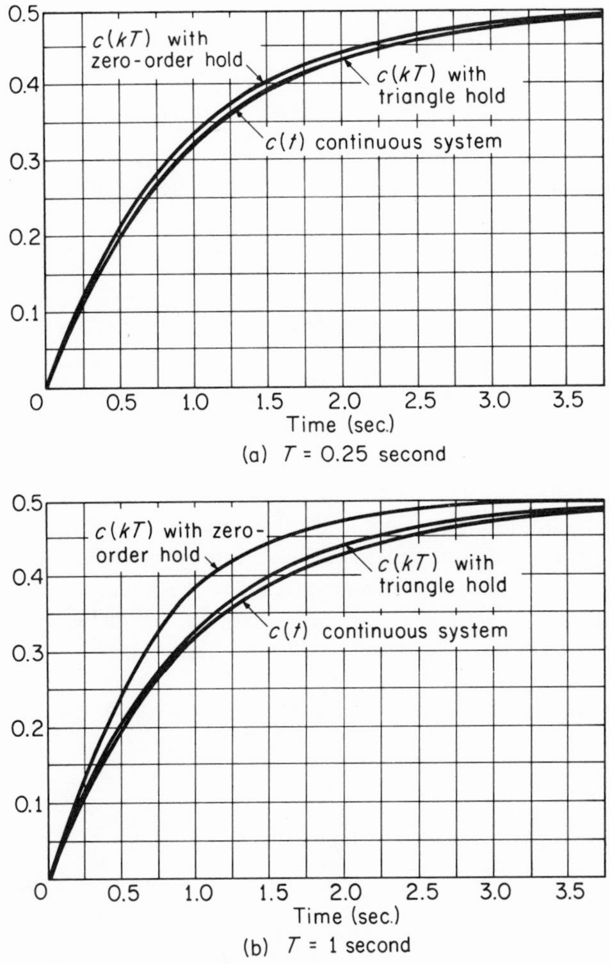

FIG. 10-9. Responses of systems by digital simulation.

sponse of the continuous-data system and the response of the digital model becomes greater as expected.

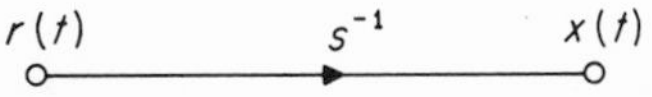

FIG. 10-10. An integrator of the state diagram.

Since the continuous-data system is of the first-order, it is stable for all finite positive values of K. However, the digital model may be stable or unstable depending on the values of K and T. For instance, with the zero-order hold and $K = 2$, the maximum value of the sampling time to insure stability in the digital system is 2.2 seconds. When K is increased to

10, the marginal value of T is 0.404 second. With $T = 2$ seconds, the z-transform of the digital system with a polygonal hold is

$$C(z) = \frac{0.632Kz}{(z - 1)[(1 + 0.368K)z + (0.264K - 0.368)]} \tag{10-43}$$

It can be shown that the system is stable for all finite positive values of K. Therefore, the polygonal hold not only gives a better approximation but also a more stable system in this case.

10.4 DIGITAL SIMULATION—NUMERICAL INTEGRATION[1]

Another popular method of digital simulation of continuous-data systems is the use of numerical integration. Since integration is the most time-consuming and difficult basic mathematical operation on a digital computer, its digital simulation plays an important role here. Instead of inserting sample-and-hold devices at strategic locations in a continuous-data system, the approach now is to approximate the continuous integration operation by numerical methods. The problem can also be stated as the simulation of the integrators s^{-1} in a continuous-data state diagram by digital models. Shown in Figure 10-10 is the integrator element of a state diagram. The input-output relationships are written as

$$x(t) = \int_0^t r(\tau)\,d\tau \tag{10-44}$$

and

$$\frac{X(s)}{R(s)} = \frac{1}{s} \tag{10-45}$$

If the input $r(\tau)$ to the integrator is as shown in Figure 10-11, the output $x(t)$ is equal to the area under the curve $r(\tau)$ between $\tau = 0$ and $\tau = t$.

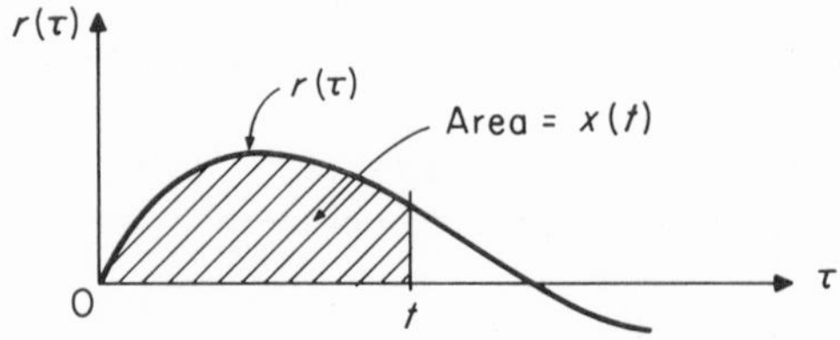

FIG. 10-11. Input and output relation of an integrator.

Rectangular Integration

One of the standard methods of numerical integration is the rectangular approximation, two types of which are shown in Figure 10-12. The approxima-

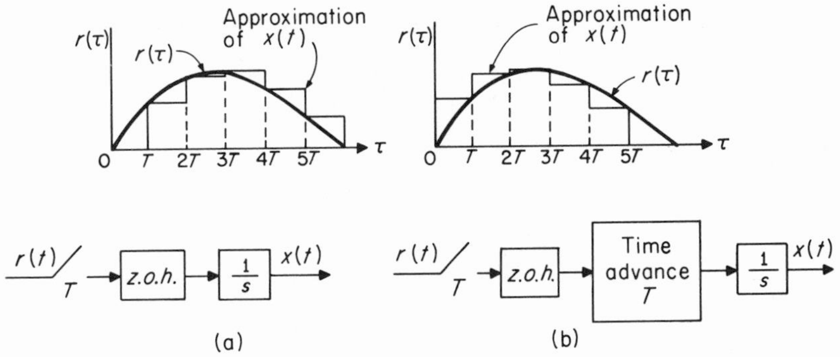

FIG. 10-12. (a) Rectangular approximation in numerical approximation; (b) advanced rectangular approximation in numerical approximation.

tion is performed by summing the rectangular areas of width T under the staircase waveforms. These rectangular approximations are equivalent to inserting a sample-and-zero-order hold device before each integrator as illustrated in Figure 10-12. The scheme shown in Figure 10-12(a) is referred to as the rectangular approximation, whereas the one shown in Figure 10-12(b) is the advanced rectangular approximation.

Referring to Figure 10-12(a), the z-transfer function of the rectangular approximated integrator is

$$\frac{X(z)}{R(z)} = (1 - z^{-1})\mathscr{Z}\left(\frac{1}{s^2}\right) = \frac{T}{z - 1} \tag{10-46}$$

The state equation is

$$x[(k + 1)] = x(kT) + Tr(kT) \tag{10-47}$$

Similarly, the z-transfer function of the advanced rectangular approximated integrator is written

$$\frac{X(z)}{R(z)} = z(1 - z^{-1})\mathscr{Z}\left(\frac{1}{s^2}\right) = \frac{Tz}{z - 1} \tag{10-48}$$

and the state equation is

$$x[(k + 1)T] = x(kT) + Tr(k + 1)T \tag{10-49}$$

As an illustrative example the continuous-data system of Figure 10-8(a) is first approximated by rectangular integration. With the integrator $1/s$ replaced by the transfer function $T/(z - 1)$, the state diagram of the digital model is shown in Figure 10-13.

The z-transform of the output of the digital model when the input is a unit-step function is ($K = 1, T = 0.25$ sec.)

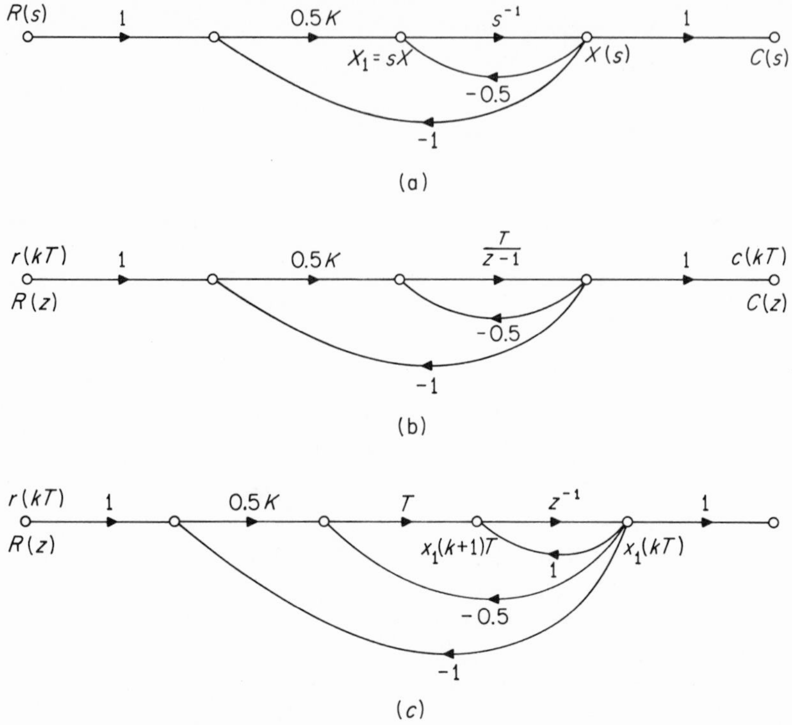

FIG. 10-13. State diagram of the control system in Fig. 10-8 with rectangular integration approximation: (a) state diagram of continuous-data system; (b) integration replaced by rectangular integrator; (c) state diagram of digital model with rectangular integration.

$$C(z) = \frac{0.125z}{(z-1)(z-0.75)} \tag{10-50}$$

Expanding $C(z)$, we get

$$\begin{aligned} C(z) = {} & 0.125z^{-1} + 0.218z^{-2} + 0.288z^{-3} + 0.34z^{-4} \\ & + 0.379z^{-5} + 0.408z^{-6} + 0.431z^{-7} + 0.45z^{-8} \\ & + 0.463z^{-9} + 0.473z^{-10} + \cdots \end{aligned}$$

The response is plotted as shown in Figure 10-14, together with the results of the sample-and-hold approximation scheme. Choosing the advanced rectangular approximation, the integrator transfer function $1/s$ in the continuous-data system is replaced by $Tz/(z-1)$. With $K = 1$ and $T = 0.25$ second, the output transfer function expression of the digital system with unit-step input is

$$C(z) = \frac{0.1z^2}{(z-1)(z-0.8)} \tag{10-51}$$

and the response is plotted as shown in Figure 10-14.

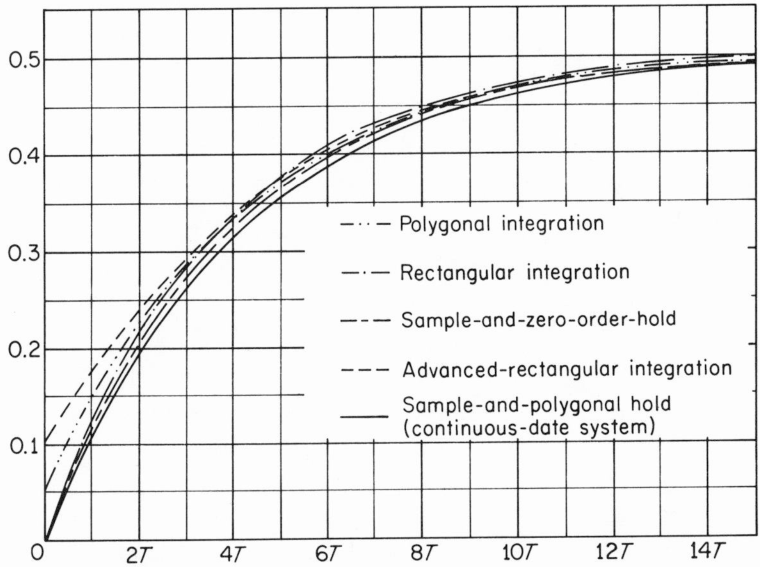

FIG. 10-14. Comparison of results of various methods of digital simulation.

Comparison of these results showed that both versions of the rectangular integration approximation give inferior results to those of the sample-and-hold approximation. The advanced rectangular approximation causes the output step response to jump at $t = 0$, although the response gets close to the actual values after the large errors for the first few sampling periods.

Polygonal Integration

A generally more accurate numerical integration scheme is achieved by using the polygonal hold concept. As shown in Figure 10-15(a), the area under the curve $r(t)$ can be approximated by summing up the areas under the polygons of widths T. It is apparent that the approximation can be made as close as possible by reducing the sampling period T. This is often referred to as the polygonal or the trapezoidal integration approximation. This type of approximation is equivalent to inserting a sample-and-polygonal hold device before each integrator as illustrated in Figure 10-15(b).

Since the transfer function of the polygonal hold is

$$G_h(s) = \frac{e^{Ts} + e^{-Ts} - 2}{Ts^2} \tag{10-52}$$

the transfer function of the polygonal integrator is written

$$\frac{X(z)}{R(z)} = \frac{z + z^{-1} - 2}{T} \mathscr{Z}\left(\frac{1}{s^3}\right) = \frac{T}{2}\left(\frac{z+1}{z-1}\right) \tag{10-53}$$

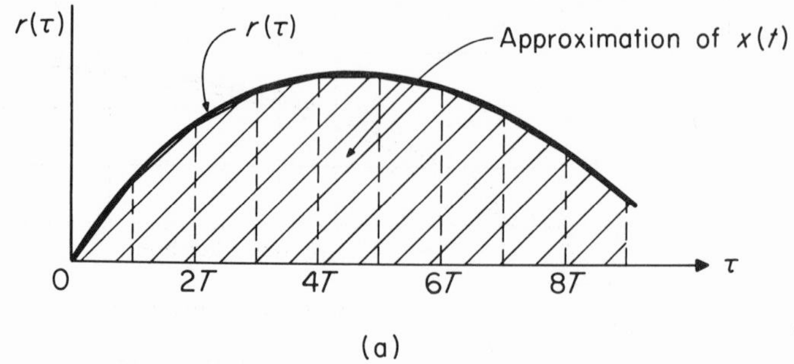

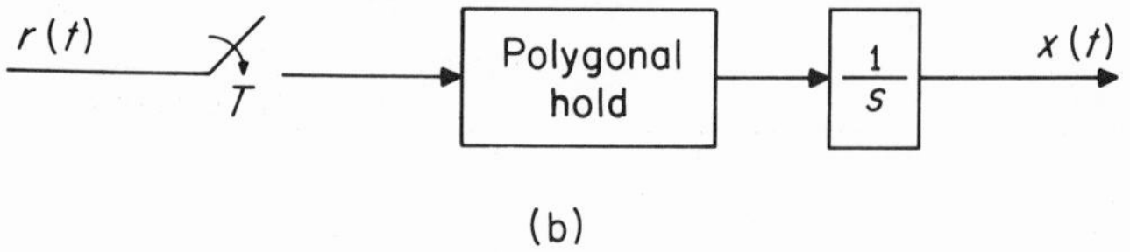

FIG. 10-15. Polygonal or trapezoidal integration approximation.

The state equation of the system is

$$x[(k+1)T] = x(kT) + \frac{T}{2}r[(k+1)T] + \frac{T}{2}r(kT) \quad (10\text{-}54)$$

The unit-step response of the system of Figure 10-8(a) using the polygonal integration approximation is shown in Figure 10-14. The fact that the polygonal integration gives an approximation inferior to that of the rectangular integration is due to the characteristic of this first-order system under consideration. Notice that the input to the integrator of the continuous-data system is written from Figure 10-13(a).

$$X_1(s) = \frac{0.5s}{s+1}R(s) = \frac{0.5}{s+1} \quad (10\text{-}55)$$

Therefore,

$$X_1(t) = 0.5e^{-t} \qquad (t > 0) \quad (10\text{-}56)$$

The numerical integration of $x_1(t)$ by means of the three schemes described above is illustrated in Figure 10-16. Since the signal $x_1(t)$ has a jump discontinuity at $t = 0$, the advanced rectangular and the polygonal integrations produce approximations which have areas prior to $t = 0$.

Stability Considerations

Since the digital model of the simulated system must be stable regardless of the method of approximation, it is necessary to investigate the stability ef-

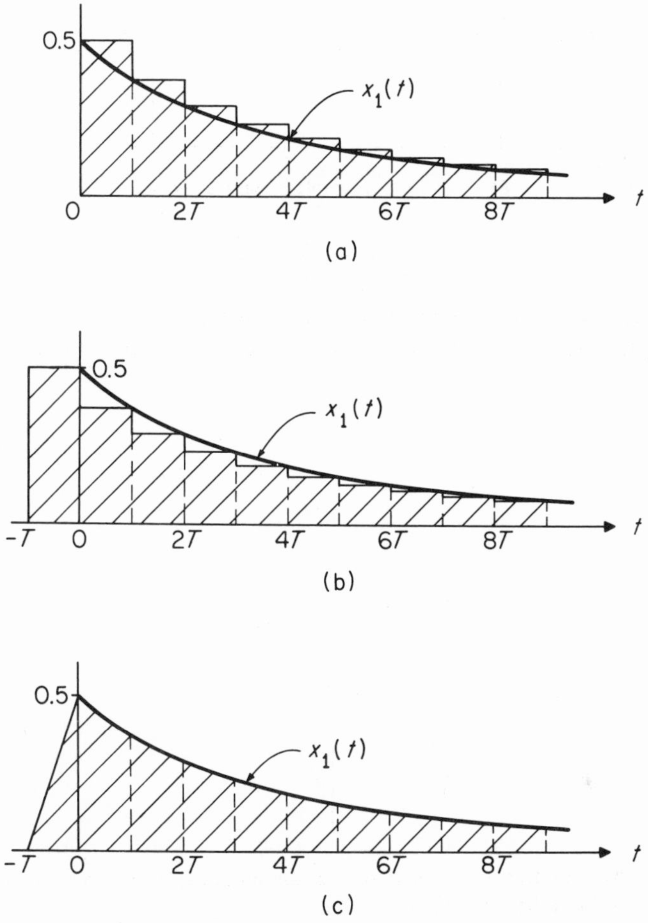

FIG. 10-16. Input to integrator $x_1(t)$ of system in Fig. 10-8(a) and areas showing the outputs of numerical integration by (a) rectangular integration, (b) advanced rectangular integration, (c) polygonal integration.

fects of the integration schemes discussed in the preceding section. The root locus method is used here for its simplicity. Let us use the closed-loop system configuration of Figure 10-17 for this study. Although in general the actual effects on stability due to the various numerical integration schemes will depend on the transfer function of the continuous-data system being simulated, it is believed that the simple integrator system of Figure 10-17 will provide a qualitative indication of simulation stability in general. For the continuous-data system, the open-loop transfer function is simply

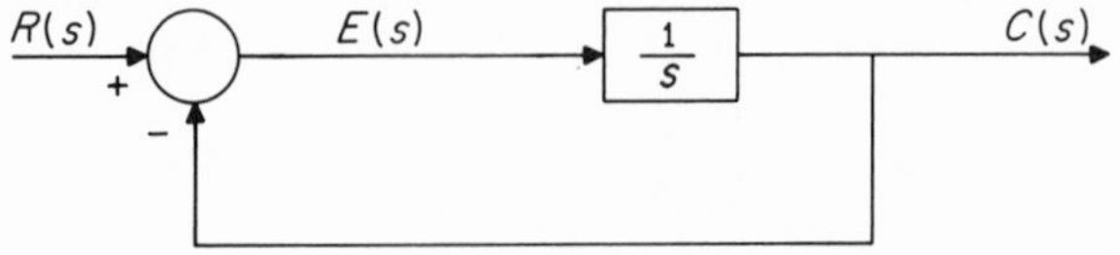

FIG. 10-17. A simple integrator system.

$$G(s) = \frac{1}{s} = \frac{T}{\ln z} \tag{10-57}$$

Therefore, when $T = 0$, the characteristic equation root is at $z = 1$, and when $T = \infty$, it is at $z = 0$. The root locus is sketched as shown in Figure 10-18.

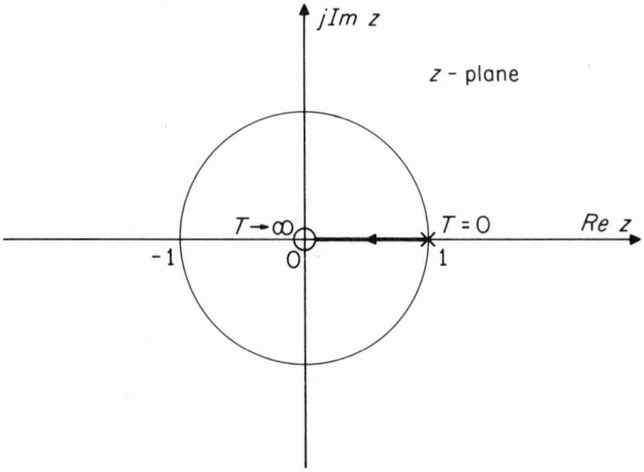

FIG. 10-18. Root locus of the continuous-data system of Fig. 10-17 in the z-plane.

Therefore, the continuous-data model is always stable. For the rectangular integration, the transfer function of the system in Figure 10-17 is replaced by $T/(z - 1)$. Then, the root locus of the digital system is drawn as shown in Figure 10-19. The system is unstable for $T \geq 2$.

For the advanced rectangular integration,

$$G(z) = \frac{Tz}{z - 1} \tag{10-58}$$

The root locus of the digital model is identical to that of Figure 10-18, although the values of T will be different along the locus. For the polygonal integration,

$$G(z) = \frac{T}{2}\left(\frac{z + 1}{z - 1}\right) \tag{10-59}$$

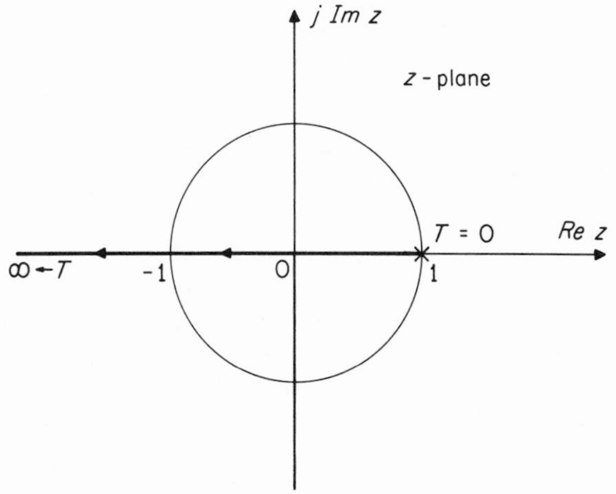

FIG. 10-19. Root locus of the system of Fig. 10-17 with rectangular integration approximation.

and the root locus is shown in Figure 10-20, which represents a stable simulation for all finite T. Therefore, from this investigation, we conclude that, in general, the rectangular integration provides the worst effect on system stability.

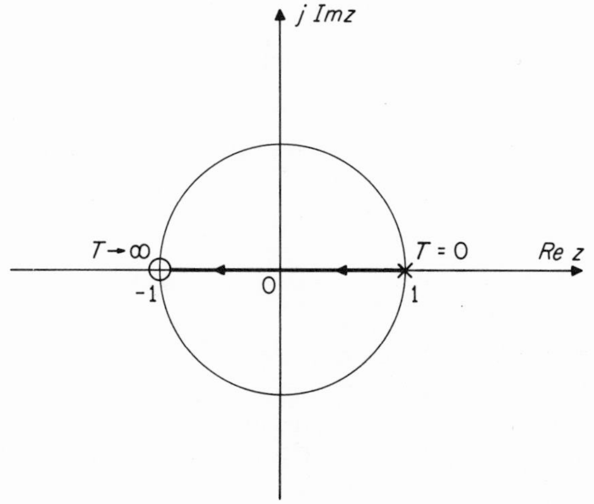

FIG. 10-20. Root locus of the system of Fig. 10-17 with polygonal integration approximation.

In general, higher-order and more complex numerical integration methods such as the Simpson's rules are available. However, these schemes usually generate higher-order transfer functions which cause serious stability problems in the simulation models. Therefore, in control system applications, these higher-order integration methods are seldom used.

10.5 DIGITAL SIMULATION BY THE Z-FORMS

The disadvantage of the foregoing methods of numerical integration for digital simulation is that the system transfer function must first be decomposed so that the integrators are identified and then replaced by the approximating numerical integrator model. A simpler procedure is made possible by use of the z-forms. The z-form method is simpler because one can work directly with the continuous-data transfer function in the s-domain and transform it into an equivalent discrete-data transfer function in the z-domain. The method is described as follows:

Let us assume that a continuous-data signal is represented by $g(t)$. Then, $g(t)$ and its Laplace transform $G(s)$ are related by

$$G(s) = \int_0^\infty g(t)e^{-st}\,dt \tag{10-60}$$

and

$$g(t) = \frac{2}{2\pi j}\int_{c-j\infty}^{c+j\infty} G(s)e^{st}\,ds \tag{10-61}$$

If $g(t)$ is sampled once every T second, the discrete-data signal is represented by

$$g^*(t) = \sum_{k=0}^{\infty} g(kT)\delta(t - kT) \tag{10-62}$$

where

$$g(kT) = \frac{1}{2\pi j}\oint_\Gamma G(z)z^{k-1}\,dz \tag{10-63}$$

where the Γ path is a circle described by $|z| = e^{cT}$ in the z-plane. The line integral of Eq. (10-61) is performed along the path as shown in Figure 10-21. Let us divide the path of integration between $c - j\infty$ and $c + j\infty$ in the s-plane into three parts as shown in Figure 10-21. Then, Eq. (10-61) can be written as

$$g(t) = \frac{1}{2\pi j}\int_{c-j\infty}^{c-j\omega_s/2} G(s)e^{ts}\,ds + \frac{1}{2\pi j}\int_{c-j\omega_s/2}^{c+j\omega_s/2} G(s)e^{ts}\,ds + \frac{1}{2\pi j}\int_{c+j\omega_s/2}^{c+j\infty} G(s)e^{ts}\,ds \tag{10-64}$$

where $\omega_s = 2\pi/T$.

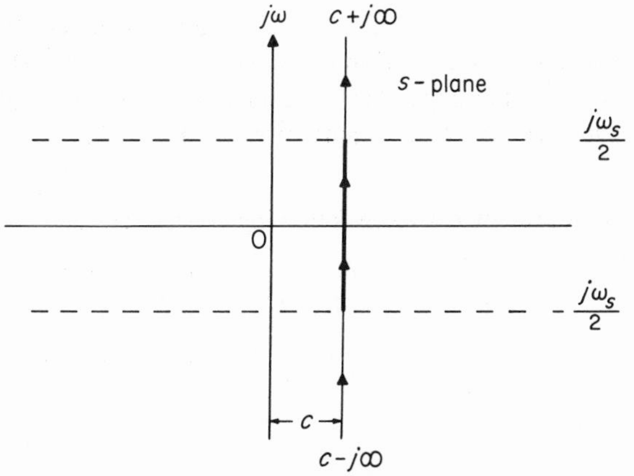

FIG. 10-21. Path of integration of the line integral in Eq. (10-61).

If the sampling period T is sufficiently small, $g(t)$ can be approximated by just the second term on the right side of Eq. (10-64). Therefore,

$$g(t) \cong \frac{1}{2\pi j}\int_{c-j\omega_s/2}^{c+j\omega_s/2} G(s)e^{ts}\,ds \tag{10-65}$$

Replacing t by kT, and letting the right side of the last equation be represented by $g_A(t)$, we have

$$g(kT) \cong g_A(kT) = \frac{1}{2\pi j}\int_{c-j\omega_s/2}^{c+j\omega_s/2} G(s)e^{kTs}\,ds \tag{10-66}$$

Since s is related to z by

$$s = \frac{1}{T}\ln z \tag{10-67}$$

substituting this relation into Eq. (10-66) for s, we get

$$g_A(kT) = \frac{1}{2\pi j}\oint_\Gamma G\left(\frac{1}{T}\ln z\right)e^{kT(1/T)\ln z}\,d\,\frac{1}{T}\ln z \tag{10-68}$$

Or

$$g_A(kT) = \frac{1}{2\pi j}\oint_\Gamma \frac{1}{T}G\left(\frac{1}{T}\ln z\right)z^{k-1}\,dz \tag{10-69}$$

Comparing Eq. (10-69) with Eq. (10-63) we notice that the two integrals are similar except for the factor of $1/T$ in the former. This simply means that $g_A(kT)$, an approximation of $g(kT)$, may be obtained by expanding $G[(1/T)\ln z]/T$ into a power series in z^{-1}. Therefore, we can initially summarize the z-form approximation method as follows:

1. Replace s^{-1} by $s^{-1} = T/\ln z$ in the transfer function $G(s)$ of the continuous-data system. This gives $G[(1/T)\ln z]$.

2. The value of $g_A(kT)(k = 0, 1, 2, \cdots)$ which approximates $g(kT)$ is obtained by expanding $G[(1/T)\ln z]/T$ into a power series in z^{-1}. However, before $G[(1/T)\ln z]/T$ can be expanded it must be in the form of a rational function in z, which means that $T/\ln z$ must first be approximated by a truncated series.

Let $\ln z$ be represented by the following power series:

$$\ln z = 2\left[\left(\frac{z-1}{z+1}\right) + \frac{1}{3}\left(\frac{z-1}{z+1}\right)^3 + \frac{1}{5}\left(\frac{z-1}{z+1}\right)^5 + \cdots\right]$$
$$= 2\left[\left(\frac{1-z^{-1}}{1+z^{-1}}\right) + \frac{1}{3}\left(\frac{1-z^{-1}}{1+z^{-1}}\right)^3 + \frac{1}{5}\left(\frac{1-z^{-1}}{1+z^{-1}}\right)^5 + \cdots\right]$$
$$|z| > 0 \quad (10\text{-}70)$$

Then,

$$\frac{1}{s} = \frac{T}{\ln z} = \frac{T/2}{u + \frac{1}{3}u^3 + \frac{1}{5}u^5 + \cdots} \quad (10\text{-}71)$$

where

$$u = \frac{1-z^{-1}}{1+z^{-1}} \quad (10\text{-}72)$$

By synthetic division, Eq. (10-71) is written

$$\frac{1}{s} = \frac{T}{\ln z} = \frac{T}{2}\left[\frac{1}{u} - \frac{1}{3}u - \frac{4}{45}u^3 - \frac{44}{945}u^5 + \cdots\right] \quad (10\text{-}73)$$

In general, for positive integral n,

$$\frac{1}{s^n} = \left(\frac{T}{2}\right)^n\left[\frac{1}{u} - \frac{1}{3}u - \frac{4}{45}u^3 - \frac{44}{945}u^5 + \cdots\right] \quad (10\text{-}74)$$

Notice that the right side of Eq. (10-73) as well as that of Eq. (10-74) is in the form of a Laurent series. If we take only the principal part and the constant term of the Laurent series, we get

$$\frac{1}{s^n} \cong \frac{N_n(z^{-1})}{(1-z^{-1})^n} = G_n(z^{-1}) \quad (10\text{-}75)$$

where $N_n(z^{-1})$ is a polynomial in powers of z^{-1}, and $G_n(z^{-1})$ is called the *z-form* of s^{-n}. It should be pointed out that since $s = 0$ corresponds to $z = 1$, the poles on both sides of Eq. (10-75) correspond. This means that although the right side of Eq. (10-75) contains only the principal part and the constant term of the Laurent series of s^{-n}, inclusion of additional terms would introduce additional poles in the z-domain and, therefore, would actually lead to larger rather than smaller errors.

Let us now demonstrate the determination of the z-forms for $n = 1$ and $n = 2$.

For $n = 1$, taking the principal part of the right side of Eq. (10-73) yields

$$\frac{1}{s} = \frac{T}{\ln z} \cong G_1(z^{-1}) = \frac{T}{2}\left(\frac{1}{u}\right)$$
$$= \frac{T}{2}\left(\frac{1 + z^{-1}}{1 - z^{-1}}\right) \tag{10-76}$$

For $n = 2$,

$$\frac{1}{s^2} = \left(\frac{T}{\ln z}\right)^2 \cong G_2(z^{-1}) = \left(\frac{T}{2}\right)^2\left[\left(\frac{1}{u}\right)^2 - \frac{2}{3}\right]$$
$$= \left(\frac{T^2}{12}\right)\frac{1 + 10z^{-1} + z^{-2}}{(1 - z^{-1})^2} \tag{10-77}$$

Following the same procedure, the z-forms for higher orders of s^{-1} are determined and are tabulated in Table 10-1.

Table 10-1

z-FORMS OF s^{-1}

s^{-n}	$G_n(z^{-1})$
s^{-1}	$\frac{T}{2}\frac{1 + z^{-1}}{1 - z^{-1}}$
s^{-2}	$\frac{T^2}{12}\frac{1 + 10z^{-1} + z^{-2}}{(1 - z^{-1})^2}$
s^{-3}	$\frac{T^3}{2}\frac{z^{-1} + z^{-2}}{(1 - z^{-1})^3}$
s^{-4}	$\frac{T^4}{6}\frac{z^{-1} + 4z^{-2} + z^{-3}}{(1 - z^{-1})^4} - \frac{T^4}{720}$
s^{-5}	$\frac{T^5}{24}\frac{z^{-1} + 11z^{-2} + 11z^{-3} + z^{-4}}{(1 - z^{-1})^5}$

Although the z-form for $1/s$ is identical to the expression for the polygonal integration, there is a basic difference between the present z-form method and the numerical integration method discussed earlier. In the polygonal and the rectangular integration approximation, the continuous-data system is first approximated by a digital model, and then the digital form of the input is applied to give the simulated output. However, in the z-form method, the Laplace transforms of the input signal are first multiplied by the transfer function of the continuous-data system, and then the z-forms are substituted into the expression of $C(s)$ to give the approximated output $C_A(z)$.

Therefore, we can now summarize the steps of obtaining an approximation to the response of a continuous-data system by the z-form method as follows:

1. Express the Laplace transform of the output of the system, $C(s)$, as a rational function in powers of s^{-1}.

2. Substitute for s^{-n} the corresponding z-forms using Table 10-1. This converts $C(s)$ into a rational function in powers of z^{-1}.
3. Divide the expression obtained from the last step by T, the sampling period, to give $C_A(z)$.
4. Expand $C_A(z)$ by synthetic division into a power series of the form:

$$c_A(0) + c_A(T)z^{-1} + c_A(2T)z^{-2} + \cdots + c_A(kT)z^{-k} + \cdots$$

where $c_A(kT)$ is the approximate value of the time response $c(t)$ at $t = kT$.

EXAMPLE 10-1

As an illustrative example of the z-form method of digital approximation, let us consider that the open-loop transfer function of a feedback control system with unity feedback is given as

$$G(s) = \frac{K}{s(s+1)} \tag{10-78}$$

where K is a constant gain factor.

The closed-loop transfer function is written

$$\frac{C(s)}{R(s)} = \frac{K}{s^2 + s + K} \tag{10-79}$$

For a unit-step input, the output transform is

$$C(s) = \frac{K}{s(s^2 + s + K)} \tag{10-80}$$

Multiplying the numerator and the denominator of the right-hand side of the last equation by s^{-3}, we have

$$C(s) = \frac{Ks^{-3}}{1 + s^{-1} + Ks^{-2}} \tag{10-81}$$

Now substituting the corresponding forms from Table 10-1 and dividing the result by T, we get

$$C_A(z) = \frac{1}{T}\,\frac{\dfrac{T^3}{2}K\left[\dfrac{z^{-4} + z^{-2}}{(1 - z^{-1})^3}\right]}{1 + \dfrac{T}{2}\left[\dfrac{1 + z^{-1}}{1 - z^{-1}}\right] + \dfrac{T^2K}{12}\left[\dfrac{1 + 10z^{-1} + z^{-2}}{(1 - z^{-1})^2}\right]} \tag{10-82}$$

Rearranging and simplifying, the last expression is written

$$C_A(z) = \frac{6T^2K(z^{-1} + z^{-2})}{(1 - z^{-1})[(12 + 6T + T^2K) + (-24 + 10T^2K)z^{-1} + (12 - 6T + T^2K)z^{-2}]} \tag{10-83}$$

The two roots of the equation

$$(12 + 6T + T^2K)z^2 + (-24 + 10T^2K)z + (12 - 6T + T^2K) = 0 \tag{10-84}$$

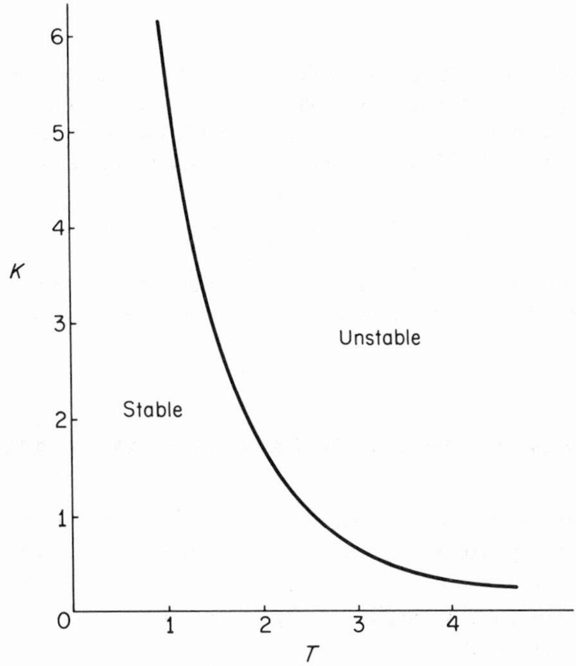

FIG. 10-22. Maximum values of K and T for a stable digital system.

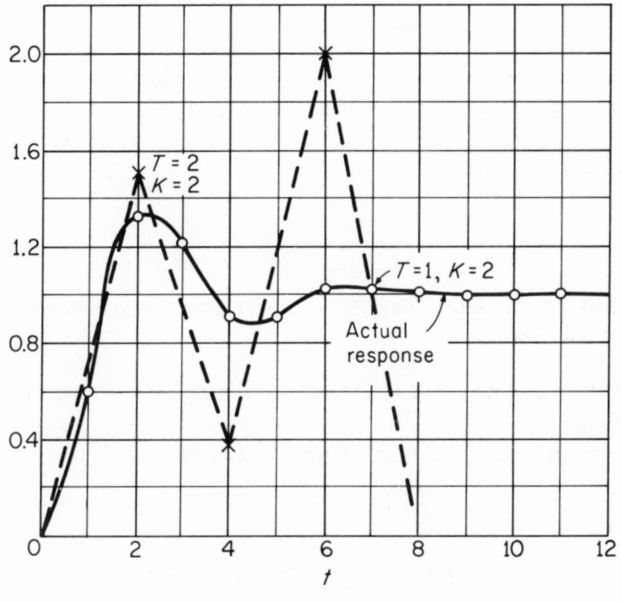

FIG. 10-23. Responses of system in Example 10-1.

determine the stability of the digital model obtained by use of the z-form. Applying the stability criterion, it can be shown that the values of K and T that correspond to a stable simulation are as illustrated in Figure 10-22. We can further show that for all values of K and T which correspond to a stable digital system, the final value of $c_A(kT)$ as k approaches infinity is unity. That is,

$$\lim_{k\to\infty} c_A(kT) = \lim_{z\to 1}(1 - z^{-1})C_A(z) = 1 \tag{10-85}$$

The unit-step responses of the digital model for $K = 2$ and $T = 1$ and $T = 2$, together with the response of the actual continous-data system are drawn as shown in Figure 10-23.

10.6 DIGITAL SIMULATION WITH MINIMUM MEAN-SQUARE ERROR

Since the problem of digital simulation is that of finding an optimal digital model of an analog system such that the error between the two systems is a minimum, we can utilize the design principle of Wiener filtering. This approach will allow us to investigate the effect of digital simulation when the input signals are described statistically.

Let the continuous-data system to be modeled be represented by the transfer function $G(s)$, and the digital model by $D(z)$. The block diagram shown in Figure 10-24 may be used for the minimum mean-square-error design of $D(z)$

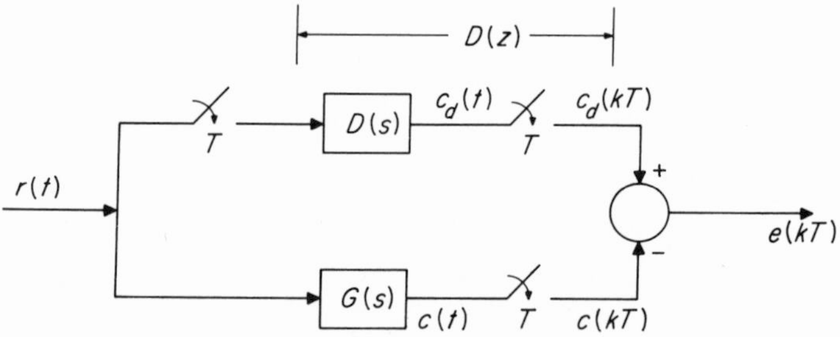

FIG. 10-24. Block diagram of digital filter for simulation of analog system.

given $G(s)$. It is assumed that the input signal $r(t)$ is a member of a stationary random process.

The error at discrete instants of time is written

$$e(kT) = c_d(kT) - c(kT) \tag{10-86}$$

where

$$c_d(kT) = \sum_{n=0}^{\infty} d(nT)r(kT - nT) \tag{10-87}$$

and

$$c(kT) = c(t)\Big|_{t=kT} = \int_0^{\infty} g(\tau)r(kT - \tau)\,d\tau \tag{10-88}$$

In the last two expressions, $d(nT)$ denotes the impulse sequence of $D(z)$ and $g(\tau)$ is the impulse response of $G(s)$.

The mean-square error is

$$\overline{e^2(kT)} = \overline{[c_d(kT) - c(kT)]^2} = \overline{c_d^2(kT)} - \overline{c_d(kT)c(kT)}$$
$$- \overline{c(kT)c_d(kT)} + \overline{c^2(kT)} \tag{10-89}$$

Or

$$\overline{e^2(kT)} + \Phi_{c_dc_d}(0) - 2\Phi_{c_dc}(0) + \Phi_{cc}(0) \tag{10-90}$$

where

$$\Phi_{c_dc_d}(0) = \sum_{n=0}^{\infty}\sum_{m=0}^{\infty} d(nT)\,d(mT)\Phi_{rr}(nT - mT) \tag{10-91}$$

$$\Phi_{c_dc}(0) = \sum_{n=0}^{\infty} d(nT)\int_0^{\infty} g(\tau)\Phi_{rr}(nT - \tau)\,d\tau \tag{10-92}$$

$$\Phi_{cc}(0) = \int_0^{\infty}\int_0^{\infty} g(\tau)g(\sigma)\Phi_{rr}(\tau - \sigma)\,d\tau\,d\sigma \tag{10-93}$$

These three correlation functions are interpreted as shown in Figure 10-25.

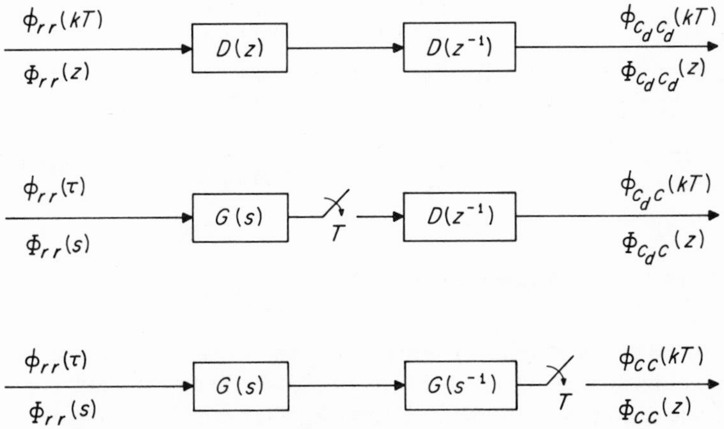

FIG. 10-25. Physical interpretations of Eqs. (10-91), (10-92), and (10-93).

Then, following the development of Chapter 8, the mean-square error can be written

$$\overline{e^2(kT)} = \frac{1}{2\pi j}\oint [\Phi_{c_dc_d}(z) - 2\Phi_{c_dc}(z) + \Phi_{cc}(z)]z^{-1}\,dz \tag{10-94}$$

where

$$\Phi_{c_d c_d}(z) = D(z)\, D(z^{-1})\Phi_{rr}(z) \tag{10-95}$$

$$\Phi_{c_d c}(z) = \Phi_{rr}G(z)\, D(z^{-1}) = \mathscr{Z}[\Phi_{rr}(s)G(s)]\, D(z^{-1}) \tag{10-96}$$

$$\Phi_{cc}(z) = \Phi_{rr}G\bar{G}(z) = \mathscr{Z}[\Phi_{rr}(s)G(s)G(s^{-1})] \tag{10-97}$$

Therefore,

$$\overline{e^2(kT)} = \frac{1}{2\pi j}\oint [D(z)\, D(z^{-1})\Phi_{rr}(z) - 2\Phi_{rr}G(z)\, D(z^{-1})$$

$$+ \Phi_{rr}G\bar{G}(z)]z^{-1}\, dz \tag{10-98}$$

Now following through the optimization procedure as described in Sec. 8.6, by perturbing $D(z)$ by $\epsilon\eta(z)$, we can show that the optimum digital filter in the sense of minimum mean-square error is given by

$$D^o(z) = \frac{z}{\Phi_{rr}^+(z)}\left\{\frac{\Phi_{rr}G(z)}{z\Phi_{rr}^-(z)}\right\}_+ \tag{10-99}$$

where

$$\Phi_{rr}(z) = \Phi_{rr}^+(z)\Phi_{rr}^-(z) \tag{10-100}$$

and $\Phi_{rr}^-(z)$ denotes the part with poles and zeros inside the unit circle.

$$\frac{\Phi_{rr}G(z)}{z\Phi_{rr}^-(z)} = \underbrace{\left\{\frac{\Phi_{rr}G(z)}{z\Phi_{rr}^-(z)}\right\}_+}_{\uparrow \text{ poles inside unit circle}} + \underbrace{\left\{\frac{\Phi_{rr}G(z)}{z\Phi_{rr}^-(z)}\right\}_-}_{\uparrow \text{ poles outside unit circle}} \tag{10-101}$$

As in Chapter 8, in order for $D(z)$ to have at least one more pole than zero so that the impulse sequence is zero at $t = 0$, a suboptimal solution is

$$D^o(s) = \frac{1}{\Phi_{rr}^+(z)}\left\{\frac{\Phi_{rr}G(z)}{\Phi_{rr}^-(z)}\right\}_+ \tag{10-102}$$

10.7 EFFECTS OF QUANTIZATION

The process of converting a physical quantity in analog form into digital or numerical form is called quantization. In digital systems, only discrete levels of amplitude of a signal are allowed. Therefore, when considering analog to digital conversion or digital simulation of analog system it is important to consider the effect of amplitude quantization of signals. The block diagram of an amplitude quantizer, together with its typical input-output characteristics, is shown in Figure 10-26. The dotted line represents the desired transfer characteristic and the staircase function is the actual characteristic of the quantizer. Notice that the input $r(t)$ to the quantizer could be of any form or magnitude, but the output $q(t)$ can take only those discrete quantizing levels which are nearest to the value of $r(t)$. The transfer characteristic shown in Figure 10-26 has uniformly

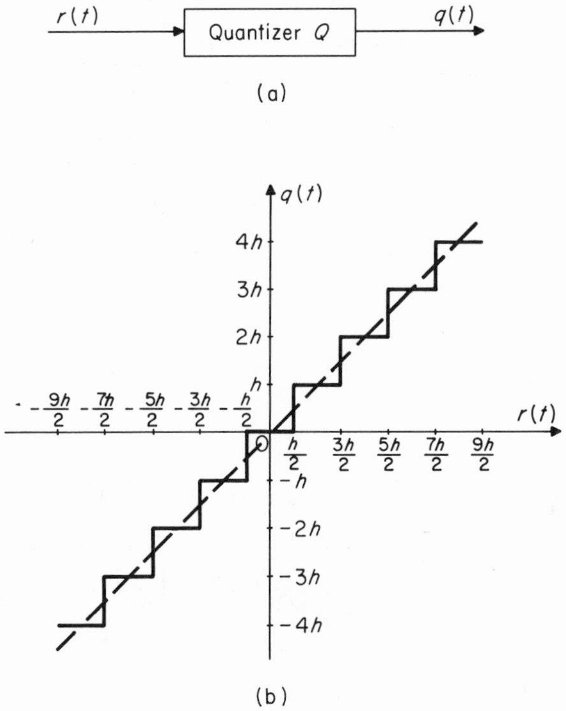

FIG. 10-26. (a) Block diagram of a quantizer; (b) transfer characteristic of a quantizer.

spaced quantizing levels, so that when the input amplitude lies between $-h/2$ and $h/2$ the output is zero. When the amplitude of the input lies between $h/2$ and $3h/2$, the output amplitude is h, and so on.

If we regard the straight line with a slope of unity as the input to the quantizer of Figure 10-26, the output is also the staircase signal as shown in Figure 10-27. The quantization error is defined as the difference between the input and the output of the quantizer; that is

$$\epsilon_q(t) = r(t) - q(t) \tag{10-103}$$

For the case considered at present the quantization error is illustrated Figure 10-27.

With the characteristic of a quantizer defined as above, the operation of an analog-to-digital converter can be replaced by the equivalent model involving a sampler and a quantizer, as shown in Figure 10-28.

In view of the transfer characteristic of the quantizer as shown in Figure 10-26, it is apparent that the device is nonlinear. In contrast to this, a sampler is a linear device. Therefore, the operation with the samplers as shown in Figure 10-29(a) holds as a test of linearity, whereas the operations on quantizers shown

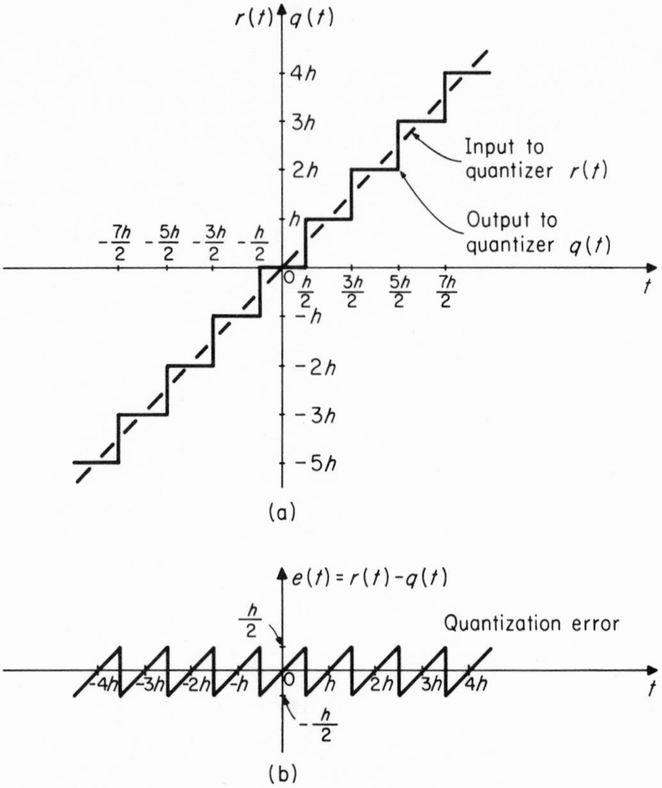

FIG. 10-27. (a) Input-output signals of a quantizer; (b) quantization error.

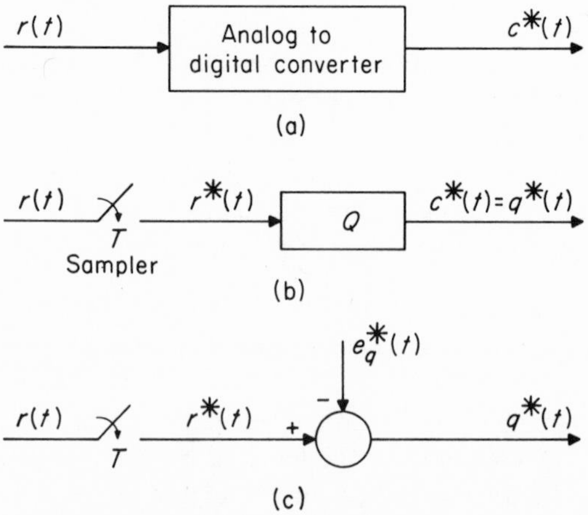

FIG. 10-28. (a) Block diagram of an analog-to-digital converter; (b) equivalent representation of an analog-to-digital converter by sample-and-quantization; (c) representation of quantizer by the algebraic sum of input and quantization error.

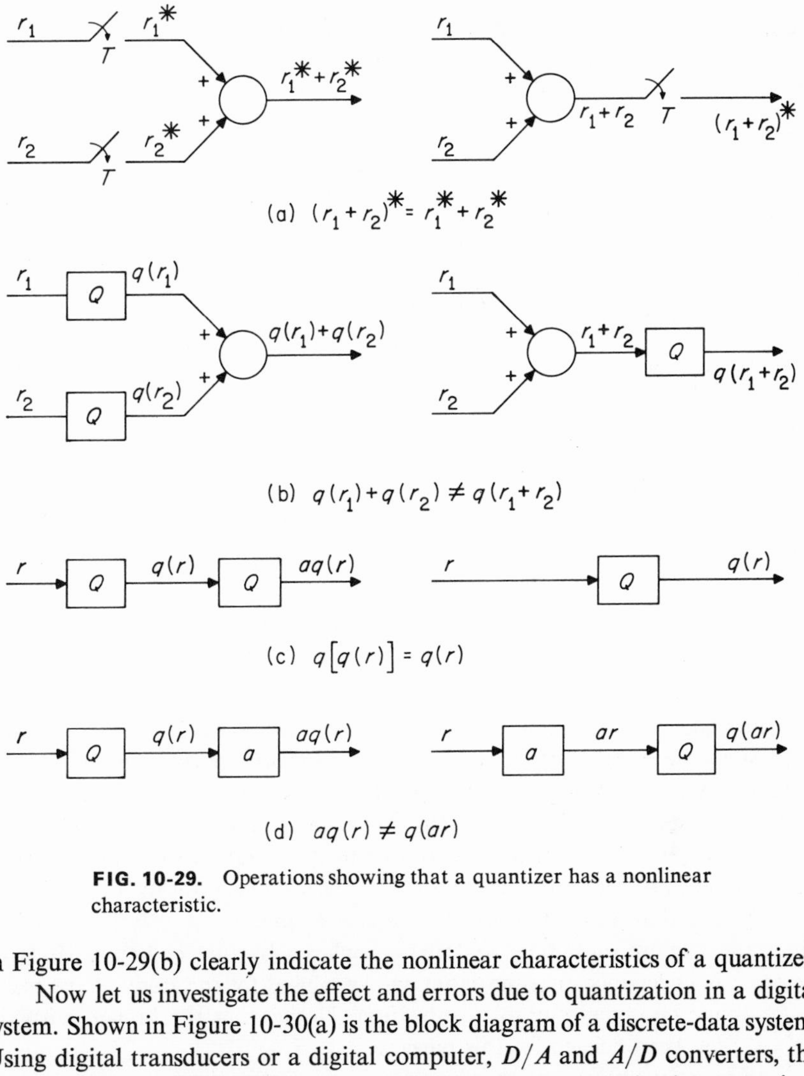

FIG. 10-29. Operations showing that a quantizer has a nonlinear characteristic.

in Figure 10-29(b) clearly indicate the nonlinear characteristics of a quantizer.

Now let us investigate the effect and errors due to quantization in a digital system. Shown in Figure 10-30(a) is the block diagram of a discrete-data system. Using digital transducers or a digital computer, D/A and A/D converters, the system is implemented as shown in Figure 10-30(b). Quantization operations are found in the A/D converters, the digital transducers, and the computer. The equivalent representation of an A/D converter by a sampler-quantizer combination has been discussed. The role that the quantizers play in the implementation of digital controllers can be investigated by means of a simple illustrative example. Consider that the digital controller of the system in Figure 10-30 is represented by

$$D(z) = \frac{M(z)}{E(z)} = \frac{1 + az^{-1}}{1 + bz^{-1}} \tag{10-104}$$

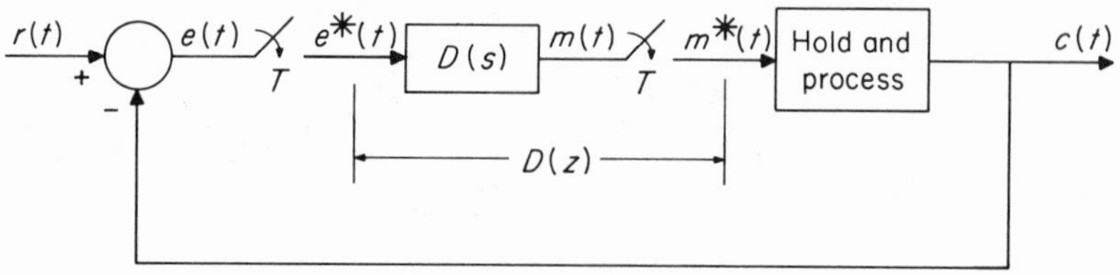

(a)

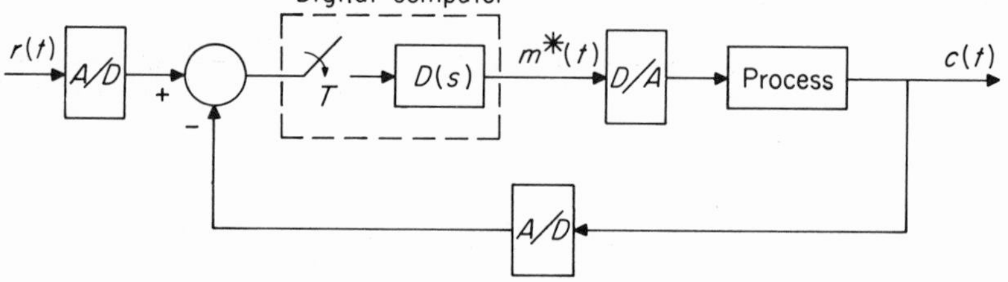

(b)

FIG. 10-30. (a) Block diagram of a discrete-data control system; (b) implementation of the system in (a) by digital transducers or digital computer and D/A and A/D converters.

A state diagram of the controller is drawn as shown in Figure 10-31(a). However, since the signal at all points in the digital transducer is in digital form, it is necessary in the implementation model to include quantizers at appropriate locations as shown in Figure 10-31(b). Since the quantization error has a maximum bound $\pm h/2$, the "worst" error due to quantization in a digital system can be studied by replacing the quantizer in the state diagram by a branch with unity gain and an external source with signal magnitude of $\pm h/2$. Therefore, the state diagram of Figure 10-31(b) is replaced by that of Figure 10-31(c), where it is understood that the latter model is used for the study of the maximum bound on errors due to quantization. Notice that the branch with the gain z^{-1} has been deleted for this purpose.

Without quantization, the dynamic equations are linear and are written from Figure 10-31(a).

$$x_1[(k+1)T] = -bx_1(kT) + e(kT) \tag{10-105}$$

$$m(kT) = (a-b)x_1(kT) + e(kT) \tag{10-106}$$

Let $\hat{x}_{1q}(kT)$ and $\hat{m}_q(kT)$ represent the system variables when the maximum quantization error $\pm h/2$ is considered. Then, from Figure 10-31(c), we have, by applying Mason's gain formula,

$$\hat{x}_{1q}[(k+1)T] = -b\hat{x}_{1q}(kT) + h + e(kT) \tag{10-107}$$

$$\hat{m}_q(kT) = (a-b)\hat{x}_{1q}(kT) + 2h + e(kT) \tag{10-108}$$

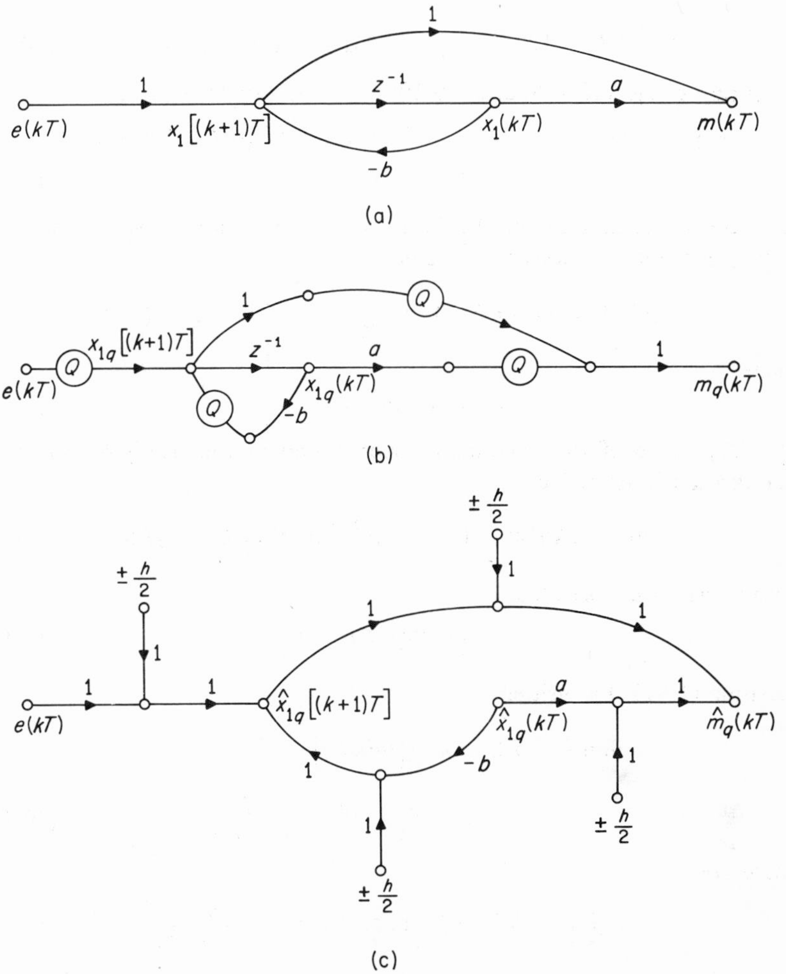

FIG. 10-31. (a) State diagram of the digital controller in Eq. (10-87); (b) state diagram with quantizers inserted at appropriate locations; (c) state diagram with quantizers replaced by unity-gain branches and inputs of $\pm h/2$ to give maximum bounds on errors due to quantization.

For the upper bound in error, the positive signs in $\pm h/2$ have been used. (As an alternative, all negative signs can be used.)

Let us define the error in the state variable due to quantization as

$$\epsilon_x(kT) = x_1(kT) - x_{1q}(kT) \tag{10-109}$$

and the upper-bound error is

$$\hat{\epsilon}_x(kT) = x_1(kT) - \hat{x}_{1q}(kT) \tag{10-110}$$

Then,

$$\hat{\epsilon}_x[(k+1)T] = x_1[(k+1)T] - \hat{x}_{1q}[(k+1)T] \tag{10-111}$$

Substituting Eqs. (10-105) and (10-107) into Eq. (10-111) yields

$$\begin{aligned}\hat{\epsilon}_x[(k+1)T] &= -bx_1(kT) + e(kT) + b\hat{x}_{1q}(kT) - h - e(kT)\\ &= -b\hat{e}_x(kT) - h\end{aligned} \tag{10-112}$$

Since the last equation is of the form of a state equation, we can solve for $e_x(kT)$ in the usual manner. Therefore,

$$\hat{\epsilon}_x(NT) = \Phi(NT)\hat{\epsilon}_x(0) + \sum_{k=0}^{N-1}\Phi[(N-k-1)T](-h) \tag{10-113}$$

where

$$\Phi(NT) = e^{-bNT} \tag{10-114}$$

The magnitude of the maximum upper bound of the steady-state error of quantization is defined as

$$|\lim_{N\to\infty}\hat{\epsilon}_x(NT)| = |\lim_{N\to\infty}(-h)\sum_{k=0}^{N-1}\Phi[(N-k-1)T]| \tag{10-115}$$

since if the system is stable,

$$\lim_{N\to\infty}\Phi(NT) = 0 \tag{10-116}$$

Equation (10-115) is written

$$\begin{aligned}|\lim_{N\to\infty}\hat{\epsilon}_x(NT)| &= h\left|\sum_{k=0}^{\infty}\Phi(kT)\right|\\ &= h\left|\sum_{k=0}^{\infty}e^{-bkT}\right| = \frac{h}{1-e^{-bT}}\end{aligned} \tag{10-117}$$

Therefore,

$$|\lim_{N\to\infty}\epsilon_x(NT)| \le |\lim_{N\to\infty}\hat{\epsilon}_x(NT)| = \frac{h}{1-e^{-bT}} \tag{10-118}$$

In a similar manner, the upper bound of the quantization error in the output $m(kT)$ is defined as

$$\hat{\epsilon}_m(kT) = m(kT) - \hat{m}_q(kT) \tag{10-119}$$

Substituting Eqs. (10-106) and (10-108) into Eq. (10-119), we get

$$\begin{aligned}\hat{\epsilon}_m(kT) &= (a-b)[x_1(kT) - \hat{x}_{1q}(kT)] - 2h\\ &= (a-b)\hat{\epsilon}_x(kT) - 2h\end{aligned} \tag{10-120}$$

Using the results of Eq. (10-117), we get

$$|\lim_{N\to\infty}\hat{\epsilon}_m(NT)| = \left|(a-b)\frac{h}{1-e^{-bT}} - 2h\right| \tag{10-121}$$

which is the maximum upper bound on the steady-state error in the output due to quantization.

EXAMPLE 10-2

As a more complex example of the error bounds on quantization, let us consider the control system shown in Figure 10-30. The transfer functions are given as

Digital Controller:

$$D(z) = \frac{M(z)}{E(z)} = 2.72\,\frac{1 - 0.368z^{-1}}{1 + 0.72z^{-1}} \tag{10-122}$$

Process:

$$G(z) = \mathscr{Z}\left[\frac{1 - e^{-Ts}}{s}\,\frac{1}{s(s+1)}\right] = \frac{0.368z^{-1} + 0.264z^{-1}}{(1 - z^{-1})(1 - 0.368z^{-1})} \tag{10-123}$$

The sampling period is one second.

A state diagram of the system is shown in Figure 10-32. Since the digital con-

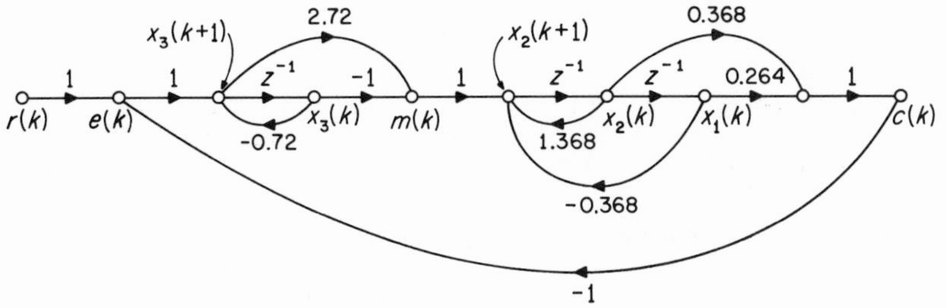

FIG. 10-32. A state diagram of the discrete-data system in Example 10-2.

troller is implemented by digital transducers or a digital computer, the effect of quantization should be included. The objective of this problem is to determine the maximum upper bound of the steady-state error due to quantization.

Shown in Figure 10-33 is the state diagram of the system with quantizers placed at appropriate places in the digital controller. For the study of the maximum error bound, we replace the quantizers in the state diagram of Figure 10-33 by a unity-gain branch and an external source of strength $\pm h/2$. Actually, maximum error is considered by using errors of the same sign.

Following this procedure, the state diagram of Figure 10-34 is obtained.

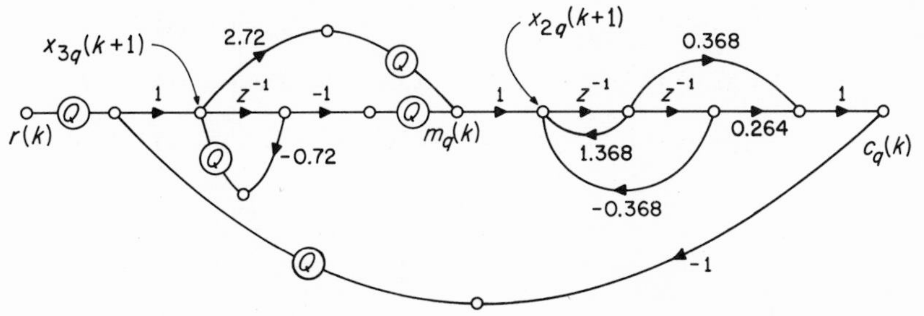

FIG. 10-33. The state diagram of Fig. 10-32 with quantizers added in the digital controller.

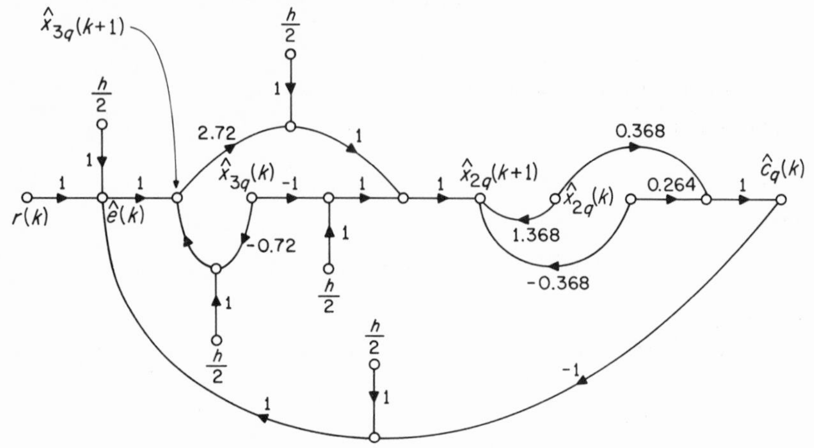

FIG. 10-34. The state diagram of Fig. 10-33 with quantizers represented by branches with unity gains and external sources of strength $h/2$.

The state equations of the system without quantization are written from Figure 10-32.

$$\begin{bmatrix} x_1(k+1) \\ x_2(k+1) \\ x_3(k+1) \end{bmatrix} = \begin{bmatrix} 0 & 1 & 0 \\ -1.086 & 0.368 & -2.958 \\ -0.264 & -0.368 & -0.72 \end{bmatrix} \begin{bmatrix} x_1(k) \\ x_2(k) \\ x_3(k) \end{bmatrix} + \begin{bmatrix} 0 \\ 2.72 \\ 1 \end{bmatrix} r(k) \tag{10-124}$$

The output equation is

$$c(k) = [0.264 \quad 0.368 \quad 0] \begin{bmatrix} x_1(k) \\ x_2(k) \\ x_3(k) \end{bmatrix} \tag{10-125}$$

For the system with quantizers, the dynamic equations considering the maximum errors due to quantization are written directly from Figure 10-34.

$$\begin{bmatrix} \hat{x}_{1q}(k+1) \\ \hat{x}_{2q}(k+1) \\ \hat{x}_{3q}(k+1) \end{bmatrix} = \begin{bmatrix} 0 & 1 & 0 \\ -1.086 & 0.368 & -2.958 \\ -0.264 & -0.368 & -0.72 \end{bmatrix} \begin{bmatrix} \hat{x}_{1q}(k) \\ \hat{x}_{2q}(k) \\ \hat{x}_{3q}(k) \end{bmatrix} + \begin{bmatrix} 0 \\ 2.72 \\ 1 \end{bmatrix} r(k) + \begin{bmatrix} 0 \\ h + 2.72\left(\frac{3h}{2}\right) \\ \frac{3h}{2} \end{bmatrix} \tag{10-126}$$

$$\hat{c}_q(k) = [0.264 \quad 0.368 \quad 0]\begin{bmatrix}\hat{x}_{1q}(k)\\ \hat{x}_{2q}(k)\\ \hat{x}_{3q}(k)\end{bmatrix} \tag{10-127}$$

The maximum upper bound of the error in the state vector due to quantization is defined as

$$\hat{e}_x(k) = \mathbf{x}_q(k) - \hat{\mathbf{x}}_q(k) \tag{10-128}$$

Thus,

$$\begin{aligned}\hat{e}_x(k+1) &= \mathbf{x}_q(k+1) - \hat{\mathbf{x}}_q(k+1)\\ &= A\hat{e}_x(k) - \mathbf{u}\end{aligned} \tag{10-129}$$

where

$$A = \begin{bmatrix} 0 & 1 & 0\\ -1.086 & 0.368 & -2.958\\ -0.264 & -0.368 & -0.72\end{bmatrix} \tag{10-130}$$

$$\mathbf{u} = \begin{bmatrix}u_1\\ u_2\\ u_3\end{bmatrix} = \begin{bmatrix}0\\ h + 2.72\left(\frac{3h}{2}\right)\\ \frac{3h}{2}\end{bmatrix} = \begin{bmatrix}0\\ 5.08h\\ 1.5h\end{bmatrix} \tag{10-131}$$

The solution of Eq. (10-129) is

$$\hat{e}_x(N) = \Phi(N)\hat{e}_x(0) + \sum_{k=0}^{N-1}\Phi(N-k-1)\mathbf{u} \tag{10-132}$$

where

$$\Phi(N) = e^{AN} = \begin{bmatrix}\Phi_{11}(N) & \Phi_{12}(N) & \Phi_{13}(N)\\ \Phi_{21}(N) & \Phi_{22}(N) & \Phi_{23}(N)\\ \Phi_{31}(N) & \Phi_{32}(N) & \Phi_{33}(N)\end{bmatrix} \tag{10-133}$$

Expanding Eq. (10-132) in scalar form, the upper bound on the steady-state error of x_i due to quantization is written

$$|\lim_{N\to\infty}\hat{e}_{xi}(N)| = \left|u_1\sum_{k=0}^{\infty}\Phi_{i1}(k)\right| + \left|u_2\sum_{k=0}^{\infty}\Phi_{i2}(k)\right| + \left|u_3\sum_{k=0}^{\infty}\Phi_{i3}(k)\right| \tag{10-134}$$

$i = 1, 2, 3$.

An alternative way of computing the quantities of Eq. (10-134) is by use of the z-transform. Taking the z-transform on both sides of Eq. (10-129), and rearranging, we get

$$\mathscr{E}_x(z) = (zI - A)^{-1}\hat{e}_x(0) - (zI - A)^{-1}\frac{1}{1 - z^{-1}}\mathbf{u} \tag{10-135}$$

Applying the final-value theorem of z-transform to the last equation, we have

$$\lim_{N\to\infty}\hat{e}_x(N) = \lim_{z\to 1}(1 - z^{-1})\mathscr{E}_x(z) \tag{10-136}$$

Since $(zI - A)^{-1}$ does not contain $(1 - z^{-1})$ in its denominator, substitution of Eq. (10-135) into Eq. (10-136) yields

$$\lim_{N\to\infty} \hat{\epsilon}_x(N) = \lim_{z\to 1} (zI - A)^{-1}\mathbf{u} \tag{10-137}$$

The matrix $(zI - A)^{-1}$ is evaluated

$$(zI - A)^{-1} = \frac{1}{z(z^2 + 0.352z - 0.265)} \times \begin{bmatrix} z^2 + 0.352z - 1.35 & z + 0.72 & -2.958 \\ -1.086z & z(z + 0.72) & -2.958z \\ -0.264z + 0.497 & -(0.368z + 0.264) & z^2 - 0.368z + 1.086 \end{bmatrix} \tag{10-138}$$

Then

$$\lim_{z\to 1} (zI - A)^{-1} = \frac{1}{1.325}\begin{bmatrix} 0 & 1.72 & -2.958 \\ -1.086 & 1.72 & -2.958 \\ 0.233 & -0.632 & 1.718 \end{bmatrix} \tag{10-139}$$

Therefore, the three components of the magnitudes of the upper bound on $\hat{\epsilon}_x(N)$ are computed as

$$|\lim_{N\to\infty} \hat{\epsilon}_{x1}(N)| = \left| \frac{1}{1.325}[0 \quad 1.72 \quad -2.958]\begin{bmatrix} 0 \\ 5.08h \\ 1.5h \end{bmatrix} \right| = 3.96h \tag{10-140}$$

$$|\lim_{N\to\infty} \hat{\epsilon}_{x2}(N)| = \left| \frac{1}{1.325}[-1.086 \quad 1.72 \quad -2.958]\begin{bmatrix} 0 \\ 5.08h \\ 1.5h \end{bmatrix} \right| = 3.96h \tag{10-141}$$

$$|\lim_{N\to\infty} \hat{\epsilon}_{x3}(N)| = \left| \frac{1}{1.325}[0.233 \quad -0.632 \quad 1.718]\begin{bmatrix} 0 \\ 5.08h \\ 1.5h \end{bmatrix} \right| = 0.475h \tag{10-142}$$

The quantization error in the output $c(k)$ is defined as

$$\epsilon_c(k) = c(k) - c_q(k) \tag{10-143}$$

and using Eqs. (10-125) and (10-127), we have

$$\epsilon_c(k) = [0.264 \quad 0.368 \quad 0]\hat{\boldsymbol{\epsilon}}_x(k) \tag{10-144}$$

Therefore, the magnitude of the upper bound of the quantization error in $c(k)$ is

$$\begin{aligned} |\lim_{N\to\infty} \epsilon_c(N)| &= 0.264|\lim_{N\to\infty} \hat{\epsilon}_{x1}(N)| + 0.368|\lim_{N\to\infty} \hat{\epsilon}_{x2}(N)| \\ &= 2.5h \end{aligned} \tag{10-145}$$

REFERENCES

1. Salzer, J. M., "Frequency Analysis of Digital Computers Operating in Real Time," *Proc. IRE*, Vol. 42, No. 2, February 1954, pp. 457–466.

2. Fryer, W. D., and Schultz, W. C., "A Survey of Methods for Digital Simulation of Control Systems," Cornell Aeronautical Lab., Cornell University, New York.

3. Halijak, C. A., "Digital Approximation of the Solutions of Differential Equations Using Trapezoidal Convolution," Bendix Systems Divisions, Report ITM-64, August 1960.

4. Boxer, R., and Thaler, S., "A Simplified Method of Solving Linear and Non-Linear Systems," *Proc. IRE*, Vol. 44, January 1956, pp. 89–101.

5. Liff, A. I., and Wolf, J. K., "On the Optimum Sampling Rate for Discrete-Time Modeling of Continuous-Time Systems," *IEEE Trans. on Automatic Control*, Vol. AC-11, April 1966, pp. 288–290.

6. Widrow, B., "A Study of Rough Amplitude Quantization by Means of Nyquist Sampling Theory," *IRE Trans. on Circuit Theory*, Vol. CT-3, December 1964, pp. 266–276.

7. Tou, J. T., *Digital And Sampled-Data Control Systems*. McGraw-Hill, New York, 1959.

8. Johnson, G. W., "Upper Bound on Dynamic Quantization Error in Digital Control Systems via the Direct Methods of Liapunov," *IEEE Trans. on Automatic Control*, Vol. AC-10, October 1965, pp. 439–448.

9. Sage, A. P., and Burt, R. W., "Optimum Design and Error Analysis of Digital Integrators for Discrete System Simulation," *AFIPS Proc.*, Vol. 27, Part 1, 1965, pp. 903–914.

10. Smith, F. W., "System Laplace-Transform Estimation From Sampled-Data," *IEEE Trans. on Automatic Control*, Vol, AC-13, February 1968, pp. 37–44.

11. Katzenelson, J., "On Errors Introduced by Combined Sampling and Quantization," *IRE Trans. on Automatic Control*, Vol. AC-7, April 1962, pp. 58–68.

12. Kramer, R., "Effect of Quantization on Feedback Systems With Stochastic Inputs," M.I.T. Electronic Systems Lab., Report 7849-R-9, June 1959.

13. Kramer, R., "Effects of Quantization on Feedback Systems With Stochastic Inputs," *IRE Trans. on Automatic Control*, Vol. AC-6, September 1961, pp. 292–305.

14. Greaves, C. J., and Cadzow, J. A., "The Optimal Discrete Filter Corresponding to a Given Analog Filter," *IEEE Trans. on Automatic Control*, Vol. AC-12, June 1967, pp. 304–307.

15. Brule, J. D., "Polynomial Extrapolation of Sampled Data With an Analog Computer," *IRE Trans. on Automatic Control*, Vol. AC-7, January 1962, pp. 76–77.

16. Monroe, A. J., *Digital Processes For Sampled Data Systems*. John Wiley & Sons, New York, 1962.

PROBLEMS

PROBLEMS FOR CHAPTER 2

2-1 A particle of mass M is connected as shown in Figure 2P-1, and is at rest in its equilibrium position at $t = 0$. The force is in the form of blows of impulses of strength F which are given to the particle at $t = nT$ sec $(n = 0, 1, 2, \ldots)$. Write the equation of motion for the system and solve for the displacement $x(t)$ of the particle by means of the Laplace transform method. $K =$ linear spring constant $= M(u^2 + \omega_s^2)$; $u =$ constant and $\omega_s = 2\pi/T$; $f =$ viscous friction coefficient $= 2Mu$.

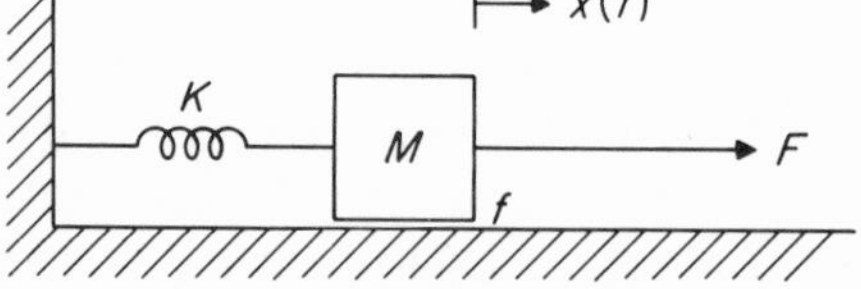

FIG. 2P-1.

2-2 The following signals are sampled by an ideal sampler with sampling period T. Determine the sampler output $f^*(t)$, and evaluate the pulse transform $F^*(s)$ by the Laplace transform method and the complex convolution method.
(a) $f(t) = te^{-at}$
(b) $f(t) = e^{-at}\sin t \qquad (a = \text{positive constant})$

2-3 Prove the "Sampling Theorem."

2-4 Prove that if a signal $f(t)$ contains no frequency components higher than ω_c rad/sec, it is completely characterized by the values $f^{(k)}(nT)$, $f^{(k-1)}(nT)$,

$\ldots, f^{(1)}(nT)$, and $f(nT)$ $(n = 0, 1, 2, \ldots)$, of the signal measured at instants of time separated by $T = (1/2)(k + 1)(2\pi/\omega_c)$ second, where

$$f^{(k)}(nT) = \left.\frac{d^k f(t)}{dt^k}\right|_{t=nT}$$

2-5 (a) Find $f^*(t)$ for $f(t) = e^{-a|t|}$, $(-\infty < t < \infty)$. Find the two-sided Laplace transform of $f^*(t)$, $F^*(s)$. Determine the region of convergence of $F^*(s)$.
(b) Find $f^*(t)$ of $f(t) = e^{-at} - e^{at}$, $(t \geq 0)$, and find $F^*(s)$.
Compare the two $F^*(s)$ functions obtained in (a) and (b). Discuss your findings.

2-6 A "delayed" sampler is considered to close for a short duration p at instants $t = \Delta, \Delta + T, \Delta + 2T, \Delta + 3T, \ldots, \Delta + nT, \ldots$, where $T > \Delta > 0$, and T is the sampling period in seconds. The input to the sampler, $f(t)$, is continuous. The output of the sampler, denoted by $f_p^*(t)_\Delta$, is assumed to be a flat-topped pulsed train, since $\Delta \ll T$. Derive the pulse transform of $f_p^*(t)_\Delta$, $F^*(s, \Delta)$. (Note that this can also be referred to as the "delayed pulse-transform" of $f^*(t)$.) For small p, assume that $1 - e^{-ps} \cong ps$. Determine the "delayed pulse-transform" of $f(t) = e^{-at}$.

2-7 The "delayed pulse-transform" described in Problem 2-6 can be applied to finite-pulsewidth considerations in sampled-data systems. As shown in Figure 2P-7(a), a pulse of width p can be considered as the resultant of N elementary

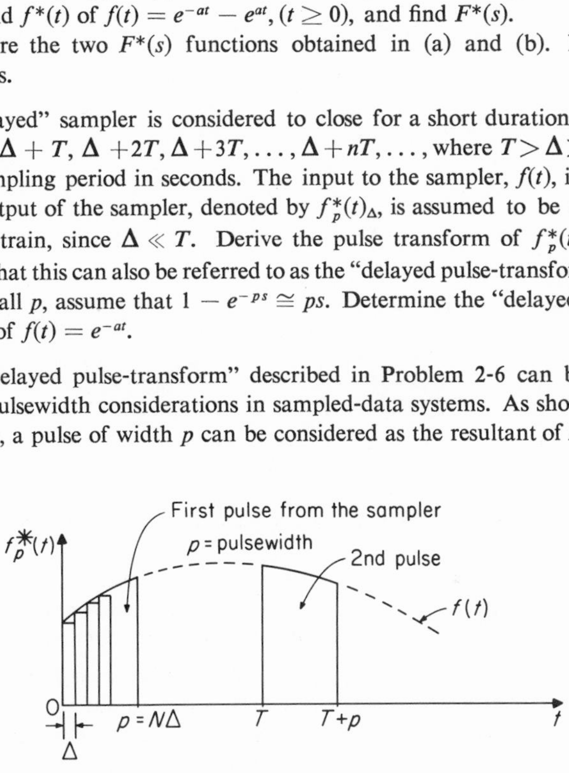

FIG. 2P-7(a).

pulses of width Δ, such that $N\Delta = p$, and Δ is very small. Therefore, a practical sampler can be represented by N samplers $S_0, S_\Delta, S_{2\Delta}, \ldots, S_{(N-1)\Delta}$, connected in parallel as shown in Fig. 2P-7(b). This implies that sampler S_0 samples at instants $0, T, 2T, \ldots, nT, \ldots$; sampler S_Δ is actuated at instants $\Delta, \Delta + T, \Delta + 2T, \ldots, \Delta + nT$; etc. Derive the pulse transform" of $f_p^*(t)$.

2-8 It is pointed out in this chapter that the frequency characteristic of a zero-order hold device has a rapid attenuation for low-frequency signals, while that of a first-order hold has an overshoot at low frequency. It is suggested that a "fractional-order hold" can be devised by extrapolating the function in any given sampling interval with only a fraction of rather than the full first difference. The process is illustrated in Figure 2P-8. In the first sampling period

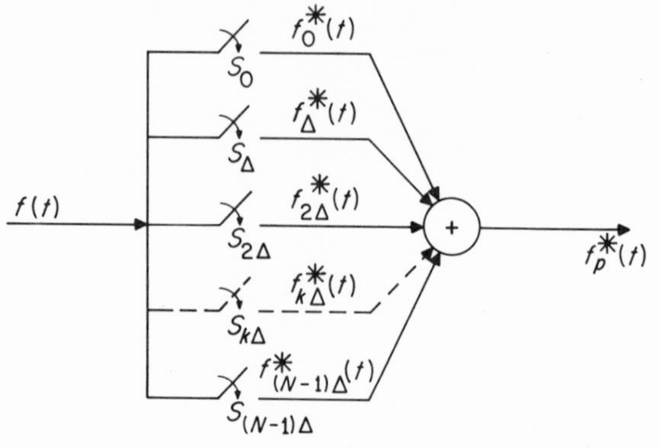

FIG. 2P-7(b).

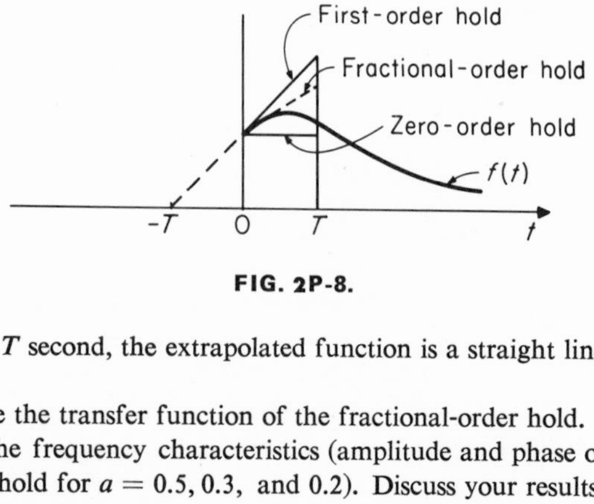

FIG. 2P-8.

from 0 to T second, the extrapolated function is a straight line whose slope is a ($a < 1$).

(a) Derive the transfer function of the fractional-order hold.

(b) Plot the frequency characteristics (amplitude and phase of the fractional-order hold for $a = 0.5, 0.3,$ and 0.2). Discuss your results.

2-9 Two different schemes of triangular hold are shown in Figures 2P-9(a) and 2P-9(b), respectively. Write the impulse responses and the transfer functions of the triangular holds. Which of these is physically realizable?

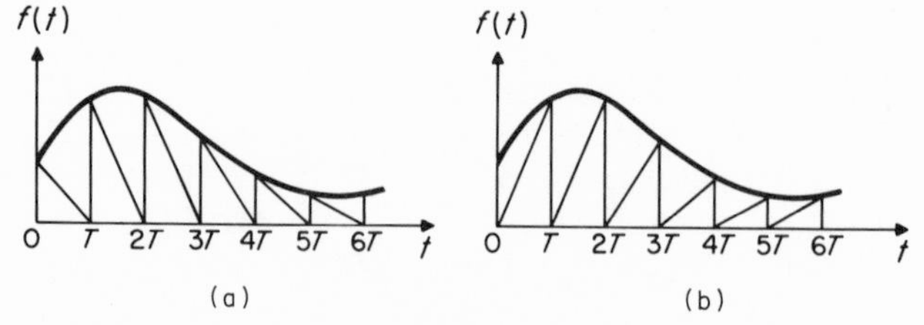

FIG. 2P-9.

2-10 Show that the polygonal approximation of a continuous signal shown in Figure 2P-10 can be made by use of the two triangular holds of Problem 2-9. Derive the impulse response and the transfer function of the polygonal hold device.

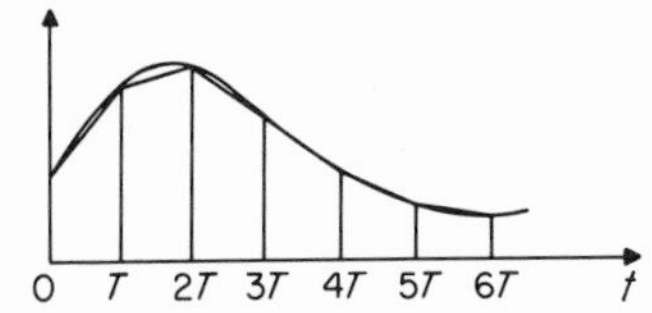

FIG. 2P-10.

PROBLEMS FOR CHAPTER 3

3-1 Determine the z-transforms of the following functions:

(a) $F(s) = \dfrac{1}{s(s^2 + 2)}$

(b) $F(s) = \dfrac{5}{s^2(s + 1)}$

(c) $F(s) = \dfrac{10}{s(s^2 + s + 1)}$

3-2 Given that the z-transform of $g(t)$ for $T = 1$ second is

$$G(z) = \frac{1 - 3z^{-1} + 3z^{-2}}{(1 - 0.5z^{-1})(1 - 0.8z^{-1})}$$

Find the values of $g(kT)$ for the first ten sampling periods.

3-3 Given $f_1(t) = t^2u(t)$ and $f_2(t) = e^{-2t}u(t)$. Find the z-transform of $f_1(t)f_2(t)$ by means of the partial differentiation theorem of z-transformation.

3-4 Given $f_1(t) = t^2u(t)$ and $f_2(t) = e^{-2t}u(t)$. Find the z-transform of $f_1(t)f_2(t)$ by means of the complex convolution theorem of z-transformation.

3-5 Find the inverse z-transform of

$$F(z) = \frac{z(z + 1)}{(z - 1)(z^2 - z + 1)}$$

by means of the following methods:
(a) the real inversion formula
(b) power series expansion
(c) partial fraction expansion.

3-6 If $F_1(s)$ is band limited, that is, $|F_1(j\omega)| = 0$ for $|\omega| \geq \omega_1$, and the sampling frequency ω_s is greater than or equal to $2\omega_1$, show that

$$\mathscr{Z}[F_1(s)F_2(s)] = TF_1(z)\bar{F}_2(z)$$

where

$$F_2(z) = \mathscr{Z}[F_2(s)]$$
$$\bar{F}_2(z) = \mathscr{Z}[\bar{F}_2(s)]$$

and

$$\bar{F}_2(s) = F_2(s) \qquad \text{for } \omega < \omega_1$$
$$= 0 \qquad \text{for } \omega \geq \omega_1$$

If $F_1(s)$ and $F_2(s)$ are both band limited with $|F_1(j\omega)| = 0$ for $|\omega| \geq \omega_1$ and $|F_2(j\omega)| = 0$ for $|\omega| \geq \omega_2$, and the sampling frequency $\omega_s \geq 2 \max(\omega_1, \omega_2)$, show that

$$\mathscr{Z}[F_1(s)F_2(s)] = TF_1(z)F_2(z)$$

3-7 The weighting sequence of a linear discrete-data system is

$$g(kT) = 5e^{-(k-1)T} \qquad \text{for } k \geq 1$$
$$= 0 \qquad \text{for } k = 0$$

Let the input to the system be $r(kT) = kT$ for $k \geq 0$. The sampling period is 0.5 second.

(a) Find the z-transform of the system output.

(b) Find the output sequence $c(kT)$ in closed form.

3-8 Find the output at the sampling instants for the system shown in Figure 3P-8. The input is a unit-step function, and the sampling period is one second.

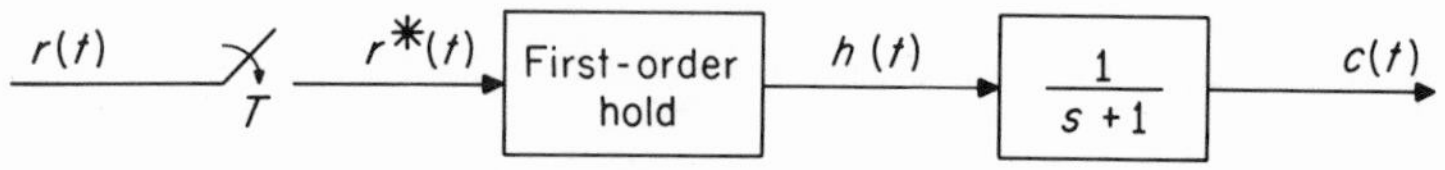

FIG. 3P-8.

3-9 Consider the sampled-data system with a nonintegral time delay as shown in Figure 3P-9. Show that the z-transform of the output is

$$C_m(z) = z^{-1} \sum_{k=0}^{\infty} c[(k+m)T]z^{-k}$$

Assume that $c(t) = 0$ for $t < 0$.

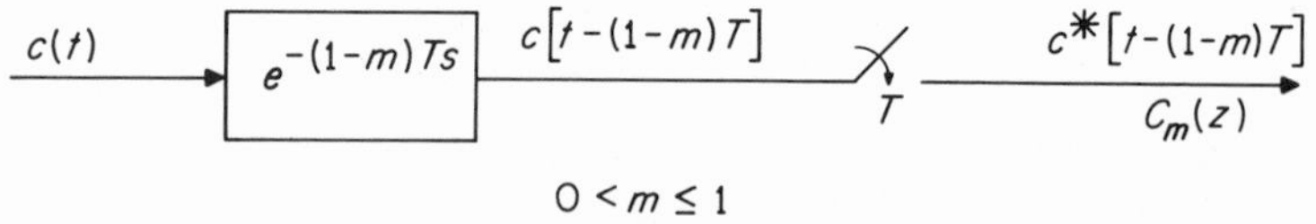

FIG. 3P-9.

3-10 The block diagram of a discrete-data control system is shown in Figure 3P-10. Given that

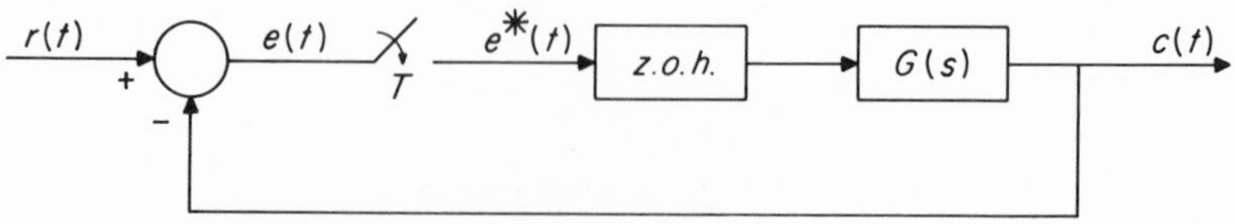

FIG. 3P-10.

$$G(s) = \frac{1}{s(s + 0.2)}$$

$T = 0.2$ second, $r(t) =$ unit-step function. Determine the following:
(a) the z-transform of the output, $C(z)$
(b) the output response at the sampling instants
(c) the final value of $c(kT)$.

3-11 The figure in Figure 3P-11 shows a discrete-data system with a pulse-width

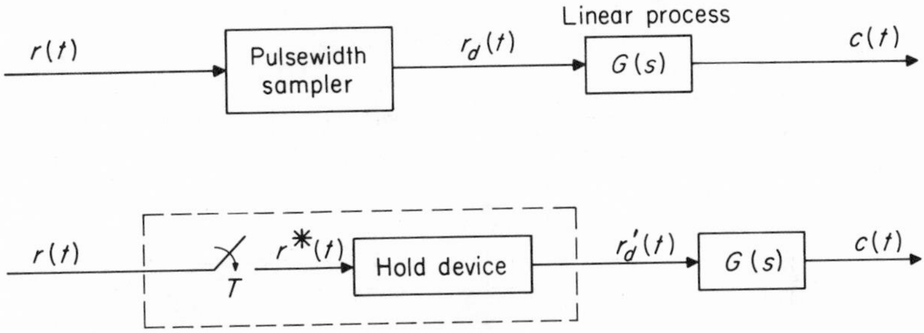

FIG. 3P-11.

sampler. If the input to the pulse-width sampler is $r(t)$, then the sampled output $r_d(t)$ is written

$$r_d(t) = E \sum_{n=0}^{\infty} [\text{sign } r(nT)][u(t - nT) - u(t - nT - p_n)]$$

where p_n, the variable pulse duration, is

$$p_n = a|r(nT)| \qquad \text{for } a|r(nT)| \leq T$$

$$p_n = T \qquad \text{for } |r(nT)| > T$$

a and E are constants, and T is the sampling period. R. E. Andeen suggested in his paper that the pulse-width sampler may be replaced by an equivalent arrangement with a pulse-amplitude sampler and a special hold device as shown in Figure 3P-11, if the principle of equivalent area is considered. In other words, it can be shown that two signals $r_d(t)$ and $r_d'(t)$ are equivalent for suitably small T if

$$\int_{(n-1)T}^{nT} r_d(t)\,dt = \int_{(n-1)T}^{nT} r_d'(t)\,dt \qquad (1)$$

where $r_d'(t)$ is now the output of the equivalent sample-and-hold device. The pulsewidth of $r_d'(t)$ is chosen to be $aR_{\max}$ where $R_{\max}$ is the largest value of $r(t)$ which is measurable.
(a) Is the pulse-width sampler a linear or a nonlinear device? Given reasons for your answer.
(b) Find the transer function $H(s)$ of the hold device so that Eq. (1) is satisfied.
(c) Find $C(z)$, if $G(s) = b/(s + b)$ and $r(t) = u(t)$.

3-12 For the discrete-data systems shown in Figure 3P-12, determine $C(s)$ and $C(z)$

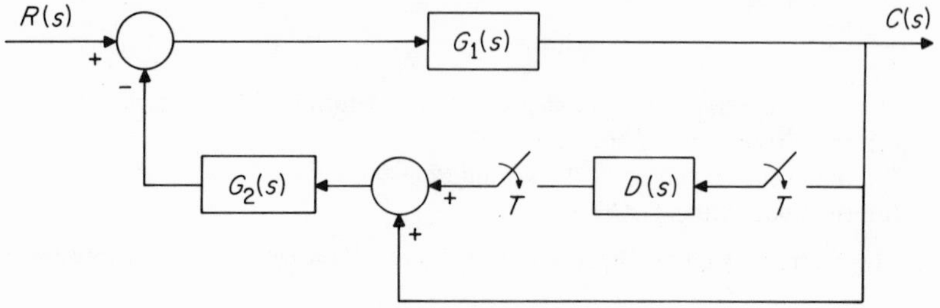

FIG. 3P-12(a).

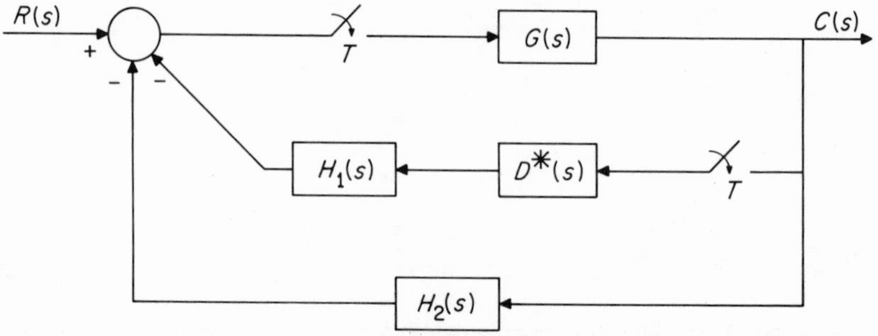

FIG. 3P-12(b).

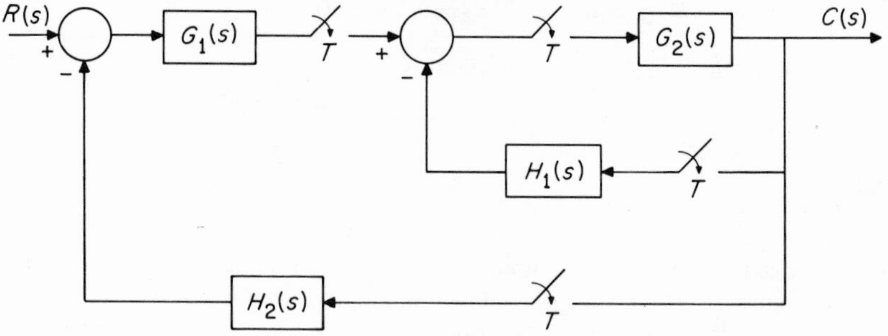

FIG. 3P-12(c).

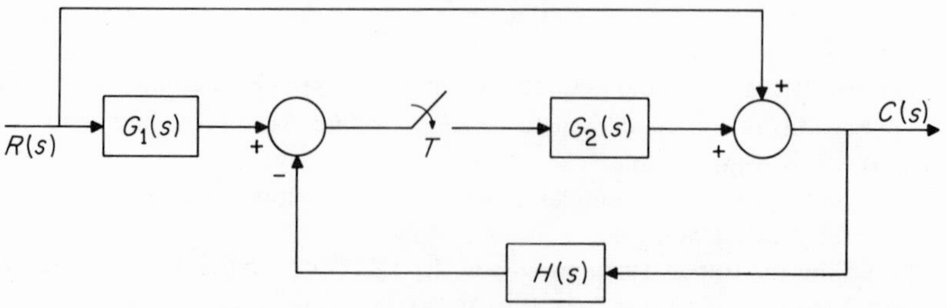

FIG. 3P-12(d).

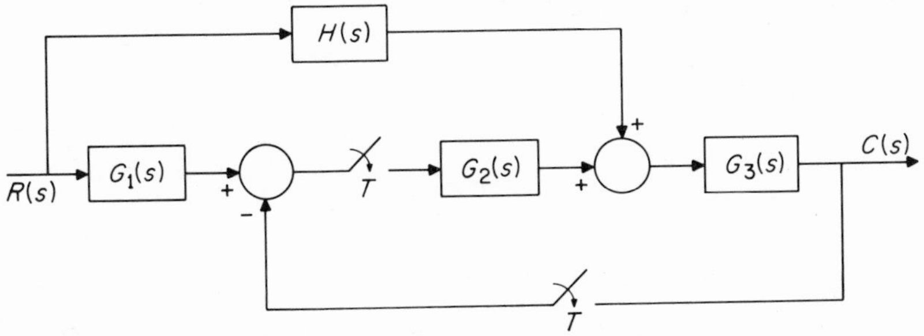

FIG. 3P-12(e).

using the sampled signal flow graph method. Check the answers with the direct signal flow graph method.

PROBLEMS FOR CHAPTER 4

4-1 A discrete-data system is represented by the following difference equation. Describe the system by a set of dynamic equations.

$$c(k+3) + 4c(k+2) + 3c(k+1) + c(k) = 2r(k+1) + r(k)$$

4-2 The state diagram of a discrete-data system is shown below. Write the dynamic equations for the system.

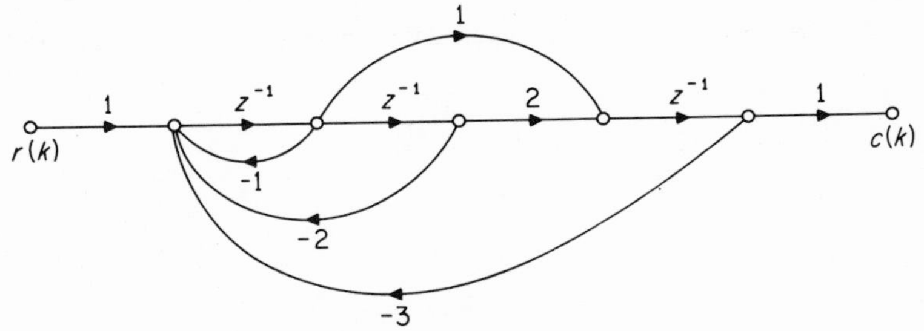

FIG. 4P-2.

4-3 Given the state equation

$$\mathbf{x}(k+1) = A\mathbf{x}(k)$$

where

$$A = \begin{bmatrix} 0.5 & 1 & 0 \\ 0 & 1 & 0 \\ 2 & 0.5 & 1 \end{bmatrix}$$

and the initial state $\mathbf{x}(0) = [1 \quad 1 \quad 0]'$. Find $\mathbf{x}(k)$ for $k > 0$.

4-4 Given the state equation

$$\mathbf{x}(k+1) = A\mathbf{x}(k) + B\mathbf{r}(k)$$

where

$$A = \begin{bmatrix} 0 & 1 & 0 \\ 0 & 0 & 1 \\ -3 & -1 & -1 \end{bmatrix} \qquad B = \begin{bmatrix} 0 & 1 \\ 1 & 0 \\ 2 & 1 \end{bmatrix}$$

Find the state transition matrix $\Phi(k)$.

4-5 (a) Draw a state diagram for the discrete-data system which is represented by the following dynamic equations:

$$\mathbf{x}(k+1) = A\mathbf{x}(k) + B\mathbf{r}(k)$$

$$c(k) = x_1(k)$$

$$A = \begin{bmatrix} 0 & 2 & -1 \\ 0 & 1 & 2 \\ 3 & 5 & -1 \end{bmatrix} \qquad B = \begin{bmatrix} 0 \\ 0 \\ 1 \end{bmatrix}$$

(b) Find the characteristic equation of the system.

4-6 A discrete-data system is described by the following transfer function:

$$\frac{C(z)}{R(z)} = \frac{1 - 0.2z^{-1} - 0.5z^{-2}}{1 + 2z^{-1} + 2z^{-2}}$$

(a) Draw a state diagram for the system.
(b) Write the dynamic equations for the system.

4-7 The block diagram of a discrete-data system is shown in Figure 4P-7.

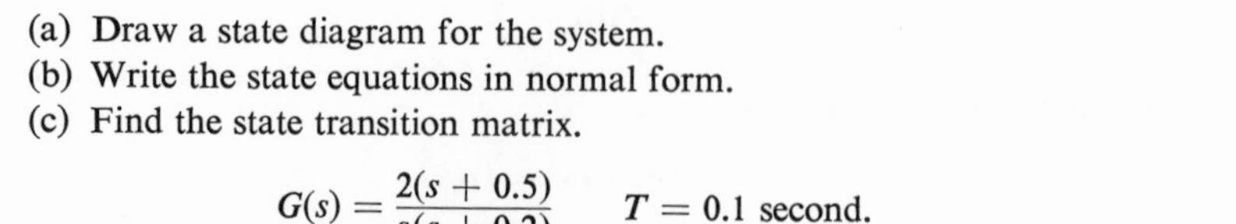

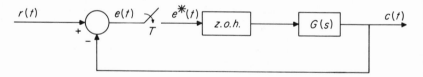

FIG. 4P-7.

(a) Draw a state diagram for the system.
(b) Write the state equations in normal form.
(c) Find the state transition matrix.

$$G(s) = \frac{2(s+0.5)}{s(s+0.2)} \qquad T = 0.1 \text{ second.}$$

4-8 The state diagram of a discrete-data system is shown in Figure 4P-8. Write the dynamic equations for the system.

4-9 Draw a state diagram for the first-order hold device.

4-10 Draw a state diagram for the system shown in Figure 4P-10. Write the discrete state equation for the system in the form of

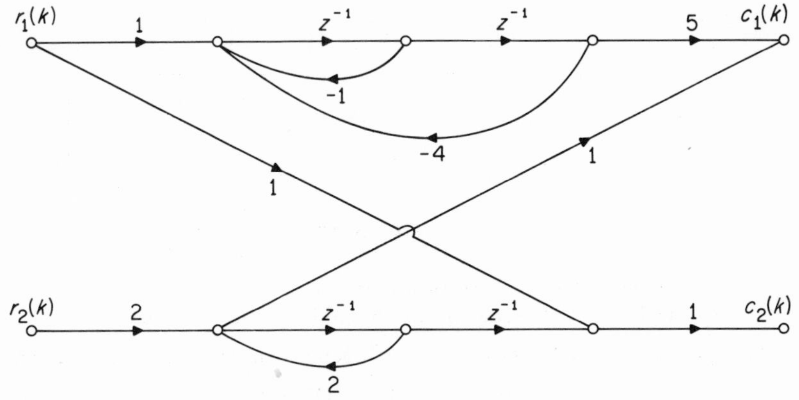

FIG. 4P-8.

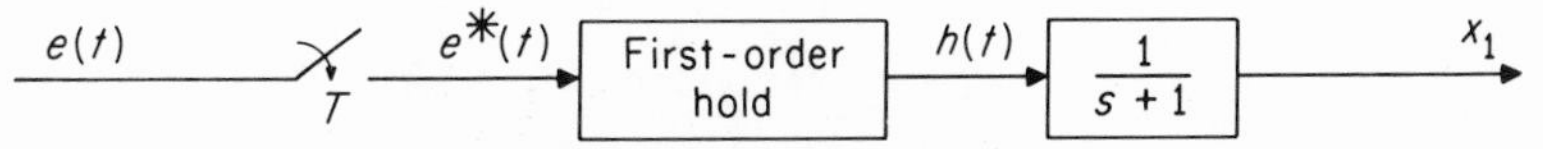

FIG. 4P-10.

$$x_1[(k+1)T] = \Phi(T)x_1(kT) + \Theta_1(T)e(kT) + \Theta_2(T)e[(k-1)T]$$

The sampling period is assumed to be one second.

4-11 The block diagram of a discrete-data system is shown in Figure 4P-11. Find the state transition equations of the system at the sampling instants.

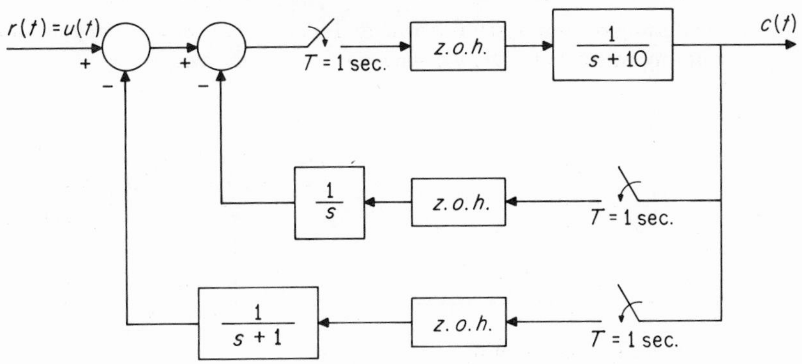

FIG. 4P-11.

4-12 The block diagram of a feedback control system with discrete-data is shown in Figure 4P-12.

$$D(z) = \frac{1 + 0.5z^{-1}}{1 + 0.2z^{-1}} \qquad G(s) = \frac{10(s+5)}{s^2}$$

$T = 0.1$ second.

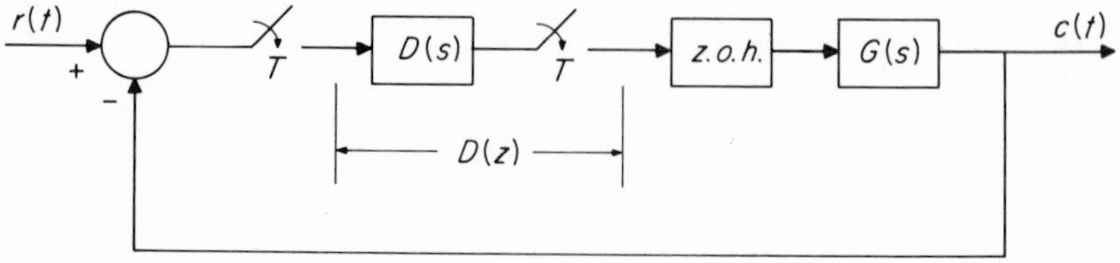

FIG. 4P-12.

(a) Draw a state diagram for the system.
(b) Write the state equations in normal form for the system.

4-13 Decompose the following transfer functions and draw the state diagrams.

(a) $$\frac{C(z)}{R(z)} = \frac{(1 - 0.1z^{-1})z^{-1}}{(1 - 0.5z^{-1})(1 - 0.8z^{-1})}$$

(b) $$\frac{C(z)}{R(z)} = \frac{z^{-1}}{(1 - 0.2z^{-1})(1 - z^{-1})}$$

(c) $$\frac{C(z)}{R(z)} = \frac{z + 0.8}{z^3 + 2z^2 + z + 0.5}$$

4-14 A discrete-data control system is described by the state equation

$$x(k + 1) = e^{-1}x(k) + (1 - e^{-1})m(k)$$

It is desired that the equilibrium state of the system be at $x = x_d$. The control $m(k)$ is given by

$$m(k) = K[x_d - x(k)]$$

where K is a constant. Determine the actual equilibrium state of the system.

4-15 The sampled-data system shown in Figure 4P-15 has a triangular hold for data-reconstruction. Draw a state diagram for the system.

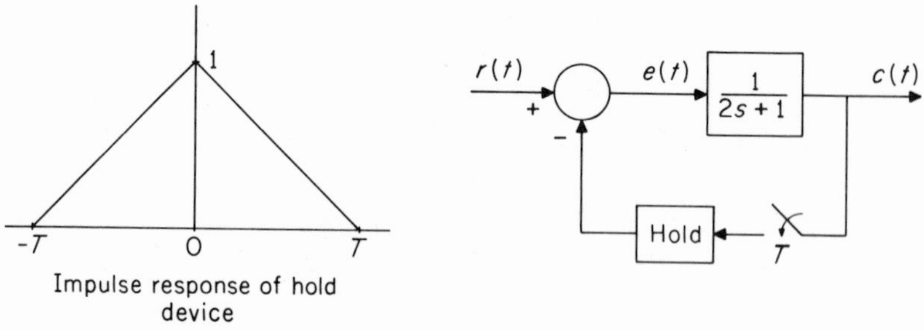

FIG. 4P-15.

4-16 The block diagram of an open-loop discrete-data control system is shown in Figure 4P-16. The input signal is a unit-step function. Write the state transition equations for the system at $t = kT_2, k = 0, 1, 2, \ldots$.

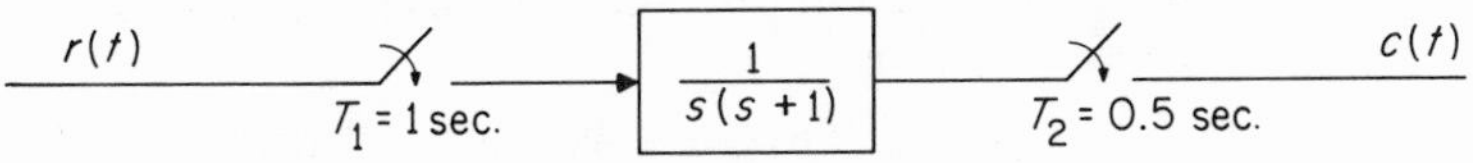

FIG. 4P-16.

4-17 The block diagram of a discrete-data system with multirate sampling is shown in Figure 4P-17. It is assumed that the samplers are synchronized.

(a) Write the state equations for the system.

(b) Find the state transition equations for the system.

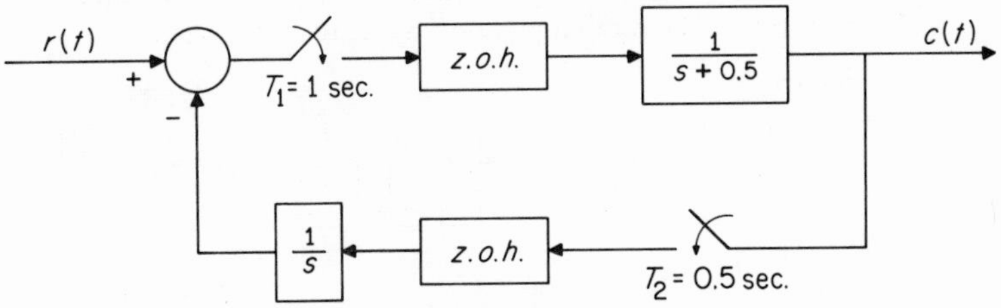

FIG. 4P-17.

4-18 Figure 4P-18 illustrates a linear process whose control signal is the output of

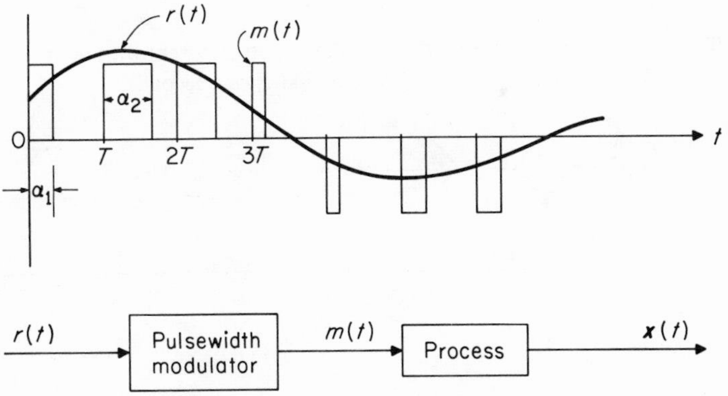

FIG. 4P-18.

a pulsewidth modulator. The control signal $m(t)$ is also shown. The pulsewidth at the ith sampling instant is denoted by α_i, where $\alpha_i \leq T$ for all i. The magnitude of $m(t)$ is designated by $m(kT)$ which is $+1$ or -1 depending upon the sign of $r(t)$. The linear process is described by the state equation

$$\frac{d\mathbf{x}(t)}{dt} = \mathbf{A}\mathbf{x}(t) + \mathbf{B}m(t)$$

Show that the system can be described at the sampling instants by the following discrete state equation:

$$\mathbf{x}[(k+1)T] = \mathbf{\Phi}(T)\mathbf{x}(kT) + \mathbf{\Phi}(T)\mathbf{\Phi}(-\alpha_k)\mathbf{\Theta}(\alpha_k)m(kT)$$

where

$$\mathbf{\Phi}(T) = e^{AT}$$

$$\mathbf{\Theta}(T) = \int_0^T \mathbf{\Phi}(T-\tau)B\,d\tau$$

4-19 The block diagram of a nonlinear discrete-data system with time delay is shown in Figure 4P-19. Sketch the state plane trajectory in x_2 versus x_1, where $x_1 = c$ and $x_2 = dc/dt$. Assume that a unit-step input is applied and the initial states are zero. The sampling period is one second.

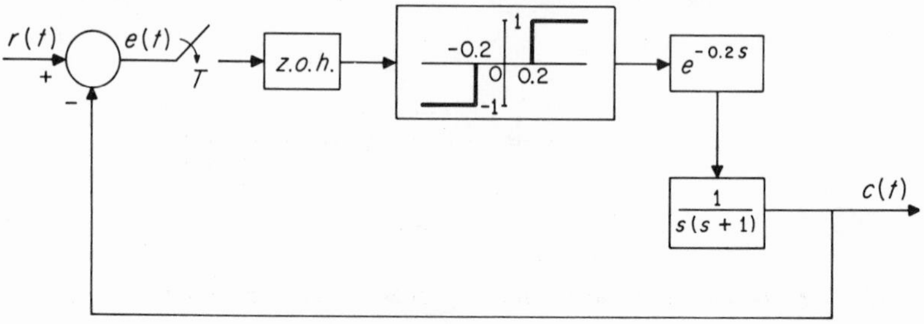

FIG. 4P-19.

4-20 The block diagram of a discrete-data control system with quantization is shown in Figure 4P-20. The sampling period is one second.

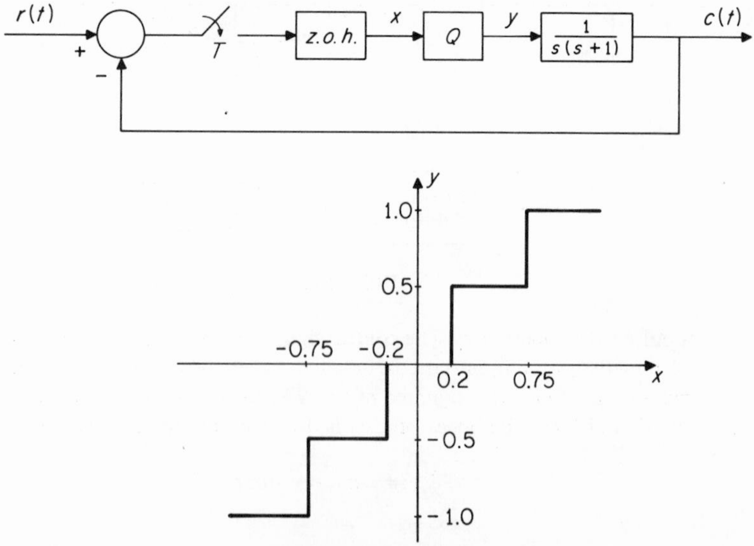

FIG. 4P-20.

(a) Evaluate the output response by the state variable method with $r(t) = 0.5u(t)$.

(b) Sketch the state plane trajectory with $x_1 = c$ and $x_2 = dc/dt$.

4-21 The block diagram of a digital tracking loop of a missile system is shown in Figure 4P-21. The sampling period is 1/12 second. Determine the maximum magnitude of the step input which will be just short of saturating the nonlinearity of the system. Let $K = 8$ and $\Delta = 0.4$.

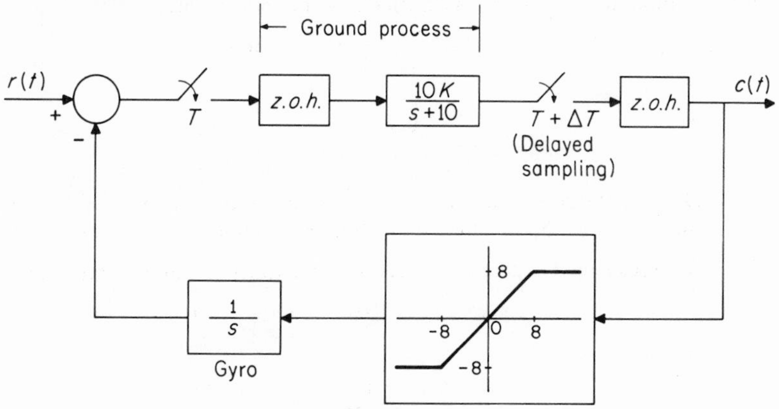

FIG. 4P-21.

PROBLEMS FOR CHAPTER 5

5-1 Determine the stability of the discrete-data systems which are represented by the following characteristic equations.

(a) $z^3 + 5z^2 + 3z + 2 = 0$

(b) $z^4 + 9z^3 + 3z^2 + 9z + 1 = 0$

(c) $z^3 - 1.5z^2 - 2z + 3 = 0$

(d) $z^4 - 1.55z^3 + 0.5z^2 - 0.5z + 1 = 0$

5-2 The characteristic equation of a linear discrete-data system is

$$z^3 + Kz^2 + 1.5Kz - (K + 1) = 0$$

Determine the values of K for the system to be asymptotically stable.

5-3 The block diagram of a sampled-data control system is shown in Figure 5P-3.

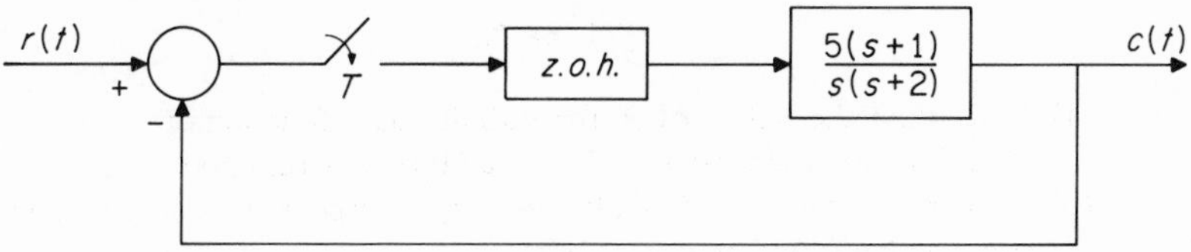

FIG. 5P-3.

Determine the values of the sampling period T so that the system is asymptotically stable.

5-4 Repeat Problem 5-3 when the hold device is a first-order hold.

5-5 A discrete-data control system is described by the state equation

$$x(k+1) = [0.368 - 0.632K]x(k) + 0.632(K+1)m(k)$$

Determine the values of K for the equilibrium state to be asymptotically stable.

5-6 A discrete-data system is described by the state equation

$$\mathbf{x}(k+1) = A\mathbf{x}(k) + B\mathbf{r}(k)$$

where

$$A = \begin{bmatrix} 0.5 & 1 \\ -1 & -1 \end{bmatrix} \qquad B = \begin{bmatrix} 1 \\ 1 \end{bmatrix}$$

Determine the stability of the equilibrium state with $\mathbf{r}(k) = \mathbf{0}$ by means of the second method of Liapunov. Find a Liapunov function $V(\mathbf{x})$.

5-7 A discrete-data system with state delays is described by the state equation

$$\mathbf{x}(k+1) = A\mathbf{x}(k) + E\mathbf{x}(k-1)$$

where

$$A = \begin{bmatrix} 0.5 & 1 \\ -1 & -1 \end{bmatrix} \qquad E = \begin{bmatrix} 0 & 0 \\ 0 & 1 \end{bmatrix}$$

Determine the stability of the equilibrium state.

5-8 The block diagram of a digital tracking loop of a missile system is shown in Figure 5P-8. The sampling period is 1/12 second.

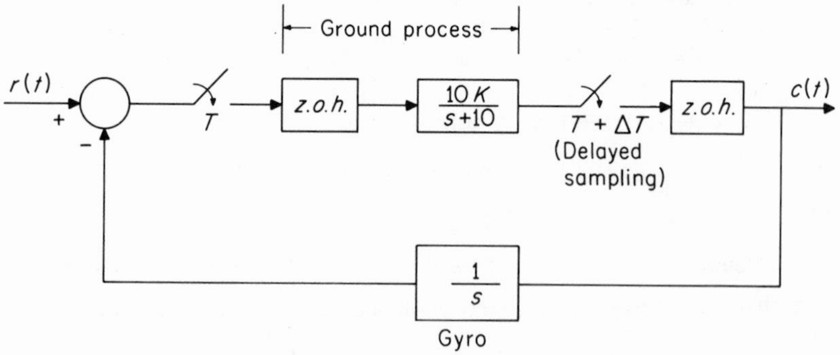

FIG. 5P-8.

(a) Determine the values of K for stability as a function of $\Delta (0 \leq \Delta \leq 1)$. Plot the maximum value of K for stability as a function of Δ.

(b) With $K = 8$ and $\Delta = 0.4$, plot the output response of the system with the input a step function, $r(t) = 2u(t)$.

PROBLEMS FOR CHAPTER 6

6-1 Consider that a digital system is described by the state equation

$$\mathbf{x}(k+1) = A\mathbf{x}(k) + B\mathbf{m}(k)$$

where

$$A = \begin{bmatrix} 1 & -2 \\ 1 & -1 \end{bmatrix} \quad B = \begin{bmatrix} 1 & 0 \\ 0 & -1 \end{bmatrix}$$

(a) Determine the controllability of the system.
(b) Find a P matrix which will diagonalize A in the form of $A = P^{-1}AP$. Verify the results of part (a) with the uncoupled state equations.

6-2 Consider that a digital system is described by the state equation

$$\mathbf{x}(k+1) = A\mathbf{x}(k) + B\mathbf{m}(k)$$

where

$$A = \begin{bmatrix} 1 & -2 & 0 \\ 3 & 2 & 1 \\ -1 & 1 & 4 \end{bmatrix}$$

$$B = \begin{bmatrix} 1 & 0 \\ -1 & 1 \\ 0 & 1 \end{bmatrix}$$

Determine the controllability of the system.

6-3 For a linear time-invariant discrete-data system with single input, show that if there is pole-zero cancellation in the input-output transfer function, the system will be either uncontrollable or unobservable.

6-4 The block diagram of a discrete-data system is shown in Figure 6P-4. Determine the controllability of the system.

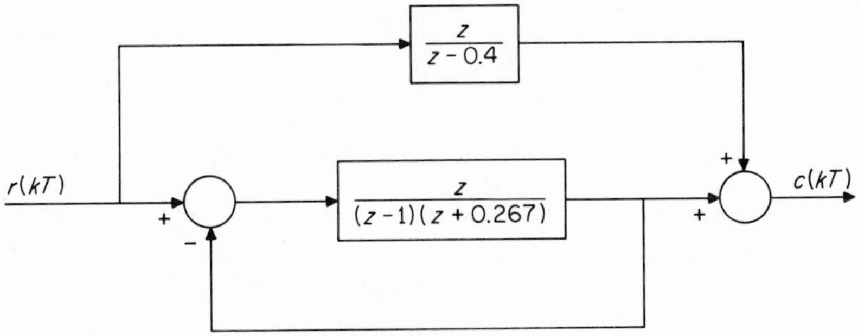

FIG. 6P-4.

6-5 The block diagram of a sampled-data system is shown in Figure 6P-5. De-

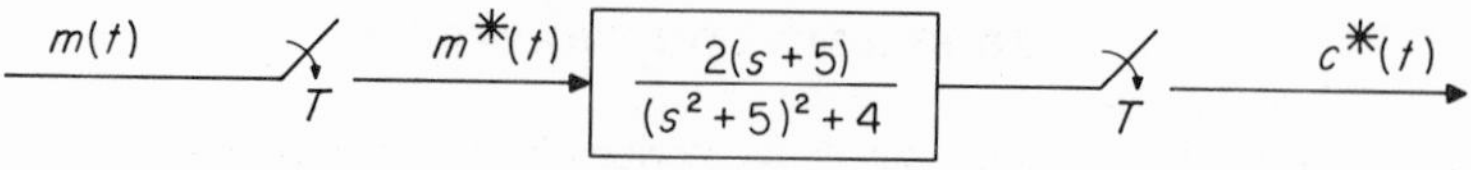

FIG. 6P-5.

termine what values of T must be avoided so that the system will be assured of complete controllability.

6-6 The linear process of a sampled-data control system is described by the following state equation

$$\begin{bmatrix} \dfrac{dx_1}{dt} \\ \dfrac{dx_2}{dt} \end{bmatrix} = \begin{bmatrix} 0 & 1 \\ -1 & 0 \end{bmatrix} \begin{bmatrix} x_1 \\ x_2 \end{bmatrix} + \begin{bmatrix} 0 \\ 1 \end{bmatrix} m(t)$$

where $m(t) = m(kT)$ for $kT \leq t < (k + 1)T$.
Determine the values of the sampling period T which make the system uncontrollable; that is, the initial state $\mathbf{x}(0)$ cannot be driven to the equilibrium state $\mathbf{x}(NT) = \mathbf{0}$ for any finite N.

6-7 Prove that if $\mathbf{x}(0) \in R'_N$ and $\mathbf{x}'(0) \in R'_N$, then the initial state $\mathbf{x}''(0)$ given by

$$\mathbf{x}''(0) = \lambda\mathbf{x}(0) + (1 - \lambda)\mathbf{x}'(0)$$

is $\in R'_N$ for all $\lambda \in [0, 1]$.

6-8 The transfer function of a linear controlled process is given by

$$G(s) = \frac{1}{s(s + 1)}$$

The input to the process is the output of a zero-order hold and is described by $m(t) = m(kT)$, $kT \leq t < (k + 1)T$. The magnitude of $m(t)$ is constrained by $|m(kT)| \leq 1$ for $k = 0, 1, 2, \ldots$. The sampling period T is one second.

(a) Write the discrete state transition equation for the system in the form of

$$\mathbf{x}[(k + 1)T] = \mathbf{\Phi}(T)\mathbf{x}(kT) + \mathbf{\Theta}(T)m(kT)$$

where $\mathbf{\Phi}(T)$ is a diagonal matrix with the eigenvalues of $\mathbf{\Phi}(T)$ as its elements. Sketch the vectors

$$S(iT) = -\mathbf{\Phi}(-iT)\mathbf{\Theta}(T)$$

for $i = 1, 2, 3$ in the (x_1, x_2) plane. What special properties do you observe from these vectors?

(b) Sketch the convex polygons R'_1, R'_2, and R'_3, where R'_N denotes the set of initial states which can be brought to the equilibrium point $\mathbf{x}(NT) = \mathbf{0}$ in N sampling periods or less by admissible control.

6-9 The vectors $S_i = -\mathbf{\Phi}(-iT)\mathbf{\Theta}(T)$ for $i = 1, 2$, and 3 for a second-order system are shown in Figure 6P-9. Sketch the convex polygons for R'_1, R'_2, and R'_3 with $|m[(i - 1)T]| \leq 1$.

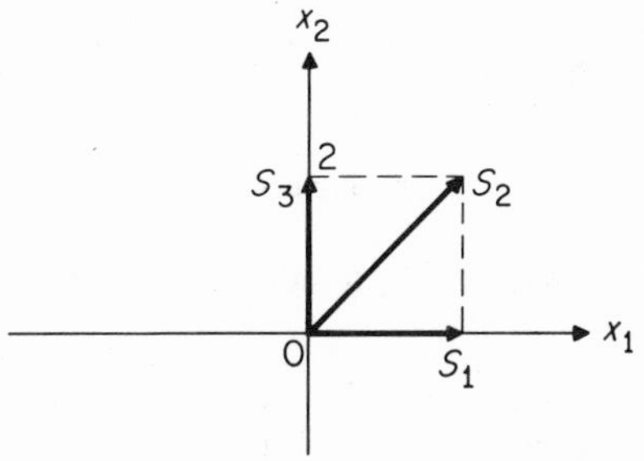

FIG. 6P-9.

6-10 The vectors, $S_i = -\Phi(-iT)\Theta(T)$, $i = 1, 2, 3$, and 4, for a third-order system are shown in the diagram of Figure 6P-10. Sketch the convex polyhedron for R_1', R_2', R_3', and R_4', with $|m[(i-1)T]| \leq 1$.

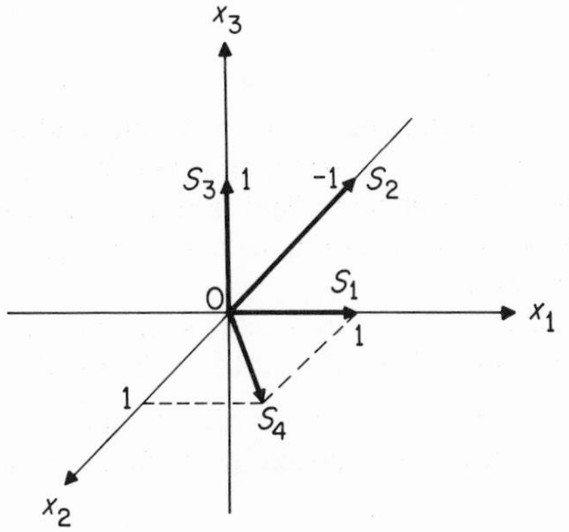

FIG. 6P-10.

6-11 The transfer function of a linear controlled process is given by

$$G(s) = \frac{1}{s(s+1)}$$

The input to the process is the output of a zero-order hold and is described by $m(t) = m(kT)$, $kT \leq t < (k+1)T$. The magnitude of $m(t)$ is constrained by $|m(kT)| \leq 1$ for $k = 0, 1, 2, \ldots$. The sampling period is one second. Find the set of initial states $\mathbf{x}(0)$ which is uncontrollable no matter how large N is.

6-12 Given the plane transfer function $G(s) = 1/s^2$. Find the equation of the *optimum switching curves* in the (x_1, x_2) plane for $+1$ forcing and -1 forcing.

6-13 The block diagram of a discrete-data system is shown in Figure 6P-13. The linear process is described by the state equation.

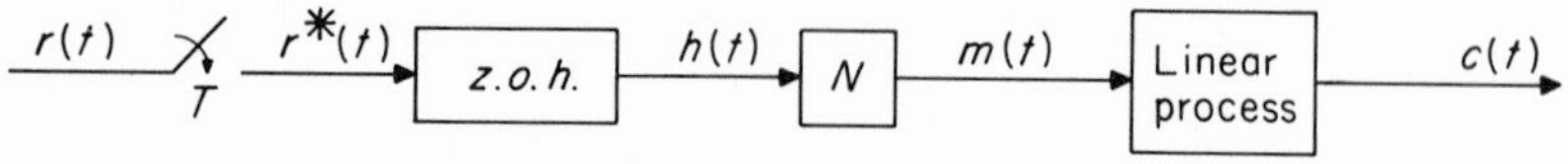

FIG. 6P-13.

$$\begin{bmatrix} \dfrac{dx_1}{dt} \\ \dfrac{dx_2}{dt} \end{bmatrix} = \begin{bmatrix} 0 & 1 \\ -1 & 0 \end{bmatrix} \begin{bmatrix} x_1 \\ x_2 \end{bmatrix} + \begin{bmatrix} 0 \\ 1 \end{bmatrix} m$$

The sampling period is $\pi/2$ seconds. Construct the regions of controllable states for $N = 1, 2, 3, 4$, with the origin of the state space as the equilibrium point. The control signal is constrained by $|m(t)| \leq 1$.

6-14 A linear process is described by the following state equation

$$\begin{bmatrix} \dfrac{dx_1}{dt} \\ \dfrac{dx_2}{dt} \end{bmatrix} = \begin{bmatrix} 0 & 1 \\ -1 & 0 \end{bmatrix} \begin{bmatrix} x_1 \\ x_2 \end{bmatrix} + \begin{bmatrix} 0 \\ 1 \end{bmatrix} m$$

where $m(t)$ is the output of a zero-order hold, and is subject to the amplitude constraint $|m(t)| \leq 1$. The sampling period is $\pi/2$ seconds.

(a) Find the regions of controllable states in the $\mathbf{x}$ state space which can be brought to the equilibrium state $\mathbf{x}(NT) = [1 \quad 0]'$ by the control signal for $N = 1, 2, 3$.

(b) Draw the regions of controllable states for the equilibrium state $[3, 0]'$. Show why a controller cannot be realized with the constraint $|m(t)| \leq 1$.

6-15 The block diagram of a nonlinear digital control system is shown in Figure 6P-15. The process G is described by the following state equation

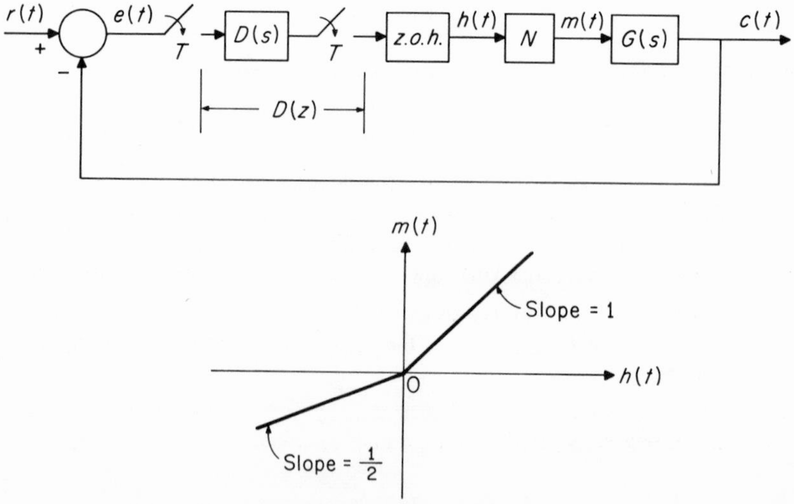

FIG. 6P-15.

$$\begin{bmatrix} \frac{dx_1}{dt} \\ \frac{dx_2}{dt} \end{bmatrix} = \begin{bmatrix} 0 & 1 \\ 0 & -1 \end{bmatrix} \begin{bmatrix} x_1 \\ x_2 \end{bmatrix} + \begin{bmatrix} 1 \\ 0 \end{bmatrix} m(t)$$

(These equations are equivalent to the transfer function $G(s) = 1/s(s+1)$.) The output equation is $c(t) = x_1(t)$. The nonlinear characteristic is illustrated as shown. Find the transfer function $D(z)$ so that the system exhibits a deadbeat response to a unit-step input. The sampling period is one second, and the initial conditions of the system are assumed to be zero.

6-16 A discrete-data control system with a nonlinear element N is shown in Fig. 6P-16. The linear process is described by the state equation

$$\begin{bmatrix} \frac{dx_1}{dt} \\ \frac{dx_2}{dt} \end{bmatrix} = \begin{bmatrix} 0 & 1 \\ -1 & 0 \end{bmatrix} \begin{bmatrix} x_1 \\ x_2 \end{bmatrix} + \begin{bmatrix} 0 \\ 1 \end{bmatrix} m$$

Determine the transfer function of the digital controller, $D(z)$, so that the system is driven to the equilibrium state $\mathbf{x}(NT) = \mathbf{0}$ for minimum N. The initial state is $\mathbf{x}(0) = [-3 \quad 0]'$. The input $r(t)$ is zero. The sampling period is $T = \pi/2$ seconds. Sketch the responses $x_1(t)$ and $x_2(t)$.

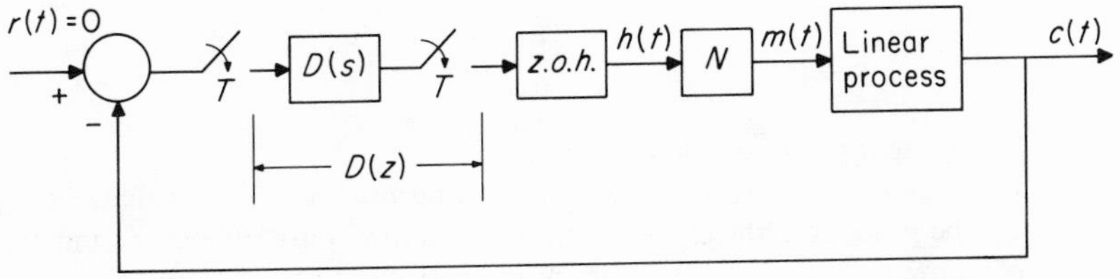

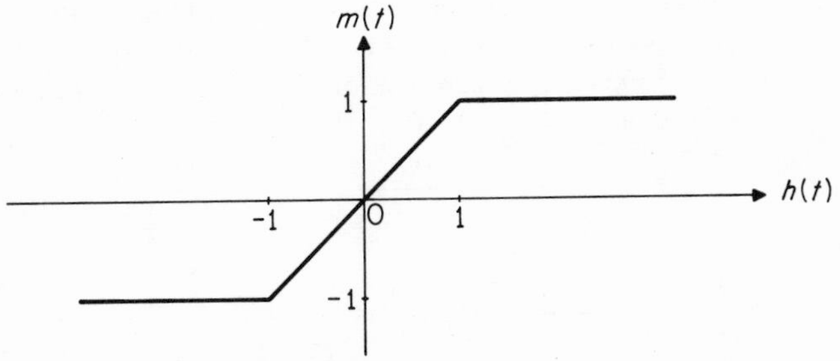

FIG. 6P-16.

6-17 The block diagram of digital tracking loop of a missile system is shown in Fig. 6P-17. The sampling period is $\frac{1}{12}$ second.

(a) Consider that the two samplers are synchronized, design an optimum digital controller $D(z)$ so that the output response is deadbeat (follows the input in the shortest possible time without overshoot) when the input is $r(t) = Ru(t)$. Sketch the output of the optimum system when $R = 1, 2, 4$, and 6. What is the maximum value of R if the system is to reach its steady state within one second?

(b) Consider that the sampler S_2 has a delay time of $T = 0.4T$, where T is the sampling period ($\frac{1}{12}$ second). What is the effect of this delay sampling to the system described in part (a)?

(c) Repeat part (a) when $r(t) = Rtu(t)$.

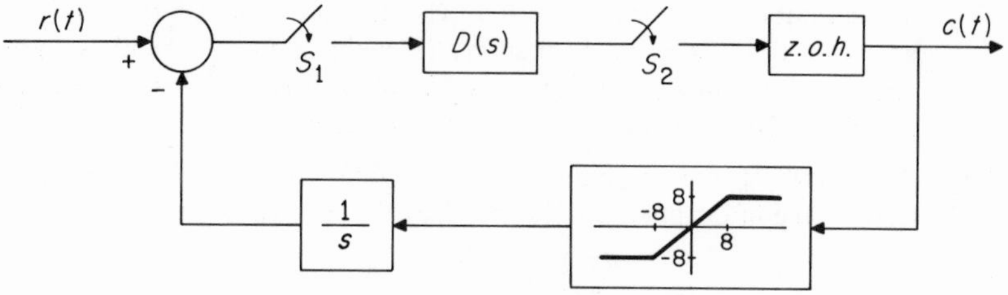

FIG. 6P-17.

6-18 A linear multivariable system with input delay is described by

$$\frac{d\mathbf{x}(t)}{dt} = A\mathbf{x}(t) + B_1 m(t - T)$$

where

$$A = \begin{bmatrix} 0 & 1 \\ -1 & 0 \end{bmatrix} \qquad B_1 = \begin{bmatrix} 0 \\ 1 \end{bmatrix}$$

The sampling period is $\pi/2$ seconds.
Find the optimum strategy for $\mathbf{m}(kT)$ in terms of $\mathbf{x}(kT)$ such that the system can be brought to the equilibrium state $\mathbf{0}$ from any $\mathbf{x}(0)$ in a minimum number of sampling periods. The values of the past input are not known in this case.

6-19 A system with input delays is shown in Figure 6P-19. Find the optimal control $\mathbf{m}^o(kT)$, $k = 0, 1, 2, \ldots, (N - 1)$, so that $\mathbf{x}(NT) = \mathbf{0}$ for minimum N. $m_0(t)$ and $m_1(t) =$ constant for $kT \leq t < (k + 1)T$. Assume that $\mathbf{m}(t) = \mathbf{0}$ for $t < 0$, and $\mathbf{x}(0) = [1 \quad 0]'$.

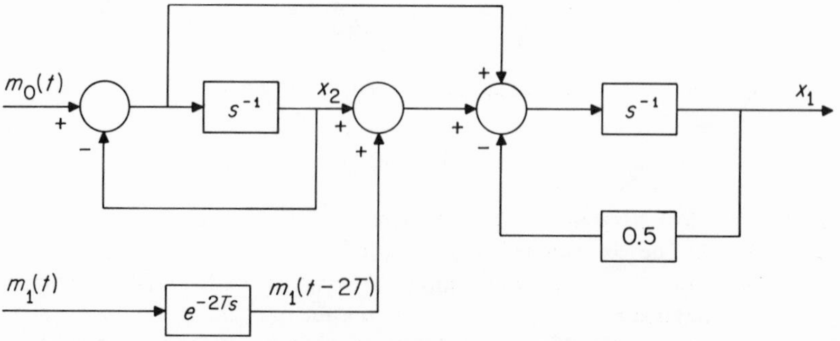

FIG. 6P-19.

6-20 Shown in Figure 6P-20 is a discrete-data system with nonuniform sampling time. The sampling scheme is described as shown in the figure. The linear process is described by

$$\frac{d\mathbf{x}(t)}{dt} = A\mathbf{x}(t) + Bm(t)$$

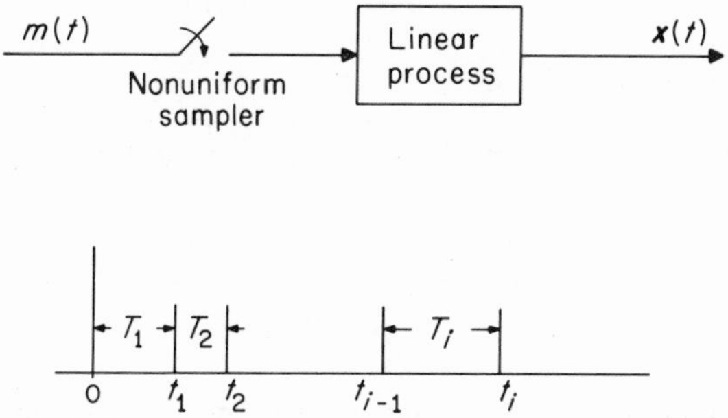

FIG. 6P-20.

(a) Show that the initial state which can be brought to the equilibrium state $\mathbf{x}(t_N) = \mathbf{0}$ in N sampling periods is represented by

$$\mathbf{x}(0) = \sum_{i=0}^{N} S_i m(i-1)$$

where

$$S_i = -\prod_{j=1}^{i} \Phi(-T_j)\Theta(T_i)$$

$$\Phi(t) = e^{At} \qquad \Theta(t) = \int_0^t \Phi(t-\tau)B\,d\tau$$

(b) Show that when the sampling rate is uniform, $T_i = T$ for $i = 1, 2, \ldots, N$, the expressions in part (a) become those for the uniform-rate sampling case.

(c) Determine the vertices of the polygon of the controllable region in terms of $\Phi(-T_j)$ and $\Theta(T_i)$ when $N = 2$.

PROBLEMS FOR CHAPTER 7

7-1 Consider the following optimal allocation problem:

c = initial amount of capital
x = amount invested in the first stage in choice A
$g(x) = 0.4\sqrt{x}$ = return from investment in choice A for one stage
$h(c-x) = 0.2\sqrt{c-x}$ = return from investment in choice B for one stage
ax = capital left at the end of the first stage from investment A
$b(c-x)$ = capital left at the end of the first stage from investment B
$a = 0.6$ and $b = 0.8$

(a) Using the dynamic programming method, find the maximum return for a three-stage process ($N = 3$). That is, find $f_3(c)$.
(b) Determine the optimum allocation at the beginning of each stage.

7-2 The block diagram of a sampled-data control system is shown in Figure 7P-2. The dynamics of the process is described by

$$\frac{dx}{dt} + 2x = m \qquad r(t) = 0$$

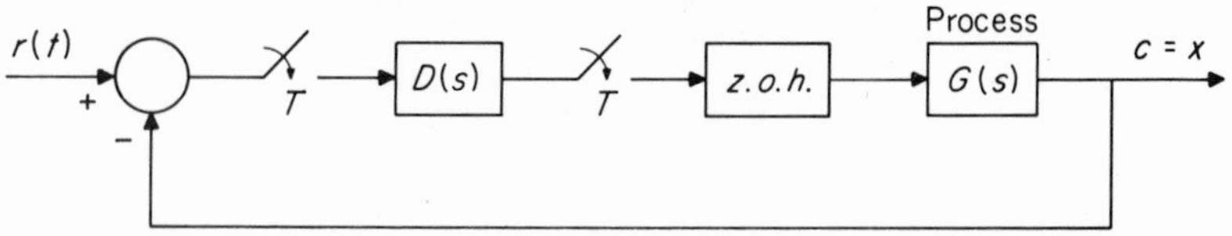

FIG. 7P-2.

The sampling period is 1 second. Determine the optimum control law $m^o(k)$ which minimizes the performance index

$$I_N = \sum_{k=1}^{N} [x^2(k) + m^2(k-1)]$$

7-3 Consider that a linear process is characterized by the state equation

$$\frac{d\mathbf{x}}{dt} = A\mathbf{x} + Bm$$

where

$$A = \begin{bmatrix} 0 & 1 \\ -4 & -5 \end{bmatrix} \qquad B = \begin{bmatrix} 0 \\ 1 \end{bmatrix}$$

The control is described by $m(t) = m(kT)$ for $kT \leq t < (k+1)T$, where $T = 0.5$ second. Determine the optimum control $m(kT)$ which minimizes the performance index

$$I_5 = \sum_{k=1}^{5} x_1^2(k)$$

Draw a block diagram for the entire system with a digital controller generating the desired optimal control.

7-4 A linear controlled process is represented by the state equation

$$\mathbf{x}(k+1) = A\mathbf{x}(k) + Bm(k)$$

where

$$A = \begin{bmatrix} 0 & 1 \\ -1 & -2 \end{bmatrix} \qquad B = \begin{bmatrix} 0 \\ 1 \end{bmatrix}$$

Determine the optimal control $m(0), m(1), \ldots, m(4)$, so that the performance index

$$I_5 = \sum_{k=1}^{5} [x_1^2(k) + x_2^2(k)]$$

is minimized.

7-5 Repeat Problem 7-4 when the performance index is

$$I_5 = \sum_{k=1}^{5} [x_1^2(k) + x_2^2(k) + 0.5m^2(k-1)]$$

7-6 A linear discrete-data process is described by the state equation

$$\mathbf{x}(k+1) = A\mathbf{x}(k) + B\mathbf{m}(k) + \mathbf{n}(k)$$

where $\mathbf{n}(k)$ denotes the disturbance vector. Determine the optimal control vector $\mathbf{m}(k)$, $k = 0, 1, 2, \ldots, N-1$, which minimizes the performance index

$$I_N = \sum_{k=1}^{N} [\mathbf{x}'(k)Q\mathbf{x}(k) + \lambda\mathbf{m}'(k-1)H\mathbf{m}(k-1)]$$

7-7 A time-varying discrete-data process is described by the state equation

$$\mathbf{x}(k+1) = A(k)\mathbf{x}(k) + B(k)m(k)$$

where $m(k)$ is the scalar control. Determine the optimal control which minimizes the performance index

$$I_N = \sum_{k=1}^{N} \mathbf{x}'(k)Q\mathbf{x}(k) + \lambda m^2(k-1)$$

where 0 is a positive definite symmetric matrix.

7-8 A linear process is described by the state equation

$$\mathbf{x}[(k+1)T] = \Phi(T)\mathbf{x}(kT) + \Theta(T)m[(k-p)T]$$

where $\mathbf{x}(kT)$ is an $n \times 1$ state vector, and $m[(k-p)T]$ is the delayed scalar control signal. The delay time pT for the input is an integral multiple of T. Given the initial state $\mathbf{x}(0)$ and the past inputs, $m(-pT), m[(-p+1)T], \ldots,$ $m(-T)$, determine the optimal controls $m(0), m(T), \ldots, m[(N-1)T]$ so that the performance index

$$I_{N+p} = \sum_{k=1}^{N+p} [\mathbf{x}'(k)Q(k)\mathbf{x}(k) + \lambda m^2(k-p-1)]$$

is minimized. $(N > p)$

7-9 Consider that a linear discrete-data process with input delay is described by the first-order difference equation

$$x(k+1) = \Phi x(k) + \Theta m(k-1)$$

where $\Phi = 0.368$ and $\Theta = 0.632$. Determine the optimal control $m(0)$, $m(1)$, $m(2)$, and $m(3)$, so that the performance index

$$I_4 = \sum_{k=1}^{4} [x^2(k) + m^2(k-2)]$$

is minimized. Assume that $x(0) = -1$ and $m(-1) = 0.178$. Determine the state variables $x(1)$, $x(2)$, $x(3)$, and $x(4)$.

7-10 A discrete-data process with state delay is described by the state equation

$$x(k+1) = x(k) + 2x(k-1) + m(k)$$

The performance index is

$$I_N = \sum_{k=1}^{N} [x^2(k) + m^2(k-1)]$$

Determine the optimal control $m(0)$, $m(1)$, so that I_N is minimized for $N = 2$. The initial states are given as $x(0)$ and $x(-1)$.

7-11 A discrete-data process with state delay is described by the state equation

$$x(k+1) = x(k) + ax(k-M) + m(k)$$

where a is a constant and M is a positive integer. The initial states are given as $S_0 = [x_0, x_{-1}, \ldots x_{-M}]'$. The performance index is

$$I_N = x^2(N) + \lambda \sum_{k=0}^{N-1} m^2(k)$$

Let

$$f_k(S_0) = \underset{m(j)}{\text{Min}} \left[\sum_{j=0}^{k-1} m^2(j) + x^2(k) \right]$$

$$j = 1, 2, \ldots, k-1$$

Show that for $k \leq M$, the minimum return over k stages is

$$f_k(S_0) = \frac{\lambda}{\lambda + k} \left[a \sum_{j=0}^{k-1} x_{-(M-j)} + x_0 \right]^2$$

and the input controls which achieve this minimum return are

$$m_k(S_0) = -\frac{\lambda}{\lambda + k} \left[a \sum_{j=0}^{k-1} x_{-(M-j)} + x_0 \right]$$

$k = 1, 2, \ldots, M-1, M.$

PROBLEMS FOR CHAPTER 8

8-1 The auto-power spectral density function of a stationary random process $r(t)$ is given by

$$\Phi_{rr}(s) = \frac{4}{0.04 - s^2}$$

Determine the discrete power spectral density function $\Phi_{r^*r^*}(z)$.

8-2 For the discrete-data system shown in Fig. 8-11(a), Chapter 8, given

$$G(s) = \frac{1 - e^{-Ts}}{s^2(s+1)}$$

$$\Phi_{r_s r_s}(s) = \frac{1}{1 - s^2} \qquad \Phi_{r_n r_n}(s) = 0.1$$

The signal and noise are uncorrelated. $T = 0.2$ sec. Determine the transfer function of the optimal digital controller in the sense of minimum sampled mean-square error. The desired $M_d(s)$ is unity.

8-3 Repeat Problem 8-2 when

$$M_d(s) = e^{-2Ts}$$

where T is the sampling period.

8-4 Repeat Problem 8-2 when $M_d(s) = 1/s$.

8-5 Repeat Problem 8-2 when $M_d(s) = s$.

8-6 Figure 8P-6 shows a sampled-data system whose mean-square error is to be evaluated. Assume that the system M is a zero-order hold and that

$$M_d(s) = e^{-Ts/2}$$

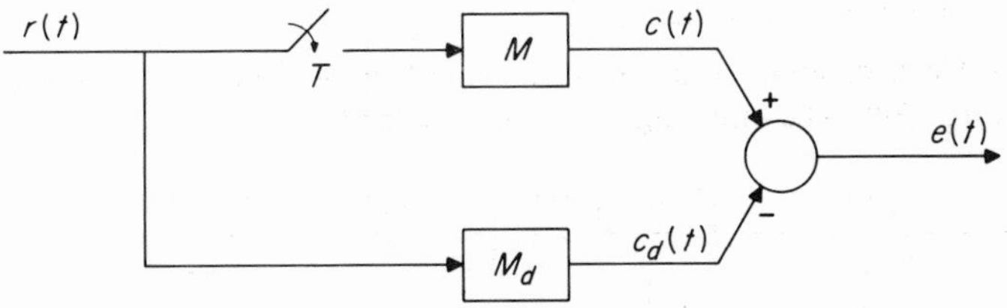

FIG. 8P-6.

The power spectral density of the input signal is

$$\Phi_{rr}(s) = \frac{2\omega_0}{\omega_0^2 - s^2}$$

(a) Find the mean-square error $\overline{e^2(t)}$ as a function of ω_0 and T.
(b) Plot $\overline{e^2(t)}$ as a function of $\omega_0 T$.

8-7 For the system block diagram shown in Figure 8P-7, show that the optimal digital filter in the sense of minimum mean-square error is

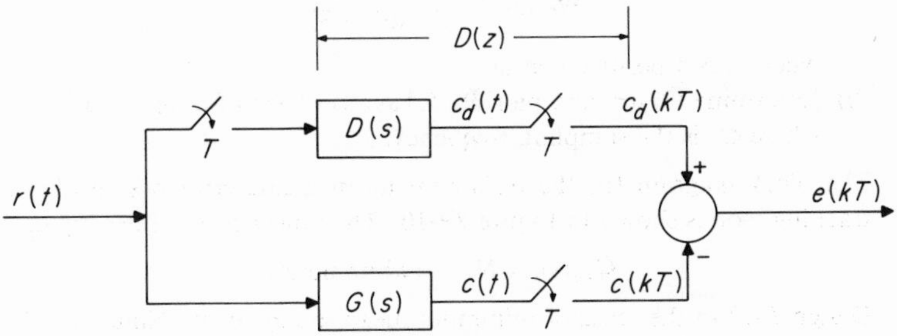

FIG. 8P-7.

$$D^o(z) = \frac{z}{\Phi_{rr}^+(z)} \left\{ \frac{\Phi_{rr}G(z)}{\Phi_{rr}^-(z)} \right\}_+$$

where

$$\Phi_{rr}G(z) = \mathscr{Z}[\Phi_{rr}(s)G(s)]$$

8-8 For the block diagram shown in Figure 8P-7, given that

$$G(s) = \frac{2(s+1)}{s^2}$$

$$\Phi_{rr}(s) = \frac{2}{1-s^2}$$

(a) Determine the transfer function of $D(z)$ so that the sampled mean-square error, $\overline{e^2(kT)}$ is minimized.

(b) Determine suboptimal solution to part (a) so that $D(z)$ represents a physically realizable function.

8-9 The block diagram of a digital tracking loop of a missile system is shown in Figure 8P-9. The sampling period is $\frac{1}{12}$ second.

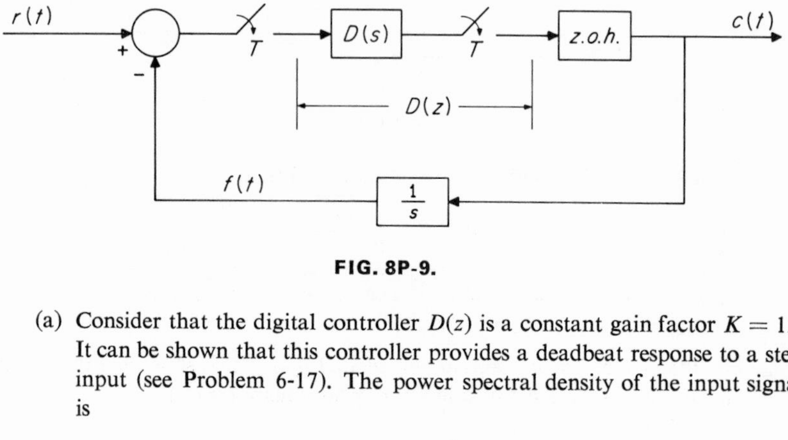

FIG. 8P-9.

(a) Consider that the digital controller $D(z)$ is a constant gain factor $K = 12$. It can be shown that this controller provides a deadbeat response to a step input (see Problem 6-17). The power spectral density of the input signal is

$$\Phi_{rr}(j\omega) = k \frac{3}{(\omega^2 + 9)}$$

where k is a positive constant.

(b) Determine $\Phi_{ff}(z)$ and plot $\Phi_{ff}(z)$ as a function of ω for $-\omega_s/2 \leq \omega_s/2$, where ω_s is the sampling frequency.

8-10 The block diagram for the minimum mean-square-error design of a digital tracking loop is shown in Figure 8P-10. The power spectral density of $n(t)$ is

$$\Phi_{nn}(s) = N \qquad \text{(white noise)}$$

Design $D(z)$ in the sense of minimum mean-square error. Note that $D(z)$ will be a function of k, N, and n. If k/N is defined as the signal-to-noise ratio, determine the optimum k/N and n (positive integer) so that $D(z)$ will approach the minimal-time controller for a step input (Problem 6-17). Consider this

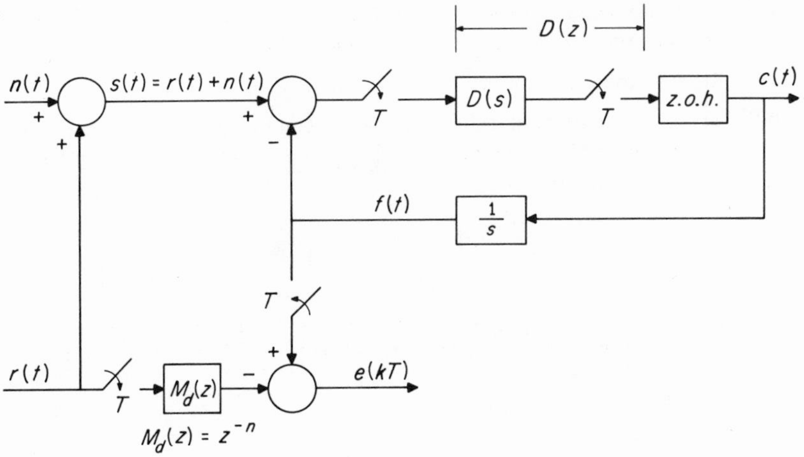

FIG. 8P-10.

to be an extension of Problem 8-9 so that the input data and sampling period are identical to those of the previous problem.
Show that for all other k/N ratios it is sometimes possible to find an integral n so that $D(z)$ will again be very close to that of the minimal-time controller.

PROBLEMS FOR CHAPTER 9

9-1 Prove Eq. (9-78).

9-2 For the system shown in Figure 9P-2, design the optimum Kalman filter so that the quadratic error $E[e^2(k+1)]$ is a minimum. It is assumed that

$$\text{Cov}\,[m(k)] = Q(k) = 0.5$$
$$\text{Cov}\,[w(k)] = R(k) = 0.1$$

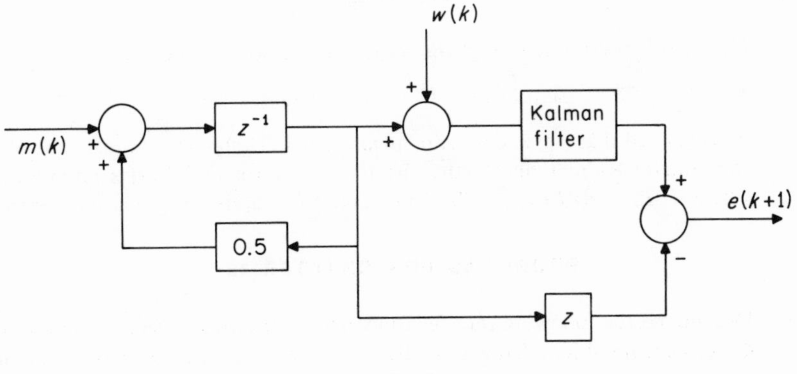

FIG. 9P-2.

9-3 For the system described in Problem 9-2, formulate an equivalent Wiener filter problem and determine the optimal Wiener filter.

9-4 Repeat Problem 9-2 when $M_d = z$ is changed to $M_d = 1$.

9-5 The block diagram of a digital tracking loop of a missile system is shown in Figure 9P-5. The sampling period is $\frac{1}{12}$ second. The power spectral density of the input signal is

$$\Phi_{rr}(j\omega) = \frac{3}{\omega^2 + 9}$$

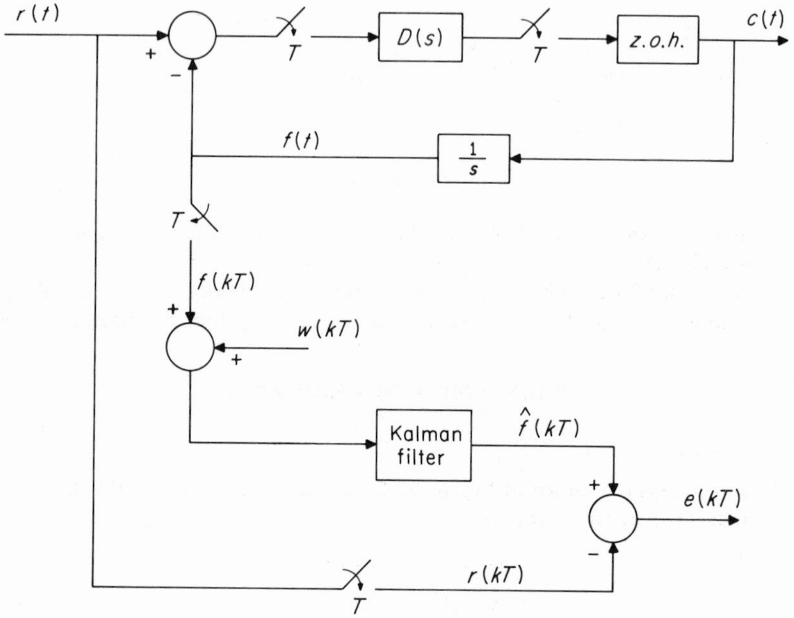

FIG. 9P-5.

The power spectral density of the measurement noise $w(t)$ is

$$\Phi_{\omega\omega}(j\omega) = W \qquad \text{(white noise)}$$

It is assumed that $r(t)$ and $w(t)$ are uncorrelated. Design the optimal controller $D(z)$ and the Kalman filter so that the quadratic error $E[e^2(kT)]$ is a minimum, where $e(kT) = r(kT) - \hat{f}(kT)$. The signal $f(kT)$ is assumed to be inaccessible.

PROBLEMS FOR CHAPTER 10

10-1 Decompose the following transfer functions by the two different methods of direct decomposition. Draw state diagrams for the system and write dynamic equations in matrix form.

(a) $$D(z) = \frac{1 - 0.5z^{-1} - 0.2z^{-2}}{1 - 0.2z^{-1}}$$

(b) $$D(z) = \frac{1 - 5z^{-1} + z^{-2}}{1 + 3z^{-1} + 2z^{-2}}$$

10-2 The block diagram of a continuous-data control system is shown in Figure 10P-2. Insert sample-and-zero-order-hold devices at appropriate locations in the system to simulate the system by a digital model. Select a suitable sampling period. Find the output of the digital model when the input is a unit-step function.

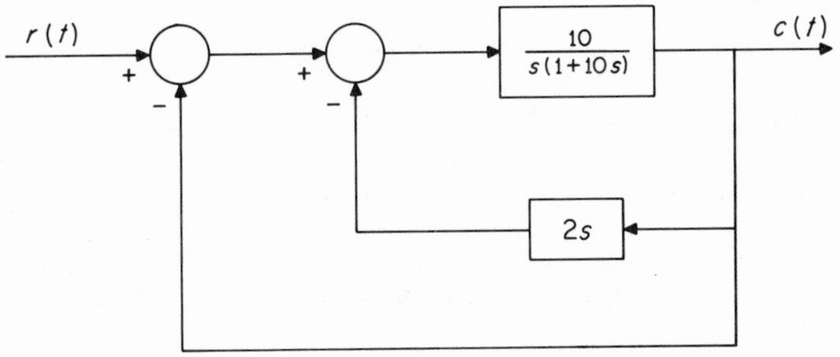

FIG. 10P-2.

10-3 The block diagram of a continuous-data system is shown in Figure 10P-3. It is desired to simulate the system by a digital model.

(a) Insert sample-and-zero-hold devices at appropriate locations in the system to form the digital model. Select a suitable sampling period with considerations on stability and accuracy of the digital simulation. Evaluate the response of the digital model when the input is a unit-step function.

(b) Replace the zero-order hold devices in part (a) by an exponential hold device whose transfer function is $T/(1 + Ts)$, where T is the sampling period, and repeat the problem. Compare results.

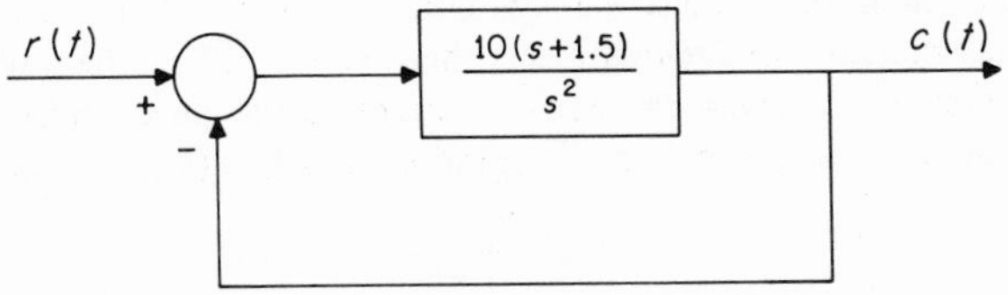

FIG. 10P-3.

10-4 Solve Problem 10-3 by means of the z-form method.

10-5 Consider the system described in Problem 10-3. Find the transfer function of the optimal digital model of the analog system in the sense of minimum sampled mean-square error with the error defined as

$$e(kT) = c_d(kT) - c(kT)$$

where $c_d(kT)$ is the sampled output of the digital model.

10-6 Solve the following differential equations using the z-form method. That is, find the output $c(t)$ at discrete intervals. Select proper sampling periods with considerations on stability and accuracy of the simulation. In each case, determine $\lim_{k\to\infty} c(k)$.

(a) $\dfrac{d^2c}{dt^2} + 4\dfrac{dc}{dt} + c = 2u(t) \qquad c(0) = 1,\ \dot{c}(0) = 0$

(b) $3\dfrac{d^2c}{dt^2} + \dfrac{dc}{dt} + 5c = tu(t) \qquad c(0) = \dot{c}(0) = 0$

(c) $\dfrac{d^2c}{dt^2} + \dfrac{d}{dt}[tc(t)] = 1 \qquad c(0) = \dot{c}(0) = 0$

10-7 The z-form method can be used to approximate the response of time-varying systems. The state diagram of a control system with a time-varying parameter is shown in Figure 10P-7. The gain $a(t)$ is a function of time and is defined to be $a(t) = t$. Compute the approximate output of the system $c_A(kT)$ for $k = 0, 1, 2, \ldots, 10$, by means of the z-form method. The input is a unit-step function. Assume that $a(kT) = kT$ for $kT \leq t < (k+1)T$. Solve the problem with $T = 0.4$ and $T = 1$ sec.

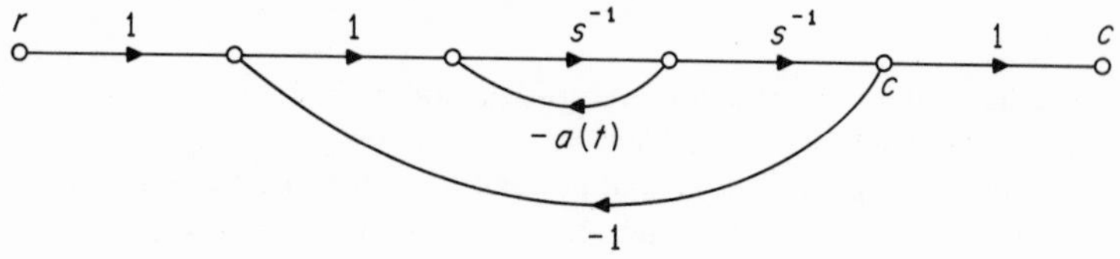

FIG. 10P-7.

10-8 Compute the magnitudes of the upper bounds of quantization errors on $x_1(k)$, $x_2(k)$, $x_3(k)$, and $c(k)$ for the control system described in Example 10-2, Figure 10-28, when the sampling period is 0.1 second.

10-9 The state diagram of a discrete-data control system is shown in Figure 10P-9. The characteristics of the quantizers that are located at the various locations in the system are given as shown. Determine the magnitudes of the upper-bound quantization errors on the variables $x_1(k)$, $x_2(k)$, and $c(k)$.

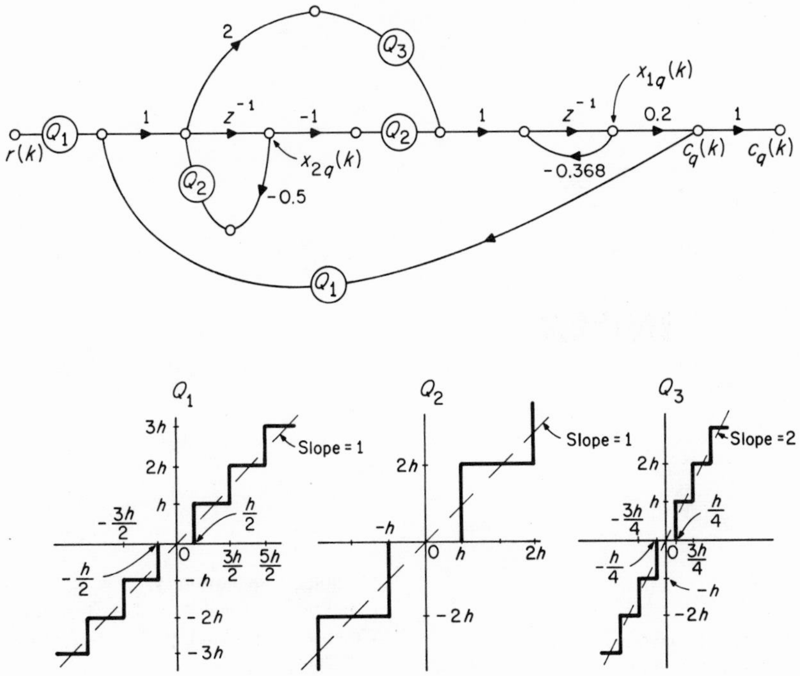
2
Q_3
$x_{1q}(k)$
1
z^{-1}
-1
Q_2
1
z^{-1}
0.2
1
Q_1
$r(k)$
$x_{2q}(k)$
-0.368
$c_q(k)$
$c_q(k)$
Q_2
-0.5
Q_1
Q_1
3h
2h
h
Slope = 1
$-\frac{3h}{2}$
$\frac{h}{2}$
$-\frac{h}{2}$
0
$\frac{3h}{2}$
$\frac{5h}{2}$
-h
-2h
-3h
Q_2
Slope = 1
2h
-h
0
h
2h
-2h
Q_3
Slope = 2
2h
h
$-\frac{3h}{4}$
$\frac{h}{4}$
$-\frac{h}{4}$
0
$\frac{3h}{4}$
-h
-2h

FIG. 10P-9.

INDEX